Environment, Biodiversity and Conservation

ENVIRONMENT, BIODIVERSITY AND CONSERVATION

Edited by
Prof. M.A. KHAN
Division of Environmental Sciences
S-K University of Agricultural Sciences & Technology
Shalimar Campus, Srinagar - 191 121 Kashmir, INDIA

With the collaboration of

Prof. S. FAROOQ
Department of Botany
University of Kashmir
Srinagar - 190 006, INDIA

A.P.H. PUBLISHING CORPORATION
ANSARI ROAD, DARYA GANJ
NEW DELHI-110 002

Published by
S.B. Nangia
A.P.H. Publishing Corporation
4435–36/7, Ansari Road, Darya Ganj,
New Delhi-110002
Phone: 011–23274050
e-mail: aphbooks@gmail.com

2026

Rs. 3995/-

Typeset by
Ideal Publishing Solutions
C-90, J.D. Cambridge School,
West Vinod Nagar, Delhi-110092

Printed at
RD DIGITAL PRINTERS
Ansari road , Daryaganj, Delhi-110002

Preface

The planet Earth is confronted with plethora of serious environmental problems. Dwindling biological diversity, one such problem, has gained paramount importance in the general debate on the sustainable use of the world's natural resources. Out knowledge of existing biological wealth is woefully inadequate which led to intensification of international efforts to assess and maintain biodiversity. A realization has dawned that global environment is at stake and a feeling felt, the worldover, to redress the balance between the man and natural environment. Consequently, the Global Biodiversity Convention was adopted at the United Nations Conference on Environment and Development at Rio de Janeiro, Brazil in 1992. It marked an important turning point addressing, among others, the major topics including biodiversity, climate change, forest principles and sustainable development.

The present book, compendium of thirty-four papers by national and international experts, provides a major review and synthesis of current state-of-art on biodiversity in conjunction with environmental impact assessment of different ecoregions of India including inhospitable, fragile arid environment of Ladakh. The book is spread over following nine sections : (i) Biodiversity, Relevance, Conservation and Sustainable Development, (ii) Biodiversity and Conservation Strategies of Inland water Ecosystem, (iii) Biodiversity of Saline and Mangrove Habitats, (iv) Diversity of Terrestrial Ecosystem: Analytical Approach, (v) Arid Environment: Biodiversity, (vi) Genetic Resources and Biodiversity, (vii) Impact of Anthropogenic perturbations on Biodiversity, (viii) Biodiversity in Relation to Ecosystem Development and Modelling, (ix) Environmental Legislation.

Section I deals with the biodiversity erosion, an appalling issue viewed as one of the major environmental challenges for 21st century, and highlights the role of biodiversity in the maintenance of life support system - so vital for the human survival. Explaining global interrelationship between biodiversity, bioproductivity and biotechnology, another study stresses the need to augment conservation efforts. Issues related to biodiversity and sustainable development are discussed with emphasise on policy-formulations on biosafety and judicious use of planet's resources.

Assessment of biodiversity in a range of aquatic ecosystems (wetlands, lakes, saline habitats and mangroves) and conservation strategies needed constitute an important segment of section II and III. Section IV deals with floristic diversity in India and evaluation of species composition and biological spectrum of selected areas (e.g. Western Ghats, North-west Himalaya etc.). Insect fauna, much ignored the worldover, have been attempted with special reference to ant-aphid associations. Little explored, but scientifically interesting, cold arid environment of Ladakh forms a core component of section V. An ethnobotanical exploration, and the varietal diversity in the germplasm of valuable crops (e.g., cherries, almonds, rice) are discussed in section VI.

The papers in section VIII are devoted to the impact of anthropogenic activities on biodiversity, and topics treated are changing environmental patterns in relation to health, effect of coal-smoke pollution on plants, man's influence on ichthyofauna, and causes of species extinction. Section VIII pertains to the role of biodiversity in ecosystem development and agroforestry models for degraded and wastelands whilst section IX focusses the environmental issues in the light of legislative and judicial response.

Many more publishable papers received posed a difficult task of choosing the one's addressing the theme which necessitated prioritizing manuscript based on their potential impact and relevance to the readership. All the manuscripts have been subject to peer review system to determine their suitability for publication. Nevertheless, profound thanks to all those who submitted their contributions. Thanks are due to colleagues and others who spared their valuable time, and painstakingly reviewed the appears. My wife Roshan was, indeed, patient and tolerant of daily long hours this arduous task purloined from the tedious family responsibility.

The Directorate of Environment and Remote Sensing, J&K, during my tenure of stay at the University of Kashmir, sponsored a seminar in June, 1997 which provided a stimulating and thought-provoking forum for undertaking this venture. The interest of Prof. M.Y. Quadri, Vice Chancellor and Prof. A. Wahid, Ex-Dean Academics, University of Kashmir is highly appreciated. At the wrapup stage Prof. M. Yusuf Kamal, Vice Chancellor, S-K University of Agricultural Sciences & Technology, Srinagar, Kashmir provided further impetus and encouragement.

The book would be useful to a wide range of environmentalists, ecologists, biologists, taxonomists, conservationists, biotechnologists, policy-makers, research scholars and students, and non-governmental organizations. The cooperation and dedication of the publisher for bringing out the book is gratefully acknowledged.

5th July, 1999

Prof. M.A. KHAN
Editor

Division of Environmental Sciences
S-K University of Agricultural Sciences & Technology
Shalimar Campus,
Srinagar - 191 121
Kashmir, INDIA

List of Contributors

Rais Akhtar
Department of Geography and Regional Development, University of Kashmir, Srinagar - 190 006

Nisar Ali
Department of Economics, University of Kashmir, Srinagar - 190 006.

S. Tanvir Ali
Department of Botany, Faculty of Science, Jamia Hamdard, Hamdard Nagar, New Delhi - 110 062.

Vijay K. Anand
Department of Biosciences, University of Jammu, Jammu - 180 006.

Syed Arshad
Centre of Research for Development, University of Kashmir, Srinagar 190 006.

L.K. Banerjee
Botanical Survey of India, 1-Sudder Street, Calcutta - 700 016.

R.C. Bhagat
Department of Zoology, University of Kashmir, Srinagar - 190 006.

G.M. Buth
Department of Botany, University of Kashmir, Srinagar - 190 006.

K.P.S. Chandal
National Faculty for Plant Tissue Culture Repository, National Bureau of Plant Genetic Resources, New Delhi - 110 012.

S. Chandrashekharan
Department of Plant Sciences, School of Biological Sciences, Madurai Kamraj University, Madurai - 625 021.

J.C. Dagar
Central Soil Salinity Research Institute, Karnal - 132 001.

G.H. Dhar
Department of Botany, University of Kashmir, Srinagar - 190 006.

G.H. Hassan Dar
Division of Environmental Sciences, S.K. University of Agricultural Sciences and Technology, Shalimar, Srinagar - 191 121.

M. Amin Dalal
Biotechnology Laboratory, Division of Pomology, S.K. University of Agricultural Sciences and Technology, Shalimar, Srinagar - 191 121.

S.A. Gangoo
Divison of Forestry, S.K. University of Agricultural Sciences and Technology, Shalimar, Srinagar - 191 121.

Brij Gopal
School of Environmental Sciences, Jawaharlal Nehru University, New Delhi - 110 067.

R.K. Gupta
Centre for Research on Ecology, Environmental Applications, Training and Education, 91/2, Vijay Park, Dehra Dun - 248 001.

R.K. Gupta
S.K. University of Agricultural Sciences and Technology, Regional Horticultural Research Station Udheywalla, Jammu 180 002.

Muhammad Iqbal
Department of Botany, Jamia Hamdard, New Delhi - 110 062.

Ulfat Jan
Department of Zoology, University of Kashmir Srinagasr - 190 006.

Azra N. Kamili
Centre of Research for Development, University of Kashmir, Srinagar 190 006.

Shashi Kant
Department of Biosciences, University of Jammu, Jammu - 180 006.

M.A. Khan
Division of Environmental Sciences, S-K University of Agricultural Sciences & Technology, Shalimar Campus, Srinagar - 191 121.

T.N. Khosshoo
Tata Energy Research Institute, Habitat Place, Lodhi Road, New Delhi - 110 003.

Fayaz A. Lone
Department of Botany, University of Kashmir, Srinagar - 190 006.

Mahmooduzzafar
Department of Botany, Faculty of Science, Jamia Hamdard, Hamdard Nagar, New Delhi - 110 062.

A.S. Makaya
Division of Forestry, S.K. University of Agricultural Sciences and Technology, Shalimar, Srinagar - 191 121.

L. Makhnova Maria
Division of Biotechnology, All Union Plant Breeding and Genetics Institue, Odessa Ukrain.

List of Contributors

A.M. Matta
Department of Law, University of Kashmir, Srinagar - 190 006.

T.A. Matta
Department of Zoology, University of Kashmir, Srinagar - 190 006.

M.S. Mir
Regional Agricultural Research Station, S.K. University of Agricultural Sciences Technology, Kargil - 194 103, J&K.

A.H. Mughal
Division of Forestry, S.K. University of Agricultural Sciences and Technology, Shalimar, Srinagar - 191 121.

Irshad A.Nawchoo
Department of Botany, University of Kashmir, Srinagar - 190 006.

B. Neog
Department of Life Sciences, Dibrugarh University, Dibrugarh, Assam - 786 004.

A. Gaffar Kuchay
S.K. University of Agricultural Sciences and Technology, Shalimar, Srinagar - 191 121.

M.Y. Qadri
Vice-Chancellor, University of Kashmir, Srinagar - 190 006.

A.Q. Rather
Directorate of Research, S.K. University of Agricultural Sciences and Technology, Shalimar, Srinagar - 191 121.

M. Ashraf Rather
S.K. University of Agricultural Sciences and Technology, Shalimar, Srinagar - 191 121.

Zafar A. Reshi
Department of Botany, University of Kashmir, Srinagar - 190 006.

M. Salman
Biotechnology Laboratory, Division of Pomology, S.K. University of Agricultural Sciences and Technology, Shalimar, Srinagar - 191 121.

A.M. Shah
Centre of Research for Development, University of Kashmir, Srinagar - 190 006.

G. Mustafa Shah
Department of Zoology, University of Kashmir, Srinagar - 190 006.

D.C. Sharma
National Faculty for Plant tisue Culture Repository, National Bureau of Plant Genetic Resrouces, New Delhi - 110 012.

Koshal Kr. Sharma
Department of Biosciences, University of Jammu, Jammu - 180 006.

Suman Sharma

Department of Biosciences, University of Jammu, Jammu - 180 006.

SM. Sundarapandian

Department of Plant Sciences, School of Biological Sciences, Madurai Kamaraj University, Madurai, T.N. - 625 021.

P.S. Swamy

Department of Plant Sciences, School of Life Sciences, Madurai Kamraj University Madurai, T.N. - 625 021

C.L. Trisal

Coordinator, Wetlands International - South Asia, C-57, Defence Colony, New Delhi - 110 024.

B.A. Wafai

Department of Botany, University of Kashmir, Srinagar - 190 006.

M.H. Wani

Section of Agricultural Economics, Faculty of Agriculture, SKUAST, Wadura, Sopore - 193 201.

A.R. Yousuf

Department of Zoology, University of Kashmir, Srinagar - 190 006.

A.H. Zargar

Division of Environmental Science, S.K. University of Agricultural Sciences and Technology, Shalimar, Srinagar - 191 121.

M.Y. Zargar

Division of Environmental Sciences, S.K. University of Agricultural Sciences and Technology, Shalimar, Srinagar - 191 121.

D.P. Zutshi

School of Enviornmental Sciences, Jawaharlal Nehru University, New Delhi - 110 067.

Contents

SECTION III : BIODIVERSITY OF SALINE AND MANGROVE HABITATS

SECTION IV : DIVERSITY OF TERRESTRIAL ECOSYSTEMS : ANALYTIC APPROACH

SECTION V : ARID ENVIRONMENTS : BIODIVERSITY

SECTION VI : GENETIC RESOURCES AND BIODIVERSITY

SECTION VII : IMPACT OF ANTHROPOGENIC PERTURBATIONS ON BIODIVERSITY

SECTION VIII : BIODIVERSITY IN RELATION TO ECOSYSTEM DEVELOPMENT AND MODELLING

SECTION IX : ENVIRONMENTAL LEGISLATION

SECTION - I

BIODIVERSITY: CONCEPT, RELEVANCE, AND SUSTAINABLE DEVELOPMENT

Chapter - 1

Biodiversity Erosion : Global Environmental Challenge

*M.A. Khan**

ABSTRACT

Widely agreeable (and uniform) meaning of biodiversity (contraction of biological diversity) is elusive because of divergent conceptualized, emanating from various organizational levels (ecosystem, generic, species and genetic) of diversity. All living systems are however, characterised by a strong linkage of natural process (e.g. ecological, biogeochemical, evolutionary and geological) which play a determinant role in structural-shape and functions of biodiversity. Considered a sustained life-support on the planet Earth, biodiversity is closely linked with the survival of human-being. Little wonder, rapid erosion of biodiversity is recognised globally as one of serious environmental problems and if not arrested, likely to gain an alarming proportion as a major environmental challenge in 21st century. Deforestation, population explosion, habitat destruction, introduction of alien species, climate changes, resource exploitation, and agricultural and industrial activities are the major causative factors of depletion of biodiversity. More vigorous world-wide measures (scientific, legislation, etc.) are needed to usher in an acceptable model of development for the sustainable use and conservation of biodiversity at the dawn of new millennium.

***Key Words** : biodiversity, conservation, depletion, rates global, biowealth, significance.*

INTRODUCTION

Biodiversity (contraction of biological diversity) is considered to have an inherent complexity of extremely high proportion, and as such no single process or theory can explain the phenomenon. A strong linkages of multiple processes (e.g. ecological, biogeochemical, evolutionary and geological) play an important and determinant role in structural–shape and functional aspects of biological diversity in a range of living system.

It is believed that as many as 500 millions species of plants, animals and micro-organisms inhabited the earth since life began more than 3.5 billion years ago. The species

* Division of Environmental Sciences, S-K University of Agricultural Sciences & Technology, Shalimar Campus, Srinagar - 191 121, Kashmir, India.

are the product of long evolutionary process. At present, the total number of species existing on the earth is said to be extremely variable (5 to > 100 million); though a conservation estimate (WCMC, 1992) suggests a figure of 12.5 million. To date, however, only around 1.4 million species have been identified and described. Thus our knowledge of comprehension of global biodiversity is amazingly very deficient.

SCALES OF DIVERSITY AND GLOBAL BIO-WEALTH

The concept of biodiversity, though known to man since he started observing living-beings around 'him, the world abruptly became in a buzz-word especially during the last few decades. The concept of biological diversity means the variety of life-forms in a given region, ecological roles they play and genetic diversity they contain (Wilcox 1984). Simply stated, biodiversity is the measure of the world's variety of genes species, and ecosystems and according to McNeely (1994) biodiversity only really came into the public vocabulary around 1988. However, widely agreeable (and uniform) meaning of biodiversity is still elusive because of its divergent conceptualized-framework emanating from four levels of biodiversity [(i.e. ecosystem diversity, generic diversity, species diversity and genetic diversity (variety in one species)]. But, common usage of species diversity has come to be, generally, well recognised. Species diversity has also been described at three spatial scales: (i) within habitat (alpha) diversity connots simply the number of species (or other components of species diversity, (ii) between-habitat (beta) diversity reflects the way in which organisms respond to environmental heterogeneity and (iii) geographical (gama) diversity refers to the number of species within a region (large-scale phenomenon).

As reported in Grumbine (1991) biodiversity refers to multiple levels of biological organisation; the genetic diversity within populations, the numbers and kind of species and populations, the diversity of ecosystem, as well as the ecological patterns and process that knit these different levels together into functional wholes through space and time.

Biodiversity is comprehensive way of approaching conservation blending facets of information, knowledge, awareness and ethics into a complex mixture of protected areas, agriculture, forestry, economics, intellectual property rights (IPRs), trade etc. It culminated into the Convention on Biological Diversity (CBD) signed in June, 1992 by around 157 governments at the Earth Summit in Rio De Janeiro, Brazil, under the aegis of the United Nations Conference on Environment and Development (UNCED). The Convention laid tremendous stress on global biological diversity and emphasized the importance of preserving it for posterity. The Convention on Biodiversity defines biological diversity to mean the variability among living organism from all sources including *inter alia*, terrestrial, marine and other aquatic ecosystems and the ecological complexes of which they are part; this includes diversity within species, between species and of ecosystems. The estimated number of species in the world and their scarcity is given in Table 1.1.

BIODIVERSITY EROSION AND NEED FOR CONSERVATION

Biodiversity is an elementary requirement for the long term sustainability of the environment, continuity of life on the earth and the maintenance of its integrity. Biodiversity is ultimately linked with the survival of human-beings since it constitutes a sustained life-support on the Earth including, among others, the fixation of solar energy, regulation of water-flow, protection of soil and recycling of neutrients. The fundamental needs of man (food, clothing, shelter) and range of products (pharmaceuticals, pesticides, fertilizes etc.) together with expansion of recently emerging scientific frontiers (genetic engineering, tissue

Table 1.1 : Estimated number of species in the world and their scarcity.

Group	*Total number of species*	*Number of identified species*	*%age of identified*	*Number of scarce species*	*%age of species identified scarce*
Mammals, reptiles and amphibians	15210	14484	95	728	5
Birds	9225	9040	98	638	8
Fish	21000	19056	90	472	3
Plants	480000	322311	67	-	-
Insects	30000000	751000	3	895	<1
Other invertebrates and microorganisms	3000000	276594	9	530	<1
Total	**33525435**	**1392485**	**4**	**-**	**-**

Source: Solbrig (1993)

culture etc.) are all met by rich and varied potential of bioresources. Thus biodiversity is the corner stone of biotechnology. Anthropogenic activities have, dismayingly, altered the natural environment and diminished biological diversity. In fact, the rapid pace of erosion of biodiversity has come to be recognized globally as one of the major and serious environmental challenges and if left unchecked, likely to gain an alarming proportion in 21st century. As reported in Wright *et al.* (1993) half of all extent species may become extinct within next 100-300 years. Extinctions of taxa appear generally to be greater in an island environment than on a continent, probably; because organisms have less area to hide or migrate on an island (WCMC, 1992). The rate of habitat loss is considered to be even greater than species extinctions. The other extremity of the scale indicates that at the molecular dimension, there is the decline in the total genetic resource due to plant and animal extinctions. Thus, anthropogenic activities leading to extinctions have caused genuine global concern. The loss of species is viewed in terms of loss of the earth's genetic diversity, as a loss of flora and fauna of great socio-economic and agricultural potential, and more importantly as a threat to the climate and environment rendering *Homo sapiens* susceptible to set of hazardous conditions. No wonder, decline in biodiversity and disappearance of species came to be identified as a grave environmental problem. Myers (1986) aptly stated: "Every time species goes extinct, we are irreversibly impoverished". Biodiversity, in essence, entails a more comprehensive approach to conserve gene, species and ecosystem. Since late 80s, such an approach gained momentum which focuses not only on merely managing the endangered species and threatened protected areas, but recognises the need to address the economic, social, ethical and even political issues related to conservatory plans. Understanding the natural process that regulate biodiversity provides the key to managing and preserving humanity's natural heritage since biological diversity most importantly, constitutes a critical component of the continued habitability of our planet.

Threats (natural and unnatural) to biodiversity (at genetic, species and ecosystem level) are numerous. Data emanating from various sources on loss of biodiversity are indeed alarming. According to Myers (1988), 50 species are being drawn to extinction per day. Reid (1992) states that 2-8% of the world's total species are likely to be lost in the next 25 years. Watt (1993) reports that on world-wide basis the genetic diversity within species (with potential value to mankind) is being eroded; many species including the yet unclassified are being destroyed or degraded. It is estimated that around 50000 species per year are likely to be lost over next decades (Raven 1994). Deforestation, population explosion, over hunting, habitat destruction, climate change, resources exploitation and activities related to expanding agricultural and industrial development are the major causative factors for loss of biodiversity. Furthermore, the introduction of alien species can be equally a major cause of extinction of native species and the global effect of increased rates of species introduction is overwhelmingly a net loss of global species diversity. The fundamental factors contributing to erosion of biodiversity are given in Table 1.2. (cf. Solbrig 1993). According to McNeely *et al.* (1995) the underlying causes of loss of biodiversity stem from changes in attitudes towards nature; growth in human population and natural resources consumption; the impact of global trade, economic systems that fail to value the environment, and flow of benefits from both the use and conservation of biological resources.

The global distribution of biological wealth is highly uneven. Khoshoo (1994) established a global relationship between biodiversity and biotechnology. Thus, such countries as Brazil, China, Ethiopia, India, Indonesia, Malaysia, Mexico are biodiversity-rich, but poor in bio-technology whilst reverse holds true for U.S.A., Japan, France, Germany and UK. This disparity, obviously results in the flow of biodiversity wealth, from developing to developed countries. The relationship between biodiversity and technology has been further expanded (*see* Khoshoo, 1999); the author stresses the need for formulation of policy in gene-rich/technology - poor (or deficient) countries to avoid what may result in "gene imperialism".

Table 1.2. Fundamental human factors contributing to erosion of biodiversity.

Factor	*Example of impact on conservation*
Population growth	Demographic pressure.
Poverty	Hunger, deforestation, trade in endangered species, lack of popular support.
Bad perception	Desire of quick results and negation of failures in the long term.
Anthropocentralism	Absence of support for non-utilitarian cause.
Cultural transitions	Unsustained management of resources during colonization and quick social changes.
Economy	Absence of planning as a result of the internationalization of markets and the erratic price of goods.
Implementation of Policies	Social crises, wars, corruption, non-fulfilment of law.

Source: Solbrig (1993)

One among top 12 mega diversity countries of the world, India accounts for over 2% of the world's landmass. The country's rich biological diversity (with about 130000 species of living organism) is due to a great diversity of environment regimes which, consequently determine wide natural vegetation ranges spanning over tropical evergreen and mangroves to dry deciduous and desert scrub. According to Gadgil (1993), India is a biomass-based civilization, and ever two-thirds of its population depend on agriculture for their substance. They gather some 150 million tons of fuel wood annually for domestic use, and free-range grazing by their livestock amounts to a fodder demand over 500 million tons. India's flora and fauna are documented to some extent, although some groups of insects and mites have hardly been studied. As indicated in Gadgil (1993), the total number of described species from the Indian subcontinent includes about 13000 species of flowering plants, some 50000 species of insects, 4000 species of molluscs, 6500 species of their invertebrates, 2000 fish species, 140 amphibian species, 420 reptile species, 1200 birds species and 340 mammal species. In all, there is a total of nearly 15000 of species of angiosperms and 80000 of animals. India is reported (cf. Gadgil 1993) to harbour 5.2% and 4.2% of all known plant and animal species respectively. This proportional representation may well hold if the yet, undescribed 30 million or so species are documented by science.

The world attention on biodiversity conservation got focussed through the Earth Summit of 1992 which devoted Article 8 on '*in-situ*' conservation and Article 9 on '*ex-situ*' conservation. CBD contains 42 articles pertaining, among others, to identification and monitoring, conservation and sustainable use of biodiversity, access to genetic resources and transfer of technology. India is a signatory to the Convention on Biological Diversity (CBD). Efforts are afoot to enact a law governing the sustainable use and conservation of biodiversity. The Ministry of Environment and Forests (Anon, 1994) has drafted an approach document for the conservation of biodiversity in India. The foundation of India's biodiversity policy with inputs from renowned scientists like Dr. M.S. Swaminathan, Dr. T.N. Khoshoo and Prof. Madhav Gadgil is indeed a step in right direction. Based on the draft, the country's biological diversity conservation bill is to be prepared and presented to Parliament. The policy enunciates that India has a sovereign right over its biological resources. Eleven areas have been identified for the draft action-legal and policy framework; survey of biodiversity and natural data base; *in-situ* and *ex-situ* conservation; sustainable utilization, indigenous knowledge system; innovations and practices and benefiting sharing; people's participation; research and international cooperation and education, training and extension. In specific terms, the legislation seeks to regulate access to biological resources, notification of areas important from the standpoint of biodiversity and protection of threatened species. Gadgil (1991) reports that no segment of Indian society is presently motivated to use biological resources in a prudent fashion, and this is the dilemma underlying India's biodiversity crises. The past efforts have been only dismal, it is time to rise to occassion (for the survival of *Homo sapiens*) since it is not too late to put in place measures (scientific and legislation etc.) to usher in a more sustainable model of development at the dawn of new millennium.

LITERATURE CITED

Anonymous (1994), Conservation of Biological Diversity of India: An approach. Ministry of Environment and Forests, Government of India, New Delhi.

Gadgil, M. (1991). Conserving India's biodiversity: the societal context. Evolutionary trends in plants 5: 3-8.

Gadgil, M. (1993). Biodiversity and India's degraded lands. Ambio 22: 167-172.

Grumbine, R. E. (1991). Coöperation or conflict? Interagency relationship and the future of biodiversity for US parks and forests. Ennvironmental Management 15: 27-37.

Khoshoo T.N. (1994). India's biodiversity: tasks ahead. Current Science 67: 577-582.

Khoshoo T.N. (2000). Biodiversity of the Himalaya. In: Khan, M. A. (Editor) Environmental, Biodiversity and Conservation.

McNeely, J. A. (1994). Biodiversity conservation and the role of global community. World Wide Fund for Nature–India, 9-16.

McNeely, J.A; Gadgil, M; Leveque, C; Padoch, C; Redford, K. (1995). Human influences on biodiversity. pages 715-733 In: Heywood, V.H. (Editor). Global Biodiversity Assessment, published for the United Nations Environment Programme, Cambridge University Press, Cambridge.

Myers, N. (1986). Biological Resources of the Tropics. Indian Council for Agricultural Research (ICAR) Publication, New Delhi, 1-9.

Myers, N. (1988). Threatened biotas: "hotspot" in tropical forestry. The environmentalist 8: 1-20.

Raven, P.H. (1994). Our planet 6: 5-8.

Reid, W.V. (1992). How many species will there be? pages 55-73 In : Whitmore, T.C. and Sayer, J.A. (Editors), Tropical Deofrestation and Species Extinction. Chapman and Hall, London.

Solbrig, O.T. (1993). All of US–1. Sept.

Watt, A. (1993). Biodiversity in Montane and Forest Ecosystem Spectrum (British Science News No. 232),.January–February Issue 11-14.

World Conservation Monitoring Centre (1992). Global Biodiversity: Status of the Earth's Living Resources. Chapman and Hall, London.

Wilcox, B. A. (1984). "In situ" conservation of genetic resources: determinant of minimum area requirements. Proceedinngs of the World National Parks Congress, Bali, Indonesia, Gland, IUCN.

Wright, F.J. Dykjova, D. and Hejny, S. (1993). Wetlands of the World, I. Inventory Ecology and Management. Handbook of Vegetation Science, Kluwer Academic Publishers, 768 pages.

Chapter - 2

Biodiversity of the Himalaya

*T.N. Khoshoo**

ABSTRACT

Himalaya is indeed a unique mountain, being the youngest, the tallest and the fragilest system on the earth. Today it is sick on account of loss of biodiversity of sorts. Therefore, conservation efforts needs to be augmented considerably. The paper discusses some of these questions together with the interrelationship between biodiversity, bioproductivity and biotechnology. The country though gene-rich, needs to be equally rich in technology so as to use the genes and biomolecules to the good, the benefit and the wellbeing of the teeming millions of India.

Key Words : biodiversity hotspot, bioproductivity, conservation, Himalaya, Technology

INTRODUCTION

If one looks at the continent of Asia from outer space, its unique feature is the Himalaya (over 2500 km long), also known as the "Third Pole", which separates the Indian Subcontinent from the northern part of Asia. The Himalaya is the youngest, the tallest and the most fragile mountain system in the world. It has considerable diversity in longitude, altitude, climate, habitat, soil, flora and fauna. Himalayan biodiversity is also reinforced by human, cultural and development diversity.

The Indian Himalaya has about 22% of area under close forest cover with 40% or more crown density. Ideally forest cover should have been 60%. Rate of deforestation is high. Out of about 8,000 species of flowering plants in the Himalaya, nearly 5,000 grow in Eastern Himalaya alone. Similar figures are not available on Himalayan animals. No account exists on microorganisms diversity of Himalaya.

Some 3,165 species of flowering plants, 17 species of mammals and 27 species of birds are endemic to the Himalaya. It has been estimated that 337 species of flowering plants and 111 species of higher animals are endangered, and must be conserved through both *in situ* and *ex situ* methods (Khoshoo, 1992, 1995b). Among the endangered plants are

*Tata Energy Research Institute, Habitat Place, Lodhi Road, New Delhi - 110 003

the herbal drugs of repute, collected and sold by the locals. Similarly, among animals, particular mention must be made of the pheasants including jungle fowl, patridges and quail. Most of these are gorgeously plumaged, thus becoming targets for poachers, smugglers and illegal exporters. At least nine species of plants and animals have become extinct in the Himalaya (Khoshoo, 1992, 1995b).

One of the 18 hotspots in the world is the Eastern Himalaya. This region is very rich in biodiversity and home to many economic plants. It also shows a very high evolutionary activity as revealed by the pioneering work of Janaki Ammal and others (vide Khoshoo, 1992).

India is an important Vavilovian Centre of agri-biodiversity having contributed 167 species to the world agriculture. Three of the eight subcentres of agribiodiversity fall in the Himalayan belt. These harbour wild relatives of cereals, pulses, fruits, oil-yielding plants, species, medicinal and aromatic plants used in indigenous systems of medicine. Himalayan animal species, as relevant to agriculture, are jungle fowl, zebu, mithun and yak.

A common feature of Himalaya is mixed cropping in different permutations and combinations, and there are five distinct agroecosystems as practised in North-Western Himalaya, Western Himalaya, Assam Plains, Eastern Himalaya and North Eastern Himalaya. Each has its own distinctive crop and livestock profiles. New economic plants are also being constantly introduced in the Himalaya.

CONSERVATION EFFORT

In situ conservation in the Himalaya involves watersheds and sacred groves, and the Protected Area Network (PAN) of biosphere reserves, national parks and wildlife sanctuaries. About 8.5% of the area of the Himalaya is under *in situ* conservation. India's six biosphere reserves fall in the Himalayan belt: Nanda Devi, Uttarkhand, Manas, Namdapha, Nokrek and Kazirang. Three States in the Himalayan belt (Himachal Pradesh, Sikkim and Arunachal Pradesh) are among the best protected states, while three others (Mizoram, Nagaland and Tripura) have no protected areas. Others fall in between. *In Situ* conservation is at macro-level (ecosystems, communities and populations), and also includes the establishment of Forest Tree Genetic Reserves so as to conserve tree germplasm. Representative lakes and wetlands with their unique flora and fauna also need documentation and conservation. There is also a need for Special Reserves so as to conserve *in situ* some of the important and unique biota given under Schedule IC endagered and/or those that are not allowed to be exported and need conservation) under Indian Wildlife (Protection) Act of 1993. Some examples are: rhododendrons, orchids, wild relatives of citrus, mango, banana, rice, important medicinal and aromatic plants, and animals like pheasants, markhor, ibex, nayan, urial, shou, chiru, musk deer, hangul, snow leopard, yak, kiang and several species of cats (*Felis*). *In situ* conservation guarantees refinement and updating of the germplasm due to natural selection.

Ex Situ Conservation deals with genetic diversity within a species. Under this are included the live collections in 11 botanical gardens and 17 zoological parks. Altogether these cover about 250 ha, but these are neither representative nor have any worthwhile management system. Furthermore, there is a well organized seed bank under the National Bureau of Plant Genetic Resources (ICAR). The Bureaus of animal, fish and microorganisms genetic resources of ICAR are at various stages of development. As yet there are no banks in operation for pollen, sperms, eggs, embroys, tissues, organs and DNA and for tree genetic resources.

Informal Community Gene Banks exists at the village level. These need to be formalized keeping in view the different agroforestry systems existing in the Himalayan belt. These system are indeed "treasure troves" of genes.

There is an urgent need to have an indepth look at the entire informal and formal conservation and utilization efforts in the entire Himalayan belt. This should include a study of the relationship between loss of biodiversity due to major developmental projects like road building, hydro-electric dams, large scale housing and fruit growing and other developmental schemes. The process of change thus set in the biodiversity regimes needs close monitoring so that timely remedial action is taken.

Conservation and domestication of wild economic plants (and some animals) that are in constant demand in trade, need priority attention. These are mostly herbal drugs of commercial value. Species yielding a host of other economic products (poles, ropes, mats, brooms, baskets, fodder, wild fruit, etc.) have also become endangered due to over-harvesting. Production through domestication alone will reduce pressure on the populations of such species. Local Technical Knowledge and the trade involving such microenterprises need not only proper documentation but must also ensure benefit to the local people and not the middle men. Last but not the least, a modern database needs to be organized on biodiversity of the Himalaya (Khoshoo, 1995b).

BIODIVERSITY, BIOPRODUCTIVITY AND BIOTECHNOLOGY

It is indeed ironical that while developing countries like India have given to the world most of the agri-biodiversity, they themselves have been areas of low productivity and high population density. This belt has also been the traditional "hunger belt", and hunger still haunts parts of Africa. The industrial countries by making use of science of genetics and breeding have raised the productivity. A perusal of Figure 2.1 shows that low biodiversity and low bioproductivity have been prevalent in harsh ecosystems. Underlying the pre-Green Revolution agriculture, was high genetic diversity and low bioproductivity. The world agriculture then moved to high productivity accompanied by low diversity and the result was Green Revolution, which has paid its dividends and in most developing countries hunger is almost absent. However, such agriculture has not been entirely ecological friendly. Thus it has also been a question of balancing immediate economic and social gains and a largely manageable level of environmental degradation and pollution. Among other aspects, sustainability in agriculture, animal husbandry, fisheries and forestry will depend on the ability to combine high productivity with high diversity. The world has to move towards such a broad goal (Fig. 2.1).

Formulation of policy in gene-rich/technology-poor (or deficient) countries is necessary in order to avoid what may result in "gene imperialism". The INBio experience, if implemented without thought and care, could prove to be the thin end of the wedge (Khoshoo, 1995a, 1996a-c). Therefore, each country must ensure that agreements, like the one reached between Costa Rica and USA, do not hurt their on going efforts in conservation, sustainable utilization, development and equity in biodiversity. Technology (including biotechnology) is inherently knowledge-intensive and having good biotech infrastructure would lead to value-addition to the products from biodiversity in the general area of agriculture, animal husbandry, fisheries, forestry and medicine. This has to be accomplished without ignoring rights and privileges of the local communities.

Transfer of biotechnology and pharmaceutics requires a certain minimum amount of technical and legal capability, which many biodiversity-rich developing countries lack at present. Therefore, considerable competenance and skills are needed in whole range of subjects starting with taxonomy, genetics, breeding, molecular biology, microbiology, pharmaceutics, biochemistry, fermentation technology, biochemical and process engineering, economics, law and last but not the least training of local communities in modern conservation skills.

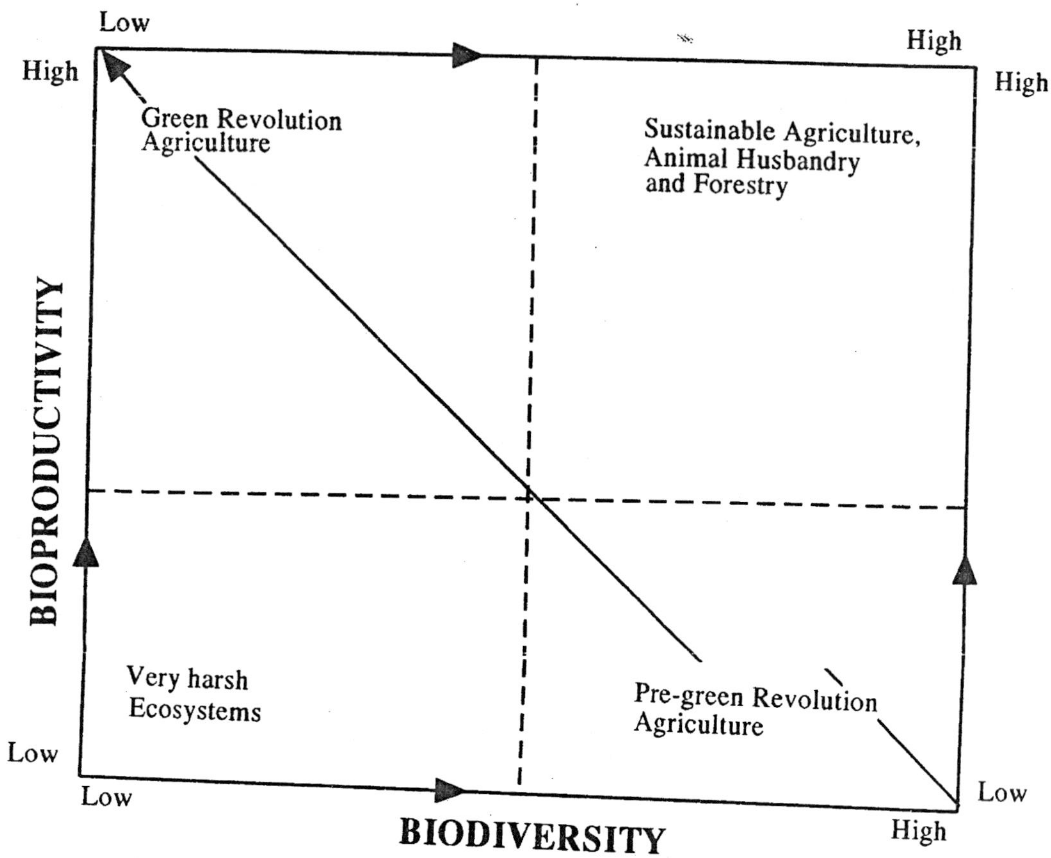

Fig. 2.1 : Interrelationship between biodiversity and bioproductivity

Although the "gene and drug rush" is inevitable, it should not take place in a policy vacuum. There is a need for an all-encompassing law on biodiversity that should take into account its deep inter-connection with technology and remove all inadequacies. This would be one tangible step towards sustainable bioprospecting.

The relationship between biodiversity and technology is depicted in Figure 2.2. The countries of the world can be divided in four groups: (a) biodiversity-poor and technology-poor, (b) biodiversity-rich but technology-poor, (c) biodiversity-poor but technology rich, and (d) biodiversity-rich and technology-rich. To the first group belong some arid-zone countries in the Middle East, the second group comprises southern countries like Indonesia, India, Malaysia, Brazil, Mexico and others in the tropical-subtropical belt, to the third group belong USA, Japan, Germany, France, Sweden, UK. etc. and there is no country which alls in the fourth group. At present there is a flow of biodiversity from second group (South) to the third group (North). The extent and nature of flow of technology from North to South is not commensurate with the flow of biodiversity from South to North. This is an unequal exchange and will remain so till such time as countries of the South become self-reliant in technology. An important factor underlying this exchange is that while some countries (like India and China) do have the capability to enter the fourth group (rich in both biodiversity and technology), the countries in North can never make it to fourth group in the real sense of the word. The reason is that they do not have any worthwhile agribiodiversity of their own growing naturally, although they do have excellent *ex situ* facilities in the form of field gene banks, seed and other banks. Contrary to countries in South, the countries in North do not have the advantage of long range ecological processes and organic evolution that operate under natural conditions in the former. These process constantly refine and update biodiversity through mutation, recombination and natural selection. In essence biodiversity stored in industrial countries is in the form of "gene morgues", because advantages conferred by exposure to the process of natural selection are not available. In contrast to *in situ* conditions, organic evolution is virtually at a stand still or halted under *ex situ* conditions. Therefore, in all bio-banks, germplasm is *preserved* (like jams and jellys) and not conserved in time and space. However, both *in situ* and *ex situ* conservation are necessary so as to complement and supplement each other (Khoshoo, 1995).

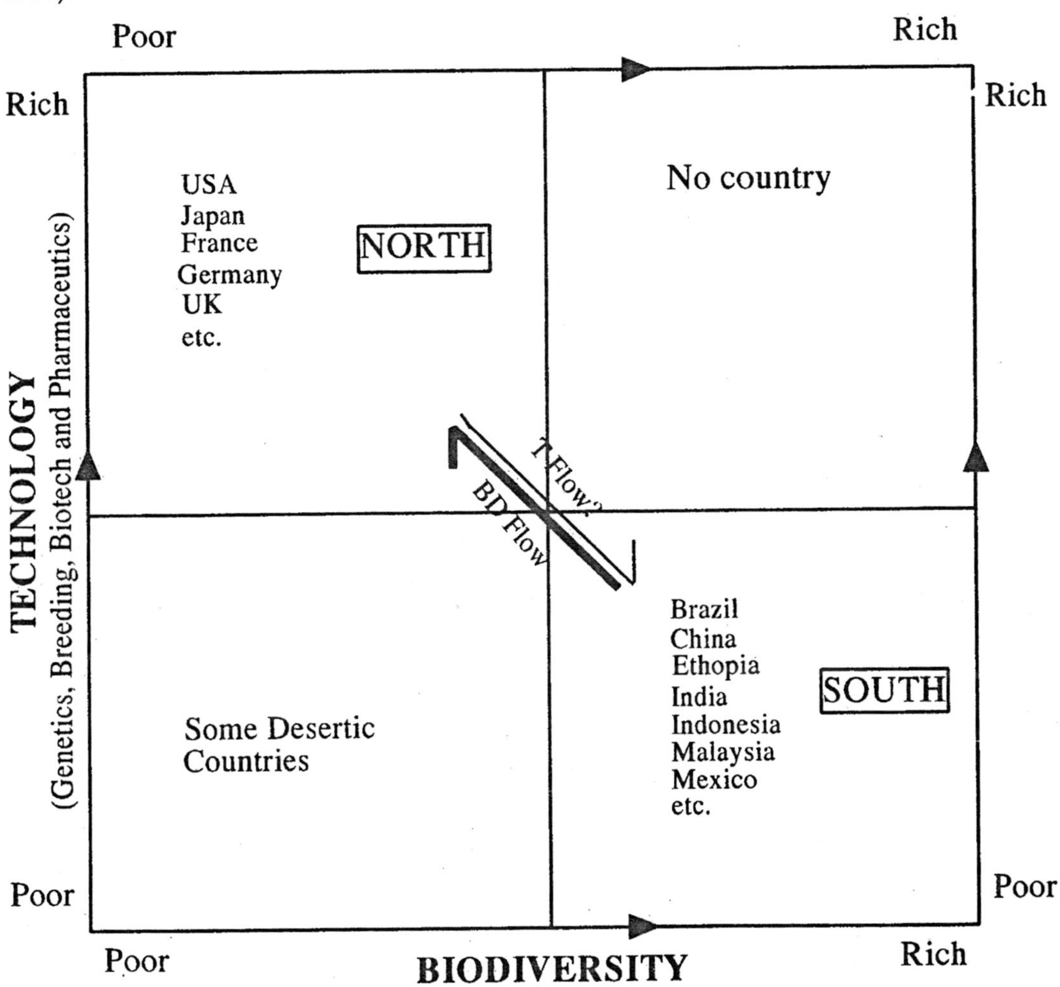

Fig. 2.2 : Interrelationship between biodiversity and technology

THE FUTURE COURSE

Gene-rich/tecnology-poor developing countries like India, Indonesia Malaysia, China, Brazil and others must, therefore, come together and reach an understanding regarding various aspects, including scientific and technological, economic, social, cultural and legal issues, and collection, supply and costing of the raw materials of biodiversity. Today the cost of biodiversity is only the cost of collection and travel involved. If these countries remain divided (as they are today) regarding their stand on prospecting for new genes and drugs and IPR, and compete among themselves, even the type of benefits that INBio has been able to get from Merck (USA), will no longer be forthcoming (Khoshoo, 1995a, 1996b).

It is, therefore, imperative that a multilateral mechanisms is created, because of the tremendous global upsurge in interest about natural products and vegetarianism. Thus there is today considerable interest on biopharmaceuticals, biopesticides, bioinsecticides, biofertilizers and biocosmetics. All these are biodegradable. Industry would be ready to make investments in these areas because gene and drug screening technology itself has reached a level of perfection (Khoshoo, 1995a 1996b-c).

The extent and nature of capability in technology will determine the capability of the country to conserve and sustainably utilize biodiversity. Conservation no longer means simply building a fence round an area, but involves considerable amount of upstream science and technology. The longer a country takes to make such a transition, the farther the country would be from reaping the harvest from its rich biodiversity (Khoshoo, 1995a, 1996b-c).

LITERATURE CITED

Khoshoo, T.N. (1992), Plant Diversity in the Himalaya: Conservation and Utilization. G. B. Pant Memorial Lecture. G.B.P. Institute of Himalayan Environment and Development. Almora, pp. 1-129.

Khoshoo, T.N. (1995a). Biodiversity, bioproductivity and biotechnology. Ambio (Sweden) 24: 251-53.

Khoshoo, T.N. (1995b), Biodiversity in the Indian Himalaya pp. 181-255. In : Banking on Biodiversity. ICIMOD, Kathmandu, Nepal.

Khoshoo, T.N. (1996a). Concern mounts on rapid erosion of Himalayan Biodiversity, Diversity (USA), 12: 25-26.

Khoshoo, T.N. (1996b). Biodiversity in the developing countries. Pages 304-311. In : Di Castri, F and Younes, T (Editor) Biodiversity, Sciences and Development : Towards a New Partnership. Cab International (London).

Khoshoo, T.N. (1996c). Biodiversity in India: Reality and Myth. pages 215-275 In Rao, UR (Editor) Science and Technology for Achieving Food, Economic and Health Security. Prism Books Pvt Ltd. Bangalore.

Chapter - 3

Relevance of Biodiversity Concept in Environmental Protection of Mined Areas

*R. K. Gupta**

ABSTRACT

The mined ecosystems experiences ecological alterations and instability as a result of environmental, biological and developmental factors leading to the total loss of biodiversity. The disturbances in these ecosystems are reflected in the physical, biological and sociological parameters as a result of overuse of the mineral, overburden disposal, irrational practices of transport of mined mineral that trigers erosion and other processes of degradation. The process of opencast mining plays a important role in modifying both faunal and floral diversity of the sites which in some places is reduced to zero, if necessary measures are not taken in time.

Mining being a commercial activity with an eye on profit, serious efforts by the miners for reclaiming the mined area at the fagend of the mines life or after its closure are sometimes doubtful. Though technology of rehabilitation of mined areas is now available it is seldom practiced for want of will and money. The process of preparing *Environmental status* in the mining sector has come into operation since 1993. The dovetailing of reclamation of mined areas with long term land use planning is still the necessity to meet the demands of land. Therefore developmental strategy has to be in harmony with overall conservation strategy keeping in view ecological zones with regard to biodiversity status. The paper discusses some of these issues in detail.

Key Words : *Biodiversity, environmental management, exclusive marketing right (EMR), intellectual property rights (IPR), land-use planning.*

*Centre for Research on Ecology, Environmental Applications, Training & Education 91/2, Vijay Park, Dehradun 248 0012.

INTRODUCTION

The term bio-diversity, an amalgam of two words, biological and Diversity, is a blanket term for the natural resources and its biological resources. The term bio-diversity refers to the number, variety and variability of all life forms, micro-organisms, plants and animals and the ecological complexes they inhabit. It denotes the life forms, the ecological role they play, the genetic diversity they contain and includes all inter/intra-specific and ecosystem diversity. Its ready availability with a wide resource base, appropriate maintenance and judicious utilisation are the pre-requisites to sustainable eco-development of any degraded ecosystem.

The mined-ecosystem experiences the ecological alterations which are the result of environmental, bio-daphic and developmental factors, which ultimately lead to their total loss of bio-diversity. Disturbances in a mined ecosystem are reflected both in the physical, biological and sociological parameters as a result of overuse of minerals, overburden disposal and irrational practices for transport of the mined mineral that triggers erosion and other processes of degradation. Surface Opencast mining, in particular, plays an important role in modifying faunal and floral diversity of the site, which infact is reduced to zero, if necessary measures are not taken for their reclamation.

GROWING CRISES OF BIODIVERSITY DEPLETION

The biodiversity is defined in terms of three fundamental and hierarchially related levels of biological organisation such as the ecosystem or ecosystem diversity, species or species diversity and genes of genetic diversity. Since genes are part of the species and the species make up the ecosystem-the concept of biodiversity reflects the interrelationship among these three components. India's geographical position in the centre of the old world, a variety of physiographic feature, bio-climates, edaphic parameters and socio-cultural systems prevailing in various parts of the country, make her as one of the most significant places with rich biodiversity. It is estimated that some 45,000 species of plants and over 77,000 wild species of animals are recorded, comprising about 6.5 percent of the world's known biodiversity. Given the fact that a number of biologically rich areas exists in India, certain regions like northeastern part and the tribal belts of central and Deccan plateau are still not fully explored. What is more striking is the uniqueness and the amount of endemisms (33 percent) of all flowering plants and 18 per cent of all plants in India. Equally important is the range of domesticated biodiversity; at least 166 species of crops, 320 species of wild relatives of crops are known to have originated in India. Within each species, diversity of species is astounding, as for example 50,000 to 60,000 local varieties of rice are reported. Other species having wide species biodiversity are Okra (Lady's finger); Citrus, Banana, Mango, Jamun, Jute, Ginger, Turmeric, Pepper, Cardamom, Cinnamon etc. Also significant is the indigenous diversity of livestock with 27 breeds of cattle, 40 of sheep and 22 of goats.

The biodiversity is not distributed uniformly and is substantially greater in some areas than in the others. Though reasons for a region's biodiversity may not be easily explained but a good rainfall, its periodic length along with temperature are fair indicators. Information on the main eco-floristic regions of India with degraded landforms and biodiversity erosion has been described earlier by the author (Gupta and Chauhan 1994).

Direct or indirect impact of man on the soil reflects overall disturbances of structural and functional properties of the soil erosion, deflation, salnization, laterization and accumulation of adverse substances and compounds that are the main processes responsible for soil degradation in tropical regions of India. Soil degradation followed with salinisation and ultimately to desertification is evident in many parts of the country. Conversion of forests to other land uses, particularly mining, urbanisation, road building etc., imbalances the ecosystem which increases surface erosion, showing "zero" biodiversity. Illegal trade and depletion of forest cover due to expanding agriculture, large scale deforestation for various reasons, both commercial and domestic use, overgrazing etc., are the main causes of bio-diversity depletion. Various estimates of loss to animal diversity (tigers, rhinoceros, pink headed duck etc.) have been recorded in spite of different projects and programmes for their protection in the form of National Parks, wildlife sanctuaries and reserves. It has however been estimated that 5.4x 10^6 tonnes (16.4 tonnes per ha per yr) of soil is being displaced which is the very basis of survival of biological resources (Dhruvanarayan and Ram Babu, 1983).

Wilson (1990) estimated that the global extinction rate due to habitat destruction in the world is about 17,500 species per year or 0.1 per cent of the existing species, though figures for Indian are not still available. No one can say for definite, so far how many species are lost, the figures estimated is about 10 per cent of the recorded wild flora and possibly a large reaction of the wild fauna are threatened (Sahni, 1967). Reports on the forest cover of India are alarming, which show that during the last few decades the forest cover (with more than 40 per cent or own density) varied from 10.99 per cent (in 1993) to 11.93 per cent (FSI reports). The domesticated flora and fauna also did not lag behind. The fact is that thousands of exotic varieties replaced countless indigenous varieties, though the indigenous varieties through broad genetic base of crops provided an insurance against pests and diseases and is a bank for desired characteristics like the nutritional value apart from being important culturally.

The earth and its diversity are dynamic and ever changing with species extinction a natural phenomenon over a geological time scale. It is likely that more species have become extinct than are living today. However, manmade alterations and destructions of environment, such as the mining activity, is extremely ironical and worrying aspect of the situation. What adds to the tragedy is our limited knowledge of the earth's biodiversity. An ever increasing number of species are being lost without even being discovered, much less studied. As a result of mining, the degradation and loss of habitat threatens a wide range of taxa which occur as a result of forest clearing, over-burden removal over large areas and the minerals in an improper way. Not only there is decrease in overall population of many species but a break up of their distribution pattern, rendering them into isolated sub-populations which are more vulnerable to extinction. The main problem of biodiversity protection in such situations is still confined to shift in development planning towards a model that respects both nature and culture which empowers the local community to participate in decision making regarding the use of natural resources (Gupta, 1971), and which is based on the twin principles of sustainability and equity. There has to be a recognised understanding of the peoples' relationship with biodiversity. Their local knowledge and technique, their philosophy and critical importance of biodiversity in their lives requires to be widely recognised in spite of the various Acts and National laws and policies on mining. Thus, the people have to be brought to centre stage of the process of development with community control over the natural resources, reviving their traditional systems and channelising the benefits of biodiversity conservation back to the local communities' social and democratic development, leading to open and transparent planning, public awareness, scientific mining technology and education. These are some of the steps required to be initiated for the management of biological resources in the mined ecosystems of the country/community.

IMPACTS DUE TO MINING AND STRATEGY FOR BIODIVERSITY PROTECTION

The probable impacts of mining are:

(a) Impacts on natural resources in terms of air, land (pollution and degradation); water (surface and subsurface), biological regime (flora and fauna); noise and vibrations, accelerated erosion due to deforestation.

(b) Socio-economic impacts: Need of the people as a result of population explosion and capacity to develop.

(c) Environmental costs associated with such activities, namely social, economic and environmental costs.

The above impacts have to be assessed and analysed with regard to both qualitative and quantitative approaches of resource management for an environmental management plan. A comprehensive management plan has to be developed for proper mining methodology, planned reclamation and restorative measures.

Since no industrial activity can be seen in isolation with respect to its impact on environment and ecology, mining in ecologically sensitive areas become more complex since the management of rejects hills is not as simple as in open-cast mining in the plains. The various processes involved are:

(a) Planning and development.

(b) Extraction of mineral.

(c) Transportation and handling of excavated material/ore benefication, and

(d) Management of rejects and reclamation of mined-out areas.

Though there are a number of laws and agreements pertaining to exploration in mining, a new concept of *Environmental statement* has come into operation which is based on environment audit and has been introduced since 1993. This concept is defined as a management tool comprising a systematic, documented, periodic and objective evaluation of the performance of environmental organisation, management and equipment. However, there is still no Reclamation Act in operation till date, though environmental management cells are being created in the mining organisations for the purpose.

Mining is essentially a commercial activity with an eye on profit, any serious effort by a mining company at the fag end of the mine's life or after the closure for reclamation may be doubtful. Adequate amount of cess, legislation, monitoring and policing may not help in the reclamation and bio-diversity restoration, because it is a labour intensive and a difficult mission to carry. Due to this reason, the mine reclamation has not been as successful as it ought to have been. There are many old mines that still lie in desolate conditions, posing a serious threat and a serious environmental hazard. The long term planning of the after use of the abandoned mines will be necessary with increasing demands of land due to population explosion.

The dovetailing of the mined area with long term land use planning in the normal development activities shall be necessary. The development strategy has to be so devised so as to be in harmoney with overall conservation, as long as mining is taken up in consonance with various zones viz. ecologically sensitive and biologically rich; restricted zone with managed land use under controls and protection measures and zones where developmental activities can be taken up. In the restricted zone a trade-off has to be worked out between the loss incurred due to destruction of ecological balance due to mining and the benefits of mineral extraction. It would, however be appropriate to evaluate not merely the biomass but the cost of biodiversity and the genetic stock that is likely to be affected.

Such an evaluation shall be necessary in various parts of the country. There are many locations where mineral locations are overlain by good forest cover, rich in biodiversity. Under such situations conservation efforts calls for both *in situ* and *ex situ* measures. For *in situ* conservation minerals extraction might have to be abandoned and the area protected by creating enclosure or declaring them as National Parks or Sanctuaries to provide effective protection cover. *Ex situ* conservation shall necessitate maintenance of genetic resources outside their natural habitats which is extremely essential for low volume and high cost genetic stock.

The intention in these pages is not to describe the technology for mined areas which is by and large available both within and outside the country (Soni *et al.* 1989; Dhruva Narayan *et. al.* 1987; Katiyar et al 1987, 1999; Misra *et. al.* 1988; Rajwar 1988, Gupta 1978, 1982, 1996), which gives hope that mining activities can be undertaken in conformity with the objective of mineral exploitation as well as natural/ biodiversity conservation.

WHAT MAKES BIODIVERSITY A FASHIONABLE WORD TODAY?

The rapid extinction of species brought issues relating to biodiversity into sharp focus. Nations and governments realised that individual species were becoming extinct, in addition to large scale habitat loss. The Convention of Biodiversity (CBD) was agreed upon at the Earth Summit in 1992. The mandate of CBD was to evolve ways and means of protecting and conserving remaining diversity of life in the world. The equitable use of this collection of living things has, as yet, not been resolved even after five years of Earth Summit. Economically poor nations are richer in biodiversity as compared to economically richer nations. Higher levels of biodiversity are in the rainforests of the tropics, tropical mountains, coral reefs of coastline. India is considered one of the 12 megadiversity centres. Pilfering genetic resources with no return to gene rich countries is taking place. Naturally economically advanced countries want to control and exploit genetic wealth of the tropics and reap benefits that acrue due to bio-technological use of these genes. The third world countries are opposing the new regime being thrust down on poor countries in the form of Intellectual Property Rights (IPR) and Exclusive Marketing Right (EMR). It is estimated that from plant based drugs in USA the revenue surpass 300 Billion US Dollars.

Efforts in India are being made to conserve biodiversity both *in situ* and *ex situ* and about 4 per cent of the total area of the land has been designated as Parks, Sanctuaries, Reserves (Biosphere), Botanical Gardens, Zoological Parks etc. The National Bureau of Plant Genetic Resources and National Bureau of Animal Genetic Resources are the nodal agencies for the collection and maintenance of genetic resources, primarily for *ex situ* conservation of diversity. The International Model of IPR protection is fundamentally opposed to Indian conditions where knowledge is jointly produced and exchanged in a traditional way to generation after generations. Yet another area of concern for India is the right of small farmers who ensure country's food security by saving, conserving and freely exchanging seeds.

Such an experimentation at the grassroots level has helped to evolve seeds suited to each ecological niche. This stands threatened today under the TRIPS (Trade Related Intellectual Property Rights) dispensation. Transferring the control of seed production to large seed companies would ensure only monopolistic control.

It is worth noting that many of the plants have traditional use in medicine. In case of NEEM (*Azadirachta indica*) out of the 26 patents 20 are in America, two in Japan, three in India and one in Germany. Similarly Haldi (*Curcuma domestica*), a traditionally used plant, has been listed for patent in USA. Species like *Tulsi* might also be in the pipeline. Locally useful plants have been collected and their shipment has become a major enterprise by many multinational companies like Ceiba Geigy, Hoechst, Sabina corporation, Phytera Inc., Ecofarm & Phyton Catalytic Inc. USA, Xenova Ltd. UK and Pharma Mar (Spain), who have set up sophisticated screening laboratories. Considering that more than two-third of the World's plant species, at least 35,000 is estimated to have medicinal value-came from the developing countries. The conservation of biological diversity for pharmaceutical use is critical to India, that is the home to some 7500 medicinal species. An intergovernmental meeting of the developing countries estimated the value of developing country's germplasm to the pharmaceutical industries at 47,000 million US Dollars by the year 2000 AD. It is primarily this that some countries continue to drag their feet when it comes to ratifying the COB. The convention, however, recognises that "plant, animal and micro-organisms are no longer mankinds heritage but a national property calling for a fair and equitable sharing of the benefits arising out of the utilisation of genetic resources". What is required today is immediate policy initiatives on how to check the piracy of the biological resources, including plants, animal and human beings.

The Economic Espionage Act passed last September by USA takes espionage from military domain and redefines intellectual Property infringement as a crime. On the contrary, globalisation is being used to force other countries to give up their national interest, national security and national sovereignty. The definition of economic espionage in terms of Intellectual Property infringement is arbitrary since IPR is being expanded to other areas like public domain of universe systems as well as collective knowledge of non-western societies. Though collective and co-operative research between nations such as Asia Pacific region could undoubtedly pay dividends, it is also urgent for the countries of the regions to reset their research and other priorities.

A draft biodiversity proposal, in India however envisages the creation of State and District biodiversity cells to regulate biodiversity use. In addition each village Panchayat is required to maintain a community biodiversity register for documenting existing knowledge on local use and practices. Foreign access to biological resources, according to this proposal, shall be permitted under strict material and transfer agreements, prior information consent (PIC) of the community and charge of license fee.

Though the need for such a legislation is well established, the government, as usual, has hardly moved forward, leaving foreign prospectors to feerly exploit the resources. Indian policy makers are grossly undervaluing the country's rich biological wealth which may be ruthlessly squandered for short term monetary gains under the enthusiasm to welcome foreign multinationals.

LITERATURE CITED

Dhruva Narayana, V.V. and Ram Babu (1983). Estimation of Soil Erosion in India. Journal of Irrigation and Drainage,109 : 409-414.

Dhruva Naranaya, V.V., Katiyar, V.S. and Dadwal, K.S. (1987). Mined area reclamation in Mussoorie hills. *Indian Journal of Soil Conservation*, 15 : 82-89.

Gupta, R.K. (1971). Planning Natural Resources. A Hand book of Aerospace Techniques for Integrated Surveys and Studies in Natural Resources Ecology. Navayug Traders, New Delhi. 264 pages.

Gupta, R.K. (1978). Some Aspects of Himalayan Ecosystem vis-a-vis Maldeota Phosphorite mine Operations. Pages 97-109 In : Deshbandhu and Beena Bharadwaj. *Ecology & Development*. New Delhi.

Gupta, R.K. (1982). Reclamation and use of Coalmined ecosystem in India. A review. Pages 510-515. In : Wali, M.K. (Editor). *Ecology & Coal Resource Development*. Pergamon Press. New York.

Gupta, R.K. (1996). Ecological Techniques for the Restoration of a minded ecosystem in the outer Himalayas. Proceedings of *Ecosummit*, 1996. (Abstract) Copenhangen, August 19-23.

Gupta, R.K. and Chauhan, Anita (1994). Biodiversity Protection and Soil Conservation in India. *Indian Journal of Soil Conservation*, 22: 201-213.

Gupta, R.K. and Chauhan Anita (1994). Biodiversity and Forest Conservation for Ecological Sustainability of the Himalayan Watersheds, 8th ISCO Conference, New Delhi, pages 408-410.

Katiyar, V.S. Shastry, G. and Adhikari, R.N. (1987). Mechanical Measures for mine spoils and landslides. Indian Journal of Soil Conservation, 15: 108-116.

Katiyar, V.s., Juyal, G.P. and Dadwal, K.S. (1990). Management of a Mined Watershed. *Third IWRS National Symposium*, Kanpur, Pages 110-117.

Maudgal, s. and Kakkar, M. (1994). Mining Operations in Ecologically sensitive regions. Third IWRSS National Symposium, Kanpur, 19-30.

Misra, R.C., Malhotra, R.C. and Joshi, M.N. (1988). Perception of Ennvironmental hazards in Dehradun Valley due to Exploitation of Natural Resources. Mining and Environment in India, *Himalayan Research Group*, Nainital. 294-298.

Rajwar, G.S. (1988). Limestone Quarrying–a Hazard to Mussoorie Mountain and their Ecosystem. Mining and Environment in India. Himalayan Research Group. Nainital. 294-298.

Sahni, K.C. (1967). Protection of Rare and Endangered Plants in the Indian Flora. Proceedings of the *IUCN/IITM Workshop*. New Delhi. Mimeo.

Soni, P., Vasistha, H.B. and Kumar Om, (1989). Biodiversity in Surface Mined Areas After Reclamation, The Indian Forestry 115: 475-482.

WII (Wildlife Institute of India) (1993). Biodiversity and Its conservation in India (A Primer) pp 1-19. Dehradun

Wilson, E.G. (1990). Threats to Biodiversity in Managing Planet Earth. Readings from Scientific American. W.H. Freeman. New York. pages 46-68.

Chapter - 4

Biodiversity and Sustainable Development: Issues in Conflict and Conservation

*Nisar Ali**

ABSTRACT

The ozone depletion, desertification and proliferation of toxic chemicals and hazardous wastes with urban phenomena are the natural outcome of sustained discoveries and massive industrialization across the globe. This reinforced the convergence of governmental agencies and green activitists to think dissemination information and policy formulation on biosafety, environment facility and judicious use of planet's resources. From Rio Summit of 1992 to Global Forum of 1994, scores of policy documents were prepared to make human habitat to realize the conflict in biodiversity and its conservation. The damages to biological diversity and environment in seventees and eightees in Europe and continuous strain on biomass and other forest resources in developing nations to make one to realize conflict resolution of biodiversity. The inadequate fresh water resources, growth of wastelands and pollution need a fresh look for economic and industrial policy at the global level. In India many streams, lakes and rivers have become outlets for sewage bringing about changes in aquatic ecosystem inconsistent with the basic norms of biosafety.

Key Words: biodiversity, biomass strain, conflict, poverty, sustainable development, toxic emissions, vulnerabilit.

INTRODUCTION

Consequent upon the sustained discoveries and industrialisation through massive use of fossil fuels on the one hand and intra and inter-regional transfer of technology and resources on the other, the world is gripped in a concomitant resultants of ozone depletion, desertification, proliferation of toxic chemicals and hazardous wastes, expansion of human habitat through urban phenomenon posing a threat to environment and biodiversity. All

*Department of Economics, University of Kashmir, Hazratbal, Srinagar-190006

this has converged the global community to search for solution-driven and result-oriented performances through sustainable development. The protection of environment and conservation of biological diversity are the two major themes that attracted global gathering of 400 thousand individuals from 18000 organisations drawn from 172 countries to serve the cause under a single banner of Rio Summit of 1992 followed by Global Forum in June, 1994. A healthy and hygienic life and quality environment for all are the major objectives of sustainable development. It simply implies achievement of improved living conditions and quality life through efficient and judicious use of planet's resources, while minimising impact of wastes on its eco-system. Most of the materials and energy resources are consumed today in the process and operation production and maintenance of human settlement amidst of emissions of toxic wastes and are serious concerns to which global community addresses. While the nations particularly the industrial ones, geared up to take off from primitive stage of growth, aimed at high per capita incomes through industrial growth, followed by equity in the distribution of gross national product, were the central concerns until fiftees and sixtees, therefore, ruled by environmental insensitive growth theories and models. It is now that macroeconomic model builders realise that economic development and environmental stewardy are interdependent if sustainable development is to achieved (Ruskin 1993 Flores 1994 Mayde 1994).

SUSTAINABLE DEVELOPMENT AND CONTEMPORARY GLOBAL POLICY

The sustainaning of unsustainable development in developing nations, as a result of reinforcing factors, like population growth, over-intensive farming, excessive use of ground water, degradation of lands, deforestation, soil erosion, farming on sub-marginal lands, energy scarcity and water pollution seems to be difficult task for conflicting objectives of take-off in development on the one hand, and conserving biodiversity, with poor technological base, on the other. Therefore, the environmental deterioration in developing countries is more serious than the developed ones. The poverty alleviation development process would more lead to environment degradation than eco-friendly development. Thanks to the decade of 90s which witnessed several groups and fora creating general awareness on consequential impact of industrial growth on environment and biodiversity.

The Intergovernmental Committee on Convention on Biological Diversity, (ICCBD) 1993, addresses to the several issues like funding biosafety issues, methods of achieving conservation and sustainable use of biological diversity. The Global Forum, 1994 invited experts and green activists at Manchester to work out "Changing Tactics in the Changing World" aiming biosafety and conserving biodiversity through innovative programmes. The Rio Summit of 1992 and Manchester meet of 1994 created International Publicity Centre, (IPC) and International Media Centre, (IMC) respectively aiming dissemination of information and knowledge through modern electronic communication technology across the globe. Thus, the Earth Summit followed by Global Forum meet at Manchester created world over concern to design national as well as global policy for saving planet on the one hand, and protecting environment and conserving biodiversity in the countries on the other. The African Ministerial Conference on Environment (AMCEN) in November, 1993 at Adis Ababa addressed themselves to the designing of direction for environmental policy and programmes at sub-regional levels. In the same year National Association for Resource Improvement (NARI) in Bangladesh held a Workshop in September on "Women, Environment and Development" and adopted an Action Plan and also created an institutional framework for implementation of environmental policy as per agenda 2 of Global Forum. Similarly, Indian Ministry of Environment and Forests, in conjunction with

Tata Energy Research Institute, New Delhi has prepared a first draft of countrys detailing the phase out schedule of ozone depleting substances, for example, cholorofluore-carbons, CFCs. The United Kingdom presented her "Strategy for Sustainable Development to United at the end of 1993. The Royal Society of Canada and Canadian Climate Board committed by report that it" would be possible and considerably beneficial to reach Canadas interim targets of stabilising greenhouse gas emissions at 1990 level by the year 2000 and to achieve absolute reduction of 20 per cent by 2010". The World Bank initiated, from early 1990s, policy discussion on replenishment process of Global Environmental Facility, (GEF). In fact participating countries reiterated their intention to contribute to reprenishment of GEF. The GEF was launched in 1991 with collaborative institutional structure involving World Bank, UNDP and UNEP with new "incremental cost" approach providing concessional financing for projects with global environmental benefits. The concern with environment and biosafety has not been realized in isolation of human habitat. Over 100 delegates all over the world met in Dhaka in July, 1993 to debate the linkage between" Poverty and Environment" as a follow up of Rio Summit setting up Global Forum on Environment and Poverty (GFEP) and mandated Bangladesh to set up its first Secretariat. The world experts, while addressing to environment-poverty-linkage deeply felt the role of poor/poverty in the regeneration of environment employment, empoverment, innovation, cropping strategies, militariarization of resources and ecosystematic planning. The Integrated Convention on Climate Change brought common issues under one umbrella in 1995 and participants first addressed themselves to methods for calculating and evaluating greenhouse gas emission criteria for joint implementation of the programme to combat "climate change". The Integrated Panel on Climate Change, (IPCC) in 1994 advocated reduction of 60 per cent or more in CO_2 emissions are required to stabilize atmospheric concentrations.

In fact the first serious meeting, after Rio, took place in October in Geneva in 1993 when Governments demonstrated concern with conservation of biodiversity. The Intergovernmental Committee for the Convention of Biological Diversity, (ICCBD) was set up and addressed to the issues like, (i) conservation of biodiversity, (ii) sustainable use of resources, and (iii) equitable distribution of profits from biodiversity.

Prior to this, Global Biodiversity Forum addressed to the critical ecological, economic, institutional and social issues for furtherence of conservation of biological diversity and Global Biodiversity Strategy.

WATER RESOURCES AND ENVIRONMENTAL FACILITY

Thus, the foregoing concern and addresses confirm that environment and biodiversity are essential pillars of sustainable development. This in turn is equally concerned with "fresh water", waste lands, water resource management, soil erosion, ravine formation and so on. The concern about fresh water was realised way back in 1977 when United Nations Water Conference was held in Argentina. The UN Commission on sustainable Development has further taken this issue and in 1992 about 500 Government experts met at Dublin at the International Conference on Water and Environment. In fact, the serious and widespread concern about fresh water has not diminished but has risen over the past 16 years. The said Conference felt that rapid population growth, expanding economic activity and poor irrigation facilities endanger water resources which are often already overloaded with toxics, other dangerous chemicals and raw sewage. The imbalance between supply and demand for fresh water in a large number of countries, uncertainty looms the devastating long run effects of the said prevailing situations.

The seriousness of the threat to global fresh water resources posed by several constraints and strains vary from region to region. In Pakistan fresh water problems pose major constraints on development, while in India demand for fresh water is far exceeding its supply. The excessive farming strain on under-ground water resources in India have lead to deepening of water table which should be a serious cause of concern. While the biological diversity can be thought to be conserved at the national and sub-regional levels, but the problem of fresh water extends the political boundries that has led to the several international water disputes. Thus, the fresh water resource management need to be effectively taken up at the international level. Much of the water resources go to the farming and its inefficient use cause waste considerable amount of it.

Global Environment Facility, GEF is committed to protecting biodiversity around the globe. The GEF funds are public fund and in the view of Newcombe (1994) the subscribing Governments support to developing nations, under the biodiversity convention are likely to be insufficient and could not realize the reduction in the loss of global biodiversity. On the other hand, "venture capital" funds under public and private partnership are explored and to that effect are underway to bring in private investment for biodiversity conservation in developing countries. It is also believed that private investment would experience new areas of activity and in that direction blodiversity investment areas could include:

(a) New timber products from forest lands/drugs, industrial products from plants/animals, food products, ethnobotinical products,

(b) timber from low impact sustainable harvests from primary forests, eco-tourism,

(c) preservation of crop varieties and development or under-used species in agriculture, and

(d) acquaculture game ranching species threatened in the world.

Initiating the financial flow would ensure sustainable activity to protect the national biodiversity.

VULNERABILITY OF BLODIVERSITY CONSERVATION

Biological diversity is susceptable to variety of degradations, particularly to human settlements, derivation of energy sources, construction of dams, and industrial complexes and above all conflict in the social objective functions between wooded wild-life and recreation. In the domain of conflict, hunting is an important activity in almost all countries, with Ireland, Turkey and United Kingdom as major exceptions. Many countries are of the view, that, "venison, rent for hunting grounds, and effects on other sectors of economy/hotel/represent an economic value equivalent to a substantial proportion of the value of wooded production" (FAO, 1985). In Austria bark damage caused by game affects 6.3 per cent of stems, and 3.7 per cent of standing value. A heavy increase in damage was observed during 1975-85. It is also claimed that "role of hunting in the national economy is positive in spite of considerable damage to forests" (FAO, 1985). In Ireland prior to 1968, the forests were regarded as private property and any entry or public use was actively discouraged except under special license for hunting until 1985. There were 70,000 licensed hunters in Ireland. In United Kingdom it was estimated that about 26 million days visits were made to forests each year and forests were thrown open to public for many activities including fishing and shooting. The European Economic Commission report observes that, "world animal damage will increase in future as reaforestation of old forests become more widespread" (FAO, 1985). In Greece and Italy hunting has been more pronounced to the extent of 300 thousand hunters were in Greece alone and in Portugal the total hunters were 250 thousand and in Spain the number of hunters was 1.2 million in 1985. This has been invariably the case in whole of the East and West Europe.

STRAIN ON BIOMASS AND FOREST RESOURCES: GLOBAL SCENARIO

If one could look at the other forest life, that is, the forestry itself, much of the energy, for domestic end-use, is produced from forests and agriculture in developing nations. In Bangladesh alone 90 per cent of energy end-use in agriculture is produced within the system, (Bhatia, 1985). In India 150 X 10^{12} Kcals of biomass was given by agriculture in 1992. Biomass continues to be a major source of energy in rural sector of developing world. It accounts for over 90 per cent of total household use in poor countries and more than 30 per cent wood fuel is needed in urban household sector in developing countries, (Ali, 1991, 1996, 1997, Alam *et al.*, 1985). The potential yield from forests has been estimated by Dunkerly *et. al.* (1981) and fall within the range of 500 million to 5 billion meteric tonnes of oil equivalent (mtoe) per year, which mainly come from forests areas. The potential annual energy from biomass varies between regions. It is estimated 370–2000 mtoe in Latin America, 280–1900 mtoe in Africa and 650-1700 mtoe in Asia, (Ramsay, 1985). According to Food and Agriculture Organisation, FAO, UN estimates of 1976, the consumption of food fuel was 1184 million cubic meters, that is, 47 per cent of total wood production (Vaelav and Knowland, 1980). Most of the developing world depend upon the biofuel for their energy requirements, particularly in household sector. Table 4.1 shows the dependence of these selected countries on biomass for their energy end-use.

Table 4.1 : Percentage of non-commercial energy consumption in some selected developing countries

S. No.	*Country*	*Percentage of traditional fuel, biomass, to total energy consumption.*
1.	Nepal	96.07
2.	Bangladesh	86.19
3.	Nigeria	62.18
4.	Sri Lanka	59.54
5.	Thailand	55.30
6.	Indonesia	54.88
7.	Phillippines	52.70
8.	Zambia	44.57
9.	Zimbabwe	40.92
10.	India	40.46
11.	Pakistan	39.91
12.	Brazil	38.36

Source: Ali, N. 1986

The above statistics reveals a considerable strain on forests and is expected to continue for decades. Thus, the conflicting objectives biological diversity conservation on the one hand and continuous dependence on the biomass as source of energy coupled with strain on wild life, intensify vulnerable and put the present ecosystem in jeorpardy. The present technological diffusion reinforce the factors for furtherence of this conflict (Ali, 1991).

BIODIVERSITY AT SUB-REGIONAL LEVEL

At the national level many non-governmental organisations and the Government have launched multiple programmes involving rural community in particular and urban population in general for conservation of biodiversity and protection of environment. Among the several constraints the emerging problems are the unabated growth of wastelands by water logging, soil erosion, the reckless management of farm fields causing infertility of top soils, growth of sand dunes, formation of ravines and marshy lands including the formation of "Usar" lands. All these put together exert strain on biomass growth and pollute environment. These problems need administrative input with community involvement. Some community based programmes were embarked upon in India towards resource management, yet the gravity of the problem continues. In order to realize objective to sustainable development one has to address to optimal water resource management for biosafety. This would contain growth of wastelands, which may be classified as under:

Shallow soils

Shifting cultivation in forests and mismanagement of farm fields cause top fertile soil to wash away with run-off rain water every year. The continuous soil erosion cause accumulation of gravel and stones in the field and moisture retention capacity of the soil decline.

Send Dunes

There are undulating heaps of sand very poor fertility and water retaining capacity. They get heated in summer and damage the crops. They are usually of two types: *First,* shifting sand dunes of desert which change the location due to physical movement of sant particiles due to wind and consequently damage vegetation, *secondly,* coastal sand dunes which do not shift from their original sites. Water table in these locations is quite high and not suitable for drinking due to high salt content and only can be used for certain specific tree species.

Ravines and River Banks

In sandy and loamy fields, ravines are gradually formed due to flow of rain water in the gullies. Timely farm operations and construction of contour bounds can restrict the formation of ravines. In attended areas farm fields adjacent to ravines, get gradually converted into ravines. The river banks too are subjected to soil erosion. The silt eroded from the banks gradually deposit at the river basin at some other locations. This affects the normal flow of water and the river is likely to change its course.

Water-logging and Marshy Fields

Wastelands are generated by waterlogging found in coastal areas or along the flat river banks where the water table is very high or where no outlet is available to drain out excess water.

There are also drought affected cultivable fields where fertile lands would remain idle due to lack of water. Water being a critical input for raising biomass, with adequate and timely supply of this input the biomass may increase 300-400 per cent. Though several programmes like gully plugging, building of percolation tanks, digging of open and bore wells have been put inn to force, yet water resource distribution in the country is extremely at sub-optimal level. There is a huge capital input required to generate optimal water resource input to convert wastelands in cultivable fields for raising huge biomass required by mankind and ecosystem. The controlled method of water distribution is important to prevent soil erosion and waste of water resources.

A considerable amount of land form wasteland of the said categories which inturn strain on the underground resources and hence reduce fertility of lands due to decrease in moisture retention capacity. Due to conflicting objectives environment protection is more talked about and less felt in terms of pursuance and biodiversity conservation what is required is policy mix to achieve sustainable development. Among several States Jammu and Kashmir, could be rich in its environment and biosafety relative to other sates, but due to certain amount of mismanagement of land, water resources, forests and to some extent wildlife, biosafety may form a challenge in years ahead if existing policy goes unbated. The maintenance of ecosystem is to be imbibed in the culture of the community. The State is in a advantageous position as compared to rest of the country as emission of toxins and other hazardous wastes is potentially low, but the strain on forests continue to be alarming.

STRAIN ON BIOMASS AND FORESTS OF J&K STATE

While the forests in the State of Jammu and Kashmir are precious to maintain ecosystem, they are equally assumed to have greater importance for the economy-mainly as revenue yielding sector. Thus, in order to avail of benefits from forests to maintain the economy, a considerable aforestation is to be made to offset the strain on ecosystem. The State has forest cover to the extent of 22193 sq. kms. out of which 91 per cent is the forest area and 9 per cent wildlife area. Among the principal forest species, deodar, chir, kail and fur account for the 43 per cet of area, while 8.5 per cent area is under non-coniferous. The Table 4.2 gives the relative position of the State in the country:

Table 4.2 : Percentage and per capita distribution of forest area in selected States in India: 1995

S. No.	*State*	*Total State forest area as percentage of geographical area of the country*	*Per capita forest area. ha*
1	2	3	4
1.	Andhra Pradesh	23.2	0.1
2.	Arunachal Pradesh	61.5	6.0
3.	Assam	39.1	0.14
4.	Bihar	16.3	0.00
5.	Gujarat	9.9	0.05
6.	Haryana	3.9	0.51
7.	Himachal Pradesh	3.9	0.74

1	2	3	4
8.	Jammu and Kashmir	9.1	0.26
9.	Karnataka	20.2	0.09
10.	Madhya Pradesh	35.0	0.23
11.	Punjab	5.5	0.01

Source: Forest Statistics (1994), Forest Deptt. Jammu and Kashmir Govt.

Since much of the land is under mountains, the forest area as percentage of geographical area is low. While in 1955-56 extraction from forest, timber, was 313.69 thousand cubic meters and that of firewood 142.97 thousand cubic meters. The same doubled in 1980-81 and reached to 609.21 thousand cubic meters of timber and 238 thousand cubic meters of firewood. The minor forest products realised were worth Rs. 10.45 lakhs in 1955-56 and the same increased to Rs. 1650 lakhs in 1995-96. The forests in the State of Jammu and Kashmir have been more taken as revenue realising sector, and the revenue realised in 1950-51 was/ Rs. 49.6 lakhs which increased to Rs. 4314 lakhs in 1992-93. This seems to be in contradiction to agenda 21 of Global Forum.

The supply sources of biomass are forests, private farms and individual plantations. While the Government source constitute 52 per cent, private farms 35 per cent and purchases from open market 8 per cent followed by other sources of 5 per cent.

A study carried out by Social Forestry Project estimated biomass in the State, and we have computed from this study the stocks and fuelwood as given in Table 4.3.

Table 4.3 : Percentage distribution of estimated growth of stocks from various sources/ broad leaved species/and fuelwood of various species in Jammu and Kashmir State./Lakh Qts, 1985

Species	*Sources of selected species*		*%*	
	Forest	*pvt farms*	*individual plantation*	*Total*
1	2	3	4	5
Willow	27.65 (60.30)	68.35 (37.51)	4.00 (2.19)	100 (100)
Papolar	0.06 (-)	98.66 (98.71)	1.28 (1.29)	100 (100)
Kiker	2.95 (1.84)	87.54 (88.09)	9.51 (10.07)	100 (100)
Fruit trees	- (-)	99.93 (99.99)	0.07 (0.07)	100 (100)
Walnut	3.13 (3.13)	96.87 (96.87)	+ (-)	100 (100)
Eeuclyputs	- (-)	94.15 (94.15)	5.85 (5.85)	100 (100)

1	2	3	4	5
Sisoo/Tahali	9.8 (4.09)	86.76 (94.46)	3.44 (1.45)	100 (100)
Bamboo	- (-)	100 (-)	- (-)	100 (-)
Total above and other species	11.30 (18.12)	86.22 (79.88)	2.48 (2.00)	100 (100)

Figures in parentheses are fuelwood percentages. Computed from Wood Balance Study, Director Economics and Statistics for Social Forestry Project, J and K Govt, Srinagar, 1985 One Lakh = 100 Thousand

The strain on forests by way of timber extraction and woodfuels are considerable. The above table gives the extent of dependence of fuelwood in the State. In urban Srinagar city fuelwood consumption is estimated 1.93 qt per capita while the state average is estimated more than 4 qts per capita.

It is estimated that by the year 2000, the fuelwood requirements alone would be 19 lakh quintals. If we take unreported rural population depending upon fuelwood, it would be alarming and should be a serious cause of concern.

The State is abounds of natural streams. Lakes and hilly torrents which offer tremendous scope for fish production. Although estimates of fish production are difficult to make, yet one could get glimpse of strain on lakes and fish breeding streams by looking at "fish caught" annually. The crude estimates show that all species put together, 43 thousand fish were caught in 1965-66, the figure went up to 1.6 lakhs in 1995-96. Most of these estimates pertain to the Valley. The water bodies are extremely important in case of biodiversity conservation. There are established views that various species in the Valley have considerably declined over the period. It may be attributed not only to human consumption but changes in acquatic ecosystems. Many streams, lakes and rivers have become outlets for sewage. Similarly there are no estimates on birds and wildlife in the State, but these have been subjected to game recreation for a long time so much so one has a reason to believe that migrant bird visits to the Valley have drastical declined. The existing 'rakhs' have extremely low seasonal density of birds and bird movement.

CONCLUSION

The concern about environment and biodiversity conservation has been institutionalised by establishing scores of forums and organisations at the international and regional levels taking into consideration location-specific problems and issues particularly in developing countries. However, contradictions still persist in the conflicting objectives when one encounters economic activity and technology transfer, on the one hand, and environment protection on the other. Such a contradiction and conflict can be minimised and overcome when biosafety culture is generated by imbibing world community on sacrifices and judicious use of resources at macro and micro level.

LITERATURE CITED

Ali, N. (1986). Energy and economic growth: a world outlook staff discussion paper. Institute of Economics of Hungarian Academy of Sciences, Budapest.

Ali, N. (1990). Technology diffusion and Indian labour movement. pages 232-242. In : Okologie Technischer Wandel und Arbeiterbewegung, Vienna, Austria.

Ali, N. (1996). Changing structure of energy demand and economic growth. pages 351-388 In : Khan, A.H and Pandit, A.K. (Editors). Ecology, Environmennt and Energy. Department of Distance Education, University of Kashmir, Srinagar.

Ali, N. (1997). Energy consumption, economic growth and disparities: a cross country analysis of developed, developing and Muslim economics. The Business Review, 3: 129-136.

Alam, M; Dunkerly, J; Ramsay, W and Elizabeth D. (1985). Fuelwood in urban markets. Concept Publishing Co, New delhi.

D.E.S. (1985). Wood Balance Study. Directorate of Economics and Statistics, Planning Department, Jammu and Kashmir State.

Dunkerly, J; Ramsay, W; Gordon, L and Cecelski, E. (1981). Energy Strategies for Developing Nations. John Hopkins Press, Baltimore.

F.A.O. (1985). The Forest Resources of European Economic Community Region, Food and Agricultural Organization of the United Nations, Rome, Italy.

Florea, E.O. (1994). Human Settlement and Quality of Life. Network (April) Geneva.

Forest Statistics (1994). Digest of Forest Statistics. Forest Department, Jammu and Kashmir Government, Srinagar.

Mayde, J. (1994). Sustainable Development and Environment. Our Common Future, Network Geneva.

Newcombe, K. (1994). Public and private partnership in protecting Global Biodiversity. Network (February-March), Geneva.

Ramsay, W. (1985). Biomass Energy in Developing Countries. Energy Policy (August), Butterworths, London 326 pages.

Ruskin, P. (1993). Sustainability and Equity. Network, Out Common Future, Geneva.

Vaclav, S and Knowlannd, W.E. (1980). Energy in Developing World. Oxford University Press, Oxford.

SECTION - II

BIODIVERSITY AND CONSERVATION STRATEGIES OF INLAND WATER ECOSYSTEMS

Chapter - 5

Conservation and Management of Wetlands in India

C.L. Trisal*

ABSTRACT

The paper gives an overview of diversity of wetland habitats in India and their ecological significance. The measures undertaken for the conservation such as, protection through protected area networking; catchment area treatment to control siltation; pollution control; wildlife conservation; control of exotic weeds, generating awareness and peoples participation have been reviewed. The institutional mechanism for implementation of management action plans at the State Government level, alongwith the existing regulatory regimes have been discussed in the paper. India has designated six wetlands for inclusion in the list of Wetlands of International Importance under the Convention of Wetlands of International Importance especially as Waterfowl Habitat (Ramsar Convention, 1971). An overview of the activities carried out by India under the Convention are discussed in the paper.

Keywords : biodiversity, ecological significance, conservation, Indian activities, Ramsar Convention, Wetlands

INTRODUCTION

India, by virtue of its extensive geographical stretch and varied terain and climate, has a wealth of wetland habitats. These wetlands are of immense importance to the people for their ecological and economic security. They are a part of India's cultural and natural heritage and provide sustenance to a large population.

The wetlands in India harbour a wide gamut of flora and fauna and carry out a crucial role as breeding grounds for waterfowl, fish and other aquatic life. These wetlands also serve as an important life support system by helping in flood control, recharging of aquifers, regulating water quality, pollution abatement and in reduction of sediment load etc.

*Coordinator, Wetlands International - South Asia New Delhi E-mail: wisaind@del2.vsnl.in

The ethos of conservation of natural resources is ingrained in Indian culture. The people in different parts of the country have evolved unique conservation and management approaches based on their culture, religion, ethics and traditions. In spite of this, our wetlands are shrinking or getting polluted at a rapid rate, as in other parts of the world. This is happening not so much due to natural reasons but more so because of incursions and encroachments. In view of this several significant steps have been initiated for conservation of wetlands.

The present paper gives an overview of wetland resources in India, their status and measures taken for their conservation.

WETLAND HABITATS IN INDIA

The Indian landmass extending over a total area of 329 million hectares, is bounded by the Himalayas in the north, the Bay of Bengal in the east and the Arabian Sea in the west. It consists of three well defined regions:

(i) the trans-Himalayan mountainous zone including the North-Eastern Hill Ranges;

(ii) the Indo-Gangetic Plain; and

(iii) the Deccan Peninsula (including the Eastern and Western Ghats).

The Himalaya, with a total area of nearly 100 million hectares is strategically important as a highway for living species linking Chinese and Malayan elements in the east and Palaeractic and Ethiopean elements in the west. Besides Himalaya also deflect the rain bearing clouds into the region thereby controlling the rainfall regime.

The greater part of Northern India and the coastal areas are characterised by the alluvial soils formed by the silt carried down by the rivers. India is fairly rich in water resources with an average rainfall of more than 130 cm. There are 14 major river systems, with their extensive network of tributaries, draining the Indian landmass in all directions. Thus, floodplains are extensively distributed throughout the country and many oxbow lakes are found in the floodplains of these river systems. There are innumerable natural wetlands ranging from temporary ponds and seasonal floodplains to permanent freshwater marshes and swamp forests; and inland saline lakes. All along the coastal line mangroves, marine wetlands and coral reefs are encountered. Man-made wetlands which are spread throughout the length and breadth of the country include shallow ponds, tanks, reservoirs etc.

The wetlands of India have been classified into nine categories in the Directory of Asian Wetlands (1989) which include:

(i) tanks, reservoirs and other water bodies of the Deccan peninsula;

(ii) backwaters and estuaries of the West Coast of the peninsula;

(iii) the vast saline expanses of Rajasthan and Gujarat (mainly Rann of the Kachchh);

(iv) freshwater lakes and reservoirs from Gujarat eastwards through Rajasthan and Madhya Pradesh;

(v) deltaic wetlands (including mangroves) lagoons and salt swamps of India's East Coast;

(vi) marshes, jheels, terai swamps and chaur lands of the Indo-Gangetic plain;

(vii) floodplain of the Brahamputra and the marshes and swamps in the hills of North Eastern India;

(viii) lakes and rivers of the montane (primarily palaearctic) region of Kashmir and Ladakh; and

(ix) wetlands (primarily mangrove associations and coral reefs) of India's island systems.

Based on the vegetational characteristics, different wetland types have been recognised which include freshwater marshes and swamps, tropical hill valley swamps, creeper swamps, seasonal swamp forest, moist temperate alder forests and mangroves.

Broadly, wetlands in India can be categorised into following:

HIMALAYAN WETLANDS

The Himalayan region is bestowed with a series of wetlands showing remarkable variations in regard to their nature and biota. The region can be further divided into three sub-regions : (i) Western Himalayas; (ii) Central Himalayas; and (iii) Eastern Himalayas.

In the Western Himalayas, Ladakh and Zanskar region (4000 meters above mean sea level) harbour a large number of high altitude wetlands. These biotopes are exposed to cold desert conditions with high solar radiation during short summer period. Both saline and freshwater wetlands are encountered in this region. The important ones are : Pangong Tso, Tso Morari, Chantasu, Noorichan, Chassul and Honlay. These wetlands are traditional breeding grounds for a number of waterfowl species. Wetlands situated in Kashmir Valley include very prominent systems viz. the Dal, Anchar, Wular, Haigam, Malgam, Hoakersar, Kranchu, etc.

In Himachal Pradesh, several natural lakes occur in tropical, sub-tropical, temperate and alpine climatic regions with altitudinal variations ranging from 400 to 5000 metres above mean sea level. Major high altitude (2500 to 5000m above mean sea level) lakes are Nako, Chander Natan, Chandratal, Surajtal, etc. In the middle altitudinal zone (1500 to 2500 metres above sea level), the prominent lakes are Khajiar, Kareri, Kumarwah, Rewalsar etc. In the low altitudinal zone (below 1500 metres above mean sea level), Renuka and Saketi are the important lakes and they experience sub-tropical climatic conditions.

The wetlands situated in the Central Himalayas are mostly located in Kumaon region. Nainital, Bhimtal and Naukuchital and a number of other small lakes are situated within this region. In the eastern Himalayas, large number of wetlands are situated in Sikkim, Assam, Arunachal Pradesh, Meghalaya, Nagaland and Manipur. These wetlands together with sanctuaries of Brahamputra Valley are internationally important wetland systems for a number of bird species like *Ardea imperialis, Leptoptilos dubius, Francollis quiaris, Perdicula manipurensis, Grus niouicollis, Houbaropsis benoalensis, Saxicola jerdoni, Pellorneum palustre, Moupinia altirostris, Prinia burnesii* and *Ploceus meoarhynchus.* Besides, these wetlands also contain rhinoceroses, tigers, marsh deer, elephants, crocodiles and water buffaloes. The Loktak lake with an area of 289 square kilometres is one of the largest freshwater lakes situated in this region. The floating swamps of Keibul Lamjao adjacent to Loktak provide refuge to the Manipur Brow-Antlered Deer, *Cervus eldi eldi,* a higher endangered species.

Floodplain Wetlands

Most of the wetlands in India are directly or indirectly associated with river systems like Ganga, Brahamputra, Narmada, Tapi, Godavari, Krishna, Cauvery etc. The largest wetland regime of India is Indo- Gangetic Flood Plain. The alluvial plains of the Ganges comprises a vast crescent of lowland, about 250 to 300 km. wide, between the Deccan peninsula and the Himalayas. The wetlands of Gangetic plain and Himalayan terrai provide a winter refuge for millions of waterfowl in the States of Uttar Pradesh and Bihar particularly. All along the course of the river Ganges and its tributaries, wetlands are scattered. Prominent among them are East Calcutta Wetlands in West Bengal, Chaurs of Bihar, the flood plains of Uttar Pradesh, Chambal wetlands in Madhya Pradesh, Keoladeo Ghana National Park in Rajasthan, etc. Various multipurpose projects launched to harness these river systems have also provided a number of man-made wetlands, for example, Harike Barrage at the confluence of Beas and Sutlej in Punjab, Bhakra Nangal Dam in Punjab and Himachal Pradesh and Kosi Barrage in Bihar, Nepal border.

Coastal Wetlands

India has a long coastline of over 7500 km. All along the coastline wetlands are dominated by mangroves, salt marshes, lagoons etc. Pulicat lake, Periyar lake, Chilka lake, Kolleru lake etc. are situated within this area. There are relatively very few natural wetlands in Deccan, the southern most section of Indian peninsula. However, there are innumerable small water storage reservoirs and tanks which have been constructed near virtually every village. These tanks provide important feeding and nesting areas for a variety of water birds. The peninsulas of Saurashtra and Kutch separated from the main peninsula by Gulf of Khambhat alongwith neighbouring Rajasthan and Madhya Pradesh are of great importance from wildlife and ecological point of view.

BIODIVERSITY

Wetlands in India due to their climatic and topographical variations and other ecological features exhibit great diversity in regard to flora and fauna. The wetland flora is represented by all taxonomic groups from unicellular algae through bryophytes, mosses and ferns to woody angiosperms. Numerous herbaceous plants including submerged, free floating, rooted floating leaf and emergents occupy different niches in wetlands. Hundreds of species of planktonic and filamentous algae are occurring in different wetlands. The macrophytic species in some wetlands form characteristic zonations of emergents, rooted floating types and submerged.

Some species having a narrow range of distribution are well represented in the North Eastern region and include *Monochoria vaginalis, M. hastata, Sagittaria trifolia, Hydrocharis dubia, Blyxa japonica, Aponogeton natans* and *Euryale ferox*. Podostomaceae, well known for its endemism is represented by *Hydrobryum polypheurum* and *H. zeylanidium* in North Eastern region (Gopal, 1990). The other endemic species found in the region are *Aponogeton echinatum, A. jacobsenii, A. microphyllus, Monochoria vaginalis, M. hastana.*

The freshwater swamps distributed in a narrow belt along the Brahmaputra and its tributaries comprise *Terminalia arjuna, Bischofia javanica, Calamus* sp., *Bambusa schizostachyoides* and *B. arundinacea*. The freshwater swamps have considerably shrunk in their area and diversity. They require special attention for conservation.

The wetlands in the western Himalaya particularly the Valley lakes in Kashmir, Renuka, Rewalsar lakes in Himachal Pradesh are profusely covered by aquatic macrophytes. The increasing concentration of nutrients in Kashmir lakes have resulted in the disappearance of some sensitive species like *Chara* sp. and *Euryale ferox*. Kaul (1983) reported on the reiterating populations of several aquatic species in the water-bodies of Kashmir e.g. *Callitriche verna* and *Utricularia minor*. Over-exploitation of some species like *Nelumbo nucifera* has also resulted in its considerable reduction in Kashmir lakes.

While on one hand some plant species are threatened, on the other hand some species particularly exotics have invaded these water-bodies at the expense of the native species. *Eichhornia crassipes* has considerably increased in the tropical and sub-tropical regions thereby posing a serious threat to the aquatic plant diversity.

The wetlands in India also exhibit great faunal diversity. All the taxonomical categories of fauna from protozoa to mammals occur in wetlands among the invertebrates, arthropods and molluscs are the most dominant components. However, many of the wetlands in the region are famous world-wide for the spectacular concentrations of wildlife. Some are the centres of rare, threatened and endangered species.

The wetlands in the Ladakh region are important transit spots and breeding grounds for a variety of waterfowl (Gole, 1982, 1987a, 1987b; Kashar, 1980). Tso Morari is an important breeding ground for the bar headed geese *(Anser indicus)* and supports significant populations of great crested grebe, Brahminy. duck, ruddy shelduck, lesser sand plover, black necked cranes *(Grus nigricollis)* and black necked grebes *(Podiceps nigricollis)*. The other important wetlands in the Ladakh region like Tso Kar, Hanie, Chusul marshes and Pangong Tso also harbour a number of migratory species.

Nearly 300 species of birds have been reported from the Manas National Park, several rare species including *Pelecanus philippensis, Leptoptilos javanicus* and *L. dubius* are found here.

The Sareswa beel supports a great diversity of resident and migratory waterfowl. Both the lesser adjutant stork *(Leptoptilos javanicus)* and greater adjutant stork (L. *dubius)* are regular visitors in winter months.

The swamp deer *(Cervus duvaucelli)* inhabits the Mahao Sanctuary. The endangered white-winged wood-duck still persists in small numbers along the lower course of the river *(Directory of Indian Wetlands, 1993)*. The swamp deer and Asian Elephant *(Elephas maximus)* utilize the valley floor wetlands of the Lali Sanctuary.

The Kaziranga National Park contains 15 species of India's threatened mammals, and has the largest population of the Indian one-horned rhinoceros *(Rhinocerus unicornis)* in the world. This population increased from only 12 members in 1908 to over 1000 in 1982. The park also provides a good habitat for wild water buffalo *(Bubalus bubalus)* in India, and supports healthy populations of Asian elephant *(Elephas maximus)* swamp deer *(Cervus duvaucelli)* and tiger *(Panthera tigris)*. The common otter *(Lutra lutra)*, and Gangetic dolphon *(Platanista gangetica)* are common in the rivers of the region.

Manas National Park supports an extremely varied mammalian fauna, including several rare and threatened species. Over 40 species listed in Schedule I of the Indian Wildlife (Protection) Act, 1972, are found in Manas. The one-horned rhino *(Rhinocerus unicornis)* is dependent on the wetlands, and there are several other species such as *Caprolagus hispidus, Lutra perspicillata, Felis viverrina, Sussalvanius* and *Cervus duvaucelli* found in the wetland. The river course may have about about 1000 ha of riparian forest of grasses and tree cover which is probably the most suitable water buffalo habitat anywhere. The residual population of the wild water buffalo *(Bubalus bubalus)* is of considerable interest genetically (Woodcock 1984)

Among mammals, *Mustela erminea* is a rare species found in Phumlen, Kharung and Ikop Lakes. *Cervus eldi eldi,* locally known as Sanghai deer is the most endangered species found in Keibul Lamjao National Park in the Loktak Lake. *Sus* scrota, Felis *chaus, Vulpes vulpes, Cuon alpinus* and *Archtonyx collaris* are other noteworthy species.

Fisheries survey carried out in some selected areas of the north-western, central and eastern Himalaya by Sehgal (1971 and 1974), Joshi *et al.* (1974) have indicated a downward trend in the fish catches of some Himalayan wetlands which is mainly attributed to the anthropogenic pressures. Some of the species like Golden Mahaseer (*Tor putitora)* have disappeared from the waterbodies of Himachal Pradesh. Even some of the valuable and rare species of the Himalayan fishes, such as Schizothoracids are disappearing fast in Kashmir, especially after the introduction of German phenotypes of the common carp - the mirror and scale carps *(Cyprinus carpio specularis* and *Carpio communis).*

THREATS TO WETLANDS

Wetlands in India are under great pressure on account of manifold causes which among others include rapid expansion of agriculture, human settlements, expansion of industries and large scale developmental projects. These activities have led to shrinking of wetlands, and their biological resource potential, pollution, siltation and weed infestation. A brief discussion of the major problems confronting the wetlands in India are :

Encroachments

Large number of people living in and around wetlands have been encroaching upon these areas and vast areas have already been drained for agriculture, urban expansion and other purposes.

Siltation

Siltation is one of the major problems in most of the wetlands in India. Deforestation and other anthropogenic activities in the catchment area have accelerated soil erosion resulting in increased sedimentation rates and resultant shrinkage of wetlands. About 13 million tonnes of silt is deposited annually in Chilka lake choking its mouth thereby leading to reduced communication with Bay of Bengal. Some other lakes experiencing this problem are Wular, Dal, Kolleru, Loktak, etc.

Weed Infestation

Almost all wetlands in the country, barring a few located in the high altitude Himalayas, are heavily infested with aquatic weeds. *Eichornea crassipes, Ipomea sp., Salvinia natans, S. molesta and Paspalm* sp., etc. are posing great threat to the wetlands hampering their ecological functions.

Pollution

Drainage and flood control programmes have long been a threat to wetlands. Chemical pollution is a more modern threat. A large number of wetlands in the country are subjected to domestic sewage, solid waste and other industrial effluents. Similarly, fertilizer and pesticide runoff from agricultural lands also augment the pollution load. All these factors are responsible for accelerating the process of eutrophication.

Aquaculture

Convertion of wetland areas indiscriminately for aquaculture particularly in the coastal areas are also posing a great threat to coastal wetland ecosystems including estuaries, lagoons and other brackishwater habitats.

The above combined threats to wetlands have led to the following issues:

- Decrease in biological diversity particularly endemic and endangered species.
- Deterioration of water quality.
- Sedimentation and shrinkage in the area.
- Decrease in biodiversity particularly of migratory bird population, fish and other faunal diversity
- Prolific growth of obnoxious aquatic weeds.

CONSERVATION MEASURES

Several regulatory and promotional measures have been adopted by Government of India to conserve wetlands and sustainablly utilise their resources. The measures taken are :

Legal and Policy Framework

The Indian Constitution enshrines the concept of environmental protection through the Section on Directive Principles of State Policy by assigning duties for the State and all citizens through Article 48 A, and Article 51 A(g), which state that the "State shall endeavour to protect and improve the environment and to safeguard the forests and wildlife in the country", and "to protect and improve the national environmental including forests, lakes, rivers and wildlife, and to have compassion for the living creatures". There are also numerous and wide-ranging policies, programmes and projects which directly or indirectly support wetland protection.

The National Conservation Strategy and Policy Statement on Environment and Sustainable Development (1992) outlines the agenda for action in regard to conservation of wetlands, coastal areas, riverine and island ecosystems. It also emphasises conservation of wetlands to ensure sustainable ecological and economic benefits and also formulation of National River Action Plan with which most of the wetlands in India are associated. The National Forest Policy as amended in 1988 and the National Wildlife Action Plan (1983) stresses the sustainable use of forests and wildlife conservation. These policies have direct bearing on wetland conservation.

Government of India have enacted several legislations for the protection of environment and conservation of natural resources. Some of these Acts which also have relevance for wetland conservation include Forest Act, 1927; the Forest (Conservation) Act, 1980; the Wildlife (Protection) Act, 1972; The Water (Prevention and Control of Pollution) Act, 1974; The Water (Prevention and Control of Pollution) Cess Act, 1977 and the umbrella provisions of the Environmental (Protection) Act, 1986. A Notification declaring the coastal stretches as Coastal Regulation Zone Notification, 1991 under the provisions of Environmental (Protection) Act, 1986 imposes graded restrictions on industrial operations and processes. This Notification is of great importance for regulating the activities in the coastal areas which are under pressure due to human activities. The Environmental (Protection) Act also specifies protection of ecologically fragile areas which include a number of wetland ecosystems in the country.

Institutional Framework

In view of the fact that wetlands are multiple value ecosystems and provide a diverse range of goods and services, a number of agencies are concerned with wetlands such as fisheries, agriculture, forestry, and irrigation. These agencies have developed some management practices over the time period to maximise the sectoral gains without understanding the inter-sectoral linkages and long term benefits of the wetlands.

Realising the importance of wetlands and developing an inter-sectoral framework for conservation of wetlands, a National Committee on Wetlands, Mangroves and Coral Reefs was constituted to advise the Government on policy guidelines, identification of priority wetlands for intensive conservation, monitoring the implementation of management action plans, research and preparation of an inventory of wetlands. At the State Government level, Steering Committees have been constituted consisting of various members drawn from the State Government Departments concerned with wetlands and experts in the subject matter area. Chief Secretary of the State Government is usually the chairman of such committees. These committees are responsible for formulation, implementation and monitoring of the progress of the management action plans prepared for each individual wetland ecosystems. At the district level, Coordination Committees have been constituted to coordinate implementation of wetland activities. The Ministry of Environment and Forest provides financial assistance to the State Governments for carrying out activities for conservation and management of wetlands, mangroves and coral reefs.

Inventory

In order to inventorise wetland resources of the country, the Ministry of Environment and Forest conducted a questionnaire survey and published a Wetland Directory in 1990. As per this survey, total wetland area of the country is about 4.1 million hectares (Table 5.1). In addition, mangroves - the coastal wetlands, occupy an area of about 6,740 sq. km. However, the periodical survey carried out by Forest Survey of India indicates that the mangrove area in the country is 4336 sq km. (Table 5.2). Major chunk of mangrove forest occur in Sunderbans of West Bengal and Andaman & Nicobar Islands which is about 80% of the total mangrove area and the rest being distributed in the coastal States of Orissa, Andhra Pradesh, Tamil Nadu, Karnataka, Maharashtra, Gujarat and Goa. However, a survey conducted by the Ministry of Environment and Forest in 1990 involving remote sensing technology, the total area of mangroves in the country has been estimated to be 2,300 sq. km. A nation-wide project on Wetlands was launched by the Ministry of Environment and Forests in 1993 to conduct survey and mapping of wetlands in the country involving remote sensing technology. The project is under implementation.

In the Asian Wetland Directory (1989) ninety three wetlands have been included from India which represent well-known sites. WWF-India and Asian Wetland Bureau have also prepared a Directory on Indian Wetlands giving a list of about 170 wetlands.

Table 5.1 : Distribution of Wetlands in India

Sl. No.	*State*	*No.*	*Natural Area (ha)*	*No.*	*Man-made Area (ha)*
1.	Andhra Pradesh	219	1,00,457	19,020	4,25,892
2.	Arunachal Pradesh	2	20,000	NA	NA
3.	Assam	1,394	86,355	NA	NA
4.	Bihar	62	1,24,788	33	48,607
5.	Goa	3	12,360	NA	NA
6.	Gujarat	22	3,94,627	57	1,29,660
7.	Haryana	14	2,691	4	1,079
8.	Himachal Pradesh	5	702	3	19,165
9.	Jammu & Kashmir	18	7,227	NA	21,880
10.	Karnataka	10	3,320	22,758	5,39,195
11.	Kerala	32	24,329	2,121	2,10,579
12.	Madhya pradesh	8	324	53	1,87,818
13.	Maharashtra	49	21,675	1,004	2,79,025
14.	Manipur	5	26,600	NA	NA
15.	Meghalaya	2	NA	NA	NA
16.	Nagaland	2	210	NA	NA
17.	Orissa	20	1,37,022	36	1,48,454
18.	Punjab	33	17,085	6	5,391
19.	Rajasthan	9	14,027	85	1,00,217
20.	Sikkim	42	1,101	2	3.5
21.	Tamil Nadu	31	58,863	20,030	2,01,132
22.	Tripura	3	575	1	4,833
23.	Uttar Pradesh	125	12,832	28	2,12,470
24.	West bengal	54	2,91,963	9	52,564
		2,164	14,49,338	65,250	25,87,965
Union Territories					
1.	Chandigarh	NA	NA	1	170
2.	Pondicherry	3	1533	2	1131
		2,167	14,50,861	65,253	25,89,266

Table 5.2: Areas of Mangroves in Different States

	State/Union Territory	*Area (km^2) (values rounded to the nearest 10 km^2)*	
		1989	1995
1.	Andaman and Nicobar groups (including Nicobar Islands)	973	966
2.	West Bengal (the Sunderbans)	2109	2119
3.	Orissa (All deltaic and coastal mangroves, including the Mahanadi area)	192	195
4.	Andhra Pradesh (the Godavari and the Krishna)	405	383
5.	Tamil Nadu (the Cauvery and adjacent coastal stretch)	47	21
6.	Karnataka (Coondapur and Maple Karwar patches)	-	2
7.	Goa	3	3
8.	Maharashtra (inclusive of Ratnagiri, Vijayadurg, Malvan, Devgad, etc. inclusive of 85 km^2)	114	155
9.	Gujarat (including the Narmada, the Tapti Gulf of Khambhayat mangroves of 105 km^2)	412	689
	TOTAL	**4,255**	**4,533**

Source: The State of Forest Report

MANAGEMENT ACTION PLANS

Based on the recommendations of the National Committee on Wetlands, Mangroves and Coral Reefs, 21 wetlands, 15 mangrove areas and 4 coral reef areas have been identified for intensive conservation (Table 5.3). As a first step towards Preparation of Management Action Plans guidelines were formulated in 1986 which were subsequently revised in 1992. Based on the guidelines prepared, management action plans have been drawn up on selected wetlands, mangroves and coral reef areas by the concerned State Governments.

Brief outlines of the promotional measures are:

Protection

An effective network of protected areas consisting of national parks, sanctuaries and biosphere reserves have been set-up in the country for the protection of wildlife. Many wetlands are included in this network. Efforts are being made to include the remaining identified wetlands under the network and the respective State Governments have been requested for the necessary steps. The State Governments concerned with these protected areas have well-structured system to protect these areas. However, assistance is also being provided to the State Governments under centrally sponsored scheme for strengthening the infrastructure facilities. Under Mangrove Programme, State Forest/Wildlife Departments have been provided financial assistance for various protective measures.

Catchment Area Treatment

Catchment area is an integral component of the wetland ecosystem. The anthropogenic activities like deforestation, overgrazing and developmental activities etc. are causative factors for the accelerated soil erosion and consequent siltation of wetlands. In view of this several activities for catchment area development including afforestation, vegetation contour bunding, construction of water harvesting structures, gully control, check dams, stream bank erosion control, etc. have been undertaken in several wetlands like Chilka, Loktak, Harike, Kanjii, Wular and Bhoj.

Table 5.3 : Management Action Plan

Policy and Legislation
Protection
Catchment Area Development
Siltation Control
Pollution Control
Wildlife Conservation
Fisheries Development
Generating Awareness
Peoples' Participation
Monitoring and Evaluation
Research

Pollution Control

As most of the wetlands are associated with the riverine systems of the country, monitoring of water quality of major, medium and minor river systems has been undertaken on extensive basis by Central and State Pollution Control Boards. Monitoring of rivers, groundwaters and lakes are also being done under the Global Environmental Monitoring Systems (GEMS), Monitoring of Indian National Aquatic Resources (MINARS) and Ganga Action Plan (GAP) Programmes. So far a network of 450 water quality monitoring stations has been established covering all the 14 major river basins, 16 m Medium rivers and 18 minor rivers, lakes and groundwater. Besides, coastal water quality is also monitored through 173 stations co-ordinated by the Central Pollution Control Board and State Pollution Control Boards of Gujarat, Maharashtra, Kerala and Tamil Nadu.

The Ministry of Environment and Forest has been assigned the responsibility for carrying out environmental impact assessment of developmental projects in various sectors such as multipurpose river valley and irrigation projects, thermal and atomic power industries, mining, ports and harbours, transport etc. in order to ascertain the impact of these developmental projects on land, water, air, flora and fauna etc., these developmental projects are required to prepare an Environmental Impact Statement (EIS) covering various issues.

Weed Control

Proliferation of the aquatic weeds like water hyacinth *(Eicchornea crassipes), Ipomea* sp. *Paspalam* in the wetlands needs to be controlled. The State Governments have taken elaborate measures to check the growth of these weeds. While biological control measures have been undertaken on experimental basis by releasing weevils and other pests, manual method is being adopted on extensive basis in Harike, Kanjii, Bhoj, Loktak, etc. There has been, however, recurrence of water hyacinth in some wetlands despite clearance due to its exponential growth. A comprehensive integrated long-term approach is required to be evolved to combat the weed menace.

Wildlife Conservation

One of the important components of management action plans is conservation of endangered and threatened species. Several programmes have been initiated by the Ministry of Environment and Forest under centrally sponsored scheme for captive breeding of some endangered species like rhinoceros (Assam), sangai *(Cervus eldi-eldi)* in Manipur, freshwater turtles and other wildlife species. Certain portions of Chilka, Kawar and Loktak wetlands have been declared as sanctuaries especially for the protection and conservation of wildlife. Construction of mounds and ponds has also been undertaken in some wetlands for developing them as suitable waterfowl habitats.

Environmental Education

One of the important components which could work as a catalyst is building of awareness about values and functions and need for conservation of wetlands. Under management action plans financial assistance has been provided by the Ministry of Environment and Forest to the State Governments, for building up awareness within all target groups through audiovisuals, posters, nature camps, films etc. Several programmes have been initiated by the Governments particularly of Punjab, Orissa, Madhya Pradesh, Kerala etc. for involving people in the conservation of wetlands.

People's Participation

For successful implementation of conservation, involvement of people and nongovernmental organisations is crucial. An effective role has been played by several NGOs in the country to save the wetlands and help in the conservation of the wetlands. A reference may be made in this case about conservation of Sukhna Lake in Chandigarh. The Environmental Society of Chandigarh, local people and school children have been participating in the conservation of lake through an environmental movement called "Shramdan". A number of activities have been taken under this movement which included desiltation of the lake, afforestation in the catchment area and generating awareness.

INTERNATIONAL COOPERATION

Ramsar Convention

The Convention on Wetlands of International Importance, especially as Waterfowl Habitat, popularly known as Ramsar Convention is one of the Intergovernmental treaties which provides the framework for international cooperation for the conservation and wide use of wetlands. The Convention entered into force in 1975 and know has Contracting Parties throughout the world. India acceded to the Ramsar Convention in October 1981, and designated Chilka Lake (Orissa) and Keoladeo National Park (Rajasthan) as Ramsar sites. Four additional sites were designated in March 1990 including Wular Lake (Kashmir), Harike Lake (Punjab), Loktak Lake (Manipur) and Sambhar Lake (Rajasthan).

The Government of India has taken several initiatives for the conservation of the Wetlands designated under Ramsar Convention. Management Action Plans have been prepared for the Wular Lake, Harike Lake, Loktak Lake and Chilka Lake. The State Governments have constituted State Steering Committees to oversee implementation of Management Action Plans. Special measures have been taken to ensure protection of Wildlife, particularly Waterfowl by bringing specific areas of these wetlands under the protection networking. Chilka Development Authority and Loktak Development Authority have been constituted to ensure effective management of Chilka Lake and Loktak Lake respectively.

Three Ramsar Sites viz. Chilka Lake, Keoladeo Ghana National Park have been put on Montreux Record. Efforts are being made to remove these wetlands from the Montreux Record by undertaking appropriate conservation measures. A Technical Advisory Committee has been constituted to look into the issues of the Loktak Lake and prepare a comprehensive Management Action Plan. A meeting in this regard was held in December 1994 with the State Government officials and the experts.

Keoladeo National Park is an important habitat for Waterfowl. There has been drastic reduction in their number of Siberian Cranes during the past few years. A detailed study on this wetland eco-system has been undertaken by Bombay Natural History Society (BNHS). This study has brought out significant alterations in the ecological character of this wetland effecting flora and fauna. International Crane Foundation and some other agencies have undertaken studies to investigate migration routes of avi-fauna particularly Siberian Cranes in order to understand the cause for their drastic reduction. Some rare birds have been released in the park and conclusions of this study are yet to be confirmed. Efforts are also being made to ensure water supply in the wetland during the lean period.

A management action plan for the Chilka Lake has been already drawn up and is being implemented by the State Government. In order to address some critical issues of water management particularly with reference to salinity regimes as a consequence of siltation, a proposal has been drawn up to seek financial and technical assistance under bilateral programme.

In order to ensure protection of Ramsar Sites efforts are being made to notify these wetlands under the provisions of Environment (Protection) Act, 1986, as ecologically fragile areas.

NGO Ramsar Committee has been set up by World Wide Fund for Nature - India to generate awareness about wetland conservation. WWF-India has published booklets on six Ramsar Sites in 1994 highlighting the issues and priorities from management perspective. A Training Workshop on Wetland Management was held in January, 1995 at Bharatpur, Rajasthan, for the wetland functionaries of the Ramsar Sites. This was supported by Ramsar Bureau.

India is also a party to other International Conventions which directly or indirectly are conducive to wetland conservation. These conventions are related to regulating trade in endangered species of wild flora and fauna, conservation of migratory species, protection of heritage sites and overall biodiversity conservation. These conventions include:-

(i) the Convention on International trade in Endangered Species of Wild Flora and Fauna.(CITES);

(ii) the Convention of Biological Diversity;

(iii) the Convention on the Conservation of Migratory Species of Wild Animals. (the Bonn Convention);

(iv) the Framework Convention on Climate Change; and

(v) the Convention concerning the protection of the world cultural and natural heritage (the World Heritage Convention).

LITERATURE CITED

Gopal, B. (1990). *Indian subcontinent and the aquatic habitats. In: Gopal, B. (Editor), Ecology and Management of Aquatic Vegetation of the Indian Subcontinent.* Kluwer Academic Publishers, Dordrecht, The Netherlands. 257 pp.

Gole, P. (1982). *Status of Anser indicus in Asia with special reference to India. Aquila,* 89: 141-49.

Gole, P. (1987a). *Future of Black-necked Crane in the Indian Sub-continent. In :Archibald,* G.A. and Pasquier, R.F. (Editor), *Proc. 1983 International Crane Workshop:* 51-54, Barabool Wisconsin: InternationalCrane Foundation.

Gole, P. (1987b). *Habitat Considerations for Black-necked Cranes in Ladakh.* Paper presented at the International Crane Workshop: 1-10 May, 1987. Qigihar, China; International Crane Foundation.

Government of India. (1990). *Wetlands of India - A Directory.* Ministry of Environment and Forests, New Delhi.

Joshi, C.B., Sehgal, K.L. and Shyam Sunder. (1974). *Observations on the fishery resources of the hill streams of Jammu province with special reference to the Mahseer and other commercially important species, Personal communication.*

Kaul, V. (1983). *Conservation of plant resources in aquatic ecosystem/with reference to some aquatic habitats of Kashmir.* In: Jain, S.K. and Mehra, K.L. (Editor), *Conservation of Threatened Plant Resources,* pp. 118-131, Botanical Survey of India, Calcutta.

Scott, D.A. (Editor) (1989). *Directory of Asian Wetlands.* IUCN, Gland, Switzerland, and Cambridge, U.K. xiv +1181 pp., 33 maps.

Sehgal, K.L. (1971). *Fisheries Survey of Himachal Pradesh and some adjacent areas with special reference to mahseerand other fishes,* J. Bombay Nat. Hist. Soc. 70: 63-71.

Trisal, C.L. and Kaul, S. (1996). *Threats and management action plan of wetlands in India.* 53-61. In: Conservation and Management of Lakes/Reservoirs in India. International Lake Environmental Committee Foundation, Shiga 525, Japan.

Directory of Indian Wetlands. (1993). WWF-India, New Delhi and AWB, Kuala Lumpur, 264pp., 32maps.

Woodcock, M. (1984). *Collins Handguide to the Birds of the Indian Sub-continent.* William Colliam Collins Sons and Co. Ltd., London.

Chapter - 6

State of Biodiversity in Lakes and Wetlands of Kashmir Valley

D.P. Zutshi and B. Gopal**

ABSTRACT

Kashmir valley in western Himalaya, spread over about 5000 sq. km at mean elevation of 1585m, is highly rich in lakes and wetlands which alone cover approx. 250 sq. km (5% of the total and area). Although there are numerous biological studies during the past more than 100 years, it is difficult to obtain a clear picture of the total biodiversity, and its changes under various anthropogenic impacts. This paper attempts to assess the aquatic biodiversity of the Kashmir Valley. The aquatic habitats in the valley support more than 250 species of marophytes, 150-200 species of phytoplankton, about 300 taxa of periphytic algae, over 50 periphytic rotifiers, about 100 species of zooplankton, about 130 species of zoobenthos, about 40 species of fish and about 100 species of waterfowl. Many species of both plants and animals are threatened or endangered, and a few have already disappeared from the valley in recent years. Several fishes are endemic to valley but the endemism among other biota is not well understood. There is also sufficient evidence that significant changes in the biodiversity (species composition, species richness as well as abundance of different species) are occurring due to growing pollution, introduction of exotics and other human disturbance. The paper calls for well-planned studies of the aquatic biodiversity, and its value and functions, for proper management of the Valley's aquatic resources.

Key Words : *biodiversity, fish, Kashmir Valley, lakes, periphyton, phytoplankton, zooplankton, zoobenthos, waterfowl, wetlannds*

INTRODUCTION

The valley of Kashmir lies to the north of the Indian subcontinent between 33°1′ and 35°0′ N Latitudes and 73°48′ and 75°30′ E Longitudes. It appears as a large 'Dun' or intermontane valley, 135 km long and 30-40 km wide, with an area of about 5000 km^2. The mean elevation of the valley is 1585 m. It is bounded by the Great Himalayan range in the

*School of Environmental Sciences, Jawaharlal Nehru University, New Delhi 110067.

north and northeast, by the Pir Panjal range in the south and southeast, and by a prominent spur of the Great Himalayan range on its west.

Kashmir valley is an alluvium-filled basin of which a substantial part is of recent formation by the River Jhelum. More than half of the valley is occupied by older alluvium known as 'udars' or 'karewa' which slope away from the high mountains surrounding the valley. The 'karewas' are supposed to be the relict basis of an ancient lake which filled the whole valley during the Pliocene-Pleistocene interglacial periods.

The valley is dotted throughout with numerous natural freshwater lakes and wetlands. According to Bacha (1996) there are 335 water bodies, of >1 ha each, covering 253 km^2 area (5.5% of the total land area) of which 186 km^2 (68.1%) is under lakes and 87 km^2 (31.9%) under wetlands. Recently Garg *et al.* (1997) reported that in Kashmir the wetlands larger than 56.25 ha each, cover 249.75 km^2 of which the maximum area lies in Baramulla district.

Although the origin of lakes and wetlands in the Kashmir valley is still controversial (for detailed discussion, *see* Zutshi and Gopal 1998), these aquatic ecosystems are unique in the entire country and play a crucial role in the socio-economic life of the State. Numerous studies on various limnological aspects of lake and wetlands of Kashmir have been undertaken in the past (Kaul 1977, Zutshi et al. 1980, Kaul and Handoo 1993) but the information on their contribution to biodiversity is still lacking. The importance of biodiversity for sustainable development is very well known and efforts are being made throughout the world to take immediate and necessary steps to conserve threatened and endangered species. Recently Gopal (1997) stated that our ignorance of biodiversity and its functions in aquatic ecosystems is overwhelmingly large. It requires serious and urgent efforts to remedy this situation.

In this paper we discuss the present state of our knowledge of the biodiversity of plants and animals in the lakes and wetlands of Kashmir valley, focus attention on areas requiring further research and identify gaps in our understanding of the problem.

THE ENVIRONMENT

The lakes of Kashmir valley are generally small and mostly shallow. The largest is Wuler Lake (59 km^2) and the deepest is Manasbai Lake (Max. Depth 12m). Over the years, not only are the lakes shrinking in their area but are also becoming shallower. For example, the area of Dal Lake has reduced by about 40% during the present century. The shape of the lake basins is prominently influenced by the land form. Most of the lakes have a saucer-shaped basin which denotes relatively gentle shores and depressions. Convexity of lake basins is a dominant feature which may be an important factor in determining high macrophyte density.

The lake waters are either turbid or moderately clear depending upon the amounts of transported silt. Biogenic turbidity (due to high plankton populations) is rare. In Dal lake, the Secchi may be visible right up to the bottom (e.g. Gagrribal basin) in some months. In Manasbal, maximum Secchi depth is 5.5 m. Hokarsar and Anchar have low Secchi transparencies (16 cm). Summer stratification is not a common feature in lakes except in Manasbal which is warm monomictic with high thermal stability. In Nagin basin of Dal Lake, thermal stratification is short lived with weak thermocline.

The lake waters are alkaline and well oxygenated. The concentration of various nutrients is moderate to high. Anoxic conditions occur in Manasbal in deep water during summer (May to October). N and P levels remain low in most of the lakes except some such as Trigam and Telwansar where water renewal is slow and inputs from catchments are high. However, the sediments are an important source of nutrients as high concentrations of N and P are locked there. Dal Lake sediments contain 0.16% of P and 0.33% of N in the upper 20 cm layer, representing an estimated quantity of 370 tons P and 720 tons N which are potentially available for plant growth under different conditions (Zutshi 1987).

Based on hydrological characteristics, the wetlands are classified as temporary and shallow, temporary and deep, permanent and shallow, and permanent and deep (Kaul and Handoo 1993). The largest wetland is Shalbogh (705 ha) and the smallest is Malangpora (127 ha). The deepest permanent wetland is Hokarsar (max. depth 1.2 m). Most of the wetlands lie in the floodplain of River Jhelum. Haigam and Hokarsar have regular inflow and outflow channels which maintain a certain level throughout the year. In comparison with lakes, the wetlands have low dissolved oxygen content but high levels of alkalinity, conductivity and phosphorus. The sediment characteristics are not significantly different.

BIOLOGICAL DIVERSITY

Macrophyte Diversity

Until mid-sixties only sporadic information was available on macrophytic getetation of Kashmir and a listing of species was conspicuously absent. Dal and Mansabal Lakes were investigated by Mukeerjee (1926, 1934). Besides commenting in a general way on the presence of *Myriophyllm*, *Potamogeton* and *Nelumbo*, Mukerjee (1926) reported the occurrence of *Chara fragilis*, *Nitella acuminata*, *Nitella dispersa* and *Nitellopsis obtusa* in Manasbal Lake. Later, Mukeerjee (1934) reported eleven species and 3 genera of Characeae to occur in Dal lake. Members of the Yale North India Expedition visited Srinagar during April-May, 1932 and provided preliminary information on the macrovegetation of Dal Lake. The dominance of *Potamogeton luceus* and *Potamogeton crispus* was reported for the first time. *Polygonum amphibium* and *Nymphoides peltata* were found to be common floating forms. In his book on Srinagar and its environs, S. Koul (1946) wrote that the surface of Dal Lake side channel water was covered with *Myriophyllum verticillatum* which had grown in profusion since 1941, *Myriophyllum spicatum* was observed as a dominant form growing under water. He found *Salvinia natas* in side channels but not in dense stands. The other specie were *Nymphaea alba*, *Nymphoides peltata*, *Hydrocharis morsus-ranae* and *Lemna* spp. *Euryale ferox* occurred freely in Hazratbal basin. Anchar lake was reported to support dense population of *Nelumbo nucifera* and Mansabal lake had fairly good population of *Myriophyllum spicatum* and Najas *graminea*. Close to the lake shore where spring water flows in the lake thick stands of *Nasturtium officinale* were common. Kaul and Zutshi (1967) were the first to provide a comprehensive list of aquatic macrophytes from lakes and wetlands of Srinagar. They recognized 117 species representing 69 genera and 42 families. The maximum number of species were collected from Dal (107), followed by Anchar (99) and Hokarsar (67) lakes. They included aquatic bryophytes and pteridophytes but not the macrolgae. Since then no attempt has been made to update the list of macrophytes. In a series of papers Kak (1984a, b, 1985) reported presence of many aquatic species from Kashmir Himalaya some of which were new records for the region. But Cook (1996) does not recognize many of these as valid Himalayan species and therefore it difficult to prepare a consolidated list of species. Subrahmanyam (1962) who published a list of 117 species of aquatic angiosperms from India made only a passing reference to the presence of *Euryale ferox* in freshwater lakes of

Kashmir besides including *Myriophyllum spicatum, Myriophyllum verticillatum* and *Hydrocharis dubia* from this region.

Lavania *et al.* (1990) who attempted a comprehensive list of 475 aquatic species of the Indian subcontinent, included only 25 species from the Kashmir region. Besides describing 800 aquatic and wetland species from India, Cook (1996) also mentioned 252 species from the Himalayan region. Taking an overall view of the published information it can be presumed that in Kashmir there are about 225 aquatic and wetland species which would constitute about 20% of the total Indian species. The figure is close to the estimation of 200 species by Kak (1978).

The diversity varies greatly between different lakes and wetlands. Zutshi (1975) recorded 25 macrophyte associations from 15 lakes and wetlands of Kashmir of which Anchar lake had the greatest diversity with 20 associations and the Ahansar had the lowest diversity with only 4 associations. Handoo and Kaul (1982) recorded 42 species of macrophytes in Malangpora, 32 species in Kranchu, 19 in Shalbogh and 19 species in Hokarsarlake. Only seven species were common to these four wetlands. The species richness in terms of number of species per square metre varied from 4.79 in Malangpora to 1.33 in Hokarsar. Kundangar and Zutshi (1987) reported low species (12) diversity in two small foodplain lakes of Jhelum. Ahansar and Waskar. They did not take into consideration the wetland species which colonize peripheral areas of the lake. Recently, Kundangar and Sarwar (1997) provided a list of 35 macrophyte species from Dal lake.

Phytoplankton diversity

In the majority of aquatic habitats phytoplankton is dominated by diatoms. Only in few valley lakes diatoms are replaced by either blue-green or green algae and in one lake it is the dominance of Dinophyceae. The early studies were essentially taxonomic in nature and reported the occurrence of either a new algal form from Kashmir (Bharadwaja 1936) or a specific group (Misra 1937). No attempts were made to provide information on all the species present in these ecosystem. Kant and Kachroo (1971) listed 157 genera from two basins of Dal lake with maximum representation of Volvocales. They presumed that the number of species in the lake would be around 350. However, no basis was given for this hazardous guess. In another publication, Kant and Kachroo (1977) reduced the number of genera to 140 and increased number of species to 360. But no lists were provided in support of their contention of high species diversity in Dal lake. Same year, Trisal (1977) reported much lower diversity in the same lake with only 71 genera and 117 species. Later studies on Dal lake by Zutshi and Vass (1982). Mufti (1982), and Zutshi and Khan (1988) identified fewer species (varying from 51 to 84). This was perhaps because of the limited area of investigation. Mir and Kachroo (1982) reported 86 diatom species and Yousuf and Parween (1992) recorded 84 genera. Recently, Kundangar and Sarwar (1997) listed 69 genera and 109 species from Dal Lake. In Anchar, Lake, the number of species in different reports varies from 90 (Zutshi unpublished) and 93 (Anonymous 1993) to only 25 (Kaul *et. al.* 1978).

Slightly higher number of phytoplankton species (125) was reported by Zutshi and Wanganeo (1984) from Manasbal lake. In Wular lake, various reports place the number of species at 27 (Kaul*et. al.* 1978), 46 (Qadri 1985) and 82 (Anonymous 1993). Again the extent of diversity may be due to the area covered during the study. In smaller lakes e.g., Bodsar (109 species), Ahansar and Waskur taken together (127 species), Pashakuri (100 species), Khanpur (75), Trigam (79) and Telwansar (70), the number of species did show significant variations. However, Pandit and Kaul (1981) reported much higher diversity in wetlannds where dominant phytoplankton was reported to be Chlorophyceae. Nowgam supported 189 species, Malgam 177, Haigam 183, Mirgund 179 and Hokarsar had only 156 species .

Thus, it can be seen that the lakes and wetlands of Kashmir do not support high diversity of phytoplankton populations. In most of lakes the species number does not exceed 100 and falls for short of the presumed number of 360 species. The upper limit in lakes may be around 150 species and in wetlands about 200 species.

The only report on species diversity (H) and evenness of species (e) in phytoplankton is that of Wanganeo and Wanganeo (1993) who found that Mansabal had higher diversity (H = 2.21-4.49) and evenness (e = 0.48-0.89) than in the smaller lake Bodsar (H = 1.16-3.9, e = 0.22-0.78).

Periphyton diversity

The earliest study on periphyton of Dal lake (Anonymous 1977) revealed that the population consists mainly of diatom, desmids and other green algae. The periphytic flora of Dal lake was investigated from two natural substrates by Pandit (1983) who recorded 264 species and 84 genera on *Potamogeton lucens* and 239 species and 77 genera on *Myriophyllum spicatum*. The population was dominated by Chlorophyceae followed by Bacillariophyceae. The distribution of species is given in Table 6.1.

Table 6.1 : Periphyton distribution on natural substrates in Dal lake.

Class/family	*Potamogeton luceus*	*Myriophyllum spicatum*
Chlorophyceae	108	107
Bacillariophyceae	104	84
Cyanophyceae	28	32
Euglenophyceae	19	9
Dinophyceae	5	3
Other groups	-	4

Sarwar (1985) recorded fewer periphytic species on five most common macrophytes in Dal lake. The maximum number of 109 species was registered on *Myriphyllum spicatum* and minimum of 89 species on *Phragmites australis*. The floating species, *Nymphoides peltata* had 107 periphytic species Bacillariophyceae was the most dominant class followed by Chlorophyceae. Sarwar and Zutshi (1987, 1989) compared the periphytic flora from three lakes but on the same natural substrate and found very little difference in the number of species. In Dal lake it was 90 species, Anchar lake 89 and Waskur lake 85. Subsequent investigations on Wular lake (Anonymous 1993) also reported the occurrence of 90 species of periphyton of which 57 were diatoms.

BACTERIAL DIVERSITY

Very few studies have been undertaken on bacterial population of lakes and wetlands of Kashmir. Although Enex (1978) reported the occurrence of coliforms and fecal streptococci in Dal lake, Kaul (1988) for the first time listed from Dal Lake 27 species of bacteria belonging to the genera: *Escherichia, Shigella, Salmonella, Staphylococcus* and *Yesima*. So far, the fungal populations of aquatic ecosystems of Kashmir have not been examined.

Zooplankton

The earliest information on zooplankton of Kashmir came from the Yale North India Expedition when Edmondson and Hutchinson (1934) reported on rotifiers and Brehm (1937) reported 21 species of cladocera. Later, S.M. Das and his associates (1969) reported zooplankton from different aquatic ecosystems but a complete enumeration has not yet been done. However, the available information indicates low zooplankton diversity in this region. Kaul *et. al.* (1978) who investigated zooplankton of 15 lakes, wetlands and ponds, reported 13 species of protozoa, 18 species of rotifers, 12 species of cladocera, 10 species of Copepoda and a single species of Ostracoda. Among the lakes, Dal with the maximum diversity of 48 species was followed by Hokarsar with 28 species and Anchar lake with only 25 species. Zutshi and Vass (1982) recorded as many as 93 zooplankton species from Dal lake with a maximum contribution from Rotifera (37 species). The copepods were few (only 8 species). Subla et al. (1992) listed zooplankton of 15 aquatic ecosystems which included information on 8 lakes from Kashmir valley. The total number of species was 151, the highest member of species was from Manasbal lake (51), followed by Panshakuri (50) and Dal (48). Very few zooplankton species (22) were reported by Kundangar and Zutshi (1985) from Ahansar and Waksur. According to Balkhi et al. (1992), in the valley lakes species diversity of cladocera was maximum (H = 0.36) with higher diversity values for Dal lake (H = 0.51). Similarly, copepod diversity was also high in valley lakes in comparison to wetlands and other aquatic habitats (H = 0.27). The values of species diversity are given in Table 6.2.

Table 6.2 : Species diversity of zooplankton in lakes and wetlands of Kashmir [after Balkhi et al. (1992)]

Habitat	*Cladocera*	*Copepoda*	*Rotifera*	*Total Zooplankton*
Valley lakes	0.36	0.27	0.67	1.31
Wetlands	0.33	0.25	0.64	1.22

In Anchar lake, Balkhi et al. (1987) recorded 87 species with 41 rotifiers, 33 cladocerans and 13 copepods. Khan (1987) compared zooplankton population of two lakes at different levels of enrichment. Trigam which is more polluted supported 40 species and Naranbagh with low trophic level had only 25 species. In both the lakes, rotifers were present in large numbers.

Crustacean plankton of one hundred freshwater bodies of Kashmir including lakes, wetlands, ponds, etc. were investigated by Balkhi and Yousuf (1992) who reported that the species diversity of cladocera and copepoda was maximum for the densely vegetated waters of valley lakes. In Anchar and Wular lakes, eight species each of copepoda was recorded in a single collection when in most of the situations it never exceeded five species. It seems that the higher presence of crustacean population of related to the density of macrovegetation in aquatic ecosystems. Raina and Vass (1993) investigated zooplankton populations of lakes and wetlands within the altitudinal range of 600 to 3800 m and observed high density of Rotifiers in the valley lakes.

Two small urban lakes, Gilsar and Khusalkar, situated between Dal and Anchar lakes were investigated for zooplankton population by Sarwar and Parveen (1996). Both the lakes had almost similar diversity (Gilsar with 51 species and Khusalsar with 53 species) with the dominance of rotifers.

Trisal et al. (1994) listed 50 zooplankton species from Wular lake with following distribution 37 species of Rotifera, 9 Cladocera and 4 Copepoda. In a recent study on Dal lake, Kundangar and Sarwar (1997) reported the presence of only 29 species of zooplankton in Dal lake.

Periphytic Zooplankton

The species composition and distribution of periphytic rotifer population of Dal lake on five submerged macrophyte species, *Ceratophyllum demersum, Myriophyllum spicatum, Potamogeton Luceus, Potamogeton crispus* and *Hydrilla verticillata* was invested by Ticku and Zutshi (1993). They reported 50 species of which only two *Lepadella ovalis* and *Monostyla lunaris* were numerically important. Other species were rare with less than 1% of occurrence. There percent occurrence of *Lepadella ovalis* on the macrophyte species is given in Table 6.3 to show its importance in the population diversity.

Table 6.3 : Distribution pattern of *Lepadella*

Macrophyte species	% occurrence
Ceratophyllum demersum	12.0
Myriophyllum spicatum	9.7
Potamogeton luceus	18.9
Potamogeton crispus	15.5
Hydrilla verticillata	12.8

In Manasbal lake, only 47 periphytic rotifer species were collected from four macrophyte species, *Ceratophyllum demersum, Potamogeton natans, Salvinia natans* and *Typha angustata* (Ticku and Zutshi 1994). The most dominant species were *Lepadella ovalis, Monostyla lunaris, Colourella obtusa* and *Monostyla bulla* Maximum diversity was registered on *Ceratophyllum* (37%) followed by *Potamogeton* (28%), *Typha* (21%) and *Salvinia* (14%). The diversity of rotifer species and the differences in the extent of population colonization were to some extent related to morphological differences of the macrophytes, in both the lakes.

Similarly, in Achar lake, 47 species were observed from four macrophytes with *Ceratophyllum* showing high diversity followed by *Potamogeton*. The highest diversity index was recorded on Typha and the lowest on Potamogeton (Zutshi and Ticku, unpublished). Table 6.4 show periphytic diversity index during different times of the year.

In Khushalsar 34 periphytic rotifer species were collected from Salvinia which forms the most dominant vegetation during summer-autumn. The diversity index is given in the Table 6.5.

Table 6.4 : Diversity index of periphytic rotifers of Anchar lake.

Season	*No. of species*	*No. of individuals*	*DI*	*Macrophyte species*
Winter	25-28	860-870	4.47-4.48	*Ceratophyllum, Potamogeton*
Spring	23.32	890-1130	4.33-4.68	*Ceratophyllum, Potamogeton, Typha*
Summer	25.37	1190-1680	4.57-4.77	*Potamogeton, Salvinia, Typha*
Autumn	24-28	880-1260	4.35-4.45	*Ceratophyllum, Potamogeton, Salvinia, Typha*

Table 6.5 : Diversity of rotifer population on Salvinia in Khushalsar.

Season	*No. of species*	*No. of individuals*	*DI*
Summer	25-26	350-360	4.48-4.49
Autumn	31-32	1050-1060	4.69-4.70

Benthic Organisms (Zoobenthos)

Studies on benthic population of lakes and wetlands have been relatively few and the data are mostly fragmentary. Zutshi and Vass (1982) reported that the benthos of Dal lake was represented by Oligochaetes, Diptera and Mollusca. No effort was made to list the species and pollution density. Data was given on seasonal basis.

Jan (1987) recognized three metazoan phyla viz., Annelida, Arthropoda and Mollusca in the Macrobenthic population of Dal lake. Within these three groups, 35 species representing 26 genera and 12 familiar were identified.

The meio benthic population of five wetlands represented 79 species belonging to Protozoa (70), Nomethelmintha (2), Rotatoria (5), Porifera (1) and Bryozoa (1). Macrobenthos comprised of only 7 taxa (Annelid 1, Arthropoda 4 and Mollusca 2). Large differences were observed in the diversity of zoobenthos of different wetlands (Pandit and Kaul 1981).

Phytobenthoss

The phytobenthos of wetlands comprised of 155 species of algae and 14 species of soil fungi (Pandit and Kaul 1981). The algal was represented by Cyanophyceae (18 species), Chlorophyceae (37), Bacillariophyceae (61), Xanthophyceae (1), Chrysophyceae (2), Dinophyceae (5), Cryptophyceae (4) and Eugleninae (27). Of the various algal classes, Chlorophyceae was the most dominant followed by Bacillariophyceae.

Sarma et al. (1983) investigated macro and meiofauna on three common macrophytes of Dal lake. *Ceratophyllum demersum, Potamogeton crispus* and *Hydrilla verticillata*. The maximum number of animals were associated with succulent smooth leaved submerged species, *Hydrilla* and *Potamogeton* with thick leaves supported the least diversity. In general, nematodes, chironomids, rotifers and cladocerans occurred in large number on the macrophytes. Population distribution and diversity are shown in Table 6.6.

Table 6.6 : Distribution and diversity of phytobenthos in Dal lake.

Plant species	*Macrofauna/100 g*	*Miofauna/100g*	*Ratio*
Hyrilla verticillata	80	15305	1 : 191.3
Ceratophyllum demersum	640	12375	1 : 19.3
Potamogeton crispus	421	7666	1 : 18.2

Fish Diversity

The ichthyofauna of Kashmir resulted from the spatial southward distribution from the interglacial periods until the recent times without much chance of Indian fishes migrating across the high Pir Panjal range in Southern Himalaya. Because of this the Palearctic fishes belonging to Schizothoracinae, Botinae and Noemachilinae are well represented in the ichthyofauna of Kashmir and over the centuries have evolved into many new species by long isolation. According to Silas (1952) fish dispersal took place in different waves of migration corresponding to the five glacial periods of Pleistocene, the pluvial periods facilitating the migration and the arid periods isolating them and thereby inducing rapid speciation.

As early as 1838 Heckel identified 16 species of fishes from the region and found them entirely different from those in the plains of India. Day (1878) listed 18 species from Kashmir but opined that the list may have included some species which may later turn out to be synonymous. Towards the middle of this century more attention was paid to the study of fishes of Kashmir. Hora *et al.* (1955) pointed out that there was great similarity between the fish fauna of River Jhelum and the valley lakes because of the continuity of drainage system and interconnection. They recorded 17 species from the valley out of which 9 were classified as of central Asiatic origin. In a list of 28 species, Silas (1960) also included two exotic forms, among the fishes of the valley. Das and Subla (1963) provided a list of 36 species found in lakes, rivers, wetlands and snow fed streams of Kashmir. A check-list of fishes of Jammu and Kashmir state was prepared by Nath (1986) who for the first time included as many as 42 species from Kashmir. Recently, Yousuf (1996) revised the list of fishes of Kashmir and brought it down to 37 species stating that the taxonomy of many a species is still confusing and therefore there is every likelihood of change in the number of species.

The number of species found in the valley lakes does not exceed 15 and in Wular lake the number is only 11. But it is interesting to find that the lakes support 5 species of *Schizothorax*, 3 species of exotic carp and some rare forms like *Noemachilus, Botia, Barbus* and *Oreinus*. In comparison to lakes the fish fauna of wetlands is limited in species diversity. Pandit (1991) reported *Cyprinus carpio communis* to be the fair fish type in Nowgam, Haigam, Mirgund and Hokarsar, while *Cyprinus carpio specularis* was limited to Mirgund and Hokarsar. The number of fishes in wetlands are restricted to 5-6 species.

Avifauna

The large number of shallow lake and wetlands afford a congenial habitat for both the resident and the migratory birds. The prime areas which attract large number of overwintering geese and duck population are Hokarsar, Haigam and Mirgund. According to Ali (1979) in the high altitude valley of Kashmir the natural freshwater biotopes such as lakes and wetlands, featuring open water for some part and a rich source of potential food, provide over-wintering resorts to thousands of waterfowl migrating from their breeding grounds in the Palearctic region extending from north Europe to central Asia.

Although some studies have been undertaken on feeding ecology (Pandit 1982), wildlife conservation (Pandit 1991) and population estimates (Bacha 1996) the census of birds has been completely neglected and therefore, exact number of species in many wetlands and lakes is not precisely known. It has been reported that till date 1950 at least, 22 avian species were regularly recorded from Hokarsar. The present number of such species has been put at 16. There include Mallard, Common teal, Blue-winged teal, Pintail, Gadwal, Widgeon, Shoveller, coot, Brahmini duck, Pochard, Greyleg goose, Common snipe, Rudly shelduck, Red crested pochard, White eyed pochard and Tufted duck. Trisal et al. (1994) reported the presence of 12 migratory and 5 resident birds from Wular lake, and recently Kundangar and Sarwar (1997) observed species of both native and summer migrants from Dal lake.

Holmes and Parr (1988) were able to identify 92 species of birds from Haigam which according to them was gross underestimation of the species and many more species will be added when more information becomes available particularly during migration phase. Haigam is an important area for autumn migrants, with 45% of species recorded as winter visitors.

RETREATING AND THREATENED SPECIES

There is growing evidence that many aquatic species are either becoming rare or are threatened due to a variety of reasons, the prominent being degradation of environment as a result of poor management in the face of mounting human pressures. There are data to show that *Euryale ferox*, once abundant in the Dal and Mansabal lakes (Lawrence1895), has virtually disappeared and is now rare (Zutshi 1987). The decline has been related to increased eutrophication of the lakes: *Euryale ferox* also has completely disappeared from Tegeleu Clay in Holland and Lachiun in Russia (Sculthorpe 1967). In the fifties seeds of *Euryale* were sold in the markets of Srinagar which goes to show that the plant was widely distributed. It is probable that besides changes in the environment of the lakes the human intervention by way of plant removal might have also been responsible for its near extinction.

There are many other macrophytes which are now confined to a very small area with much smaller populations *e.g., Typha laxmanii, Acorus calamus, Monochoria vaginalis.* Kaul (1983) provided a list of 15 macrophyte species which he classified as rare or retreating species from the Kashmir lakes. He attributed their decline to the process of the filing up of lakes by silt, mud etc. which resulted in their fast degeneration. Mukerjee (1926, 1934) reported the occurrence of 3 genera and 4 species of charophytes from Mansabal lake and three genera and 11 species from Dal lake. Over the years, there is significant decline both in the member of genera as well as in the number of species of charophyte. The decline may be related to the changing lake environment induced by flow of waste waters which contain high P and N concentration. In sixties, we often came across *Potamogeton perfoliatus* in Anchar lake but during later years this species was rarely recorded (Zutshi, unpublished). However, *Potamogeton perfoliatus* is quite abundant in snow-fed streams of Achabal area in the south of Kashmir were waters are clear and well oxygenated with moderate flow. It is mostly likely that in the past *Potamogeton Perfoliatus* was an important constituent of lake vegetation in Kashmir when waters were not polluted.

Although among aquatic macrophytes more than 60 species are known to be endemic to India (Gopal 1977), there is no information available from Kashmir on this aspect. It is most likely that there are endemic species in Kashmir which are not found in other parts of Himalaya as is the case with terrestrial plants.

Yousuf (1988) observed that two calanoid copepods, *Allodiatomus merabilipes* and *Pseudodiaptomus lobipes* which were reported by Akhtar (1973), have completely disappeared from the Mansabal lake. He ascribed this to gradual eutrophication of the ecosystem. Similarly, Raina and Vass (1993) reported significant decline in the number of *Daphnia* species since 1937 due to eutrophication.

Only one zooplankton species, *Arctodiaptomus michaeli* has been reported to be endemic to Kashmir and that too in high altitude lake, Alipather (Reddy *et al.* 1990). The plankton by and large are cosmopolitan in nature and widely distributed and therefore the likelihood of coming across many endemic species is rather remote. However, there are data to show the decline of species such as *Ceratium hirundinella* from small Karewa lakes (Zutshi, unpublished) in south of Kashmir.

One of the interesting features of biodiversity is high level of endemism in fish population of Kashmir valley. Silas (1960) lists 8 endemic species which according to him constitute 42% of the total population. Recently, Nath (1986) reported that there are 10 endemic fish species in Kashmir out of which 7 belong to genus *Schizothorax* and 3 to *Noemachiilus*. According to Das (1966) the fish fauna of Kashmir have definite palearctic elements, among which the endemic fishes of palearctic regions appear to have come from central Asia to the southern slopes of the Himalayas, to which they are now restricted in a region extending from the Kashmir valley to Ladakh.

The lakes and wetlands of Kashmir valley, featuring open water for some part, provide overwintering reports of thousands of waterfowl migrating from their breeding grounds in Palearctic region extending from north Europe to Central Asia (Ali, 1979). The littoral zone of lakes and wetlands are among the most important habitats which are visited by the birds. Twelve species have been identified as the most common waterfowl that over-winter in Kashmir.

It has been reported that till late 1950 at least, 22 avian species were regularly recorded from Hokarsar. The present number of species has reduced to 16. Many other wetlands also used to receive birds till 1950 which now only act as occasional feeding grounds (Pandit 1991).

Recently, Bacha (1996) report that in many aquatic habitats Greylag goose, Shoveller, Red Crested Pochard are rare and Ruddy Shelduck white-eyed Pochard and tufted duck could be put under vulnerable status. Special mention may be made of considerable drop in Greylag goose population in Hokarsar where this species used to visit in large numbers a few years back. It has also been noted that the common crane (*Grus grus*) has discontinued its stop-over in the lakes and wetlands of Kashmir.

ALTERATION IN ENVIRONMENT/LOSS OF BIODIVERSITY

It is now well documented that the habitat of many lakes and wetlands of Kashmir has undergone far reaching changes due to undesirable human impacts and poor management of catchment area. One of the major causes of the loss of macrophyte biodiversity is cultural eutrophication. The lake waters are more enriched now with higher concentration of phosphorus and nitrogen. This perhaps is the reason for gradual disappearance of many macrophyte species. Not only this but in many lakes there is a clear cut tendency to develop monospecies stands which form a continuous mass of vegetation affecting other denizens of the ecosystem e.g., *Cratophyllum demersum* in Manasbal lake, parts of Dal lake, Naranbagh lake and Bodsar cover large water area at the expense of other species. Another cause for macrophyte decline is siltation of water bodies due to large scale tree cutting in the watershed. The waters of Dal, Anchar, Hokarsar, Hygam and Wular are more turbid in comparison to past years. These changes become quite apparent when topographic maps different years are compared. The most significant change is seen in Anchar lake where northern portion is now completely silted up. The entire topography of western shore has undergone significant modification. The turbid water support less of submerged vegetation and more of floating forms. In Hokarsar during early sixties a tributary of Jhelum, spill channel was diverted into the lake as a populist measure by the State Government. The channel brought high silt load and deposited it along the lake periphery which was later reclaimed for paddy cultivation. The perennial turbid waters had a far reaching impact on the macrovegetation with disappearance of submerged species ad proliferation of monospecies floating form such as *Trapa natans*. The habitat change is not only characteristic of Hokarsar but also found in many water bodies.

The impact of eutrophication has been reported on plankton population also. Khan (1985) observed much higher phytoplankton population in Trigamsar in comparison to Naranbagh which he attributed to higher nutrient concentrations. Zutshi and Khan (1988) reported predominance of cyanophyceae in lake area, which are close to sewage disposal sites.

The introduction and subsequent proliferation of certain exotic species has had considerable impact on biodiversity. The invasion of *Salvinia natans* in recent years is interfering with the aquatic population of lake channels, sheltered areas and even in open waters (Zutshi and Wanganeo 1984). Similarly Chinese Capr, *Cyprinus Carpio* which was introduced in the valley lakes during late fifties have been responsible for dwindling populations of endemic snow trout. The near extinction of some waterfowl species may be attributed to the changing environment of water bodies with subsequent loss of food plants. The other factors like proximity of human settlements; poaching and legalized hunting and vagaries of climatic conditions are also effecting bird diversity.

CONCLUSIONS

It is apparent from the foregoing account that our knowledge about the biodiversity of lakes and wetlands of Kashmir Himalaya is grossly inadequate. The first and foremost effort should be directed towards preparation of reliable lists of both fauna and flora of the lakes. Once the listing is complete, attempts should be made to update these on the basis of published data and from further field observations. This would remove a certain degree of redundancy found in some published works. There is no source material which could be used for identification of plankton. Wherever possible keys for identification need to prepared to facilitate future work. There are no data available on aquatic fungi and this needs immediate attention. Studies on aquatic amphibians and reptiles is wholly lacking.

Since early eighties the State Government has been making efforts to protect Dal lake mainly with the idea of improving its environment to attract more tourists. With the establishment of Lakes and Waterways Development Authority there is a possibility of bringing more water bodies under programme of conservation. Unfortunately under these programmes the status of biodiversity is completely neglected and there is no scope to take steps for its protection. The introduction of mechanical deweeding in Dal lake has been responsible for loss of large quantities of indigenous fish (Ticku and Zutshi 1991). The suction dredging which is now in operation must be depleting the lake of its benthic population. These and other impacts need to be assessed before hand to avoid long term effects on the food chain.

Today when giant weed harvesters, aerators, suction dredges and other mechanical contraptions are freely used in lake rehabilitation programme, the knowledge drawn from traditional approaches has been completely overlooked. The presence of excessive macrovegetation is not a recent phenomenon although the problem has magnified due to pollution and discharge if waste water. In 1932 when Yale Nath India expedition visited Dal lake the members found entire lake basin carpeted with aquatic vegetation (Edmondson and Hutchinson 1934). The peculiar morphometric conditions, warm waters during major part of the year coupled with high lake fertility are some of the factors that help prolific growth of vegetation in Dal lake. The traditional approach to keep the vegetation under check and use it profitability is to harvest it manually and use the mix of weed and mud on floating islands to grow vegetables. This practice reduced biodiversity on a very small scale and managed the system using experience gained over many centuries. Times have changed and with it has changed the entire scenario of traditional management. The large scale weed harvesting must have effected the biodiversity as in case of fish and now dredging has added another dimension. These intervention by man need to be addressed immediately in terms of loss of biodiversity. The protective management of certain beneficial macrophytes such as *Nelumbo, Nymphaea* etc., also interferes with overall biodiversity which needs to be investigated.

Further studies are needed to evaluate the precise impact of carp on the indigenous fish. Normally the ecological niche does not overlap and higher numbers of carp in fish catches does not necessarily point to its impact on snow trout. The situation of inflow streams must be restricting the upstream movement of *Schizothorax* for breeding and therefore one of the causes for population decline.

LITERATURE CITED

Ali, Salim. (1979). The Book of Indian Birds. Oxford University Press, Bombay. 187 pages.

Akhtar, S. (1973). Studies on ecology an zooplankton of freshwater lakes of Kashmir. Ph.D. thesis, University of Kashmir, Srinnagar, (unpublished)

Anonymous (1993). Ecology and Conervation of Wetland of Wular Lake (Kashmir) Final Report Project No. J2203/38/88 W. Hydrobiology Research Laboratory, S.P. College, Srinagar. 89 pages

Anonymous (1977). Report on Dal lake, Srinagar, Kashmir with Suggestions for Development of its Fishery. Central Inland Fisheries Research Institute Barrackpore, India (Mimeographed), 13 pages.

Anonymous (1993). Ecological monitoring and conservation of Anchar lake, Kashmir (Report) Hydrobiology Research Laboratory, S.P. College, Srinagar, 15 pages.

Bacha, M.S. (1996). Waterfowl census in Kashmir Wetlands. Department of Wildlife Protection, Jammu and Kashmir Government, Srinagar. 21 pages.

Balkhi, M.H., Yousuf, A.R. Pandit, A.K. and Gazala F. (1992). Species diversity of zooplankton in Kashmir water. pages 245-253, In: Yousuf, A.R., Raina, M.K. and Qadri, M.Y. (Eds) Current Trends in Fish and Fishery Biology and Aquatic Ecology. Kashmir University, Srinagar, J & K 365 pp.

Balkhi, M.H. and Yousuf, A.R. (1992). Community structure of crustacean plankton in relation to tropic conditions. International Journal of Ecology and Environmental Sciences 18: 155-163.

Balkhi, M.H., Yousuf, A.R. and Qadri, M.Y. (1987). Hydrobiology of Anchar lake. Comparative Physiology and Ecology 12: 131-139.

Bhardwaja, Y. (1936). On two forms of *Hydrarus* ag. From Kashmir. Proceedings of the Indian Academy of Sciences 3(B): 278-281.

Brehm, V. (1937). Freshwater Fauna of India. Part II. Records of Indian Museum 48: 9-28.

Cook, C.D.K. (1996). Aquatic and Wetland Plants of India. Oxford University Press, Oxford. 385 pp.

Das, S.M. (1966). Palearctic elements in the fauna of Kashmir. Nature, London 212: 1-6.

Das, S.M. and Subla, B.A. (1963). The ichthyofauna of Kashmir, Part I. History, topography, origin, ecology and general distribution. Ichthyologica 2: 68-106.

Das, S.M., Daftari, S., Singh, H.B., Akhtar, S., Chaudhary, S. and Ahmad, N. (1969). Studies on the organic production in high altitude lakes of Kashmir I. The general ecology, and zooplankton of Kashmir lakes. Kashmir Science 6: 1-22.

Day, F. (1878). The Fishes of India: Being a natural history of the fishes known to inhabit seas and freshwaters of India, Burma and Ceylon. William Dawson and Sons, London. 778 pages.

Edmondson, W.T. and Hutchinson, G.E. (1934). Yale North India Expedition Article IX. Report on Rotatoria. Memoirs Connecticut Academy of Arts and Science 10: 153-186.

Enex (1978). Study of the Pollution of Dal Lake, Srinagar, Kashmir, Enex Consortium, New Zealand.

Garg, J.K., Singh T.S. and Murthy, T.V.R. (1997). Wetlands of India. Space Application Centre (ISRO) Ahmedabad.

Gopal, B. (1997). Biodiversity in Inland Aquatic Ecosystems in India: An Overview. International Journal of Ecology and Environmental Sciences 23: 305-313.

Handoo, J.K. and Kaul, V. (1982). Phytosociological and standing crop studies in wetlands of Kashmir. pp. 187-195, in: Gopal, B., R.E. turnner, R.G. Wetzel and D.F. Whigham (Eds) Wetlands Ecology and Management. International Scientific Publications, Jaipur.

Holmes, P.R, and Parr, A.J. (1988). A checklist of the birds of Haigram Rakh, Kashmir. Journal of the Bombay Natural History Society 85(3): 465-473.

Hora, S.L. (1995). Place of Kashmir in the fish geography of India. Everyday Science 3(1): 36-45.

Jan, U. (1987). Studies on macrobenthos of Dal lake. Ph.D. thesis of Kashmir University. Srinagar (unpublished).

Kak, A.M. (1978). Taxonomic studies on aquatic angiosprms of Kashmir. Ph.D. thesis, Kashmir University, Srinagar. (unpublished)

Kak, A.M. (1984a). Aquatic and Wetland Flora of the Northwestern Himalaya, XXII Aquatic cyperaceae of North western Himalaya. Journal of Econnomic and Taxonomic Botany 5(3): 687-703.

Kak, A.M. (1984b). Potamogetonn in northwestern Himalaya. Journal of the Bombay Natural History Society 8(2): 312-319.

Kak, A.M. (1985). Aquatic and wetland vegetation of the northwestern Himalaya. XXl Family Nymphaeceae in the northwestern Himalaya. Journal of Economic and Taxonomic Botany 7(3): 391-395.

Kant, S. and Kachroo, P. (1971). Phytoplankton population dynamics and distribution in two adjoining lakes in Srinagar, 1. Macroflora in relation to phytoplankton. Proceedings of the Indian National Science Academy 37: 163-188.

Kant, S. and Kachroo P. (1977). Limnological studies in Kashmir Lakes. I. Hydrological features, composition and periodicity of phytoplankton in the Dal and Nagin lakes. Phykos 16: 77-97.

Kaul, K. (1988). Study on bacterial population of Dal lake. Ph.D. thesis, University of Kashmir, Srinagar (unpublished).

Kaul, V. (1977). Limnnological survey of Kashmir lakes with reference to trophic status and conservation : International Journal of Ecology and Environmental Sciences 3: 29-44.

Kaul, V. (1983). Conservation of plant resources in aquatic ecosystems with reference to some aquatic habitats of Kashmir. pages 118-131, In : Jain, S.K. and Mehra, K.L. (Editors) Conservation of Tropical Plant Resources BSI, Calcutta.

Kaul, V., Fotedar, D.N., Pandit, A.K. and Trisal, C.L. (1978). A comparative study of plankton populations of some typical freshwater bodies of Jammu and Kashmir state. pages 249-269. In : Sen, D.N. and S.P. Bansal (eds) Environment Physiology and Ecology of Plants. Bishen Singh Mahendrapal Singh, Dehradun.

Kaul, V. and Handoo, J.K. (1993). Ecology and management of some typical wetlands of Kashmir. Pages 112-130. In: Mishra, P.C. and Trivedy, R.K. (Editors) Ecology and Pollution of Indian Lakes and Reservoirs. Ashish Publishing House, New Delhi.

Kaul, V. and Zutshi, D.P. (1967). A study of aquatic and marshland vegetation of Srinagar Proceedings of the National Institute of Sciences, India 33(B): 111-127.

Khan, M.A. (1985). Spatio-temporal phytoplankton variability in two Himalayan lacustrine ecosystem Acta Hydrochimica Hydrobiologia (Berlin) 13: 249-257.

Khan, M.A. (1987). Observations on zooplankton composition, abundance and periodicity in two flood-plain lakes of the Kashmir Himalayan Valley. Acta Hydrochimica Hydrobiologia (Berlin) 15: 167-174.

Koul, S. (1946). Srinagar and its Environs. Mysore

Kundangar, M.R.D. and Sarwar, S.G. (1997). Dal lake. Environ. Engineering Dept. Government of Jammu and Kashmir, Srinagar. 37 pages.

Kundangar, M.R.D. and Zutshi, D.P. (1987). Ecology and production of some important macrophyte species in two rural lakes of Kashmir. Journal of Hydrobiology 3(6): 57-62.

Lavania, G.S., Paliwal, S.C. and Gopal, B. (1990). Aquatic vegetation of the Indian subcontinent pages 29-78. In: Gopal, B. (Editor) Ecology and management of the aquatic vegetation of the Indian subcontinent. Kluwer Acad. Publishers, Dordrecht.

Lawrence, R.W. (1895). The Valley of Kashmir. Oxford University Press, Oxford.

Mir, A.M. and Kachroo, P. (1982). Limnology of Kashmir lakes VII the ecology of Bacillariophyceae in two lakes of Srinagar. Proceedings of the Indian National Science Academy B 48(2): 378-390.

Misra, J.N., (1937). Zygnemaceae of Kashmir I. Proceedings of the Indian Academy of Science 5B: 110-117.

Mufti, N.A. (1982). Measurements of phytoplankton production using C-14, plankton populationn and environmental factors of two inter-connected basins of Dal lake. M.Phil. Dissertation, University of Kashmir (unpublished).

Mukherjee, S.K. (1926). Aquatic and marshland vegetation of Dal lake of Kashmir. Proceedings of the Linnaean Society, London: 55.

Mukerjee, S.K. (1934). The charophyts of the Dal lake, Kashmir. Proceedings of the Indian Science Congress 21: 295.

Mukerji, D.D. (1936). Yale North India Expedition. Report on fishes. Part II. Sisoridae and Cyprinidae. Memoirs, Connecticut Academy of Arts and Science 10: 323-359.

Nath, S. (1986). A checklist of fishes of Jammu and Kashmir State (India) with remarks on the Ichthyogeography of the state. Journal of the Zoological Society of India 38: 83-98.

Pandit, A.K. (1982). Feeding ecology of breeding birds in fie wetlands of Kashmir. Indian Journal of Ecology 9: 181-190.

Pandit, A.K. (1991). Conservation of wildlife resources in wetland ecosystems of Kashmir, India. Journal of Environmental Management 33: 143-154.

Pandit, A.K. and Kaul, V. (1982). Trophic structure of some typical wetlands. pages 55-82, In: Gopal, B. Turner, R.E., Wetzel, R.G. and Whigham, D.F. (Editors) Wetlands: Ecology and Management. Vol. II. Internationnal Scientific Publications, and National Institute of Ecology Jaipur.

Pandit, S.N. (1983). Ecological studies of periphyton in two lake ecosystem of Kashmir. M. Phil. Dissertation. University of Kashmir. (unpublished).

Qadri, SA. (1985). Hydrobiology of Wular, Kashmir. M.Phil. Dissertation, University of Kashmir (unpublished).

Raina, H.S. and Vass, K.K. (1993). Distribution and species composition of zooplankton in Himalayan ecosystems. Innternationale Revue der gesamten Hydrobiologie 78: 295-307.

Reddy, Y.R., Blkhi, H.H. and Yousuf, A.R. (1990). *Arctodiaptomus* (Rhabdodiaaptomus) *Michaeli* n. sp (Copepoda: Calanoida) from Kashmir India. Hydrobiologia 190: 223-231.

Sama, A.L.N., Rao, D.G. and Durani, P.K. (1983). The phytal macro and meiofaua of Dal Lake (Srinagar, Kashmir). Journal of Environmental Biology 4: 117-126.

Sarwar, S.G. (1985). Community structure, species composition and seasonal variation in Aufwuchs population on natural and artificial substrates. Ph.D. thesis, University of Kashmir. (unpublished)

Sarwar, S.G. and Zutshi, D.P. (1987). Studies on periphyton population of Himalayan lakes. I. Species composition and community structure on natural and artificial substrates. Proceedings of the Indian National Science Academy B53: 239-243.

Sarwar, S.G. and Zutshi, D.P. (1989). Periphytic algal flora of Phragmites communis Trin. Journal of the Indian Botanical Society 69: 275-283.

Sarwar, S.G. and Parveen, A. (1996). Community structure and population density of zooplankton in two interconnected lakes of Srinagar, Kashmir. Pollution Research 15: 35-38.

Sculthorpe, C.D. (1967). Biology of Aquatic Vascular Plants, Edward Arnold, London.

Silas, B.G. (1952). Further studies regarding Hora's Satpurahypothesis. 2. Taxonomic assessment and levels of evolutionary divergence of fishes with the so-called Malayan affinities in Peninsular India. Proceedings of the National Institute of Sciences, India 18: 423-448.

Silas, B.G. (1960). Fishes from the Kashmir Valley. Journal of the Bombay Natural History Society 57: 66-67.

Subla, B.A., Wanganeo, A., Raina, R., Vishen, N. and Zutshi, D.P. (1992). Studies on zooplankton of Jammu and Kashmir State pages 33-49. In: Nath, S. (Editor) Recent Advances in Fish Ecology and Conservation .

Subrahmanyam, K. (1962). Aquatic Angiosperms. Botanical Monograph 3. Council of Scientific and Industrial Research, New Delhi.

Ticku, A. and Zutshi, D.P. (1991). Effects of weed harvestinng on the fish population of Dal Lake Interational Journal of Ecology and Environmental Sciences 17: 217-224.

Ticku, A. And Zutshi, D.P. (1993). The distribution and abundance of epiphytic rotifer populations on submerged macrophytes in Dal Lake, Srinagar. Journal of the Indian Institute of Science 73: 237-245.

Ticku, A. and Zutshii, D.P. (1994). Species composition and community structure of periphytic rotifers in Manasbal Lake, Kashmir. Proceedings of the National Academy of Sciences, India 64 B 175-180.

Trisal, C.L. (1977). Studies on primary production in some Kashmir lakes. Ph.D. thesis, University of Kashmir (unpublished).

Trisal, C.L., Robson, I. and Kundangar, M.R.D. (1994). Wular Lake. WWF-India, New Delhi.

Wanganeo, A. and Wanganeo, R. (1991). Algal population in valley lakes of Kashmir Himalaya. Archiv for Hydrobiologie 121: 219-233.

Yousuf, A.R. (1988). Copepod plankton of Manasbal lake. Journal of the Indian Institute of Science 68: 307-313.

Yousuf, A.R. (1996). Fishey resources of Kashmir, pages 87-128, In : Khan, A. H. and Pandit, A.K. (Editors) Ecology, Environment and Energy, Kashmir University, Srinagar.

Yousuf, A.R. and Parveen, M. (1992). Ecology of polluted waters of Kashmir. Brarinambal basin of Dal lake In : Yousuf, A.R., M.K. Raina and M.Y. Qadri (Editors), current Trends in Fish and Fishery Biology and Aquatic Ecology. Kashmir University, Srinnagar, J&K.

Zutshi, D.P. (1975). Associations of macrophytic vegetation of Kashmir lakes. Vegetation 30: 61-66.

Zutshi, D.P. (1987). Impact of human activities on the evolution of Dal lake environmennt. pages 565-577, In: Pangtey, Y.P.S. and Joshi, S.C. (Editors) Western Himalaya: Environment, Problems and Development. Nainital.

Zutshi, D.P. and Gopal, B. (1998). Ecology and Management of Lakes and Wetlands of Himalaya (Manuscript).

Zutshi, D.P. and Khan, A.U. (1988). Eutrophic gradient in Dal lake, Kashmir. Indian Journal of Environmental Health 30: 348-354.

Zutshi, D.P. and Vass, K.K. (1982). Limnological studies on Dal lake III. Biological features. Proceedings of the Indian National Science Academy B 48(2): 234-241.

Zutshi, D.P., Subla, B.A., Khan, M.A. and Wanganeo, A. (1980). Comparative limnology of nine lakes of Jammu and Kashmir Himalayas, Hydrobiologia 80: 101-112.

Zutshi, D.P., Wanganeo, A. (1979). *Salvinia natans*–a new menace to Dal lake. Geobios 6: 217-218.

Zutshi, D.P. and Wannganeo, A. (1984). The phytoplankton and primary productivity of a high altitude subtropical lake. Verhandlungen, Internationale Vereingung der Limnologie 22: 1168-1172.

Chapter - 7

Wetland Biodiversity in the Kashmir Himalaya : Assessment and Conservation Strategies

*M.A. Khan**

ABSTRACT

Wetlands support highly valuable pools of biodiversity and genetic resources, but unsustainable development is threatening the biowealth, and even causing species extinction. Data on Kashmir wetlands and contained biodiversity are inadequate. This paper attempts to augment and present a consolidated assessment of biodiversity potential covering plant life (algae, macrophytes) and faunal composition (invertebrates, fish and birds). The functional role of wetlands and wetland biodiversity are discussed and strategies needed for their conservation and wise use highlighted.

Key words: aquatic biodiversity, avifauna, conservation, fish diversity, functional role, invertebrates, Kashmir Himalaya, macrophytes, wetlands.

INTRODUCTION

Freshwater biodiversity is increasingly threatened by unsustainable development. Water system regulations and drainage for agriculture and urban development have been the major causes of the loss of over 50% of the wetlands in several countries including USA, Australia and India. Wetlands, after tropical rain forests, are among the richest ecosystems on the planet Earth, and display high levels of biodiversity. As per the WWF Atlas of the Environment 1990, wetlands distributed globally in different climates cover about 6% of the earth's land surface. They support highly valuable pools for biodiversity and genetic resources, and prove a livelihood to millions of people. Destruction of habitats accompanied by loss of species due to anthropogenic pressures, constitute threat to wetland biodiversity; many suffer significant and lamentably irreversible change. According to Deny

*Division of Environmental Sciences, S-K University of Agricultural Sciences & Technology, Shalimar Campus, Srinagar - 191121, Kashmir, India.

(1994), the extinction of species in wetlands is higher than for any other ecosystem; 66% of the continental extinctions are aquatic taxa. Scott and Poole (1989) state that Asian wetland scenario is distributing with 85% of studied sites threatened. It is believed that in Asia alone > 2 billion people depend on wetland crops and fish as their staple and protein resource.

Wetland conservation receive fillip at the global level when 160 countries joined the Convention on Wetlands (Ramsar Convention), originally signed on 2 February 1971 in the Iranian city of Ramsar, on the Caspian Sea. The main objective was to identify "Wetlands of International importance", so as to provide special attention to their conservation and sustainable use. Almost 900 such sites, covering some 7500000 ha (=75000 km^2) have been placed so far in the Ramsar Convention List.

India supports a rich diversity of inland and coastal wetland habitats, spread over 4.1 million ha of wetlands (excluding paddy-fields and mangroves) of which 1.5 million ha are natural and 2.6 million is man-made (Anon 1992). Mangroves are estimated to cover 0.6 million ha. On the contrary, a Director of Asian Wetlands (IUCN) shows total area of wetlands 58.2 million ha, of which paddy cultivation constitutes 40.9 million ha. Broadly, the wetlands in India can be categorized in to (i) Himalayan (ii) Indo-Gangetic, and (iii) Coastal wetlands. The Himalayan region, characterized by a series of wetlands, is distinguishable into (i) Western (ii) Central, and (iii) Eastern Himalaya.

Kashmir valley comprising a part of Western Himalaya with varied relief and dense drainage network has unique distinction of richness of natural waters. However, scanty literature is available on diverse biotic components and their functional role in ecosystem dynamics. Despite their age old recognition of socioeconomic importance, the existence of wetlands as ecological entities and associated rich biowealth has been threatened at an unacceptably high rate due to varied anthropogenic perturbations.

The present paper deals with the assessment of biological diversity of some selected Kashmir wetlands and outlines strategies for their conservation.

WETLANDS: DEFINITION AND MAJOR FUNCTIONAL ASPECTS

Wetlands are ecotones (edges), transition zones between dry land and deeper water; environments that are not always wet nor obviously dry. The unique plant and animal communities in them depend on environmental conditions created up alternating denundation and drying during different seasons or different years. The species diversity is reported (Gopal 1994) to be generally higher in transition zones than in adjacent systems. Gopal (1994) holds that wetlands can be interpreted as ecotones and, by implication, the severe should be true also. Tiner (1991) recognises wetlands "as distinct ecosystems along the wetter end of the soil moisture gradient" on the basis of distinctive characteristics of microflora invertebrate fauna and soils.

The Convention of Wetlands of International Importance (Ramsar) defines wetlands "as areas of marsh, fen, peatland or water weather natural or artificial, permanent or temporary, with water that is static or flowing, fresh, breackish or salt, including areas of marine water the depth of which at low side does not exceed six metres". Winderoxhel (1995) revised the definition to make it reflective of tropical examples such as seagrass beds and coral reefs. However, common usage centers on a modified IUCM definition of wetlands and reads "All submerged or water-saturated lands, natural or man-made, inland or coastal, permanent or temporary, static or dynamic, vegetated or non-vegetated, which necessarily have a land-water interface are wetlands".

Wetlands are extremely important as they are not only discrete ecological units, but cut across a range of boundaries from upland to lowland, terrestrial to marine, freshwater to salt water. Thus, multiplicity of interests involved extend from fisherman to farmer, herders to hunters, bird-watchers to commercial interests. Wetlands also provide great benefits by way of natural flood-control or buffering for downstream areas by slowing the flow of floodwater, desynchronizing peak contributions of tributary streams, and reducing peak flows on main rivers. Some wetlands may function as ground water recharge areas. They can effectively remove or convert large quantities of pollutants form point or non-point sources. Many nutrients are held in the wetlands system and recycled through successive seasons of plant growth, death and decay. Because of their role in limiting the damage-effects (e.g. store floodwater, trap sediment, reduce pollutants) the wetlands are known as "the kidneys of the planet". Table 7.1 presents summarized major functional aspects of the wetlands.

Table 7.1 : Functional aspects of wetlands

I. Hydrologic

* Natural flood-control or buffering for downstream areas
* Function as ground water recharge areas

II. Pollution abatement

* Remove or convert large quantities of pollutants from point and non-point sources
* Sequester heavy metals and certain poisonous chemicals in anaerobic zones of soil

III. Biogeochemical

* Play pivotal role in nutrient recycling
* Regulate nutrient inputs and organic matter flux to open waters

IV. Organic production

* Considered richest ecosystems (after tropical rain forests) on the planet
* Contribute high primary and secondary productivity
* Support fisheries resources

V. Biological

* Support highly valuable pools for biodiversity and genetic resources
* Acts as suitable habitat for residential and migratory birds including rare and endangered ones
* Yield aquatic plants of high economic utility to local population including fodder
* Provide breeding sites for variety of fauna

VI. Others

* Constitute favourable places of recreation and relaxtion

(contd....)

* Provide scenic beauty to nature-lovers (especially bird-watchers)
* Provide site for bird-shooting
* Revenue generation to concentrated agencies
* Provide sites for ecological monitoring and environmental impact assessment to scientific personnel and researchers.

WETLAND SITES AND THEIR ENVIRONMENT

Kashmir Valley is replete with diverse types of freshwater bodies. Majority of these natural aquatic systems are shallow-basined and classified under western Himalayan wetlands. Zutshi and Khan (1978) in a classic typology recognized three major categories of waterbodies: Valley lakes (1580-1600 m) of Kashmir, Forest lakes (2000-2500 m) of Pir Panjal Range, and Glacial or High altitude (3000 m) lakes. Hydrologically, valley lakes have been sub-categorized into drainage (e.g. Dal, Wular), semidrainage (Naranbaghsar, Khanpursar), and non-drainage (Trigamsar) types. The valley lakes/wetlands are mostly located in the floodplain of River Jhelum and River Sind, and are characterized by varying hydroedaphic features (Table 7.2). A brief description of eight wetlands treated is given as under:

Hokersar: The wetland has fluvial origin and is located close to Srinagar (about 10 km away). It is permanent but relatively shallow waterbody with water source mainly from Doodganga flood channel. The utilization ranges from agricultural activity to fisheries, grazing, irrigation. The wetland is well known game/wildlife sanctuary but subject to varied anthropogenic pressures including encroachment, siltation and eutrophication posing threat to its survival.

Migrund: It is a shallow wetland of fluvial origin about 15 km southwest of Srinagar. Although a temporary wetland, it at times attain maximum depth of 1m with the water supply mostly from Sukhnallah and channels irrigating paddy-fields. The waterbody is extensively used for harvesting reeds and grazing is common. The horticulture activity is witnessed in close vicinity of the waterbody. The ecological problems include encroachment, siltation, eutrophication. The wetland is also used as a game reserve.

Shallabough : Commonly known as 'Shallbough Rakh', the wetland is temporary and shallow, with main source of water from River Sind. The waterbody is about 15km southwest of Srinagar and its importance lies in its use as game reserve. The wetland is exposed to ravages so siltation and encroachment.

Trigamsar: Located in the flood plain of River Jhelum, Trigamsar is about 20km in the west of Srinagar. The waterbody is permanent, nondrainage, shallow, and is very turbid. It has been under the impact of heavy biotic interference due to harvesting of food and fodder and plants and fishing. Ecologically, the waterbody is highly polluted. Recently, J&K Fisheries Department has developed the waterbody as a fish-farm.

Naranbaghsar: Located about 30km from Srinnagar, the waterbody is semi-drainage type with major part of its drainage basin being calcareous. It is essentially an oxbow type with alluvial sediments. Water supply is mainly from springs present within its basin. Activities related to agriculture and horticulture besides raising of aquatic fruits are the main sources of human interference.

Table 7.2. Some important morphoedaphic and ecological characteristic of wetlands

Wetland	*Physical status**	*Area (ha)*	*Fluctuation in depth (a)*	*Water temp. (°C)*	*pH*	*Dissolved Oxygen (mgl^{-1})*	*Importance*
Hokersar	Permanent; shallow	500	0.5-1.6	6.5-28	7.2-8.5	4.5-8.0	Game Sanctuary
Mirgund	Temporary; shallow	400	0.05-1.0	7-27	7.1-8.4	3.2-8.0	Game Reserve
Shallabough	Temporary; shallow	700	0.05-1.0	7-27.5	7.0-7.6	4.0-7.8	Game Reserve
Trigamsar	Permanent; shallow	11	1.5-2.25	7-30	8.1-9.0	7.6-14.5	developed fish-farm
Naranbaghsar	Permanent; deep	24.3	3-5	7.5-29	7.6-8.7	7.3-12.6	Source of fodder
Khanpurar	Permanent; deep	12.5	2.5-3.8	6.5-29	7.4-8.6	7.0-12.5	-do-
Haigam	Permanent; shallow	1400	0.2-1	6-29	7.4-8.3	4.5-6.7	Game Sanctuary
Wular	Permanent; shallow	19000	0.5-4.4	4.5-30.5	7.2-9.2	4.1-11.3	Ramsar Site

* *all wetlands are natural*

Khanpursar: A semidrainage waterbody, Khanpursar is relatively deep and permanent ecosystem. A few ephermeral inflow channels constitute the main source of water. Submerged vegetation dominates the waterbody. Paddy cultivation and willow plantation in some parts of catchment makes it prone to eutrophication. The utilization of the waterbody includes fishing, harvesting of aquatic plants of varying economic importance.

Haigam: This wetland is an important game sanctuary, about 50 km west of Srinagar. The various channels irrigating the paddy fields constitute the main source of water. The activities linked with the sanctuary are reed-harvesting, fishing. Like other major wetlands, siltation and unabated encroachment are potential problems.

Wular: Lake Wular, considered to be one of thee largest-area natural freshwater lakes in Asia, is internationally important Ramsar site. Fed and drained primarily by river Jhelum, water-input is also from several springs and streams. The waterbody is permanent but relatively shallow having multiple economic utility; fishing, grazing, reed-harvesting, rice cultivation and irrigation, besides serving as a breeding habitat for a large number of resident and migratory birds. Siltation, enrichment and pollution are the main problems of the wetland.

BIODIVERSITY IN WETLANDS

Floristic diversity:

Algal flora: Composition and distribution:

The species diversity reported in some Kashmir wetlands is quite variable. Generally, the taxa are represented by various classes; Cyanophyceae, Chloophyceae, Bacillariophyceae, Chrysophyceae, Dinophyceae, and Euglenophyceae. Table 7.3 shows that green algae were remarkably more diverse in species composition followed by diatoms. Blue-greens are fairly represented but the occurrence of chrysophytes, dinoflagellates and euglenoids are rare in terms of number of species. Pandit and Kaul (1982), without documenting identification list, reported a total of 198 species of phytoplankton in some Kashmir wetlands; greens had maximum (72) taxa followed by diatoms (66), euglenoids (29), blue-greens (20), whilst chrysophytes, xanthophytes and dinoflagellates contained only 3 taxa each. The number of species inhabiting various wetlands was <150 but> 200. The distribution pattern showed the following sequence: Nowagam - 189, Haigam - 183, Migrund - 179, Malgam - 177, and Hokersar - 156. The studies reported in Anon (1993) show the presence of 82 phytoplankton species in Lake Wular.

Table 7.3 : List of algae found in some wetlands of Kashmir

1	2
I. Cyanophyceae	
Anabena circinalis Raben	*Anabaena* sp.
Aphanocaspa sp.	*Arthrospira* sp.
Calothrix sp.	*Chroococcus limneticus* Lemm.
Coelosphaerium naegelianum Ung.	*Coelosphaerium* sp.
Merismopedia elegans Braun	*M. Glauca* (Ehr.) Kutz.
M. Punctata Meyen	*Microcystis aeruginosa* (Kurtz.) Kurtz.

1	2
Microcystis sp.	*Oscillatoria limnetica* Lemm.
Oscillatoria sp.	*Spirulina* sp.
Trichodesmium lacustre Kleb.	—
II. CHLOROPHYCEAE	
Actinastrum hantzschii Lagerh.	*Ankistrodesmus convolutus* Corda.
A. falcatus (Corda.) Ralfs.	*A. spiralis* (Turner) Lemm.
Chlorella vulgaris Beijer.	*Closterium acerosum* (Schrank) Ehr.
C. moniliferum (Bory) Ehr.	*Closterium* sp.
Coelastrum microporum Nag.	*Cosmarium reniforme* Archer
Cosmarium sp.	*Crucigena tetrapedia* (Kirch.) W. and G.S. West
Crucigeniella crucifera (Wolle) Kom.	*Dictyosphaerium ehrenbergianum* Nag.
Gonium Pectorale Muller	*Kirchneriella obesa* Schmidle
Kirchneriella sp.	*Micrasterias pinnatifida* (Kütz.) Ralfs.
M. raidata Hass	*Oedogonium* sp.
Oocystis lacustris Chodat	*Pandorina morum* (O.F. Müller) Bory
Pediastrum boryanum (Turp.) Meneghini	*P. duplex* Meyen
P. simplex Meyen	*P. tetras* (Ehr.) Ralfs
Scenedesmus acuminatus (Lagerh.) Chod.	S. *arcuatus* Lemmermann
S. armatus Chodat	*S. Bijugatus* (Turp.) Kürtz.
S. carinatus (Lemm). Chodat	*S. dimorphus* (Turp.) Kütz.
S. obliquus (Turp.) Kurtz.	S. quadricauda (Turp.) Bréb.
Scenedesmus spp.	*Selenastrum* sp.
Sphaerocystis schroeteri Chod.	*Spirogyra* sp.
Staurastrum lunatum Ralfs.	*Staurastrum* sp.
Tetraedron minimum (A Braun) Hansg.	*Tetraedron* sp.
Ulothrix sp.	*Volvox globator* Ehr.
Volvox sp.	
III. Bacillariophyceae	
Achannthes coarctata (Breb). Grun.	*A. minutissima* Kütz.
Amphora ovalis Kütz.	*A. veneta* Kütz.
Aulocosiraa granulata (Ehr.) Ralfs.	*Coocconeis placentula* Ehr.
Cyclotella sp.	*Cymatopleura elliptica* (Breb.) W. Smith
Cymbella affinis Kütz.	*C. prostrata* (Berk.) C1.
C. ventricosa Kütz.	*Diatoma elongatum* (Lyngb.) Agardh

1	2
Diatoma spp.	*Epithemia turgida* (Ehr.) Kütz.
Epithemia sp.	*Fragilaria capucina* Desm.
F. crotonensis Kitton	*Fragilaria* sp.
Gomphonema anustatum (Kütz.) Rabh.	*G. constrictum* Ehr.
G. gracile Ehr.	*Gomphonema* sp.
Gyrosigma sp.	*Hantzschia amphioxys* (Ehr.) Grun.
Navicula cuspidata (Kütz.) Kütz.	*N. mutica* Kütz.
N. radiosa Kütz.	*Navicula* sp.
Nitzchia acicularis W. Smith	*N. Angustata* (W. Smith) Grun.
N. apiculata (Greg.) Grun.	*Nitzschia* spp.
Pinnularia borealis Ehr.	*Surirella ovata* Kütz.
Synedra acus Kütz.	*S. ulna* (Nitzsch) Ehr.
Tabellaria fenestrata (Lyngb.) Kütz.	
IV. Chrysophyceae	
Dinobryon sp.	*Synura* sp.
V. Dinophyceae	
Ceratium hirundinella (O.F. Müller) Duj.	*Gymnodinium* sp.
Peridinium inconspicuum Lemm.	
VI. Euglenophyceae	
Euglena acus Ehr.	*E. acus v. longissim* Defl.
E. pedunculata Gojdics	*E. spirogyra* Ehr.
Euglena sp.	*Lepocinlis ovum* Lemm.
Lepocinclis sp.	*Phacus caudatus* Hubner
P. longicauda (Ehr.) Duj.	*Phacus* sp.
Trachelomonas volvocina Ehr.	*Trachelmonas* sp.

Quantitative dominance of different algal groups in wetlands is highly variable as exemplified by the work of Khan (1985) who reported that surface water of Trigamsar was nearly 90 times and bottom water 40 times denser in algal population as compared to Narabaghsar. Furthermore, the seasonal oscillations in the Wetlands indicate that algal densities start increasing during spring, reach maximum in summer and start declining from autumn onwards. This unimodal pattern was governed by high temperature and better light conditions during summer. Trigamsar had blue-greens dominating the annual algal population followed by greens and diatoms whilst in Naranbaghsar, different sequence of algal stock prevailed, with greens constituting a dominant group followed by diatoms and blue-greens. There was insiginificant contribution from other groups. Kaul and Pandit (1982) reported dominance of green algae in some wetlands including Haigam, Mirgund and Hokersar, though on relative basis, Hokersar had minimum biomass production of phytoplankton. Zutshi (1996) evaluated the phytoplankton population of three waterbodies (Table 7.4). Diatoms were found to be numerically higher in Khanpursar as compared to other wetlands. However, the number of species was more in Trigamsar wetland.

Table 7 4. Percent phytoplankton population (cellsl^{-1}). The number is given in parenthesis.

Class	*Khanpursar*	*Telwansar*	*Trigamsar*
Bacillariophyeae	88 (34)	60.2 (33)	57.12 (37)
Cyanophyceae	5.3 (10)	27.5 (7)	19.2 (8)
Chlorophyceae	1.9 (26)	3.7 (25)	13.2 (28)
Euglenophyceae	0.16 (3)	6 (3)	9.7 (3)
Chrysophyceae	5.4 (1)		0.1 (1)

Source: Zutshi (1996)

Pandit and Kaul (1981) state that the phytobenthos of wetlands include 155 algal species and 14 soil fungal species. The distribution of algae was Chlorophyceae (37), Cyanophyceae (18), diatom (61), Xanthophyceae (1), Chrysophyceae (2), Dinophyceae (2), Crytophyceae (4) and Eugleinae (27).

Thus it is obvious that our knowledge on algal flora of wetlands is far from complete and merits attention for adequate census and taxonomic documentation. Studies gain added importance in view of occurrence of red-bloom flagellate, *Euglena pedunculata* constituting mono-specific population in Kashmir lakes and Wetlands (Khan 1993 a, b; 1994 a, b; 1996).

Macrophytic vegetation and associations

The macrophytic vegetation in Kashmir wetlands presents a great deal of diversity in composition, number and extent of colonization. Broadly, four district groups recognized are : emergents, rooted floating-leaf types, free-floating types and submersed ones. The occurrence, frequency and relative abundance of each type varies considerably in various Wetlands affecting the organic production. No attempt is made here to discuss ecophysiological aspects; instead study is restricted here to document the pattern of macrophytic diversity. Table 7.5 lists the distribution of important macrophytes (arranged alphabetically in their respective groups) in selected wetland sites.

Table 7.5 : Distribution of some important macrophytes in Kashmir wetlands

*Name of aquatic plant**	*Wetland Site*
1	2
I. Emergents	
Acorus calamus	2, 3, 7
Alisma plantago-aquatica	1, 5, 6
Carex articulata	3, 7,

1	2
C. serotinuus	1, 2, 7
Carex spp.	4, 7
Cladium mariscus	3, 5, 6, 7
Cyperus serotinus	1, 7
Cyperus spp.	1, 7
Eleocharis palustris	1, 2, 7
Juncus articulatus	1
Menyanthes trifoliata	3, 7
Myriophyllun verticillatum	1, 3, 5, 6, 7
Phragmites australis	1, 2, 3, 4, 5, 6, 7
Polygonum hydropiper	1, 7
Rannunculus spp.	2, 7
Sagittaria trifolia	1, 5, 6, 7,
Scirpus lacustris	1, 2, 3, 5, 7,
S. palustris	7
Sium latijugam	7
Sparganium ramosum	1, 2, 3, 7,
Spiranthes sinensis	1, 7
Typha angustata	1, 3, 4, 5, 6, 7, 8
II. Rooted floating-leaftype	
Hydrocharis dubia	2
Marsilea minuta	1, 2
Nelumbo nucifera	5, 6, 8
Nymphaea Alba	5, 6
N. stellata	1, 2, 3, 7
Nymphoides peltata	1, 2, 3, 4, 5, 6, 7, 8
Polygonum amphibium	1, 2
Potamogeton natans	1, 2, 3, 4, 5, 6, 7
Trapa natans	1, 2, 3, 7, 8,
III. Free-floating type	
Lemna aequinoctialis	4, 5

1	2					
Lemna spp.	1,	4,	5,	6,	7,	8
Salvinia natans	1,	2,	5,	6,	7,	8
IV. Submersed						
Ceratophyllum demersum	2,	3,	5,	6,	7,	8
Myriophyllum spicatum	2,	5,	6,	8		
Potamogeton crispus	1,	2,	5,	6,	7	
P. filiformis	5,	6				
P. lucens	1,	2,	4,	5,	6,	7
Potamogeton spp.	2					
Ranunculus aquatilis	1,	2				
Utricularia aurea	7					

Site: Hokersar (1) Mirgund (2) Nowgam (3) Trigamsar (4) Naranbaghsar (5) Khanpursar (6) Haigam (7) Wular (8).

*Arranged alphabetically in respective categories

A generation allows to state that most wetlands support emergents (e.g. *Pharagmites australis, Typha angustata*), rooted floating-leaf types (e.g. *Nymphoides peltata, Nympheae stellata, Potamogeton natans*), free-floating forms (e.g. *Lemna* spp., *Salvinia natans*) and submersed ones (e.g. *Ceratophyllm demersum, Potamogeton lucens.*). However, in turbid wetlands (e.g. Hokersar), the submersed macrophytes are very sparse, but extensive belts of reeds dominate the littorals. More comprehensive treatment of aquatic and marshland aquatic vegetation of Srinagar is provided in Kaul and Zutshi (1967) listing 117 species distributed in 69 genera. Pandit (1980) compared macrophytes in five wetlands of Kashmir and observed that four species (*Phragmites australis, Nymphoides peltata, Nympheae stellata, Scirpus lacustris*) were common to the sites; Haigam wetland recorded highest number (29) of macrophytes. According to Handoo and Kaul (1982), the composition of macrophytes in four wetlands studied was : Melangpora (42), Kranchu (32), Shallabough (19), and Hokersar (19) - only four species were common to these four wetlands. Very recent analysis (*see* Zutshi and Gopal 1999) reports the occurrence of about 225 aquatic and wetland species in Kashmir constituting about 20% of the total Indian species.

Zutshi (1975) provided associations of macrophytic vegetation in Kashmir lakes (including wetlands) and recognized 26 associations. The highest number of associations (10) is recorded in Potametea whilst Phragmitetea had 6 associations. The author opined that, on the whole, Kashmir aquatic vegetation shows a higher floristic affinity to Europe a than to Tropical-Indian aquatic communities. Observations indicate that myriophyllids occur in combination with ceratophyllids and also nymphaeids. In relatively deep wetlands (e.g. Naranbaghar and Khanpursar) the myriophyllid and ceratophyllid growth-forms are found together in the depth zone of 2.5 to 4.0 m, but in deeper zones (even in lakes-Dal, Nagin, Mansbal) ceratophyllids form pue stands. Furthermore, Zutshi (1975) showed that with regard to the distribution of plant communities. Anchar Lake wetland had the great density with 20 associations and half of it occurred in Hokersar and Haigam wetlands and minimum of 2 associations were recorded in Trigamsar. Recently, some urban Kashmir waters (e.g.

Dal Lake witnessed shifts in vegetation association as indicated by formation of wide mats of monospecific water-fern (*Salvinia natans* covering extensive areas including Nagib basin (Khan 1997).

FAUNISTIC DIVERSITY

Kashmir wetlands sustain a myriad of faunal diversity representing various taxonomic groups spread from simple unicellular protozoa to relatively large animals including varied ichthyofauna and spectacular resident and migratory bird species.

Invertebrate fauna : Composition and distribution

The zooplankton fauna seems quite diversified in Kashmir wetlands. The present study records a total of 101 species (including 18 unidentified) belonging to 56 genera. Rotifera showed rich species richness (51 in 27 genera) followed by Cladocra (28 species in 15 genera) and Copepods (20 species in 14 genera) *Brachionus* (Rotifera) and *Alona* (Cladocera) revealed high species diversity (8 species in each) whilst *Cyclops* (Copepoda) contained 6 species (Table 7.6). Pandit and Kaul (1982) reported the presence of 147 zooplankton species in five wetlands of Kashmir; the faunal distribution indicated highest number in Protozoa (61) and Clacocera (35) whilst Rotifera had only 29 species, Copepoda 13, and Ostrocoda 3 species. The same authors report that among the sites, Haigam harboured highest number (116) of taxa followed by Nowgam and Malgam (109 each), Hokersar (100) and Mirgund (97).

Table 7.6 : List of zooplankton in some wetlands of Kashmir.

*Zooplankton**	*Wetland site*
1	*2*
I. Rotifera	
Anuraeopsis fissa	8
A. navicula	4
Ascomorpha sp.	8
Asplanchna brightwelli	4, 5, 6
A. priodonta	1, 2, 3, 7
Brachinus aculeatus	4
B. angularis	4
B. bidentata	1
B. calciflorus	1, 2, 3, 4, 7
B. caudatus	4
B. falcatus	4
B. forficula	4
B. quadricornis	2

1	2
B. quadridentata	8
B. rubens	4
Brachious spp.	1, 2, 3, 7
Callotheca ornata	4
Cephalodella gibba	4
Cephalodella spp.	1, 2, 3, 7
Colurella bicuspidata	1, 2, 3, 7
C. oblongo	4
Diplois sp.	1, 3, 3, 7
Euchlanis dilata	1, 2, 3
Filinia longiseta	1, 2, 3, 4, 5, 6, 7, 8
F. opoliensis	1, 2, 3, 7
Gastropus stylifer	1, 2, 3, 7
Hexarthra intermedia	4
Horaella bremi	3, 4
Keratella cochlearis	1, 2, 3, 4, 7
K. quadrata	2, 4, 5, 6
K. tropica	4
K. valga	1, 2, 3, 7
Lecane bulla	4
L. ohioensis	1
L. pyriformis	4
Lecane spp.	1, 2, 3, 7
Lepadella ovalis	2, 4
L. patella	4
Monostyla spp.	1, 2, 3, 7, 8
Mytilina mucronata	1, 2, 3, 7
Notholca acuminata	1, 2, 3, 7
Philodinna sp.	3
Platyias sp.	1, 2, 3, 7
Polyarthra major	4, 5, 6
P. vulgaris	1, 2, 3, 7, 8

1	2
Pompholyx complanatus	4, 5, 6
Resticula melandocus	4
Synchaeta pectinata	2, 3
Trichocrc cylindrica	1, 2
T. longiseta	1
T. similis.	4
Trichocerca sp.	1, 2, 3, 7
Trichotria tetractis	2, 4
II. Cladocera	
Alona affinis	1, 2, 3, 7
A. bukobensis	4, 5, 6
A. costata	4, 8
A. guttata	2, 8
A. ladakensis	5, 6
A. Rectangula	1, 2, 4, 5, 6, 8
A. verrucosa	5, 6
Alona sp.	4
Biapetura verrucosa	5, 6
Bosmina coregoni	1, 2, 3, 7
B. longirostris	1, 2, 3, 4, 7, 8
Ceraiodaphnia quadrangula	5, 6
C. reticulata	1
Ceriodaphnia spp.	1, 2, 3, 7
Chydorus sphaericus	2, 4, 5, 6, 8
Daphnia magna	1, 2, 3, 7
D. pulex	1, 2, 3, 7
Daphnia sp.	1, 2, 3, 7
Diaphannosoma brachyurum	4, 5, 6
Graptoleberis testudinaria	1, 2, 3, 5, 6, 7
Macrothrix rosea	1, 2
Moina micrura	4

1	2						
Pleuroxus denticulatus	1,	2,	3,	7			
Pseudosida bidentata	1,	2,	3,	7			
Scapholeberis sp.	1,	2,	3,	7			
Sida crytaallina	1,	2,	3,	5,	6,	7	
Simocephalus exspinosus	1,	2,	3,	7			
S. vetulus	1						
III. Copepoda							
Acanthodiaptomus denticornis	5,	6					
Bryocamptus sp.	1,	2,	3,	7			
Canthocamptus sp.	1,	2,	3,	7			
Copepod nauplii	1,	2,	3,	5,	6,	7,	8
Cyclopds bicolor	8						
C. Bicuspidatus	1,	2,	3,	7,	8		
C. ladakensis	4						
C. scutifer	1,	2,	3,	7			
C. vicinus	4,	5,	6				
Cyclops sp.	1,	2,	8				
Diptomus spp.	1,	2,	3,	7,	8		
Encyclops agilis	1,	2,	3,	7			
Macrocyclops albidus	1,	2,	3,	8			
Megacyclopds viridis	8						
Mesocyclopds hyalinus	1,	2,	3,	5,	6,	7	
M. leuckarti	4,	5,	6				
Macrocyclopds varicans	4,	5,	6				
Neodiaptomus schmackeri	5,	6					
Paracyclopds affinis	8						
Phyllodiaptomus blanci	5,	6					
Thermocyclopds crassus	4,	5,	6				

Sites: Hokersar (1), Mirgund (2), Nowgam (3), Trigamsar (4), Naranbaghsar (5), Khanpursar (6), Haigam (7), Wular (8)

*Arranged alphabetically

Not much information is available on snail population of aquatic systems of Kashmir including wetlands. Kaul *et al.* (1980) studied the snails (Gastropod molluscs) in Haigam wetland and recorded a rich population of pulmonate snails belonging to exclusively family Lymnaeidae and Planoribidae whilst operculate snails belonging to Bulimidae family are restricted to the bordering paddy fields (temporary wetlands). Among the Lymnaeidae the records exist of *Lymnaea stagnalis* Linne, *Lymnaea* (Radix) *auricularia* Linne, *Lymnnaea (Pseudosuccinea) Columella* Say and the sporadically occurring *Lymnaea (Bulimnaea) meageasoma* Say and *L. abrussa* Say. The family Planorbidae, as reported in Kaul *et al.* (1980), is represented by *Promenetus exacuous* Say and *Planorbis* sp. whilst the family Bulimidate is represented by only *Amnicola limosa* Say.

Icthyofaunal diversity : Distribution in Kashmir Valley water

Kashmir aquatic systems, including famous wetlands (Lake Wular-Ramsar site) sustain good crop of commercially valuable ichthyofauna which supports and augments the livelihood of several thousand fishermen. Kashmir region accounts for >80% of the total fish production (1600 tonnes during 1995-96) of the state with L. Wular contributing 60% followed by River Jhelum (15%) while input from other lakes is just 5% only.

Despite long research interest in the ichthyofauna of Kashmir, an authentic census of existing fish populations is far from complete. Thus, while Hora *et al.* (1955) recorded 17 species from the Valley, Das and Subla (1963) listed 36 fish species in Kashmir waters including snow-fed streams. However, Nath (1986) reported 42 species from Kashmir. Further, revised version by Yousuf (1996) shows the presence, of only 37 species with some having uncertain taxonomy. The fish fauna, generally, can be grouped into three major categories: Cypriniformes, Siluriformes and Salmoniformes. However, fish found in Kashmir wetlands belong to mostly Cypriniformes and limited extent to Siluriformes. On the other hand, Salmoniformes occur in oxygen-rich clear waters and upstreams. Table 7.7 lists the fish species largely found in wetlands, though occurrence of some species is of sparse nature.

Table 7.7 : Occurrence of fish species in Kashmir wetlands

Cypriniformes:
Carassius Carassius (Gang)
Cyprinus carpio var. communis (Punjaeb-gaad)
Cyprinus carpio var. specularis (Punjaeb-gaad)
Puntius conchonius = *Barbus conchonius* (Safed-bacha)
Schizthorrax curvifrons (Sater-gadd)
S. esocinus (Chhuroo)
S. hugelli
S. longipinnis
S. micropogan (Ram-gaad)
S. nasus = *S. intermediatus*
S. niger (Ale-gaad)
S. labiatus
Crosocheilus diplocheilus

Botia birdi (Ram-gurun)

Siluriformes :

Gambusia affinis = *G. affinis .holbrooki* (Mahe-gaad)

*Local names given in parentheses

The number of fishes in wetlands is highly restricted. According to Pandit (1991) only 5-6 species occur in such biotopes with *Cyprinus carpio* var: *communis* fairly represented in Hokersar, Mirgund, Nowgam and Haigam wetlands. Wular lake wetland is believed to contain 11 species of fish. Some more have been recently introduced and particular mention may be made of grass carp (*Ctenopharyngodon idella*) and silver carp (*Hypopthalmichthys molitrix*) which have been introduced at Trigam Fish-Farm during 1981 by the J&K Fisheries Department.

Avifauna: Species Composition

The diversity in Kashmir avifauna is quite considerable, and varies seasonally. The waterbodies (lakes, wetlands, marshes) provide ideal breeding grounds for many species including globally threatened ones. Although rich in wetlands, Hokersar and Haigam in Kashmir are very important duck-shooting sites. These provide food and refuge to a number of overwintering water-fowl migrating from as far as north Europe to central Asia (Ali 1979) and breeding grounds to a host of summer migrants and resident birds. There is a lot of variation in avifauna data generated on Kashmir wetlands. Pandit (1982) studies feeding ecology of breeding birds in some selected wetlands of Kashmir. In another contribution in wildlife resources in Kashmir wetlands, Pandit (1991) documented 13 species of common waterfowl. More updated version by the World Wide Fund for Nature India (*see* Ahmedullah 1997) includes species whose distribution bias is aquatic (Table 7.8). However avifaunal studies covering 2-years period (1982-83) on Hokersar wetland (see Shah *et al.* 1999) yields a total of 90 birds species. Qadri (1989) observed only 13 species in Hokersar during 1985-87. Furthermore during other months of overwintering season, the populations of Red-head Pochard (*Aythaya ferina*) and Common Merganser (*Mergus merganser*) are reported to have been observed in small numbers. Similarly during 1986-88 in Haigam wetland, only 6 species have been recorded (cf. Qadri 1989).

Table 7.8 : Birds linked with water habitats in Kashmir

Common Name (Local Name)	*Scientific Name*	*Some Important Remarks*
1	2	3
Great Created Grebe	*Podiceps cristatus cristatus*	Jheels and marshes of Kashmir.
Indian Little Grebe (Pindh or Pinditch)	*Podiceps ruficollis capensis*	Not uncommon around lakes and marshes.
Indian large Cormorant (Moong)	*Phalacracorax carbo sinensis*	Not uncommon around
Pond Heron or Paddybird	*Ardelola grayii*	L. Wular and Dal.
Large Egret	*Egretta alba*	

1	2	3
Night Heron	*Nycticorax nycticorax nycticorax*	
Little Bittem	*Ixobrychus minutus minutus*	
Bittem	*Botaurus stellaris stellaris*	
Greyleg Goose	*A. anser rubrirostris*	
Whooper Swan (Safed Anz)	*Cygnus eygnus*	Recorded before 1951. No recent report.
Ruddy Shelduck (Czakuq or Tsakao)	*Tadorna ferrugina*	
Pintail	*Anas acuta*	Common passage migrant.
Common Teal	*Anas crecca crecca*	Recently reported from Kashmir
Mallard (Male = Neluj) (Female = Thuj)	*Anas platyrhynchos platyrhynchos*	
Gadwall (Buden)	*Anas stepera*	
Wigeon (Buden)	*A. penelope*	
Gargaey (Narru	*A. querquedula*	
Shoveller (Honke)	*A. clypeata*	
Red-crested Pochard (Toor)	*Neeta rufina*	
Common Pochard (Khroh)	*Aythya ferina*	
Ferruginous Duck	*A. nyroca*	Rare and no longer found breeding Formerly bred in Hokersar, Anchar Haigaun.
Pallas's Fishing Eagle	*Haliaeetus leucoryphus*	Scar near L. Wular and uncommon.
Turkestan Water Rail (Tichi or Gill)	*Rallus aquaticus korejewi*	Rare.
Eastern Baillon's Crake	*Porzana pusilla pusilla*	
Indian Moorhen (Tich)	*Gallinulla chloropus indica*	
Indian Purple Coot (Watich)	*Porphyrio porphyrio poliocephalus*	
Coot (Kayput or Kolar)	*Fulica atra atra*	
Pheasant-tailed Jacana	*Hydrophasianus chirurgus*	
Painted Snipe (Painter chah or Rahchah)	*Rostratula benghalenss*	

1	2	3
Black-winged Stilt	*Himantopus Himantopus himantopus*	Rare and no recent report.
Lapwing or Peewit (Greese Mohraz)	*Vanellus vanellus*	Rarely seen
Redwattled Lapwig	*V. indicus*	
Eurupean Little ringed Plover	*Charadrius dubius curonicus*	
Common Sandpiper	*Tringa hypoleucos*	
Fantail Snipe	*Gallingo gallingo gallingo*	
Indian Whiskered Tern (Kreshi)	*Chlidonias hybrida indica*	
Indian Lesser Pied Kingfisher	*Ceryle rudis leucomelanura*	No recent report.
Central Asian Kingfisher	*Alcedo atthiis pallasii*	
Indian Great Reed Warbler (Kurkech)	*Acrocephalus stentoreus brunescens*	
Witherby's Paddy-field Warbler	*A. concinens horingtoni*	
Himalayan Goldcrest	*Regulus regulas himalayensis*	

Source: Adopted from Ahmedullah, (1997)

Studies conducted during February-March, 1993 by Zargar and Naqash (1993) under the auspices of the department of Wildlife Protection J&K Govt. concerning nine wetlands show that the total number of birds belonging to 11 species peaked (78790) on 18 February, with successive observations registering steady decline in populations, reaching a minimum of 38374 individuals on 9 March. Computations indicate that Common Teal and Mallard comprise >24% each of the total waterfowl population whilst Pintail represents 16.7%. This trend implies that the birds migrant northward to their breeding sites. Earlier records given in Pandit (1991) document high estimates of waterfowl, 139000-158000 in mid-November in Hokersar and 30000 - 33000 in mid-December 1976 in Haigam wetland. According to Bacha (1996), in Kashmir aquatic habits, Greylag goose, Shovller, Red Crested Pochard are now rare whilst Ruddy Shelduck, White-eyed Pochard and Tufted Duck are vulnerable. Greylag goose frequented Hokersar in large numbers but now a significant decline in its population has been recorded. Bacha (1996), in his wetland census (January - April, 1994) in Kashmir wetlands reports that a small flock of 7 Carmorants (*Phalacrocorax carbo*) splendoured at Bathi (Wular) on 20 February while 2 - 5 vagrants were also recorded at Hokersar wetland on specific dates during March - April, which according to the author showed a seasonal influx of extralimital migration in spring to the Valley. Moorhens (*Gallinula chloropus*) where observed to overwinter at some local wetlands of Kashmir Valley. Three Pheasant-tailed Jacanas (*Hydrophasianus chirugus*) also landed at Hokersar on 18 April, reaching a peak of 28 within a week (Bacha 1996). The extralimital spring seasonal movement of varying population of Black Winged Still. (*Himantopus himantopus*) has been observed at Kranchu wetland during April, 1994. A small population of Black headed Gulls (*Larus *ridibundus*) sighted, among few others, in Dal/Hokarsar and as reported in Bacha (1996), the species behaviour indicates a short transitory stopover in Valley at the fag end of their wintering stay in the sub-continent.

ECONOMIC ASPECTS OF WETLANDS

Wetlands, in developed countries, have evoked surging interest because of their economic and aesthetic values. However, in developing countries including India, the importance of these highly productive ecosystems has not been much appreciated, though realization about their utility has started emerging. Kashmir, since long, has had intimate socio-economic links with wetlands. Famine that struck Kashmir Valley during the end of 19th century due to floods (which destroyed the main crop) was averted by use of the fluor of 'Singhara' (Trapa natans), a common wetland plant (cf. Lawrence (1895).

Wetlands are source of several economically important edible plants (*e.g.* Nelumbo, Trapa). *Euryale ferox*, one thriving well has sharply declined. In a report (UEED, 1997) produced under the aegis of the Urban Environmental Engineering Department, Kundangar and Sarwar erroneously reported that *E. ferox* is extinct and exists only on herbarium sheets. Very recently Khan (1999) dealt with the occurrence and disturbution of *E. ferox* in Kashmir waters which shows that the plant is not extinct. Kashmir has peculiar tradition of extensive culture of rice, an important life - sustaining crop for the human populations. A major utility lies by way of harvesting practices in numerous temporary and permanent wetlands. The plants usually harvested are *Phragmites australia, Typha angustat, nymphoides peltata, Scirpus lacustris, Cyperus sp. Carex serotinus, Hydrocharis dubia, nymphae alba, Trapa natans, Sparganium ereciurn*. Several wetlands are a source of plant materials of high nutritive value. Pandit (1980) has shown that aquatic weeds are a potential source of food in the food chain of fishes (27 - 36%) of their total volume of food), water fowl (73.4 - 98.2%) of the food volume of wild ducks and 100% of that of wild geese. The diverse rich flora and fauna of wetlands constitute a biowealth of immense economic importance contributing to high productivity of these ecosystems. Thus, they provide a suitable ecological niche for fish spawning and habitat for waterfowl and ducks. Hokersar is a famous bird-shooting wetland in Kashmir. The assessment–value of wetlands varies widely with respect to the type of wetland and the available resources (both living and non-living) in different localities including sociological, ecological and economic segments. The broader implications of economic aspects, however, merit adequate treatment. Wetland plants have ability to accumate nutrients in considerable amounts through what, is know as 'luxury consumption'. Efforts have been made to use constructed wetlands for treatment of wastes (Billore *et. al.* 1997). Use of constructed wetland (CW) is now recognized as an accepted low-cost technology. The waste water pollutants in such systems are recovered through a combination of physical, chemical and biological processes including sedimentation, precipitation, absorption to soil particles, assimilation by the plant tissue and microbial transformations. The key role of vegetation in these systems is said to be as a substrate microbial growth and transmission of oxygen to the root systems in the grave bed. The ability of emerge plants [*e.g. Phragmites Karka* (reed grass)] according to Billore *et. el* (1997) to transport oxygen from their leaves to their root systems support a population of aerobic micro-organisms in the rhizosphere is one of the key machanisms for efficient BOD, COD and Nitrogen removal. According to Denny (1994), wetlands may be evaluated under four main headings viz. Global, Functional, Habitat and Anthropogenic values.

Global values are best reflected in large continents such as Africa, where lakes and water ways provide refugia for organisms and because of geographical barriers and climate zones, become isolated in to centres of endemism. The fish population of L. Malawi and distribution of Podostemacea (water-plants whose specie are often restricted to particular waterfalls) are good examples (cf. Denny 1994). The functional values are mainfold (see Table 7.1) and the utility of wetlands (natural and constructed) for wastewater treatment is one of the most-promising clean up-bio-technologies, specially for less developed countries. Wetlands provide habitats for some of the rarest flora and fauna besides attracting fascinating wildlife diversity. Anthropogenic values, accordingly to Denny (1994) fall in two groups extrisic and itrisic values, the former refers to governments and private organizations who exploit the wetlands for major commercial purposes, the latter concerns those which are of direct value to the people who live around and in the wetlands.

WETLAND CONSERVATION STRATEGIES

Despite India being signatory to the Ramsar convention on wetlands of International Importance, these ecosystems are under increasingly greater threat due to poor and unscientific management Kashmir has great potential to conserve and develop its wetland resources in a sustainable manner. To achieve it, detailed account of sensitive, fragile and critical species and habitats needs to be developed and integrated with development of protected areas and their wise-use conceptual framework. Besides, it, a comprehensive work on Wetlands is needed in view of realization of importance of wetlands in supplementing human dietary requirements and their ecological significance in terms of flood control, water purification, aquatic productivity and microclimate regulation etc. However, the deterioration of Kashmir aquatic habitats has, of late, increased at an alarming rate causing their shrinkage (even threatened extinction) due to environmental stresses imposed by anthropogenic perturbations. The major threats of Kashmir wetlands include the menace of increased siltation, entrophication due to run-off from catchment (including fertilizers), encroachment from agriculture land and willow plantations, receding open water areas as a result of expanding reed belts. Such dangers calls for integrated management practices for wetland conservation and their contained biological diversity. The emphasis on various conservation aspects are outlined below:

1. Accumulation of silt-load along the shoreline areas has caused 'terrestrialization' in several wetlands. The construction of settling basins near inflow channel(s) could arrest the siltation menace.

2. Checking flood-water entry by its diversion in to large reservoirs and river basin (s). For example, there is an urgent need to divert the Dudhganga flood-channel at Hokrsar to check siltation problem.

3. Large-scale drive for aforestation of catchment areas to hold and conserve the soil which would mitigate the siltation menace.

4. Moratorium on indiscriminate exploitation of wetlands/lakes by way of drainage, filling, land cleaning, cattle-grazing, dumping of garbage, refuses, hazardous pollutants from agricultural fields.

5. Careful regulation of outflow channels by erection of bunds and sluice-gates to prevent floods and maintain proper and desired water-level in different wetlands to suit the specific requirements of different waterfowls. Dabbling ducks (*e.g.* Mallard, Common Teal, Pintail) usually frequent wetlands flooded to the depth of 0.2 - 0.5 m whilst diving ducks (*e.g.* Pochard) and Railes are attracted by those with a depth range of 0.5 - 1.0 m. This is related to the food resources exploitation (seeds and rhizomes of macrophytes) in sediments.

6. Construction of bunds (embankments) along wetland periphery besides regulating water-level can additionally help in checking the agricultural runoff and abate the eutrophication.

7. Effective fencing would check encroachment and reclamation of wetland areas for expanding farm-activities especially in rural areas.

8. Land-use and management practices viz-a-viz rural development (esp. Wular) deserve scientific-based restructuring in the light of 'ecodevelopment' programme.

9. Restriction on spread of human settlements in the vicinity of the waterbodies.

10. Ban on wholesale harvesting of useful food plants and destruction of nests and egg-laying sits of aquatic avifauna.

11. Allowing the growth and availability of suitable food plants (e.g. *Cyperus serotinus, Polygonm sp, Carex sp. Sagittari trifolia, Nymphoides peltata, Potamogeton crispus,* and *Trapa natans*) for wild ducks, and other aquatic plants (e.g. *Phragmites australis, Oryza sativa and Trapa natans*) for wild geese (Pandit 1980).

12. Need for institutional measures to meet dietary requirements of birds during severe winter causing occasional freezing of water-bodies (e.g. during December, 1985 - February 1986). The Department of Wildlife Protection supplied birds artificial food in the form of paddy (*Oryza sativa*) spread over large areas of selected wetlands.

13. Inventorizing the authentic available fauna especially wildlife including the threatened ones and serious scientific plans to breed the endangered life. Thus during 1986 attempt was made to rear in Hokersar, a small population (25) of endangered species of Bar-headed Goose (*Anser indicus*) raised from a consignment of 30 eggs from IWBB England (cf. Qadri 1987). Artificial feed, comprised of paddy (*Oryz sativa*), *Phragmites australis* and *Trapa natans* was provided to birds. Such a step was deemed necessary because the change in vegetation (disappearance of *Phragmites australis* accompanies with the fast spread of *Sparganium erectum*) was related (Pandit 1988) to the signs of decline in members of Greylag Geese visiting Kashmir wetlands.

14. Need to arrest and combat the spread of recently emerging monospecific red-bloom menace in wetlands/lakes of Kashmir (Khan 1993 a, b; 1994 a, b) having potential for adversely affecting the aesthetic value and biodiversity of water resources.

15. Restriction on bird-hunting (especially during their initial stages) would go a long way as an important conservation steps.

16. Looking afresh on the existing laws and regulations regarding bird-shooting. Poaching and over-hunting have harassed the bird population. The recent incident (6-7 December, 1997) of shooting of migratory water fowl in Hokersar is a glaring example of the sorry-state of affairs. The shooting was made possible by invoking the provisions of the J&K Wild Protection Act, 1987. There was an unspoken convention not to grant permission of to the hunters despite existence of the law on the statute book. The law allows hunting of birds. J&K is the only State which has such a law on its statute book. It has not been able to amend that law despite directives from the Union Environment Ministry.

17. Drive for public awareness to educate masses of the socio-economic and environmental role of wetlands and need to preserve these for posterity.

18. Promote an interdisciplinary approach to wetland management for they involve multiplicity of interests ranging from fisherman to farmers, herders to hunters, bird watchers to tourist activities. As such, wetlands cannot be managed in isolation from other ecosystem.

Concluding the overall objective is to pursue the sustainable development of Kashmir wetalands by actively seeking ways to apply the wise-use concept towards the conservation and exploitation of wetlands whilst maintaining biological diversity and safeguarding the interests of the human population.

LITERATURE CITED

Ahmedullah, M. (Editor) (1997). Biodiversity of Jammu and Kashmir–A profile. World Wide Fund for Nature–India, New Delhi, 195 pages.

Ali, Salim (1979). The Book of Indian Birds, 11th edition, The Bombay Natural History Society, India xxxxiv - 142 pages.

Anonymous (1992). Conservation of Wetlands in India. Ministry of Ennvironment and Forest, New Delhi, 48 pages.

Anonymous (1993). Ecology and Conservation of Wetland of Wular Lake (Kashmir). Final Report Project No. J2203/38/88W. Hydrobiology Laboratory S.P. College, Srinagar, 89 pages.

Bacha, M.S. (1996). Waterfowl Census in Kashmir Wetlands. Department of Wild Life Protection, Jammu and Kashmir Govt., Srinagar, 21 pages.

Billore, SK; Sharma J. K. and Thakhr, N.S. (1997). Wastewar treatment through constructed wetlannd system: a promising sustainable ecotechnology, Environment Management 1: 27 (Abstract)

Denny, (1994). Biodiversity and wetlands. Wetlands Ecology and Mannagement (The Hague). 3: 55-61.

Das, S.M. and Subla, B. A. (1963). The ichthyology of Kashmir Part I. History, topography, origin, ecology and general distribution. Ichthyologica 2: 87-106.

Gopal, B. (1994). The role of ecotones (transition zones) in the conservation and management of tropical inland waters. Mittelungen international a Vereingung Limnologie 24: 17-25.

Handoo, J. K. and Koul, V. (1982). Phytosociological and standing crop studies in wetlands of Kashmir pages 187-195. In: Gopal, B, Turner R. E, Wetzeld, R.G. and Whigham, D. F. (Editors) Wetlands Ecology and Management, International Scientific Publications, Jaipur.

Hora, S.L. Malik, G. M. and Khajuria, H (1995). Some interesting features of the aquatic fauna of Kashmir Valley. Journal Bombay National History Society 53: 140-143.

Kaul, V. and Zutshi D. P. (1967): A study of aquatic and marshland vegetation in Srinagar. Proceeding of the Nationnal Institute of Science, India. 33: 111-127.

Kaul, V. and Pandit AK (1982).: Biotic factors and food chain structure in some typical wetlands of Kashmir. Pollution Research 1: 49-54.

Kaul, V. Pandit, AK and Fotedar, D.N. (1980). Ecology of freshwater snails (Gastropod Molluscs) in Haigam–a typical wetland of Kashmir. Tropical Ecology 21: 32-46.

Khan, M. A. (1984). Studies on primary productivity and some hydrochemical aspects of a polluted waterbody of the Himalayan Jhelum Valley. Acta hydrochimica et hydrobiologica (Berlin). 12: 539-548.

Khan M.A. (1985). Spatio-temporal phytoplankton variability in two Himalayan lacustrine ecosystems. Acta hydrochimica et hydrobiologica (Berlin), 13: 249-257.

Khan, M.A. (1986). Hydrobiology and organic production in a marl lake of Kashmir Himalayan Valley. Hydrobiologia. (The Netherlands) 135: 233-242.

Khan, M.A. (1987). Observations on zooplankton composition, abundace and productivity on the flood-plain lakes of Kashmir Himalayan Valley. Acta hydrochimica et hydrobiologica (Berlin) 15: 167-174.

Khan, M. A. (1993 a). Occurrence of a rare euglenoid causing red-bloom in Dal lake waters of the Kashmir Himalaya. Archiv furhyrobiologic (Stuttgart). 127 : 101-103.

Khan, M. A. (1993 b). Euglenoid red-bloom contributing to the environmental pollution of Dal Lake, Kashmir Himalaya Environmental Conservation (Switzerland) 20: 352-356.

Khan, M.A. (1994 a). Red tide flagellate rediscovered in the Himalayan Valley. Inter-governmenntal Oceanographic Commission (HAN) UNESCO-Paris 9 : 5.

Khan, M. A. (1994 b). Red tide menace spreads to Indian sub-continent's largest freshwater lake. Ennvironmental Conservation (Switzerland) 21: 78.

Khan, M.A. (1996). The phytoplankton periodicities of two warm-climate lakes subject to marked seasonal variability. Journal of Tropical Ecology (Cambridge, UK). 12: 461-474.

Khan, M.A. (1997). Endangered ecosystem: eutrophication and conservation status of Dal Lake, Kashmir Himalaya. 7th International Conference on the Conservation and Management of lakes (Buenos Aires).

Khan, M.A. (1999). *Euryale ferox* Salisb–a threatened aquatic plant species in the Kashmir Himalaya. Indian Journal of Environmental Sciences. (in press).

Lawrence, R.W. (1895). The Valley of Kashmir. Oxford University Press, Oxford, 478 pages.

Nath, S. (1986). A check-list of Jammu and Kashmir State (India) with remarks on the ichthyogeography of the State. Journal of the Zoological Society of India 38: 83-98.

Pandit, A.K. (1980). Biotic factor and food chain structure in some typical wetlands of Kashmir, Ph. D. Thesis, University of Kashmir, Srinagar, India.

Pandit, A. K. (1988). Threats to Kashmir wetlands and their wild life resources, Environmental Conservation 15: 266-268.

Pandit, A.K. (1989). Feeding ecology of wild ducks and geese overwitering in wetlands of Kashmir. Indian Journal of Ecology 16: 199-202.

Pandit, A.K. (1991), Conservation of wild life resources is wetland ecosystem of Kashmir, India. Journal of Environmental Management 33: 143-154.

Pandit, A.K. Kaul V. (1982). Trophic structure of some typical wetlands, pages 55-83 In : Gopal, ; Turner RE, Wetzel RG, Whigham, DF (Editors), Wetlands, Ecology and Management. International Scientific Publication, New Delhi.

Qadri, S. E. (1987). Attempted introduction of bar-headed goose through a new habitat. Environmental Conservation 14: 264-256

Qadri, S. (1989). Ecological factors afecting waterfowl in the wetlands of Kashmir. Ph. D. Thesis, University of Kashmir, Srinagar, India.

Scoot, D.A. and Poole, C.M. (1989). A status overview of Asian wetlands No. 53. Asian Wetland Bureau, Kualalumpur, Malaysia, 140 pages.

Shah, Mustafa, G; Qadri, M. Y. and Jan, U. (2000). Species composition and population dynamics of birds of Hokersar wetlands, Kashmir. pages. In : Khan, M. A. (Editor). Environment, Biodiversity and Conservation.

Tiner, R. W. Jr. (1991). The concept of a hydrophyte for wetland identification. Bioscience 41: 236-247.

UEED. (1997). Dal Lake, Kashmir. Urban Environmental Engineering Department, Jammu & Kashmir Government. 37 pages.

Windevoxhel, N. (1995). In search of a definition for "Wetlands". IUCN wetlands Programme Newsletter, 11: 22-23.

Yousuf, A.R. (1996). Fishery resources of Kashmir. Page 87-105 In : Khan, A. H. and Pandit, A. K. (Editors). Ecology, Evironmennt and Energy, Department of Distance Education, University of Kashmir, Srinagar, India.

Zargar, A.R. and Naqash, R. Y. (1993). Waterfowl Census Report, March and Feb. 1993. Department of Wild Life Protection, Government of Jammu and Kashmir.

Zutshi, D. P. (1975). Association of Macrophytic vegetation in Kashmir Lakes. Vegetation 30: 61-65.

Zutshi, D. P. and Khan, M. A. (1978). On lake typology of Kashmir. Environmental Physiology and Ecology of Plants. pages 465-472.

Zutshi, D. P. (1996). Biological population of three Himalayan lakes. pages 11-24 In : Kohli, R.K. Jerath, N and Batish. D. L. (Editors) Some facets of Biodiversity, M/S Research Periodicals and Book Services, Inc, of India, New Delhi.

Zutshi D. P. and Gopal, B. (2000). State of biodiversity in lakes and wetlannds of Kashmir Valley. pages in: Khan M. A. (Editor) Ennvironment, Biodiversity and Conservation.

Chapter - 8

Interrelationship and Impact of Lake Sediment Texture on the Productivity and Diversity in Genus *Potamogeton*

*V.K. Anand and Suman Sharma**

ABSTRACT

Biomass production, occurrence and distribution of various species of *Potamogeton* have been studied in two Shivalik lakes of Jammu. It has been observed that sediment texture greatly influenced the productivity and diversity of the Genus *Potamogeton*. Various other factors that alter the sediment texture which inturn affected the growth of different species of *Potamogeton* have been discussed in order to conserve these limnologically important plants and lake ecosystem as a whole.

INTRODUCTION

Potamogeton, the most diversified genus among the aquatic and marshy plants grows in different water bodies of Jammu (Sharma, 1990; Anand and Sharma, 1991; 1993 Anand *et. al.*, 1993) especially in the lacustrine habitats i.e. Lakes Mansar and Surinsar which inhabit five species of *Potamgeton* (*P. natans, P. nodosus, p. crispus, P. lucens* and *P. salicifolius*). *P. natans* is purely a floating leaved species; *P. nodosus* being heterophyllous bears both floating and submerged leaves while rest of the three species are purely submerged. Of these five species, *P. crispus* and *P. nodosus* are common to both the lakes. *P. nodosus* exhibits very little growth in lake Surinsar. The diversification of this genus in these lakes of Jammu with respect to species composition, habit form, distribution, occurrence, biomass production is tremendous and is in turn being influenced greatly by their sediment texture.

Physical characteristics (texture, density) and nutrient content of bottom sediments have been linked to the distribution, density and viability of submerged macrophytes (Barko et al., 1982, 1986; Barko & Smart, 1979, 1981, 1983, 1986; Anderson & Kalff, 1986; Chambers & Kalff, 1985; Aiken & Pikard, 1980; Lillie & Barko, 1990). In India, though the bottom

*Department of Biosciences, University of Jammu - 180004, India

sediments have been analysed for physico-chemical characteristics (Ramakrishnan 1964; Gopal 1969 Kulshreshtha & Gopal, 1982; Trisal & Kaul, 1983; Handoo, 1986; Gopal, 1990), yet very little information is available regarding the impact of bottom sediment texture on the productivity and diversity of submerged macrophytes (Mitra 1955 ab; 1960; 1966). In the paucity of the information available, the objective of the present studies is to evaluate the impact of the bottom sediment texture on the overall diversity and productivity of the genus *Potamogeton* occurring in two lakes of Shivalik Himalayas.

STUDY AREA AND CLIMATE

Lakes Mansar and Surinsar of Jammu (J&K) are situated in the North-East of Jammu city (Fig. 8.1) with subtropical type of climate (Fig. 8.2). For other details, regarding the morphometry and physical features of these lakes refers to Table 8.1. For performing regular studies, four research stations have been fixed at lake Mansar and three stations at Lake Surinsar.

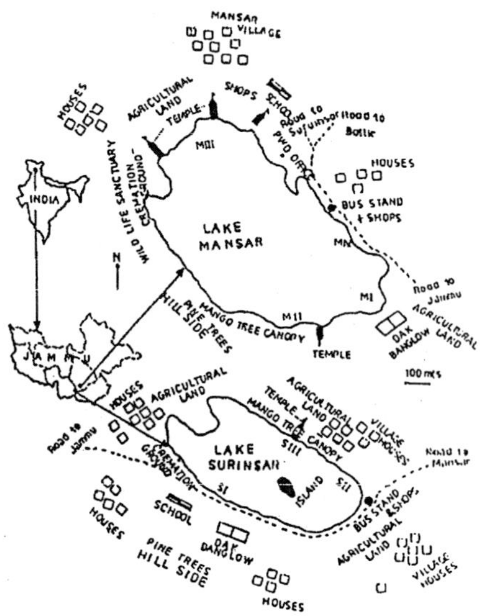

Fig. 8.1 : Location & topographical details of Lake Mansar & Lake Surinsar.

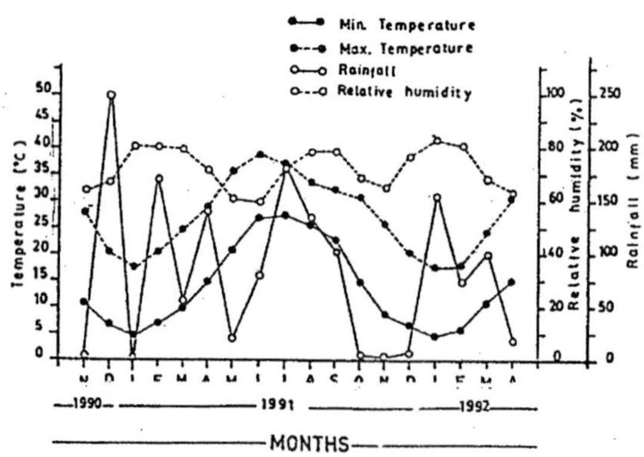

Fig. 8.2 : Meterological data in the study area.

Table 8.1 : Morphometric and physical features of the lake Mansar and lake Surinsar

Lake parameters	*Lake Mansar*	*Lake Surinsar*
Altitude (mts)	664.0	604.0
Area (hect.)	45.0	29.0
Maximum depth (mts.)	37.0	22.0
Transparency (mts.)	2-7	1.7 - 3.0
Maximum surface temperature (^{0}C)	31.0	33.0
Stratification type	Warm monomictic	Warm monomictic
Thermocline depth (mts)	3-12	3-9
Bottom type	Silty & sandy	Silty & sandy
Lake type	Non-drainage	Non-drainage

MATERIAL AND METHODS

From each fixed Station, 5 samples of the bottom sediments were collected, mixed thoroughly to form a composite sample which has been analysed mechanically for percentage of particle size distribution following U.S. Department of Agricultural Handbook No. 18 (1969). Distribution, occurrence and biomass production of *Potamogeton* has been studied regularly every month by quadrate method (Ambasht, 1971). Identifications have been made using standard literature (Hagstrom, 1916; Biswas & Calder, 1954; Cook, 1974; Subramanyum 1974; Sainity & Jacobs, 1981; Clampham et al., 1987; Burshe, 1991). Herbarium specimens have been deposited in herbarium section of Department of Biosciences, University of Jammu (J&K) India. For Statistical analysis two way annova variance analysis for each fraction of soil, coefficient correlation between biomass production of each species and each fraction of sediment texture (sand, silt, clay and loam) has also been derived using computer (HI-TECH SYSTEMS).

Similarity and dissimilarity indices between the two lakes with respect to the Potamogeton species and among different Stations of each lake have also been calculated (Odum, 1971).

RESULTS AND DISCUSSION

Figures 8.3 (a-d) & 8.4 (a-c) indicate seasonal fluctuations in various fractions of soil texture (sand, silt, clay & loam) at different Stations of the lake Mansar (MI-MIV) and lake Surinsar (SI-SIII) respectively. Percentage particle size distribution revealed that total sand content is the most dominating fraction of the soil texture at Stations MII, MIII and SI to SIII. At Station MI silt fraction dominated only for some part of the year whereas at Station MIV, it (silt) dominated for most part of the year.

Coefficient of correlation between various fractions of soil texture (Tables 8.2 & 8.3), reveals that sand is positively correlated with loam (R=0.306 & 0.171) but negatively with ilt (r=-0.927 & 0.900) and clay (r=-0.840 & -0.884). Similarly, silt is positively correlated with clay (r=0.816 & 0.696) and negatively with loam (r=-0.426 & -0.396) at both the lakes– Mansar and Surinsar respectively. Thus, in general, sand and loam fractions are negatively correlated with silt and clay fractions of sediment texture. A similar relationship of different fractions of the sediment texture has also been reported by Lillie & Barko (1990) in Devil's lake, Wisconsin.

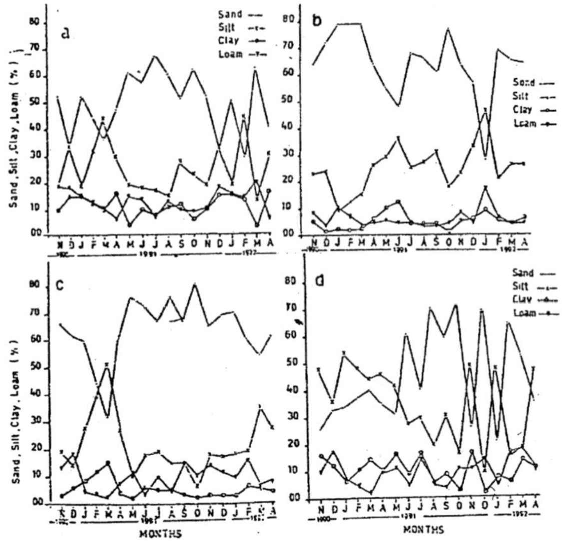

Fig. 8.3 : Seasonal fluctuations in various fractions of sediment texture at Lake Mansar.

(a) at Station MI; (b) at Station MII; (c) at Station MIII; (d) at Station MIV.

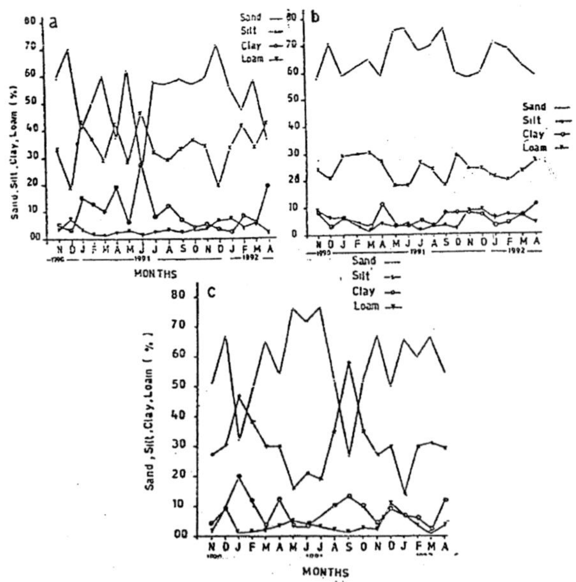

Fig. 8.4 : Seasonal fluctuations in various fractions of sediment texture at Lake Surinsar. (a) at Station SI; (b) at Station SII; (c) at Station SIII

Table 8.2 : Annova variance analysis (F) of various fractions of sediment texture of Station M I to M IV of Lake Mansar and S I to SIII of Lake Surinsar during November 1990 to April 1992.

Sediment texture (%)	*Lake*			
	(F) Mannsar		*(F) Surinsar*	
	Bet. Station	*With in Station*	*Bet. Station*	*With in Station*
Sand	11.3*	1.35***	5.6*	1.15***
Silt	6.15*	1.38***	29.36*	1.89***
Clay	10.86*	0.652***	3.03*	1.62***
Loam	11.15*	2.03**	4.04*	4.07**

* Significant at 1% level.
** Significant at 5% level.
*** Insignificant.

Table 8.3 : Correlation matrix indicating coefficient of correlation between various fractions of Soil texture at Lake Mansar

Parameter	*Depth*	*Sand*	*Silt*	*Clay*	*Loam*
Depth	-	0.028	-0.269	-0.291	0.672
Sand	-	-	-0.927	--0.840	0.306
Silt	-	-	-	0.816	-0.426
Clay	-	-	-	-	-0318
Loam	-	-	-	-	-

Table 8.4 : Correlation matrix indicating coefficient of correlation between various fractions of Soil texture at Lake Surinsar

Parameter	*Depth*	*Sand*	*Silt*	*Clay*	*Loam*
Depth	-	0.033	0.029	0.017	0.027
Sand	-	-	-0.900	0.884	0.171
Silt	-	-	-	0.696	-0.396
Clay	-	-	-	-	-0.146
Loam	-	-	-	-	-

Two way annova variance analysis reveals that the fluctuations in various fractions of soil texture (sand, silt, clay & loam) among different Stations during same month are more significant at lake Mansar than at lake Surinsar (Annova Table 8.4). So, at lake Mansar, marked zonation (erosion zone, neutral zone and sedimentation zone) of the bottom sediment with respect to its texture have been observed, but such zonation of sediment texture is not distinct at lake Surinsar (Fig. 8.5). This may be attributed to the depth, slope of bottom, wind, wave action and biotic interference (Spence, 1967; Hutchinson, 1975). Lake Mansar is very deep and steeper (0-12 mts.) even near the shores (Littoral region) at certain sites than the lake Surinsar which is shallower (0-3 mts) for a longer distance from the shore. Slope of bottom is the major factor controlling the physical characteristics of sediment by affecting sediment stability and deposits the nutrient rich fine materials (Pearsall, 1917; Spence, 1967; Hutchinson, 1975; Hakansen, 1977). In the present lakes, sediment distribution is extremely heterogenous because of combined effects of morphology, hydrology and imbalance in nutrient load entering into the water body (Kiorboe, 1980; Durate & Kalff, 1986).

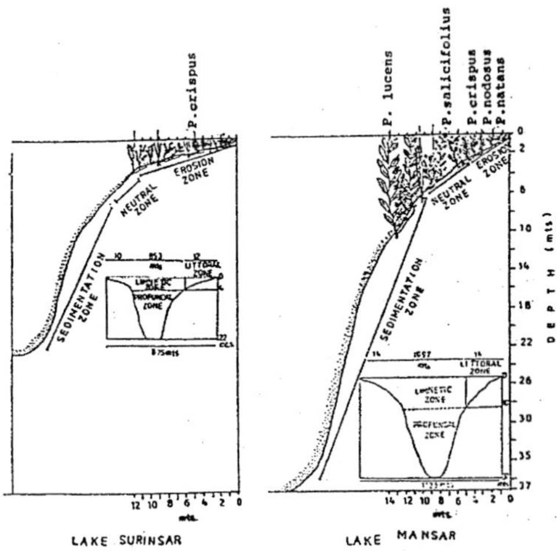

Fig. 8.5 : Cross sections of Lake Mansar & Lake Surinsar exhibiting various zones of the sediment depending upon its texture.

Wind and wave action is much more pronounced at lake Mansar than the lake Surinsar since there is an Island in the centre of lake Surinsar which acts as a potential wind breaker. Moreover, lake Mansar being developed as tourist spot and is under constant biotic interference than the lake Surinsar. All these factors collectively cause periodic agitation and results in suspension of colloidal fraction of the sediment which gradually settles in the deeper layers where the effects of all these factors are negligible. This leads to form distinct zonation of the bottom sediment at lake Mansar than the lake Surinsar where they are comparatively more uniformly distributed.

Table 8.5 reveals the distribution of various species of *Potamogeton* in two lakes as well as at different Stations of the lake Mansar (MI-MIV) and surinsar (SI-SIII). Similarity and dissimilarity indices between the two lakes are 0.57 and 0.43 (Table 8.5) indicating sufficient deal of diversity between the two Lakes so far as species composition of *Potamogeton* is concerned. However, there exists 100% dissimilarity among the various Stations of both the lakes (Mansar & Surinsar) since no one species is common to all the four Stations of the lake Mansar or to all the three Stations of the lake Surinsar (Table 8.6). It is also evident through the Table 8.7 that different species of *Potamogeton* exhibited differential rate of biomass production not only in two different habitats (Lakes) but also at different research Stations of the same lake exhibiting highest biomass production at a particular stations with specific zone of the sediment texture (Fig. 8.5). *P. natans, P. nodsus* and *P. crispus* occur in the erosion zone. *P. salicifolius* occupies the neutral zone while *P. lucens* occurs in the sedimentation zone only. *P. natans* and *P. nodosees* occupy erosion zone (0-3) mts) with higher sand contents. Biomass production in case of *P. nodosus* is positively correlated with sand and loam fractions (r=0.568, 0.584) at lake Mansar, but at lake Surinsar it exhibited positive correlation with the loam fraction only (r=0.638) where it showed poor growth that too in the ditches along the marshy shores of the lake (Station-SI). *P. crispus* occurred at the fully exposed shores (MII, MIII & SI) of both the lakes, where lot of siltation takes place especially during rainy season forming heterogenous mats with species like *C. demersum, H. verticillata, V. spirallis* and species of charophytes. Correlation matrix (Table 8.8) reveals that biomass production of *P. crispus* is negatively correlated with sand, clay & loam (r=0.653, -0.587, -0.429) but positively with silt (r=0.753) at lake Mansar. However, at lake Surinsar it also exhibits negative correlation with sand (r=-0.440) but showed positive correlation with silt, clay and loam content. (r=0.129, 0.393, 0.131) respectively. All these fractions i.e. silt, clay and loam are aluvial in origin at these particular sites (Station MII, MIII & SI) of both the lakes. On the contrary, *P. lucens* occupies the deepest, most sheltered and protected regions of the lake Mansar (Station MIV) the sedimentation zone with higher silt and low sand content. Correlation matrix reveals that biomass production in case of *P.. lucens* is positively correlated with silt (r=0.614) but negatively with sand (r=0.543). At these sites the silt is of fluvial origin. Similarly, biomass production in *P. Salicifolius* is also negatively correlated with sand (r=-0.412) and positively with silt (r=0.219) but these correlations are comparatively less significant, since it occupies the intermediate zone (3-6 mts) i.e. neutral zone, where sand remained higher for some part of the year followed by the silt. Thus, it is not only the quantity but quality of silt, clay and loam is equally important in influencing the distribution, occurrence and biomass production of the species of *Potamogeton*. Despite of the severe interspecific and intraspecific competition among these species, they continue to grow successfully together within the same lake basin since each species occupies its own specialised niche with respect to the sediment texture (sand, silt, clay and loam) in different combinations. Rath (1987) described the development of diverse vegetation corresponding to different successional stages on aluvial deposits. According to him, sporadic occurrence of submerged plants is due to less advanced eutrophication of the water body. This is in consistency with the present findings since the species of *Potamogeton* form pure monospecific stands at different depths exhibiting a peculiar zonation pattern especially at lake Mansar. But such type of zonation pattern of *Potamogeton* species has not been observed in lake Surinsar, thereby indicating that lake Surinsar is at an advanced stage of eutrophication than the lake Mansar. Patnaik (1973), Koul *et al.* (1978) categorically emphasized the importance of soil particle size on the distribution and occurrence of macrophytes in Chilka lake (Orissa) and Kashmir lakes respectively. Despite of the fact that both the habitats exhibited almost same range of physico-chemical characteristics of water yet *P. natans, P. lucens* and *P. salicifolius* do not occur in lake Surinsar probably due to the lack of suitable substrate since coarse substrate with higher sand and loam contents are not only nutrient poor but also provide feable anchoring capacity. Consequently, it supports only the species with creeping habit having few branches and smaller leaves. On the contrary, substrate rich in fine silt (autochthonoous) becomes too flocculent to support rooted submerged species with larger branches and vitiate leaves, thereby, enabling these species (*P. lucens* and *P. salicifolius*) to occupy more steep but deeper regions of the lake Mansar. Thus substrate texture also influences the various habit forms of plants within the lake basins (Mcfarnald and Barko, 1987) besides influencing distribution, occurrence and biomass production of the macrophytes (Kiorboe, 1980; Nichols & Shaw, 1986; Macfarnald and Barko, 1987; Koul et al. 1978).

Table 8.5 : Distribution of various species of *Potamogeton* in Lake Mansar & Surinsar

Name of Species	*Habitat*	
	Lake Mansar	*Lake Surinsar*
P. natans	+	-
P. nodosus	+	+
P. crispus	+	+
P. lucens	+	-
P. salicifolius	+	-

Similarity Index = 0.57 Dissimilaity Index = 0.43

Table 8.6 : Distribution of various species of *Potamogeton* at different stations of Lake Mansar and Surinsar

Name of species	*Lake Mansar*				*Lake Surinsar*		
	MI	*MII*	*MIII*	*MIV*	*SI*	*SII*	*SIII*
P. natans	-	-	-	-	-	-	-
P. nodosus	+	+	+	-	+	-	-
P. crispus	-	+	+	-	+	-	-
P. lucens	+	-	-	+	-	-	-
P. salicifolius	+	-	+	+	-	-	-
Total	3	2	4	2	2	-	-

Similarity Index = 0Dissimilarity Inndex = 1.0

Table 8.7 : Average productivity values (g dry wt/m²/day) of various species of *Potamogeton* at different stations of lake Mansar and Surinsar.

Name of species	*Lake Mansar*				*Lake Surinsar*		
	MI	*MII*	*MIII*	*MIV*	*SI*	*SII*	*SIII*
P. natans	-	-	-	-	-	-	-
P. nodosus	-	0.479	2.790	-	0.061	-	-
P. Crispus	-	0.250	0.217	-	0.471	-	-
P. lucens	0.649	-	-	4.55	-	-	-
P. salicifolius	1.122	-	0.245	0.383	-	-	-

Table 8.8 : Coefficient of Correlation between biomass production of various species of *Potamogeton* with various fractions of soil texture at lake Mansar and Surinsar

Name of species	*Lake Mansar*				*Lake Surinnsar*			
	Sand	*Silt*	*Clay*	*Loam*	*Sand*	*Silt*	*Clay*	*Loam*
P. natans	-	-	-	-	-	-	-	-
P. nodosus	0.568	-0.472	-0.381	0.584	-0.432	-0.083	-0.083	0.638
P. crispus	-0.653	0.753	0.587	-0.429	-0.44	0.129	0.393	0.131
P. lucens	-0.546	0.614	0.154	0.045	-	-	-	-
P. Salicifolius	-0.412	0.291	-0.005	0.363	-	-	-	-

It is interesting to point out that though these lakes of Jammu (Mansar and Surinsar) are situated at the same altitude, under uniform environmental conditions and with similar geological substratum yet exhibit remarkable diversity with respect to the occurrence, distribution, biomass production and habit forms of the species of *Potamogeton*. Sediment texture to a greater extent have been observed to be the consecutive factor in maintaining the diversity within the lake basin particularly at lake Mansar.

During the recent years, a gradual decline in the distribution, occurrence and biomass production of submerged macrophytes have been observed (Sharma, 1994) which is due to various developmental activities all along the shores of the lake Mansar. Some of the important factors causing such a decline are higher rate of erosion, siltation, wind and wave action, unscientific and unplanned deweeding within the lakes, biotic interference by local inhabitants, tourists and animals, boating, inpour of nutrient rich water and eroded silt from the surrounding agricultural lands. Inpour of soil and ashes from the cremation ground, pumping out of water for supply to the surrounding villages for drinking purposes and to the agricultural lands for irrigation. Construction of new shopping complexes, roads, side tracks, bathing sites, new pumping stations along the shores. All these factors collectively effect the bottom sediment texture making it more uniform which inturn lead to the decline not only in the diversity of limnologically important species of *Potamogeton* but also in other macrophytes as well. It is, therefore, suggested that some remedial policies should be formulated well in time to conserve and protect these species of *Potamogeton* and the lake ecosystems as a whole.

ACKNOWLEDGEMENT

Authors are thankful to the Head, Department of Biosciences for providing the facilities and to the UGC and CSIR, New Delhi for financial assistance.

LITERATURE CITED

Aiken, S.G. and Pikard, R.R. (1980). The influence of substrate on the rowth of morphology of *Myriophyllum exalbescens* and *Mycriophyllum spicatum*. Canadian Journal of Botany 58: 1111-1118.

Ambasht, R.S., 1971. A Text Book of Plant ecology. Students Friends & Co. Varanasi.

Anand, V. K. and Sharma, S. (1991) Studies on the macrophytic vegetation of lake Mansar (Jammu). Journal of Phytol Research 4 : 67-72.

Anand, V. K. and Sharma S. (1993a). Occurrence, distribution and composition of aquatic and marshy plants of Jammu province. Environment and Ecology, 1 : 109-114.

Anannd, V.K., Sharma S. and Anand, K. (1993). Aquatic and marshy plants of Jammu Province; II. Systematic enumerations. Advances in Plant Science 6 : 10-19.

Anderson, M.R. & Kalff, J. (1986). Regulationn of submerged aquatic plant distribution in a uniform area of a weed bed. Journal of Ecology., 74 : 953-961.

Barko, J.W. and Smart, R.M. (1979). The nutritional ecology of *Cyperus esculentus*, an emergent aquatic plant grown on different sediments. Aquatic Botany 6 : 13-28.

Barko, J. W. and Smart R.M. (1981). Sediment based nutrition of submerged macrophytes. Aquatic Botany 10 : 339-352.

Barko, J.W. and Smart, R.M. (1983). Effects of organic matter additions to sediment on the growth of aquatic plants. Journal of Ecology., 71 : 161-175.

Barko, J.W. and Smart, R.M. (1986). Sediment related mechanism of growth limitation in submerged macrophytes. Ecology, 67 : 1328-1340.

Barko, J.W., Harden D.G. and Mathews, M.S. (1982). Growth and morphology of submerged plants in relation to light and temperature. Canadian Journal of Botany., 60 : 877-887.

Barko, J.W., Adams M.S. and Clesceri, N.L. (1986). Environmental factors and their consideration in the management of submerged aquatic vegetation. A review, Journal of Aquatic Plant, Management 24: 1-10.

Biswas, K. and Calder, C.C. (1936). Handbook of common water and marsh plants of India and Burma. Health Bull No. 24, Govt. Press of India, Calcutta (1954 Ed. turn)

Burshe, E.M. 1991. A handbook of water plants. Oriental Enterprises 25-A, Kalidas Road, Dehradun.

Chambers, P.A. and Kalff, J. (1985). The influence of sediment compositions and irradiance on the growth and morphology of *Myriophyllum spicatum* L. Aquatic Botany 22 : 253-263.

Clampham, A.R., Tutin, T.G. Moore D.M. (1987). Flora of the British Isles Cambridge University Press, Cambridge.

Cook, C.D.K. (1974). Water plants of the world. Dr. W. Junk. D.V. Publishers. The Hague, IBH Press Roaman, Awn England.

Durate, C.M. and Kalff, J. (1986). Littoral slope as a predictor of the maximum biomass of submerged macrophytic communities. Limnology and Oceanography., 31 : 1072-1080.

Gopal, B., 1969. Ecological studies on the genus *Marsilea* II. Edaphic factors. Tropical Ecology 10 : 278-291.

Gopal, B., 1990. Ecology and Management of aquatic vegetation in the Indian sub-continent. Kluwer Academic Publishers, The Netherland.

Hagstrom, J.D. (1916). Critical researchers on the *Potamogeton* GI. Svenska Ventenskapsaka demiens handlinger. Band 55, No. 5, Stockholm.

Hankanson, L. (1977). The influence of wind, fetch and water depth on the distribution of sediments in lake Vanern, Sweden. Canadian Journal of Earth Sciences. 14 : 397-412.

Handoo, J.K., 1986. The nature of sediments in some wetlands. Acta Hydrochimica. Hydrobiologica 14 : 485-493.

Hutchinson, G.E., (1975). A Treatise in Limnology. Vol. III. John Willey & Sons, Inc. New York.

Kaul, V., Trisal C.L. and Handoo, J.K. (1978). Distribution and production of macrophytes in some water bodies of Kashmir, pages. 313-334. In' J.S. Singh & B. Gopal (Editors), Glimpses of Ecology. Internationnal Scientific Publications, Jaipur.

Kiorboe, T., (1980). Distribution and productionn of submerged macrophytes in Tipper Ground (Ringdobing Fgord, Den mark) and the impact of waterfowl grazing. Journal of Applied Ecology 17 : 675-687.

Kulshreshtha and Gopal, B. (1982). Observations on nutrient removal by freshwater macrophytes under different habitat conditions in Jaipur, India. Internationale Revue der gesamten Hydrobiologie., 67 : 543-553.

Lillie, R.A. and Barko, J.W. (1990). Influence of sediment and ground water on the distribution of biomass of *Myriophyllum spicatum* L. in Devil's lake, Wisconsin. Journal of Freshwater Biology 5 : 417-426.

Mcfarnald, D.G. and Barko, J.W. (1987). Effects of temperature and sediment type on the growth and morphology of monoecious and dioecious *Hydrilla*. Journal of Freshwater Ecology 4 : 245-252.

Mitra, E. (1955a). Cotribution to our knowledge of Indian freshwater plants. 1. On some aspects of the structure and life history of *Hydrilla verticillata* Presl with notes on its autecology. Journal of Asiatic Society of Science. 21 : 1-17.

Mitra E., (1955b). Contribution to our knowledge of Indian freshwater plants. 2. On some aspects of habit, structure, life history annd autecology of *Limnanthemum cristatum* Griseb & *L. indicum* thw. Proceedings of Natural Institute of Sciences. India, 212 : 170-187.

Mitra, E., (1960). Contribution to our knowledge of Indian freshwater plannts. Behaviour of *Hydrila verticillata* presl in nature and under experimental conditions. Bulletin Botanical Society of Bengal 14 : 73-75.

Mitra, E., (1966). Contribution to our knowledge of Indian freshwater plants 5. On the morphology, reproduction and autecology of *Pistia stratoites* Linn. Journal of Asiatic Society of Science, 8 : 115-135.

Nichols, S.A. and Shaw, B.H. 1986. Ecological life histories of the three aquatic nuisance plants *Myriphyllum spicatum, Potamogeton crispus* and *Elodea canadensis*. Hydrobiológia, 131 : 3-21.

Odum, E.P., (1971). Fundamentals of Ecology. W.B. Saunders Company Philadelphia London. Toronto.

Patnaik, S., (1973). A study of the aquatic plants of Chilka lake. Proceedings of National Academy of Sciences, India. 43 (b), I & II.

Pearsall. W.H. (1917). The aquatic and marsh vegetation of Esthwaite water. Journal of Ecology., 5 : 180-202.

Pearsall, W.H., (1920). The aquatic vegetation of English lakes. Journal of Ecology 8 : 163-201.

Ramakrishnan, P.S. (1964). Soil complex in relation to aquatic plant communities. Procedings of National Innstitute of Science, India, 30B : 348-353.

Rath, B., (1987). The macrophyte vegetation of a small branch system of the Danube at Dunaremate (Szigetkoz, River Km 1826). Acta Botanica Hungarica 33(3-4) : 187-197.

Sainty, G.R. and Jacobs, S.W.L. (1981). Water plants of New South Wales, Published by Water Resources Commission N.S.W. pages. 550.

Sharma, S. (1990). Ecomorphological studies on some aquatic and marshy plants of Jammu (J&K) M. Phil dissertation, University of Jammu, Jammu (J&K), India.

Sharma, S. (1994). Productivity studies on submerged macrophytes in relation to physico-chemical characteristics of Lakes-Mansar & Surinsar, Jammu. Ph.D. Thesis, University of Jammu, Jammu.

Sinha, A.K., Singh, D.K. Baruah A. and Sharma, U.P. (1992). Seasonal variations of physico-chemical properties of bottom sediments of Kawar Lake Begusarai (Bihar). Journal of Freshwater Biol., 4 : 249-254.

Spence, D.H.N. (1967). Factors controlling the distribution of freshwater macrophytes with special reference to the loches of scotlan. Journal of Ecology., 55 : 147-169.

Subramanyum, K. (1962). Aquatic Angiosperms. Botanical monograph No. 3, CSIR New Delhi.

Trisal, C.L. and Kaul S. (1983). Sediment composite mud-water inter changes and the role of macrophytes in Dal Lake, Kashmir. Int. Rev. Ges. Hydrobiol., 68 : 671-682.

Chapter - 9

Changes in Vegetation Pattern of Hokarsar (Wetland Reserve), Kashmir

S.A. Gangoo and A.S. Makaya**

ABSTRACT

Hokarsar wetland, one of the best resorts of migratory waterfowl in Kashmir, is under anthropogenic pressure and siltation due to floods. This has resulted in the loss of habitat for varied aquatic flora and fauna. The number of macrophytic plant species have been decreasing from 90 to 25 and now there are only 24 species. The species of high economic importance which have disappeared from the area during recent years include *Nelumbium nuciferum*, *Euryale ferox* and *Acorus calamus*.

Key Words: *biodiversity, conservation, macrophytes, sanctuary, wetland*

INTRODUCTION

Wetlands are among the most important ecosystems on earth. In the great scheme of nature, the swampy environment of the carboniferous period preserved many of the fossil fuels on which we now depend. In more recent biological and human time periods, wetlands are valuable as sources, sinks and transformers of a multitude of chemicals, biological and genetic materials. Wetlands have been called "biological super markets" for the extensive food chain and rich biodiversity they support. They play major role in the landscape by providing unique habitats for a wide variety of flora and fauna. Now that we have become concerned about the health of our entire planet, wetlands are being described by some as carbondioxide sinks and climate stablizers on a global scale.

In the western Himalaya, a large number of diversified types of freshwater wetland form an important component of the beautiful ecological spectrum of the high altitude valley of Kashmir. The series of the wetlands, spread all along the flood plain of the River Jhelum and Sind at the base of Kashmir Himalaya (34°5′ - 34°10′ N, 74°6′ - 74°9′ E and 1580 - 1600 m Altitude) in the northern fringe of the Indian Sub-continent, are of great aesthetic,

*Division of Forestry, S.K. University of Agricultural Sciences and Technology, Shalimar Campus Srinagar 191 121

cultural, socio-economic and ecological value, besides playing, an important role in the conservation of genetic resources of both plants and animals.

Hokarsar is one of the most important wetlands of Kashmir valley. It is a permanent eutrophic lake, surrounded by fresh water marshes on the flood plain of the Jhelum River. The lake is drained by a channel to the Jhelum River at Sozeth Narbal village. The lake reaches a maximum depth of 2.5 m in Spring with snow melt water, and a minimum of 0.7 m in Autumn (Pandit and Qadri, 1990). This lake is protected as a game reserve by the Jammu and Kashmir Department of Wildlife Protection. Waterfowl hunting is allowed on a controlled basis in winter and the harvesting of reeds is permitted in Summer. Ninety species of resident and non-resident birds have been recorded in Hokarsar (Shah and Qadri, 1990). These comprise of 25 species of resident birds, 11 species of local migrants, 33 species of summer migrants and 21 species of Winter visitors. Out of the winter visitors Mallard (*Anas platyrhyuchos*) and Greylag goose (*Anser anser*) are most important visitors and their one time population in the wetland was recorded as 4000 and 10000 respectively. In the past, the waterfowl population was still larger and is on a continuous decline because of change of habitat conditions of the wetland. Bates and Lowther, (1952) have reported that thick stands of emergents, namely *Phragmites Australias* and *Typha angustata*, at this site formed the most suitable breeding sites for mallard, white-eyed pochard and coots until 1924. To assess the habitat condition of the wetland, the change in the vegetation structure of the lake has been studied, and compared with earlier research works.

METHODS

For the purpose of the study of the present status of macrophytes of the wetland the lake was divided into three zones viz; Open water, floating garden, and marsh edges. Seven quadrats of 100 x 50 cm were laid randomly as suggested by Haynes (1982), in each of the three zones of the lake and macrophytes were counted species wise to determine their abundance during May 1997.

RESULTS

Table 9.1 presents the data on the dominant macrophytic species. 24 species of dominant macrophytes are present in the wetland at present. Not a single colony of *Phragmites australis* and *Typha augustata* could be found in the wetland. *Sparganium ramosum* is spreading fast in the wetland, particularly in the shallow waters and floating gardens. Out of 21 quadrats laid, *Sparganium ramosum* was dominant in 7 quadrats. Thus one can conclude that this species occupies 33% of the wetland at present.

Table 9.1 : Enumeration of dominant macrophtic plant species

Quadrat No.		*Name of species* *OPEN WATER*	*No. of Plants Recorded* *in 5000 Sq. cm.*
1		*2*	*3*
1	(i)	*Trapa natans*	75
	(ii)	*Ceratophyllum demersum*	8
2.	(i)	*Nymphoides peltatum*	16
	(ii)	*Nymphaea tetragona*	14

1		2	3
	(iii)	*Trapa natans*	3
	(iv)	*Ceratophyllum demersum*	4
3.	(i)	*Nymphoides peltatum*	58
	(ii)	*Ceratophyllum demersum*	10
4.	(i)	*Nymphoides peltatum*	58
	(ii)	*Batrachium trichophyllum*	2
5.	(i)	*Nymphaea tetragona*	23
	(ii)	*Potamageton crispus*	3
	(iii)	*P. lucens*	4
6.	(i)	*Trapa natans*	2
	(ii)	*Lemna spp.*	300 dots
	(iii)	*Nymphoides peltatum*	8
	(iv)	*Potamageton crispus*	2
7.	(i)	*Trapa natans*	80
		Floating Gardens	
1.	(i)	*Phragmites australis*	13
	(ii)	*Sparganium ramosum*	17
	(iii)	*Scirpus maritimus*	18
	(iv)	*Potamageton lucens*	6
	(v)	*Rumex patientia*	3
2.	(i)	*Sparganium ramosum*	19
	(ii)	*Potamogetonn lucens*	7
	(iii)	*Carex spp*	5
3.	(i)	*Phragmites australis*	2
	(ii)	*Sparganium ramosum*	18
	(iii)	*Sagittaria trifolia*	4
	(iv)	*Alisma plantagoaquatica*	4
4.	(i)	*Sparganium ramosum*	10
	(ii)	*Rumex aquaticus*	8
	(iii)	*Scirpus spp*	5
5.	(i)	*Typha angustata*	17
	(ii)	*Sagittaria trifolia*	8

1		2	3
	(iii)	*Scirpus maritimus*	10
6.	(i)	*Sparganium ramosum*	10
	(ii)	*Carex spp*	6
	(iii)	*Scirpus spp*	7
7.	(i)	*Sparganium ramosum*	35
		Marsh Edges	
1.	(i)	*Phragmits australis*	3
	(ii)	*Typha elephantina*	8
	(iii)	*Sparganium ramosum*	10
2.	(i)	*Scirpus triqueter*	102
	(ii)	*Lemna spp*	80
3.	(i)	*Sagittaria trifolia*	15
	(ii)	*Cyperus rotundus*	35
	(iii)	*Scirpus spp*	10
4.	(i)	*Scirpus spp*	35
	(ii)	*Spargannium ramosum*	25
	(iii)	*Phragmites australis*	7
	(iv)	*Ceratophyllum demersum*	6
5.	(i)	*Nymphoides peltatum*	3
	(ii)	*Sparganium ramosum*	21
	(iii)	*Scripus spp*	8
6.	(i)	*Epilobium hirsutum*	20
	(ii)	*Batrachium spp*	10
	(iii)	*Trapa natans*	4
7.	(i)	*Sparganium ramosum*	42
	(ii)	*Scirpus spp*	3

DISCUSSION

Accelerating anthropogenic pressures, coupled with recurring floods brought a series of changes in the biotic composition of Hokarsar. Hokarsar wetland has been a rich source of economically important plant species like *Trapa natans* (Gaur), *Nelumbium nuciferum* (Nadur), *Nymphoides peltatum* (Khur), *Euryale ferox* (Jewer) and *Acorus calamus* (Vai Gunder) (Lawrence 1967, Kak 1990, and Pandit 1989). Out of the above species *Nelumbium nuciferum, Euryale ferox* and *Acorus calamus* have completely disappeared from this wetland. Not a single specimen of these plants was observed while travelling through nook and corner of this wetland. According to Kaul and Zutshi (1967), and Pandit (1980), the number of macrophytic species has considerably decreased from 85 in 1963 to 24 in 1980. However, Kak (1990) has reported 58 species of macrophytes in this wetland out of which 44 are rare.

Macrophytes have an important role in aquatic ecosystem. The most important role of macrophytes is in maintaining an ecological balance, for example, the rooted aquatic often recover nutrients from the benthic zone in anerobic conditions and thus provide a useful "nutrient pump" for the ecosystem organisms. Aquatic plants have an intimate and important relationship in one way or other with almost all the animals in a aquatic ecosystem. In addition to their use as food, macrophytes also provide suitable breeding and sheltering places for varied aquatic fauna. Macrophytic plants provide basic source of food (energy) in the food web. The macrophytes, constitute the principal source of food of fish (27-367%) of the total volume, waterfowl (73.4-98.2%) of the food volume of wild ducks and 100% for wild geese (Pandit, 1980), and other animals, which in turn from the food of human beings. Besides, being the base of trophic chains, macrophytes like *Nymphoides peltatum, nympheae sp., Phragmites australis, Carex spp.* and *Scirpus* sp. are directly used as cattle feed, and duck weeds (*Lemna* spp.) are the most favourable food of domestic ducks. Man also consumes large quantities of some macrophytic plants (seed) of *Trapa natans*, rhizomes and seeds of *Nelumbium nuciferum*, petioles and fruits of *Nymphaea sp.* and stem tubers of *Sagittaria trifolia* as food.

To restore the habitat conditions of the wetland the plant species which have disappeared from the area should be reintroduced and species like *Phragmites australis* and *Typha angustata* which are fast declining should be planted. To protect the wetland from poachers and to ensure nesting success of birds, this wetland should be declared as a bird sanctuary. The sanctuary may have a core and a buffer zone. The core area should be left undisturbed, while as the rights of the people for reed cutting and hunting may be allowed in a regulated way in the buffer zone.

To preserve the biodiversity of this wetland the management programme should be based on following principles.

(i) Long-term productivity under most natural conditions possible, by maintaining natural values through the use of natural process.

One of the areas in which wildlife management emphasizes natural diversity as an objective is in maintenance of wilderness areas. Wilderness areas are protected from nearly all forms of human destruction, particularly in terms of habitat disturbance. As such they serve as biological reserves for all sorts of organisms, including those commonly called "wildlife". This objective will be fulfilled in the proposed core area of the reserve.

(ii) Long-term benefits should be superimposed over short-term benefits.

One thorny question is whether or not sport hunting should be permitted in keeping with the pristine and natural conditions of wilderness. Considering hunting and other potentially controversial issues, sport hunting with conventional weapons may be continued to be allowed in the buffer zone, with the core zone declared as the restricted area. Long term benefits include protection against; species extinction, species association or the loss of genetic variability in terms of species and population.

(iii) Conservation programme that serve the greatest range of public interests.

People living around the reserve, are greatly depend on the lake for fodder, food plants and fish. People from cities visit the wetland for sport hunting, recreation etc. Conservation programmes should not restrict these uses. Ecodevelopment of village around the lake and environmental awareness of masses in general should be the thrust of the conservation programmes.

ACKNOWLEDGEMENTS

This study is a part of the M.phil dissertation of Ist author under the title "Ecological Studies of Hokarsar (Wetland Reserve) Jammu and Kashmir" submitted to Indian Institute of Forest Management, Bhopal and completed under the guidance of Dr, P.C. Kotwal.

LITERATURE CITED

Bates, R.S.P. and Lowther, E.H.N. (1952). Breeding birds of Kashmir, London: Oxford University Presss.

Raynes, Robin (1982). Environmental Science Methods, Cambridge: University Printing House.

Kak, A. Majeed (1990). Aquatic and wetland vegetation of the Kashmir Himalaya. Journal of Economic. Taxonomic Botany, 4: 1-14

Kaul, V. and Zutshi, D.P., (1967). A study of aquatic and marshland vegetation of Srinagar. Proceedings of the National Institute of Science, India 33b: 111-127

Lawrennce, R. Walter (1967). The valley of Kashmir 2nd edition. Kesar Publishers, Sherwani Road, Srinagar (Kashmir).

Pandit, A.K. and Qadri, S.S., (1990). Floods threatening Kashmir wetlands. Journal of Environmental Management 31: 299-311.

Pandit, A.K., (1980). Biotic factors and food chain structure in some typical wetlands of Kashmir. Ph.D. thesis, University of Kashmir, Srinagar, India.

Shah, G.M., and Qadri, M.Y., (1990). Avifauna of Kashmir Himalayas. Species composition and population dynamics of birds of Hokarsar wetland. Pages 229-241. In : Himalaya: Environment Resources and Development. Shree Almora Book Depot, Almora.

Chapter - 10

Species Composition and Population Dynamics of Birds of Hokarsar Wetland Kashmir

G. Mustafa Shah, M.Y. Qadri** and Ulfat Jan**

ABSTRACT

The paper documents the birds of Hokarsar- a typical wetland of Kashmir. A total of 90 species of birds were recorded during the course of two years study programme of 1982 and 1983. These comprised of 25 species of resident birds, 11 species of local migrants, 33 species of summer migrants and 21 species of winter migrants. The data are presented on species composition, periodicity, relative abundance and habitat use of each species.

Key Words: birds, habitat use, periodicity, species composition, relative abundance, wetland

INTRODUCTION

Wetlands are by nature dynamic ecosystems and are often subject to drastic changes resulting from biotic or abiotic factors. Their avifauna population in turn fluctuate with the habitat conditions. The degradation of Hokarsar wetland by way of illegal encroachment and heavy siltation has influenced the abundance and distribution of bird species. Because of these changes several species of birds which according to Bates and Lowther (1952) were abundantly seen in the wetland premises have drastically declined in number and have been listed in various categories based on their current status. Five species are very rare, 6 species rare and 11 species uncommon.

To understand man's impact on the wetland habitat, there is a need for the accurate measurement of its bird population. Through use of these data, it will be possible to detect the average response of these populations to generalized environmental changes caused by diverse and widespread land use practices. As this kind of information becomes available,

*Post Graduate Department of Zoology, University of Kashmir, Srinagar - 190 006 (India)

**Vice Chancellor, University of Kashmir.

it will serve as the basis for developing appropriate management programmes for the purpose of maintaining and increasing overcall bird population. There do not appear to be any through systematic study on the avifauna of Kashmir Himalaya and it was with this aim that the present study on the population dynamics, periodicity and relative abundance was undertaken in one of the important wetlands of Kashmir- Hokarsar.

Hokarsar (34°06' N lat., 74°05' E long., 1584 m.a.m.s.l.) is an important wetland situated in the centre of the valley, about 10 km. to the west of Srinagar and is managed by the Department of Wildlife Protection, J&K Government chiefly for the purpose of winter visiting ducks and geese. The wetland is more or less semicircular in outline, extending in the east-west direction with a mere leftover area of 4.5 sq.km. It is fed by perennial Doodganga watershed in the Pir Panjal range of the Himalaya.

Athick willow belt (*Salix* sp.) with scattered stands of Populus trees extends along the sides of the wetlands. The dominant macrophytes of the wetland include *Saccharum spontaneum, Typha angustata, Butomus umbellatus, Sparganium ramosum, Trapa natans, Phragmites australis, Eleocharis palustris, Potamogeton* sp. and *Polygonum* sp. which provide cover, food and nesting sites for birds.

METHODS

For purpose of present investigations the population of birds was estimated once every two weeks by visual census and transect methods (Watson, 1965, Gaston, 1975).

From November to March, when the wetland lacks almost all emergent vegetation that otherwise conceal the birds, visual counting was very conveniently employed. An estimate was obtained without disturbing the birds, by scanning each part of the study unit from better view points, with the help of 20 × 50 X field binoculars.

From April to October, the transect method was used for all counts. The method entailed moving along a series of transacts to and fro to cover the units, attempting to identify the species and count the number of birds of each species (cf Jenkins and Watson, 1962).

Terms used in defining the relative abundance of each species are those used by Bull (1964) and McCaskie (1970).

Very abundant	Over 1 000 individuals per day.
Abundant	201 - I 000 individuals per day.
Very common	51 - 200 individuals per day.
Common	21 - 50 individuals per day.
Fairly common	7 - 20 individuals per day.
Uncommon	1 - 6 individuals per day.
Rare	1 - 6 individuals per season.
Very rare	Infrequent occurrence.

The birds were identified by studying their characteristic features in accordance with the identification keys evolved by Bates and Lowther (1952), Whistler (1963) and Ali (1979).

OBSERVATIONS

SPCIES ACCOUNT

Family : Podicepidae

1. Dabchick, Podiceps ruficollis capensis. Local Name: "Pind"

A common resident bird found all over the wetland. A total of 612 birds constituting 0.40% (of the total bird population) were recorded during the period of study. Although the birds were recorded throughout the year, however, the greater population density was observed during July-September. The maximum density of 100 was recorded in August.

Family : Phalacocoracidae

2. Indian Large Cormorant, Phalacocorax carbo sinensis. Local Name: "Mong"

Rare to fairly common and was found only from November through February. During these months the number of birds observed was 2, 3, 8 and 15 respectively. The bird is migratory and only a winter visitor to the valley. No nest or nest formation was recorded during the period of their stay in the wetland.

Family : Ardeidae

3. Eastern Grey Heron, Ardea cinerea. Local Name: "Breg"

The most conspicuous bird among the resident population. An almost stable population of the heron was always noticed, sitting on the floating islands, fields and edges of the water solitarily. The largest number of herons recorded at a time was 27 in March and 30 in April. At other periods of the year, their number was mainly between 15-1 8.

4. Little Egret, Egretta garzetta

Little Egret constituted an important and common group of resident birds. Their number varied between 8 in July to 93 in November, with an average of 41 for the whole year.

5. Indian Cattle Egret, Bubuicus ibis coromandus

Together with grey heron and little egret, the Indian cattle egret is a common resident bird. Although they were always fairly in large numbers but in the month of August, an abrupt spurt was noticed when they formed one of the dominant bird groups with the highest population of over 500.

6. Indian Pond Heron, Ardea grayii. Local Name:"Broku"

Resident, fairly common to very common bird that stays in the wetland all through the year. In March and April, they were very common when upto 180 individuals were found in the wet fields of the wetland. During the mid day, the birds were found in less numbers but congregated in large flocks at sun set.

7. Night Heroti, Nycticorax nycticorax nycticorax. Local Name:"Bor"

Resident, though common they occurred in small numbers from 11-19 only. On one occasion in September, 23 birds were counted.

8. Little Biftern, lxobrychus minutus minutus. Local Name:"Goi"

The Little Biftern is a breeding summer resident that arrives in Hokarsar in early May for a brief stay till end of August or early September. During their brief period of stay a total population of 422 only was actually recorded and the bird classified as common.

Family : Anatidae

9. Greylag Goose, Anser anser. Local Name:"Anz"

A typical winter visiting goose of the valley. Characteristically, they prefered mostly Hokarsar, where they formed large flocks. As many as 10, 000 were recorded during one count in November, 15th 1982. It is the only migratory goose that starts reaching Hokarsar in the middle of September. A population of 1500 were noticed in September and over 5,000 in October. They were very abundant throughout the late autumn and early winter.

10. Mallard, Anas platyrhynchos. Local Name: Male-"Niluj", Female-"Thuj".

Winter migrant, very abundant in December and February when upto 4,000 individuals were found in a single day. This gradually decreased to 2,000 and 1,000 in March and April before they left. Prior to this in October and November the ducks were abundant being upto 1,000.

11. Pintail, Anas acuta. Local Name: "Sukh Pachan"

Also a winter visitor, very abundant during October, November and December and again in February with a brief disruption in January. The duck was one of the first migratory visitors that started coming to the wetland as early as August.

12. Common Teal, Anas crecca. Local Name:"Keus"

The smallest of the winter visiting ducks but in respect of its population it was the most common. Together with the Pintail, it arrived the wetland in August. The population was built up to maximum of 6,000 in November. With a sudden decline in population in January, there was a second spurt of population in February and thereafter it gradually decreased with their outward migration which was completed by the end of May.

13. Garganey, Anas querquedula. Local Name:"Nour"

Winter visitor and passage migrant, fairly common to common till December, when upto 50 individuals were found in a day. In January, only 1 or 2 individuals were seen but from February to ending April they were abundant.

14. Gadwall, Anas strepera. Local Name:"Budun"

Winter visitor. The first flock of 1 1 birds arrived in the wetland in September. This continued till a maximum of 973 was recorded in December. Only 43 individuals were noted in January. But in February the population againg increased (575) and a second peak (713) was recorded in March. Thereafter the ducks started leaving the wetland and this outbound migration was completed till the end of May.

15. Shoveller, Anas clypeata. Local Name:"Honk"

Like Gadwal, Shoveller started collecting in September (22) and continued so during the winter months except January. The stable population was found during the spring months of February-April. In May, only small numbers (158) were found and the bird was completely gone by the end of the month.

16. Wigeon, Anas penelope. Local Name:"Shierni Budu"

Wigeon also formed an important contributor to the total population of winter visiting ducks. The large flocks of these ducks were seen to arrive in the wetland, this inward migration continued till December, when 975 ducks were recorded on the day of observation which dropped to 130 in January as was the case with other ducks. In February, the ducks started regrouping and the peak population was recorded in March which then fell to 1100 in April. During April, the ducks took to their wings for the outbound migration which was completed by the end of May.

17. Brahminy Duck, Tadorna ferruginea. Local Name:"Tsakow"

Winter migrant. The first batch of 82 ducks was recorded in October which increased to 212 by December following a general decrease in January (4). They started regrouping in February (453), attaining a peak population of 752 in March and fell to 255 in April.

18. Common Pochard, Aythya ferina. Local Name:"Thauthur"

Winter visitor, fairly common to common during the first four months (October - January). In February and March there was a sudden influx and a population of 161 and 191 was built up respectivelly in the two months. In April only 72 ducks were recorded and in May their number was only 12.

19. Tufted-Duck Aythya fuligula. Local Name:"Thauthur"

Winter migrant, fairly common to very common, arrived in October, when a flock of 8 ducks was recorded that grew to 15 by December. During February-March, the ducks became very common 121 and 151 in these two months respectively. From March onward, the bird started moving back and only 4 in early May.

20. Red-crested Pochard, Aythya rufina. Local Name:"Tsarow"

The bird was recorded for a period of 9 months, from September to May. The duck was fairly common from September through January with a maximum flock of 19 individuals in December. From February to April the duck was very common and the maximum of 92 were seen in a single day count in March.

21. White-eyed Pochard, Aythya nyroca. Local Name:"Harwat"

Winter visitor and passage migrant. Small numbers of the bird arrived in September and it was fairly common till December. During February, March and April these ducks became very common.

22. Common Merganser. Mergus merganser orientalis.

A rare to fairly common late winter visitor and passage migrant. During the entire course of study only 38 birds were recorded from November to April. In November and December the bird was rare and was completely absent in January. However, the birds were fairly common in February to April.

23. Smew, Meegus albellus. Local Name:"Gagur"

A rare winter visitor, only two birds were found each in the month of October and December.

Looking on over all population of waterfowl at Hokarsar, it is seen that in January, maximum number of ducks and geese leave the wetland for a brief period. Intensive competition for food and high expenditure of energy during periods of cold weather and snow cover (January) are probably dominant factors governing the distribution of waterfowl and fidelity to wintering grounds. Cains (1973), Bennet and Bolen (1978), Shah (1984) and Shah and Qadri (1986) also showed that severe weather conditions especially low temperatures of January are known to cause stress and effect on the distribution and movement of waterfowl. However, it is not known which wintering areas become suitable for them. Probably they move further south where they could find climatically suitable places and food. Albright (1981) also showed that during winter, migrating American black duck (*A. rubripes*) continued southward rather than settling in more northerly wintering sites.

Family: Accipitridae

24. Himalayan Griffon Vulture, Gyps himalayansis. Local Name: 'Grad'

Resident, but uncommon, generally not more than two individuals were noticed on any day of observation for most part of the year.

Family: Falconidae

25. Pallas Fishing Eagle, Haliactus leucoryphus. Local Name: "Goda Grad", "Loz"

Largely resident but uncommon, like the vulture not more than two birds were observed at any one time.

26. Common Pariah Kite, Milvus migrans govinda. Local Name: "Gaunt"

Resident but uncommon. A stable population of the kite was always found in the wetland.

Family: Gruidae

27. Common Crane, Grus grus. Local Name: "Hillaw"

A rare or casual visitor, only four birds were seen in January.

Family: Rallidae

28. Coot, Fulica atra. Local Name: "Kollur" "Kavput"

Winter visitor and was seen from September - May. In September and October, the coot was abundant but became very abundant from November to April.

29. Indian Purple Coot, Porphyrio policephalus policephalus. Local Name : "Wan Tech"

A rare winter visitor. It was observed only for the months of September and November to February. On no occasion more than two birds were recorded.

30. Turkestan Water Rail, Rallus aquaticus korejewi

Summer migrant, very rare bird only 2-3 individuals were observed during the summer months of June - August.

31. Eastern Baillon's Crake, Porzana pusilia pusilia. Local Name : "Grettu", "Gralu"

Summer migrant. Like water rail it was very rare, only 1-2 birds were seen during the months of June - August.

32. Indian Moorhen, Gailinula chloropus indicus. Local Name "Tech"

Mostly summer visitor. Very common to abundant in the wetland. Although the young ones were present throughout the year, the adults were found to start coming during April and stayed till the end of August. During this period, birds built their nests, layed eggs and raised the chicks and then undertook their migration to Indian plains before autumn; leaving most of their chicks here to undertake the migration in the next season. However, the chicks that hatched early in the season migrated the same season with their parents.

Family: Jacanidae

33. Pheasant-tailed Jacana, Hydrophasianus chirurgus. Local Name: "Gund kav/Gair kav"

Summer visitor that stayed in the wetland from May to September and were commonly found on reeds and other floating vegetation.

Family: Scolopacidae

34. Common Sandpiper, Actitis hypoleucos. Local Name: "Kola kavin" "Tont kon"

Resident, common during spring and very common during summer.

35. Wood Cock, Scolopax rusticola. Local Name : "Zar Batchi"

Resident, though fairly common but not more than 20 were recorded in any observation.

36. Fantail Snipe, Capella gallinago gallinago

Resident, fairly common in summer and fall and very common in winter and spring.

Family : Charadriidae

37. European Little-ringed Plover, Charadrius dubius curonicus. Local Name : "Kola katij"

Summer migrant that was fairly common from April to September on the floating islands and along the peripheral areas of the wetland.

38. Jack Snipe, Capella minima,

Winter migrant, uncommon to fairly common from December to March.

Family: Rostratulidae

39. Painted Snipe, Rostratula benghalensis. Local Name : "Gug", "Gugu"

Mainly a summer visitor, found from mid March to September, when it was common. It was mainly found on the floating islands and in large grasses adjacent to these.

Family: Recurvirostridae

40. Black-winged Stilt, Himanotopus himanotopus himanotopus. Local Name:"Longzeyet

Although resident but not more than 6 birds were seen on any day of observation.

Family: Laridae

41. Indian Whiskered-Tern, Chlidonias hybrida indica. Local Name:"Kreshi"/Krind/ Krew

Mainly a summer visitor that started arriving towards the middle of April, becoming very common in June. In July, the bird became abundant.

Family: Charadriidae

42. Eurasian Lapwing, Vanellus vanellus. Local Name : "Greese Mahraz"

Very rare, only one bird was found each in the month of May and August.

Family: Columbidae

43. Hume's Blue Rock Pigeon, Columba livia neglecta. Local Name:"Jungli Kotur"

Resident, very common throughout the year but more so from July - October. It mostly preferred the paddy fields on the southeast of the wetland.

44. Indian Ring Dove, Streptopelia decaocta decaocta. Local Name : "Kukil"

Summer migrant that was uncommon to fairly common from April to August around the wetland.

Family: Psittacidae

45. Himalayan Slaty-headed Parakeet, Psittacula himalayana himalayana. Local Name: "Shoga"

Summer visitor, mainly common from May to September on the Populus and Chinar trees on the sides of the wetland. The birds seen in April were amongst the early arrivals.

Family: Cuculidae

46. Asiatic Cuckoo, Cuculus canorus telephonus. Local Name:"Shah Kuk" or "Kuku"

A summer visitor, fairly common in June and July. During April and May, they were rare and only 2-4 birds were seen during this period.

Family: Strigidae

47. Indian Barn Owl, Tyto alba stertens. Local Name: "Rata Mogul"

Although resident but uncommon. The number of birds recorded ranged from 1-5 and in the month of March, July and October none of them was visible on the days of observation.

48. Turkestan Great-horned Owl, Bubo bubo turcomanas. Local Name: "Rata Mogul"

Also a resident and uncommon owl. Only 1-5 of these were seen during a single day of observation. In April and July they were not sighted.

49. Scully's Wood Owl, Strix aluco biddulphi. Local Name: "Rata Mogul"

A local migrant, rare and sighted only on three occasions- February, March and December and their numbers were only 2, 3 and 1 respectively.

Family: Alcedinidae

50. Indian Pied Kingfisher, Ceryle rudis leucomelanura. Local Name: " Hor Kola Tonch", "Sufaid Tont", "Duddru"

Resident, fairly common to common. Slightly less numerous during winter.

51. Central Asian Kingfisher, Alcedo athis pallasi. Local Nami : "Lokut Kola Tonch"

Resident, uncommon to common, slightly more common than the Indian Pied Kingfisher.

52. White-breasted Kingfisher, Halcyon smyrnensis. Local Name: "Bud Kola Tonch"

Resident of the three types of kingfishers, it was least numerous being mostly 5-9 and only once in April, 11 birds were seen in a single day of observation.

Family: Meropidae

53. European Bee- Eater, Merops apiaster. Local Name: "Tuieri Kav", "Rodabubru"

Summer visitor, rare to uncommon, found only from May to August in the vicinity of water.

Family : Upupidae

54. European Hoopoe, Upupa epops epops. Local Name: "Satut", "Hud-hud"

Summer visitor, uncommon in February and April and fairly common in June and July.

Family: Picidae

55. Kashmir Pied Woodpecker, Dryobates himalayansis albescens. Local Name: "Koel Ku-kir", "Hor Koel"

A local altitudinal migrant that descends from higher altitudes and forests to the lower reaches and is found in the wetland in winter and spring. On no occasion more than five woodpeckers were recorded.

56. Brown-fronted Pied Woodpecker, Dryobates brumofrons

Also a local altitudinal migrant, uncommon to fairly common. Like Kashmir Pied Woodpecker it was also found around the wetland during the winter and spring period.

Family: Hirundinidae

57. Common Swallow, Hirundo rustica rustica. Local Name: "Katij"

Spring and summer visitor that was very abundant during spring and early summer and common towards late summer till it was gone towards the end of September.

Family: Dicrijridae

58. Indian Grey Drongo, Dicrurus leucophaeus longicaudatus. Local Name: "Gunkots", "Telakots"

Summer migrant, uncommon seen singly from May to September. Their number seen on any day of observation was 1-3 only.

Family: Oriolidae

59. Indian Oriole, Oriolus oriolus kundoo. Local Name: 'Pshnul"

Also a summer visitor, uncommon to fairly common from April to August.

Family: Corvidae

60. House Crow, Corvus splendens zugmayeri. Local Name: "Kav"

Resident, common, though more common in summer when the egg laying and hatching in breeding birds of the wetland was at its peak.

61. Eastern Jackdaw, Corvus monedula soemeringii. Local Name: "Kavin"

Resident, common to very common. They were more common than the House Crow.

62. Himalayan Jungle Crow, Corvus macrorhynchos intermedius. Local Name: "Diva Kav" or "Pontsol Kav".

Local altitudinal migrant. Due to severe cold and snow, the bird migrates downwards from the higher reaches of the valley and was fairly common in winter.

Family: Paridae

63. Kashmir Grey Tit, Parus major cashmirensis. Local Name: "Ranga Tsar"

A permanent resident, common throughout the year especially in spring.

Family: Certhidae

64. Wall Creeper, Tichodroma muraria. Local Name: "Lamba Dider"

An uncommon to fairly common winter visitor, found mostly in the thick population of Willow trees on the north-east of the wetland.

65. Himalayan Tree Creeper, Certhis himalayana limes. Local Name: "Koel Dider"

Local altitudinal migrant, fairly common in winter during the severe cold.

Family: Timaliidae

66. Simla Streaked Laughing Thrush, Trochalopteron lineatum griseicentior. Local Name: "Sheen-a-pi-pin"

Local altitudinal migrant that descended from higher altitudes. Small numbers of these birds were seen during the winter months of December and January, never exceeding 9.

Family: Pycnonotidae

67. White-cheeked Buibul, Holpastes leucogenys leucogenys. Local Name : "Bulbul", "Bil-bi-chur".

A resident, common bird found throughout the year.

Family: Trogiodytidae

68. Kashmir Wren, Trogiodytes troglodytes neglectus.

A winter visitor otherwise local altitudinal migrant. They were found only during late winter and early spring months- December to March.

Family: Turdidae

69. Tickel's Thrush, Turdus unicolor. Local Name : "Kastoor"

Mainly summer visitor, common to fairly common in May - July when upto 30 were recorded in a day. Smaller numbers were also seen in March and April.

70. Western Spotted-Forktail, Enicurus maculatus maculatus. Local Name: "Shakel Lot"

A winter and spring visitor of wetland. Small numbers of these were noted from December-March.

71. Western Himalayan Rubythroat, Calliope pectoralis pectoralis. Local Name: "Yaquat Hot"

This also was a winter visitor otherwise local altitudinal migrant, 1-3 birds being found only during the months of January-March.

72. Himalayan Whistking Thrush, Myiophoneus caeruiens temminckii Local Name : "Hazar Dastan", "Kav Kunur"

Local altitudinal migrant, 1-2 of the thrushes were seen during January and February only.

73. Northern Indian Pied Bushchat, Saxicola caprata bicolor,

Summer visitor. The bird was uncommon and stayed in the vicinity of the wetland from April to October.

74. Indian Stonechat, Saxicola tarquata indica. Local Name : "Dofa Teriv"

Summer visitor, again uncommon from April to October, excepting the month of May, when no bird was noticed.

Family: Sylvidae

75. Indian Great Reed Warbler, Acrocephalus stentorius brunnescens. Local Name : "Kurkech"

A summer migrant, fairly common to very common from April to September. The six birds found in March were the early arrivals.

76. Witherby's Paddy-field Warbler, Acrocephalus concinens haringtoni,

Summer migrant, fairly common to common from March to September. They prefered willow trees, marsh reeds and softer tall grasses.

77. Pale Bush Warbler, Homochiamys pallidus pallidus. Local Name: "Dofa Pich".

Also a summer migrant but uncommon to fairly common found from April to September. 12-18 birds were noticed in the month of June, July and August. At other periods of their stay in the wetland they were only 2-3.

78. Large-billed Bush Warblev, Tribura major

A winter visitor found from December - March. It was uncommon, being 1-3 in number.

Family: Muscicapidae

79. White-browed Blue Flycatcher, Muscicapula supercilliaris superçilliaris,

A rare summer migrant, only 2-3 birds were seen during the months of May and June on willow trees around the wetland.

80. Himalayan. Paradise Flycatcher, Terosiphone paradisi leucogaster. Local Name: "Phambasir" or "Ranga Buibul" or "Shah Bulbul".

Arather uncommon summer visitor from May to August, singly and in pairs among the thick growth of willows and chinar trees on the south-east of the wetland.

Family: Motacillidae

81. Hodgson's Pied Wagtail, Motacilia alba alboides. Local Name:"Dobbai" or "Pinskanni"

An early summer migrant that stayed in the wetland for a brief period of April - July. In April and May, they were common but became very common in June and July when upto 132 individuals were found in the wetfields.

82. Eastern Grey Wagtail, Motacilia cinerea melanope. Local Name :"Khak Dobbai"

Summer visitor, stayed for a much longer period (March - September) than the Pied Wagtail. In its abundance it was common to very common on the floating vegetation of *Trapa natans, Nymphoides peltatum* usually in association with little bittern and moorhen.

83. Hodgson's Yellow-headed Wagtail, Motacilla citreola calcarata. Local Name: "Lidar Dobbai".

Summer visitor, fairly common to common from April - August and abounded on the floating vegetations.

Family: Laniidae

84. Rufous-backed Shrike, Lanius schach erythronotus. Local Name:"Harwatij"

A summer visitor, fairly common to common from April to September.

Family: Sturnidae

85. Common Myna, Acridotberes tristis tristis. Local Name: "Hor".

A resident bird, common in and around the wetland all through the year, especially on the exposed land masses.

86. Himalayan Starling, Sturnus vulgaris humii. Local Name: "Tsinihangur".

An early summer migrant, common to very abundant from March-September. The first flock arrived in March, this migration continued till May and were dispersed throughout.

Family: Friingillidae

87. Kashmir House Sparrow, Passer domesticus griseigularis. Local Name: Male - "Kantur", Female - "Tsar"

Resident, common in all parts all through the year.

88. Himalayan Goldfinch, Carduelis caniceps caniceps. Local Name:"Sehara, "or "Sera"

A rare to uncommon summer visitor, only 1-4 birds were noticed during May-July.

89. Gold-Fronted Finch, Metoponia pusilia. Local Name: "Tiok" or "Taer".

A summer visitor, rare to uncommon, not exceeding 1-4 in number from May to August.

Family: Camephagidae

90. Indian Short-billed Minivet, Pericrocotus brevirostris brevirostris. Local Name: "Wazul Mini".

Summer visitor, fairly common on the branches of willow trees and never exceeding 12 in number on any day of observation from April to September.

In all 90 species of birds Table 10.1 including residents and non-residents have been recorded. These comprised of 25 species of resident birds that were found in the wetland throughout the year and 11 species of local migrants, 33 species of summer migrants and 21 species of winter visitors. The number of species obtained at any time ranged between 44-67. The maximum number of species (67) were recorded in the month of May, when on one hand a number of winter visiting waterfowl each with a small population were present and on the other a large number of summer migrants also made their appearance in the wetland.

The population density of all the bird species ranged from 2, 887 (representing 58 species) in June to 28, 181 (representing 46 species) in November. The highest population was recorded during the winter months of December-February totalling to 54, 372 followed by autumn, 50,322. The least population (9836) was recorded in summer. During winter the main contributors of the population were the winter visiting ducks and geese.

LITERATURE CITED

Albright, J. J. (1981) Behavioural and physiological responses of coastal wintering Black Ducks (*Anas rubripes*) to changing weather in Maine. M.Sc. Thesis, University of Maine Orons.

Ali, S. (1979) *'The Book of Indian Birds'*. Bombay Natural History Society, Bombay.

Bates, R. S. P. and Lowrher, E. H. N. (1952) *Breeding birds of Kashmir:* Oxford University Press, London.

Bennett, J. W. and Bolen, E. G. (1978) Stress response in wintering Green-winged Teal. *Journal Wildlife Management*, 43, 81-86.

Bull, J. (1964) *Birds of the New York area* . Harper and Row, New York. 540 pp.

Cains, B. W. (1973) Effect of temperature on energy requirement and northward distribution of the Black-bellied Tree Duck. *Willson Bulittin*. 85, 308-317.

Gaston, A. J. (1975) Methods for estimating bird population. *Journal Bombay Natural History Society*, 72(2), 271-283.

Jenkins, D and Watson, A. (1962) Fluctuation in red grouse (Lagopus scoticus) population 1956-1969. *The exploitation of natural animal population*. Ed. Le Cren, E.D. and Hoidgate , M. W. Oxford, 96-117.

McCaskle, G. (1970) Shorebirds and waterbirds use of the Salton Sea. *California Fish and Game*, 66(2), 87-95.

Shah, G. Mustafa (1984) *Birds of Hokarsar: Food, feeding and breeding biology of some resident and non-resident birds*. Ph.D. Thesis, University of Kashmir.

Shah, G. Mustafa and adri, M. Y. (1988) Food of Mallard, *Anas platyrhynchos* at Hokarsar, Kashmir, *Journal Bombay Natural History Society, Bombay*, 85(2): 325-31.

Watson, A. A. (1965) A population study of the Ptarmigan (*Lagopus mutus*) in Scotland. *Journal Animal Ecology*, 34,

Whistler, H. (1962) *Popular Hand book of Indian Birds*. Oliver and Boyle, London.

Table 10.1 : Species composition of Avifauna of Hokarsar Wetland, Kashmir

Common Name		*Scientific Name*	*Local Name*
1	*2*	*3*	
Family : Podicepidae			
1.	Dabchick or Indian Little Grebe	*Podiceps Ruficollis Capensis*	"Pind"
Family : Phalacrocoracidae			
2.	Indian Large Cormorant	*Phalacrocorax Carbo Sinensis*	"Mong"
Family : Ardeidae			
3.	Eastern Grey Heron	*Ardea Cinerea Rectirostris*	"Breg"
4.	Little Egret	*Egretta Garzetta Garzetta*	
5.	Indian Cattle Egret	*Bubulcus Ibis Coromandus*	
6.	Indian Pond Heron	*Ardea Grayii*	"Broku"
7.	Night Heron	*Nycticorax Nycticorax Nycticorax*	"Bor"
8.	Little Bittern	*Ixobrychus Minutus Minutus*	"Goi"
Family : Anatidae			
9.	Greylag Goose	*Anser Anser*	"Anz"
10.	Mallard	*Anas Platyrhynchos*	Male "Niluj", Female "Thuj"
11.	Pintail	*Anas Acuta*	"Sukh Pachan"
12.	Common Teal	*Anas Crecca*	"Keus"
13.	Garganey or Blue-Winged Teal	*Anas Querquedula*	"Nour", "Noer"
14.	Gadwal	*Anas Strepera*	"Budun"
15.	Shoveller	*Anas Clypeata*	"Honk"
16.	Wigeon	*Anas Penelope*	"Shierni Budun"
17.	Brahminy Duck	*Tadorna Ferruginea*	"Tsakow"
18.	Common Pochard	*Aythya Ferina*	"Thauthur"
19.	Tufted-Duck	*Aythya Fuligula*	"Thauthur"
20.	Red-Crested Pochard	*Aythya Rufina*	"Tsarow"
21.	White-Eyed Pochard	*Aythya Nyroca*	"Harwat"
22.	Common Merganser	*Mergus Merganser Orientalis*	—
23.	Smew	*Me Rgus Albellus*	"Gagur"

1		2	3
Family: Falconidae			
24.	Himalayan Griffon Vulture	*Gyps Himalayansis*	"Grad"
25.	Pallas Fishing Eagle	*Haliactus Leucoryphus*	"Goda Grad", "Loz"
26.	Common Pariah Kite	*Milvus Migrans Govinda*	"Gaunt"
Family: Gruidae			
27.	Common Crane	*Grus Grus*	"Hillaw"
Family: Rallidae			
28.	Coot	*Fulica Atra*	"Kollur", "Kavput"
29.	Indian Purple Coot	*Alus*	"Wan Tech"
30.	Turkestan Water Rail	*Rallus Aquaticus Korejewi*	
31.	Eastern Baillons Crake	*Por Ana Pusilla Pusilla*	"Grettu", "Gralu"
32.	Indian Moorhen	*Gallinula Chloropus Indicus*	"Tech
Family: Jacanidae			
33.	Pheasant-Tailed Jacana	*Hydrophal Urgus*	"Gund Kav"
Family: Scolopacidae			
34.	Common Sandpiper	*Actitis Hypoleucos*	"Kola Kavin", "Tont Kon"
35.	Wood Cock	*Scolopax Rusticola*	"Zar Batchi"
36.	Fantail Snipe	*Capella Gallinago Gallinago*	"Chah"
Family: Charadriidae			
37.	European Little Ringed Plover	*Charadrius D Nicus*	"Kola Katij"
38.	Jack Snipe	*Capilla Minima*	
Family: Rostratulidae			
39.	Painted Snipe	*Rostratula Benghalensis*	"Gug", "Gugu"
Family: Recurvirostridae			
40.	Black-Wingewd Stilt	*Himanotopus H Himanotopus*	"Longzeyet"
Family: Laridae			
41.	Indian Whiskered-Tern	*Chlidonias Hybrida Indica*	"Kreshi, Krend, Krew"
Family: Charadriidae			
42.	Eurasian Lapwing	*Vanellus Vanellus*	"Greese Maharaz"

1		2	3
Family: Columbidae			
43.	Hume's Blue Rock Pigeon Slaty-Headed Parakeet	*Columba Livia Neglecta*	*Jungle Kotur*
44.	Indian Ring Dove	*Streptopelia Deca Deta Decaocta*	"Kukil"
Family: Psittacidae			
45.	Himlayan Slaty-Headed Parakeet	*Psittacula Himalayana Himalayana*	"Shoga"
Family: Cuculidae			
46.	Asiatic Cuckoo	*Cuculus Canorus Telephonus*	"Shah Kuk/Kuku"
Family: Strigidae			
47.	Indian Barn Owl	*Tyto Alba Stertena*	"Rata Mogul"
48.	Turkestan Great Horned Owl	*Bubo Ubo Turcomanas*	"Rata Mogul"
49.	Scully's Wood Owl	*Strix Aluco Biddulphi*	"Rata Mogul"
Family: Alcedinidae			
50.	Indian Pied Kingfisher	*Ceryl Rudis Leucomelanura*	"Hor Kola Tonch", "Sufaid Tont", "Duddru"
51.	Central Asian Kingfisher	*Alcetdo Athis Dallasi*	"Kola Tonch"
52.	White-Breasted Kingfisher	*Halcyon Smyrnensis*	"Bud Kola Tonch"
Family: Meropidae			
53.	European Bee Eater	*Merops Apiaster*	"Tuleri Kav", "Rodabubru"
Family: Upupidae			
54.	European Hoopoe	*Upupa Epops Epops*	"Satut", "Hud-Hud"
Family: Picidae			
55.	Kashmir Pied Woodpecker	*Dryobatis Himalayensis Albescens*	"Koel Kir", "Hor Koel"
56.	Brown-Fronted Pied Woodpecker	*Dryobatis Brumifrons*	—
Family: Hirundinidae			
57.	Common Swallow	*Hirundo Rustica Rustica*	"Katij"

1		2	3
Family: Dicruridae			
58.	Indian Grey Drongo	*Dicrurus Leucophaeus Longicaudatus*	"Gunkots", "Tlekots"
Family: Oriolidae			
59.	Indian Oriole	*Oriolus Oriolus Kundoo*	"Poshnul"
Family: Corvidae			
60.	House Crow	*Corvus Splendens Zugmayeri*	"Kav"
61.	Eastern Jackdaw	*Corvus Monedula Soemmeringii*	"Kavin"
62.	Himalayan Jungle Crow	*Corvus Macrorhychos Intermedius*	"Diva Kav", "Pontsol Kav"
Family: Paridae			
63.	Kashmir Grey Tit	*Parus Major Cashmirensis*	"Ranga Tsar"
Family: Certhidae			
64.	Wall Creeper	*Tichodroma Muraria*	"Lamba Dider"
65.	Himalayan Tree Creeper	*Certhis Himalayana Limes,*	"Koel Dider"
Family: Timaliidae			
66.	Simla Streaked Laughing Thrush	*Trochalopyeron Lineatum Griseicentior*	"Sheen-A-Pi-Pin"
Family: Pycnonotidae			
67.	White-Cheeked Bulbul	*Molpastes Leucogenys Leucogenys*	"Bulbul", "Bil-Bi Chur"
Family: Troglodytidae			
68.	Kashmir Wren	*Troglodytes Troglodytes Neglectus*	—
Family: Turdidae			
69.	Tickel's Thrush	*Turdus Unicolor*	"Kastoor"
70.	Western Spotted Forktail	*Enicurus Maculatus Maculatus*	"Shakhel Lot"
71.	Western Himalayan Rubythroat	*Calliope Pectoralis*	"Yaquat Hot"
72.	Himalayan Whistling Thrush	*Myiophoneus Caerulens Temminckii*	"Hazar Dastan/ Kavkunur"
73.	Northern Indian Pied Bushchat	*Saxicola Caprata Bicolor*	—
74.	Indian Stone Chat	*Saxicola Torquata Indica*	"Dofa Teriv"

1		2	3
Family : Sylviidae			
75.	Indian Great Reed Warbler	*Acrocephalus Stentorieus Brunnescens*	"Kurkech"
76.	Witherby's Paddy Field Warbler	*Acrocephalus Concinens Haringtoni*	—
77.	Pale Bush Warbler	*Homochlamys Pallidus Pallidus*	"Dofa Pich"
78.	Large-Billed Bush Warbler	*Tribura Major*	*Haringtoni*
Family: Muscicapidae			
79.	White-Browed Blue Flycatcher	*Muscicapula Supercillaris Supercillaris*	*Haringtoni*
80.	Himalayan Paradise Flycatcher	*Terpsiphone Paradisi Leucogaster*	"Phambasir", "Rangabulbul" or "Shah Bulbul"
Family: Motacillidae			
81.	Hodgson's Pied Wagtail	*Motacilla Alba Alboides*	"Dobbai", "Pinskánni"
82.	Eastern Grey Wagtail	*Motacilla Cinerea Melanope*	"Khak Dobbai"
83.	Hodgson's Yellow Headed Wagtail	*Motacillacitreola-Calcarata*	"Liderbobbai"
Family: Laniidae			
84.	Rufous-Backed Shrike	*Lanius Schach Erythronotus*	"Harwatij"
Family: Sturnidae			
85.	Common Myna	*Acridotheres Tristis Tristis*	"Hor"
86.	Himalayan Starling	*Sturnus Vulgaris Humii*	"Tsinihangur"
Family: Fringilidae			
87.	Kashmir House Sparrow	*Passer Domesticus Griseigularis*	Male- "Kantur", Female- "Tsar"
88.	Himalayan Goldfinch	*Carduelus Caniceps Caniceps*	"Sehara',"Sera"
89.	Gold-Fronted Finch	*Metoponia Pusilla*	"Tiok", "Taer"
Family: Campephagidae			
90.	Indian Short-Billed Minivet	*Pericrocotus Brevirostris Brevirostris*	"Wazul Mini"

SECTION - III

BIODIVERSITY OF SALINE AND MANGROVE HABITATS

Chapter - 11

Biodiversity of Saline Habitats

J.C. Dagar*

ABSTRACT

In India, about 8.6 million ha land is salt-affected. In most of the arid and semi-arid regions, the combination of salinity and waterlogging is a duel problem. The inland saline lands are either devoid of any vegetation or support very meager cover. Survey of highly saline inland habitats in relation to soil salinity indicated that only restricted number of species are found distributed in highly saline regions like Ran of Katchh in Gujarat and Pachpadra in Rajasthan. Contrary to inland saline habitats the coastal tidal areas are rich in biodiversity. In India, 37 species of exclusive mangroves and many species of associate mangroves are found distributed in about 6740 sq. km mangal formations which in turn are inhabited by several interesting animals such as salt water crocodiles, turtles, water monitor lizards, snakes, wild pigs, monkeys, deer, even tigers (in Sunderbans), several indigenous and migratory birds, and skippers, crabs, molluscus, insects and several crustaceans. These are nursery and breeding ground and source of food for variety of animals and micro-organisms. The aerial roots, pneumatophores and branches serve as shelter and food for large number of organisms including many beautiful orchids, ferns lichens, mosses, jungerminiales, algae, fungi, bacteria, ciliates, nematodes and amphipods which colonize and take part in complex food chain of this uniqe ecosystem. But regretfully mentioned that all these habitats are now under tremendous anthropogenic pressure. Many species are at the verge of extinction. As a result of the unsustainable exploitation of the natural saline habitats many mangrove species and animal species which have been largely or exclusively dependent on them, are dwindling at a fast rate. Entire biodiversity of these habitats is at stake.

Key words : biodiversity conservation, life-support system, mangal formation, mangroves, salinity, sustainability, waterlogging, wild life.

INTRODUCTION

Biodiversity is often defined as the variety of all life forms–the different plants, animals and micro-organisms; the genes they contain and the ecosystems of which they

*Central Soil Salinity Research Institute, Karnal - 132001

form a part. In recent past it has been widely discussed. It is not a fixed entity, but constantly changing and increased by genetic changes and by evolutionary processes and reduced by extinction and habitat degradation (Naveh, 1995). The concept emphasises the inter-relatedness of the biological world covering the terrestrial, marine and other aquatic environment. Without going in details of types of biodiversity like species (alpha), habitat (beta), landscape (gamma) and genetic diversity a brief general account of biodiversity of saline habitats of India has been given in this paper which includes both inland and coastal (mangrove) habitats.

Salt-affected soils occur under different environmental conditions and have different morphological, physical, chemical, physico-chemical and biological properties. These are universally low in fertility and difficult for rational utilisation. Though there are many classification systems for salt-affected soils, they are commonly placed under five groups (Table 11.1) based on their different chemicals types of electrolytes (Szabolcs 1989). Saline and alkali lands have existed in India since ancient times. The Indians of Harrapan and Vedic periods were aware of this soil malady and could distinguish such lands (*Usara*) from the productive lands (*Urvara*). In the present times, a perceptible increase in area under salt-affected soils, increasing pressure on agriculture lands, and concerns for ensuring sustained agricultural productivity of soils have necessitated a greater focus on the problem. There is a growing public concern about secondary salinization and waterlogging of soils in major canal commands. The salt-affected soils in India have been grouped in saline or alkaline groups and cover an area of about 8.6 million ha (Singh 1994) existing in different physiographic and climatic situations in the country (Table 11.2). It is seen that major alkali area is situated in the Indo-Gangetic plains. The soil salinity problem is mainly confined to low rainfall regions and the coastal belt of the country with a wide spectrum of rainfall. In nature, only some highly specialized plants which have great tolerance to salts, are found growing on these lands. Those plants which occur naturally on soil or in water too salty (usually more than 0.5% NaC1 content) for the average plants are generally designated as halophytes (Sen et al. 1982). Those species which attain their optimal growth at moderate to high salinity (above 0.5% NaC1) are designed as obligate halophytes where as the plants which achieve their optimal growth on non-saline soils but can also grow on saline soils are called *faculatative halophytes*. Aronson (1989), however, has set the criterion of 7-8 dsm[1] salinity for a species to be designated as salt-tolerant. In present paper plants commonly found both on inland and coastal habitats have been discussed.

Table 11.1. Classification of salt affected soils.

Type of salt affected	*Environment*	*Electrolyte(s) causing salinity or alkalinity*	*Main adverse effect on production*	*Method(s) of reclamation*
1	2	3	4	5
Saline soils	Arid, semi-arid	Sodium chloride and sulphate (in extreme cases nitrate). ECe at 25^{0}C=4.0 or more dS m^{-1}, pH = less than 8.2	High osmotic pressure of soil solution (toxic ions)	Leaching (removal of excess salt)

1	2	3	4	5
Alkali soils	Semi-arid, sub-humid, humid	Sodium salts capable of alkaline hydrolysis, ECe at 25°C generally less than 4 dS m^{-1}, ESP generally exceeds 15.	Adverse soil physical condition, low nutrient availability	Lowering excess exchangeable Na with Ca amendments
Gypsiferous soils	Semi-arid, arid	Calcium salts	Adverse soil physical condition, osmotic effects	Leaching excess salts
Magnesium soils	Semi-arid, sub-humid	Magnesium salts	Osmotic and nutritional effects, adverse soil physical condition	Leaching excess salts
Acid Sulphate soils	Sea shores, lagoons with heavy sulphur containing sediments	Ferric and aluminium salts (mainly sulphates)	Very low pH, toxic-levels of Al and Fe	Liming, judicious water management

Table 11.2 : Extent and distribution of Salt affected soils and waterlogged area in India ('000ha)

State	*Waterlogged area*			*Salt affected area*			
	Canal Command	*Unclassified*	*Total*	*Canal Command*	*Outside Command*	*Coastal*	*Total*
1	2	3	4	5	6	7	8
Andhra Pradesh	266.4	72.0	339.0	139.4	390.6	283.3	813.3
Bihar	362.6	NA	362.6	224.0	176.0	Nil	400.0
Gujarat	172.6	311.4	484.0	540.0	327.1	302.3	1214.4
Haryana	229.8	45.4	275.2	455.0	NA	Nil	455.0
Karnataka	36.0	NA	36.0	51.4	266.6	86.0	404.0
Kerala	11.6	NA	11.6	NA	NA	26.0	26.0

1	2	3	4	5	6	7	8
Madhya Pradesh	57.0	NA	57.0	220.0	22.0	Nil	242.0
Maharashtra & Goa	6.0	105.0	111.0	446.0	NA	88.0	534.0
Orissa	196.3	NA	196.3	NA	NA	400.0	400.0
Punjab	198.6	NA	198.6	392.6	126.9	Nil	519.5
Rajasthan	179.5	168.8	348.3	138.2	983.8	Nil	1122.0
Tamil Nadu	18.0	109.9	127.9	256.5	NA	83.5	340.0
Uttar Pradesh	455.0	1525.6	1980.0	606.0	689.0	Nil	1295.0
West Bengal	NA	NA	NA	Nil	NA	800.0	800.0
Total	**2189.4**	**2338.1**	**4527.5**	**3469.2**	**3027.0**	**2069.1**	**8565.2**

NA refers to data not available. The area from other category may spill to this column.

Source: Singh (1994)

DISTRIBUTION OF SALT TOLERANT PLANTS

The salt-tolerant plants may be distinguished mainly into the following six different kinds of plant communities inhabitating the (1) inland salt marshes, (2) alkali or sodic soils, (3) saline sand formations, (4) coastal salt marshes, (5) littoral woodlands, and (6) mangrove swamps.

(1) Inland salt marshes

Large parts of Gujarat and Rajasthan are arid and comprise of saline unproductive land in the form of salt basins like Ran of Katchh, Pachpadra, Thob, Sanwarla, Bap, Pokran, Lun Sarkar, Lanela, Sakhi and Khajuvalla. In these basins the soil surface upon drying is covered with patches of a white crust of salts mainly in the form of evaporite deposits of chlorides of sodium and potassium; sulphates of sodium, calcium and magnesium; and nitrates and other salts. Following are the main plant communities distributed in various saline regions:

Suaeda fruticosa (pure)
Aeluropus lagopoides-Cressa creatica-S. fruticosa
Halaxylon salicornicum-Salsola baryosma-Sporobolus marginatus
Salsola baryosma-Suaeda fruticosa-Cressa .cretica
Tamarix troupii-Suaeda fruticosa-C. cretica
S. fruticosa-A. lagopoides-Urochondra setulosa (in Rann of Katchh only)
A. lagopodies-Trianthemum triquetra-Soorobolus marginatus-S. fruticosa
Sporobolus marginatus-Dactyloctenium sindicum-Cyperus sp.
Sporobolus helvolus-Cyperus rotundus
Peganum hermale--Elensine compressa-S. marginatus

Other species which are occurring frequently in these basins in isolated communities or as associates of some communities include *Haloxylon recurvum, Euphorbia granulata, Gisekia pharnaceoides, Trianthema portulacatrum Calotropis procera, Artemisia maritima, Sphaeranthus indica, Chloris virgata, C. barbata, Cynodon dactylon, Dichanthium annulatum, Salicornia brachiata, Atriplex sp., Kochia indica, Suaeda nudifiora, S. Maritima, Croton bonplandianum, Portuleca oleracea, P. guadrifia, Solanum surattense, Fagonia cretica, Tribulus terrestris, Zygophyllum simplex,* and many others. In most of salt basins *Prosopasis julifora* has spread as a brush wood and has influenced other plant communities and has become a prominent species. Some of the soil characteristics under selected plant communities have been presented in Table 11.3. The distribution of some salt tolerant species in different salt basins is given in Table 11.4.

(2) Alkali (Sodic) soils

The sodic soils of Gangetic alluvial plains support very limited natural flora comprising of only a few salt tolerant species. *Prosopis Juliflora,* although an exotic, has spread forming gregareous patches in many salt affected areas in the country. *Acacia nilotica,, A. leucophloea, Prosopis cineraria, Salvadora persica, S. oleoides, Capparis decidua, C. sepiaria, Maytenus emerginatus, Lycium barbatum, Calotropis procera and Zizyphus nummularia* are some common woody species. At some low pH (about 9) *Balanites roxburghii, Desmanthus virgatus, Tecomella undulata and Butea monosperma* are also found. Among herbaceous species grasses like *Sporobolus marginatus, S. diander, Desmostachya bipinnata, and Chloris barbata, and other herbs like Trianthema triquetra, Suaeda fruiticosa and Kochia indica* are common. During rainy season, due to low salt contents in upper layers of soil herbaceous species such as *Cynodon dactylon, Chloris virgata, Echinochloa colonum, Croton bonplandianum, Cassia tora, C. occidentallis., Eclipta prostrata, Euphorbia thymifolia, E. hirta, Indigofera sppp. Abutilon indicum, Achyranthes aspera and Amaranthus viridis* appear and grow on these soils. Along bushes *Adhatoda vasica* also forms gregareous colonies. Among bushes of *Capparis* and *Salvadora,* climbers like *Asparagus racemosa, Cocculus pendulus, C. hirsutus, Pupasllia lappacea, Cayratia trifolia, Momordica dioica, Mukia maderaspatana and Ichnocarpus prutescens* are quite common.

Table 11.3 : Soil characteristics of prominent plant communities.

	Plant Communities	*Soil depth cm*	*pH_2*	*EC dS^c/m*	*Org.c %*	*Na*	*K*	*Ca=MG*
							meq/1	
1	2	3	4	5	6	7	8	9
Rann of Katchh								
1.	*Aeluropus logopoides* (pure)	0-15	8.4	60.2	0.01	538.0	28.6	30.5
		15-30	8.4	21.9	0.22	164.4	17.3	33.0
		30-90.	8.3	13.6	0.16	74.2	7.6	39.4
2.	*A. lagopoides, Cressa cretica, Suaeda fruticosa*	0-15	8.4	33.5	0.51	294.0	15.2	60.2
		15-30	8.3	37.5	0.44	334.3	12.2	46.0
		30-90	8.8	43.9	0.30	368.1	11.5	51.5
3.	*Cressa cretica, Cyperus sp. Sporobolus marginatus*	0-15	8.8	30.0	0.26	242.0	7.0	45.5
		15-30	8.2	38.6	0.24	255.4	9.0	120.0
		30-90	8.1	34.5	0.18	209.6	7.0	118.0
4.	*Prosopis juliflora, Sporobolus marginatus, Crotalaria burhia, Aeluropus lagopoides*	0-15	8.3	42.5	0.15	178.0	153.8	81.0
		15-30	8.6	30.9	0.12	158.1	115.0	33.5
		30-60	8.7	29.6	0.08	167.0	89.8	37.6
		60-90	9.2	29.3	0.05	170.9	76.3	40.5
5.	*Salsola baryosma, Suaeda fruticosa, Cressa cretica*	0-15	8.2	31.9	0.26	181.0	10.6	145.0
		15-30	8.5	33.5	0.21	164.3	12.1	155.0
		30-90	8.7	36.2	0.15	193.0	15.3	152.0

1	2	3	4	5	6	7	8	9
6.	*Suaeda fruticosa*	0-15	8.7	25.2	0.32	192.0	11.8	43.8
		15-60	8.6	23.9	0.21	171.6	9.6	75.6
7.	*Tamarix troupii*	0-15	8.3	27.4	0.26	230.0	5.0	37.0
	Suaeda fruticosa	15-30	8.2	24.1	0.10	146.4	10.0	83.5
	A. lagopoides	30-90	8.0	19.0	0.06	123.2	13.7	51.4
8.	*Phragmites australis,*	0-15	8.2	36.8	0.31	236.4	8.0	69.5
	Carex sp., Cyperus sp.	15-30	8.3	28.9	0.26	214.3	7.0	64.0
	(Lowlying area)	30-90	8.2	27.6	0.29	185.3	7.0	86.0
9.	*Suada fruticosa, A. lagopoides*	0-15	8.2	59.0	0.31	456.5	13.5	58.0
	Urochondra setulosa	15-30	8.6	53.0	0.22	524.5	15.7	51.0
		30-90	9.0	67.9	0.10	626.1	16.5	33.0
10.	*A. lagopoides-Trianthema*	0-15	8.4	14.3	0.24	21.0	8.8	39.4
	portulacastrum, Solanum	15-30	8.7	21.6	0.18	148.4	13.7	48.4
	Incanum, Spharanthes Indica, Spopobolus, Vrochondria setulosa							
Near Bhachau								
11.	*S. persica, Tamarix*	0-15	8.3	28.9	0.45	236.4	7.0	44.5
	troupii, Suaeda	15-30	8.2	32.0	0.49	241.8	7.8	65.4
	fruticosa	30-60	8.4	41.4	0.38	281.4	8.2	116.5
		60-90	8.6	49.7	0.27	350.6	11.2	126.7
12.	*Salvadora persica, Suaeda*	0-15	9.0	40.2	0.30	342.4	5.0	53.5
	fruticosa, S. nudiflora	15-30	8.4	34.8	0.21	274.5	6.0	60.0
		30-60	9.1	46.3	0.17	405.0	6.1	48.0
		60-90	9.0	49.6	0.09	416.7	8.6	57.4
13.	*S. persica, Tamarix troupii*	0-15	7.8	117.5	0.98	1049.0	6.0	111.0
		15-30	8.2	91.3	0.82	745.0	7.0	148.5
		30-120	8.6	78.3	0.53	585.0	8.0	187.6
14.	*S. fruticoosa, Oleochaete,*	0-15	9.2	69.1	0.48	598.0	4.0	74.5
	Sphaeranthes indica,	15-30	9.0	55.1	0.29	437.6	7.8	96.6
	Dichanthium annulatum, A. lagopoides							
15.	*Spersica, Tamarix broupii,*	0-15	8.34	89.9	0.64	790.8	28.5	76.0
	A. lagopoides	15-30	8.59	67.2	0.49	548.6	31.2	87.5
		30-60	8.4	46.6	0.27	315.4	37.4	105.6
16.	*P. juliflora, S. persica,*	0-15	8.8	39.4	0.94	210.0	11.1	68.0
	S. fruticasa, A. lagopoides	15-30	8.7	35.4	0.79	239.7	12.9	73.5
	(Near Mahi river)	30-60	8.6	26.2	0.40	160.0	10.6	88.0
		60-90	8.6	24.6	0.32	138.4	9.7	101.3
Gadoi, Nasar, 22km from Nosari								
17.	*Anacla nilotica, Ipomoea*	0-15	10.5	55.2	0.12	524.5	4.0	20.0
	carnea, Desmostachya	15-30	10.6	50.2	0.10	467.4	5.0	26.0

1	2	3	4	5	6	7	8	9
	binata, Typha, Fimbristilis,	30-60	10.7	45.1	.08	456.5	4.0	17.0
	Bacopa monerii	60-90	10.5	57.3	.08	548.9	4.0	15.5
	(Alakali Quake)							
Nalsarovar, Gujarat								
18.	*Scirpus littoralis,*	0-15	9.9	25.2	0.25	169.0	4.0	57.5
	Cynodon datylon,	15-30	9.8	15.2	0.18	109.8	3.0	30.0
	Cyperus sp. (low lying)	30-60	10	12.1	0.08	74.5	3.6	35.0
Sampla, Haryana								
19.	*Heliotropium curassavicum,*	0-15	8.4	30.6	0.18	74.5	3.6	43.2
	S. fruticosa, Cynodon dactylon	15-30	8.3	14.31	0.15	68.5	2.5	31.5
		30-60	8.5	19.7	0.08	75.6	2.3	30.7
		60-90	8.5	20.9	0.06	78.8	1.5	37.5
20.	*Tamarix troupii,*	0-15	8.5	35.4	0.30	93.5	11.2	75.0
	Plucheria, lanceolata, Cynodon	15-30	8.1	29.1	0.28	100.5	11.8	51.5
	dactylon, Sporobolus	30-80	8.5	28.7	0.20	102.8	12.3	63.4
	marginatus,	60-90	8.3	28.5	0.11	109.8	12.9	57.4
	S. fruticosa							
21.	*H. curassavicum, S. fruticosa,*	0-15	9.0	29.6	0.28	91.3	10.2	79.0
	Launea aspleniifolia,	15-30	8.5	19.1	0.15	51.1	2.9	1.5
	Blumea lacera	30-60	8.0	19.6	0.11	43.5	2.6	58.4
Pachpadra, Rajasthan								
22.	*Prosopis juliflora, Aeluropus*	0-15	8.1	43.0	0.25	262.2	2.8	-
	lagopoides Suaeda fruticosa,	15-30	7.9	34.0	0.21	189.4	2.4	-
	Sporobolus marginatus	30-90	7.8	37.8	0.16	102.5	2.0	-
23.	*S. fruticosa, A. lagopoides*	0-15	8.6	49.9	0.27	947.2	6.2	-
	Cressa cretica, Cyperus	15-30	8.4	40.8	0.21	805.1	5.8	-
	bulbosus, Trianthema triquetra,	30-60	8.2	36.0	0.19	368.0	2.5	-
	S. marginatus,	60-90	7.8	30.8	0.16	437.0	2.4	-
	Sphaeranthus Indica							
Lunsarkar, Rajasthan								
24.	*Cressa cretica, A. lagopoides,*	0-15	8.5	48.9	0.31	1060.0	8.3	-
	Tamarix sp., Suaeda fruticosa	15-30	8.4	45.6	0.26	839.4	8.2	-
		30-60	8.4	41.0	0.21	947.4	5.7	-
Hanumangarh, Rajasthan								
25.	*Species of Haloxylon,*	0-15	-	63.4	0.29	1311.1	5.5	-
	Salsola, Tamarix,	15-30	-	29.8	0.21	485.9	4.7	-
	Suaeda, Sporobolus	30-60	-	22.1	0.18	329.8	4.1	-
26.	*Acacia tortalis, Calligonum*	0-15	7.8	29.8	0.26	220.6	4.3	-
	Polygonoides, Salsola baryosma,	15-30	7.6	26.3	0.18	185.7	4.0	-
	Leptadenia pyrotechnica	30-60	8.0	21.9	0.14	78.8	3.0	-

1	2	3	4	5	6	7	8	9
27.	*Capparis decidua, Prosopis Juliflora, P. cineraria, Salvadora oleoides, Zizyphus nummularea*	0-15	7.8	20.8	0.19	189.3	3.0	-
		15-30	8.0	17.0	0.15	105.6	2.0	-
		30-90	8.5	15.4	0.12	40.9	1.0	-
Sambar Lake, Rajasthan								
28.	*Species of Suaeda, Sperobolus, Trianthema, Salsola, Dactyloctenium, Xanthium, Sphaeranthus, Baunaea*	0-15	8.1	57.0	0.29	708.4	3.0	-
		15-30	7.8	40.8	0.19	368.0	2.0	-
		30-90	8.3	27.8	0.15	56.6	1.0	-
On way from Jodhpur to Ajmer								
29.	*Species of Suaeda, Cressa, Trianthema, Sporobolus, Aeluropus, Pagonia*	0-15	7.9	48.9	0.18	619.6	2.0	-
		15-30	8.0	39.7	0.11	406.7	1.0	-
		30-90	8.7	30.1	0.08	216.6	1.0	-
On way back from Bharatpur to Mathura (U.P.)								
30.	*S. persica Diospyros montana, Capparis decida*	0-15	7.6	33.1	0.21	149.0	3.0	-
		15-30	8.0	29.4	0.17	108.1	2.1	-
		30-60	8.2	25.5	0.09	95.4	0.9	-
		60-90	8.6	18.4	0.07	79.0	0.9	-
31.	*Species of Tamarix, Salvadora, Prosopis, Suaeda, Capparis*	0-15	8.0	51.1	0.19	619.8	4.0	-
		15-30	8.2	22.4	0.09	295.4	3.0	-
		30-90	8.3	29.9	0.06	198.6	2.0	-
32.	*Species of Sporobolus-Suaeda and Sphaeranthus*	0-15	8.5	41.0	0.20	564.0	2.0	-
		15-30	8.3	37.4	0.13	460.0	2.0	-
		30-60	8.0	25.2	0.08	195.0	1.0	-

In *Vertisols Acacia nilotica, A. eburnea, Balanites roxburghii, Butea monosperma and Desmenthus virgatus* are common woody species. *Alhagi pseudoalhagi* is common herbaceous legume. Some interesting grasses like *Cymbopogon martinus, Heteropogon contortus, Themeda triandra, Chrysopogon fulvus, Isschaemum rugosum, Dichanthium annulatum, Bothriochloa pestusa, Apluda aristata, Tragus biflorus, Iseilema laxum and Sehima nervosum* along with legumes like *Alysicarpus vaginalis, A. rugusum, A. longifolius, Rhynchosia minima, Phaseolus trilobus, Indigofera linifolia and I. glandulosa* appear after closing from grazing. *Azadirechta indica* when grown performs very well in these soils.

(3) Coastal salt marshes

Although typical salt marshes are not found along Indian coasts some saline marshes develop in, areas which are flooded with marine water for long periods. Soils are sandy loam to silty loam with moderately alkaline reaction (pH 7.8 to 8.6), high organic matter (1.4 to 2.6%), and NaCl content 1.5 to 4.0%. Along Gujarat coast *Salvadora persica, Suaeda fruticosa, S. nudiflora, Atriplex stocksii, Salicornia brachiata, Tamarix troupii, Sesuvium portulacastrum, Aeluropus lagopoides and Urochondria setulosa* arer common species. *Suaeda maritima, S. fruticosa, Heliotropium curassavicum, Myriostachya wightiana, Fimbristylis cymosa, Aeluropus lagopoides and Cressa cretica* are found nearer to the mouth of Krishna river. The grass *Porteresia coarctata* grows abundantly in newly formed tidal mudflats particularly nearer Sunderbans. *Scirpus littoralis, Fibristylis spp., cyperus spp, and Acrostichum aureum* are common in Bay Islands. *Arthrocnemum indicum, Suaeda nudiflora S. monoica, Salicornia brachiata, Sesuvium portulacastrum and Heliotropium curassavicum* are common in Pichavaram.

(4) Saline sandformations along beaches

These formations are influenced by tidal inundations but because of sandy nature of the soil, the salinity in root zone is not very high. *Ipomoea pes-caprae* is most common creeper throughout the coasts. In coastal regions of Andhra Pradesh and Orissa, *Spinfex littoralis* is most common grass bordering beaches. *Ipmoea gracilis, Vigna marina, Clitoria ternatea, Canavalia maritima, Ischaemum muticum, Thuareainvoluta, Cassytha filiformis, Eclipta prostrata, Launea sarmentosa* and many species of *Fimbristylis, Scirpus and Cyperus* are other important species found distributed along beaches particularly in the Bay Islannds. *Triumfetta repens* is found forming mats in Car Nicobar. In Pichavaram, *Cyperus arenarius, Fimbristylis cymosa, Zoysia matrella, Aeluropus logopoides, and Geniosporum tenuifolium* make their appearance. Near Bombay, halophytes such as *Cyperus bulbosus, Paspalum distichum, Aeluropus lagopoides, Fimbristylis ferruginea, Sporobolus viriginicus, S. helovolus, S. marginatus and. S. glaucifolius* are common. Sometimes *Suaeda fruticosa, Sesuvium portulacastrum and Porteresia coarctata* are also found. All along Orissa coast, *Casuarina equisetifolium* has been raised along beaches successfully.

Table 11.4 : Classification and relative distribution of species found naturally in inland salt affected localities.

Species	*Adaptability class*	*Raunkiaers Life form*	*Relative distribution in location of*								
			P	*L*	*GH*	*S*	*T*	*Gc*	*M*	*HP*	*R*
1	2	3	4	5	6	7	8	9	10	11	12
Trees & shrubs											
Acacia farnnesiana	G	N	-	-	-	-	-	-	-	r	r
Acacia jaquemontii	G	N	o	-	o	-	-	-	-	r	r
A. leucophloea	G	M	-	-	-	-	-	-	-	o	-
A. nilotica	G	M	-	-	o	o	-	f	f	f,la	r
Balanites aegyptiaca	G	N	-	-	-	-	-	-	-	r	o
Butea monosperma	G	M	-	-	-	-	-	-	-	o	-
Calotropis procera	G	M	o	-	o	o	-	-	f	o	o
Capparis decidua	F.H.	M	-	-	o	la,f	f	f	f	a,f	o
C. sepiparia	FH	N	-	-	-	-	-	-	o	f	-
Carissa spinarum	FH	N	-	-	-	-	-	-	-	o	-
Clerodendrum phlomidis	G	N	-	-	r	-	-	-	r	o	-
Haloxylon recurvum	H	N	f	-	f	-	f	o	-	-	-
H. salicornicum	F.H.	N	f	-	f	-	-	o	-	-	-

1	2	3	4	5	6	7	8	9	10	11	12
Holoptelea integrifolia	G	N,M	-	-	-	-	-	-	-	r	-
Leptadenia pyrotechnica	G	N	r	-	-	-	-	-	-	r	-
Lycium barbatum	G	M,N	o	-	-	-	-	-	-	r	-
Maytenus emerginnatus	G	N	-	-	-	-	-	o	-	o	-
Mimosa hamata	G	N	-	-	-	-	-	-	-	o	-
Prosopsis cineraria	G	M	o	r	o	-	o	o	o	o	-
P. juliflora	G	M	f	f	f	f	f	f	f	o	f
Salsola baryosma	H	N	f	o	f	f	f	f	o	r	f
Salvadora oleoides	FH	M	f	-	o	f	f	f	a	f	o
S. persica	FH	M	o	-	o	o	o	o	f	o	o
Securingea leucoppyrus	G	N	r	-	-	-	-	-	-	r	-
Tamarix dioica	H	M	o	-	-	-	o	o	-	o	-
T. ericoides	H	N	o	o	-	o	-	-	o	-	o
T. troupii	H	N	-	-	f	r	r	r	o	-	o
Tecomella undulata	G	M	-	-	-	-	-	o	-	r	-
Zyzyphus nummularia	G	M	o	-	f	-	-	-	f	o	o
Herbs, Grasses & Sedges											
Abutilon indicum	G	Ch	-	-	-	-	-	-	o	o	o
Achyranthes aspera	G	N	o	-	-	r	o	o	f	f	o
Adhatoda vasica	G	Ch	-	-	-	-	-	-	f	f	-
Aeluropus lagopoides	H	He	f	f	o	f	f	f	o	r	a
Aerva persica	FH	N,Ch	o	o	o	r	-	-	-	r	-
A. pseudotomentosa	G	Ch	o	-	o	-	-	-	-	r	-
Alysicarpus monilifer	G	Th	-	-	-	-	-	-	-	o	-
Amaranthus gracilis	G	Th	-	o	-	o	o	o	f	f	o
A. hybridus	FH	Th	r	-	-	-	-	-	-	-	-
A. spinosus	G	Th	-	-	-	-	-	-	f	r	r
Anticharis senegalensis	G	Th	-	-	-	-	-	-	-	-	-
Argemone mexicana	G	Th	-	r	-	-	-	-	f	r	-
Aristida funiculata	G	Th	o	-	-	o	-	-	-	-	-
A. setacea	G	Th	r	-	-	-	-	-	-	r	-
Asparagus racemosus	G	CP,N	-	-	-	-	-	-	r	r	-

1	2	3	4	5	6	7	8	9	10	11	12
Blepharis sindica	G	Ch	r	-	-	-	-	-	r	-	-
Boerhaavia diffiusa	FH	Th	o	-	-	o	o	o	o	o	r
B. repanda	G	Cp	r	-	-	-	-	-	-	r	-
Brachiaria ramosa	Fh	Th	o	-	-	-	-	o	o	o	-
Bulbostylis barbata	FH	Ge	o	r	f	o	o	r	r	r	-
Cassia italica	FH	N	r	-	-	-	-	-	-	-	-
C. occidentalis	G	Th	-	-	-	-	-	-	-	r	-
C. tora	G	TH	-	r	-	-	-	-	f	o	o
Cenchrus biflorus	G	Th	f	-	-	o	o	o	f	f	r
C. ciliaris	G	Ch	o	-	-	-	-	-	-	o	r
C. prieurii	G	Th	r	-	-	-	-	-	-	r	-
C. setigerus	G	Ch	o	-	-	r	r	r	o	o	-
Chenopodium album	G	Th	-	-	-	-	-	-	f	r	-
C. ambrosioides	FH	Th,Ch	-	-	-	-	-	-	r	f	-
C. murale	G	Th	-	-	-	-	-	o	f	o	-
Chloris barbata	G	Th	-	-	-	-	-	-	-	r	-
Chloris gayana	FH	Th	-	-	-	-	-	-	o	o	-
C. virgata	FH	Th	-	-	-	-	-	r	r	o	-
Citrullus colocynthis	G	He	o	-	-	-	-	-	o	r	o
Cleome vahliana	FH	Th	o	-	-	r	o	o	-	-	-
C. viscosa	G	Th	r	-	-	r	-	-	o	o	-
Coccinia grandis	G	Cp	o	-	-	-	-	-	o	o	-
Convolvulus hirsutus	G	Cp	o	-	-	-	-	-	o	o	-
Convolvulus microphyllus	FH	He	o	-	-	o	-	-	o	o	-
Corchorus depressus	G	He,Th	r	-	-	o	-	r	o	o	-
C. tridens	G	Th	r	-	-	o	-	r	o	o	-
Cresa cretica	H	Ch	f	f	f	f	f	f	r	r	a
Crotalaria burhia	H	N	f	-	-	-	-	-	-	r	-
C. medicaginea	G	Th	-	-	-	-	-	-	r	r	-
Croton bouplandianum	G	Th	-	-	-	-	-	-	o	f	-
Cucumis callosus	FH	Th	o	-	-	r	-	r	o	o	-
Cynodon dactylon	G	Cr	o	o	o	r	o	r	f	f	o

1	2	3	4	5	6	7	8	9	10	11	12
Cyperus arenarius	G	Cr	r	-	-	-	-	-	-	-	-
Cyperus bulbosus	FH	Cr	o	-	-	r	r	r	o	r	o
C. compressus	FH	Cr	-	-	-	o	-	-	o	r	-
C. flavidus	FH	Cr	-	-	-	r	-	-	-	-	-
C. iria	FH	Cr	o	-	-	o	o	r	o	o	-
C pengurei	G	Cr	-	-	-	r	-	-	-	-	-
C. rotundus	FH	Cr	o	o	o	o	o	o	f	f	-
C. triceps	FH	Cr	-	-	-	r	-	-	o	-	-
C. tuberosus	FH	Cr	r	-	-	o	o	o	-	r	-
Dactyloctenium aegyptium	FH	Th	o	o	o	o	o	o	o	f	-
D. sindicum	G	Ch,Cr	o	-	o	-	-	r	o	o	-
Desmodium trifloru	G	Cr	-	-	-	-	-	-	o	-	-
Dichanthium annulatum	FH, G	Cr	o	-	-	r	o	o	f	f	-
Digera muricata	G	Th	r	-	-	o	r	r	o	f	-
Digitaria ciliaris	FH	Th	o	o	r	-	r	r	o	f	-
Echinochloa colonum	FH	Th	o	-	o	o	o	o	o	o	-
Eclipta prostrata	FH	Th	o	o	o	o	o	o	o	f	o
Echinops echinatus	G	Th	o	-	-	r	o	o	r	o	o
Eleusne compressa	G	Ch	o	-	-	o	o	o	o	o	-
E. indica	G	Ch	-	-	-	-	-	-	o	f	-
Enicostema hyssopifolia	G	Th	o	-	-	-	-	-	-	-	-
Eragrostis ciliaris	FH	Th	o	r	r	o	r	r	-	o	-
E. pilosa	FH	Th	r	-	-	o	o	o	-	o	-
E. tremula	FH	Th	o	-	-	r	-	o	-	-	o
Erigeron canadensis	G	Th	o	o	o	r	o	o	o	o	o
Eupphorbia granulata	G	Th	o	r	o	o	o	o	f	f	-
E. hirta	G	Th	o	r	o	o	n	o	f	f	-
E. Jodhpurensis	G	Th	r	-	-	-	-	-	-	-	-
E. prostrata	G	Th	o	r	r	r	o	o	r	o	o
E. thymifolia	G, FH	Th	o	o	o	-	o	-	f	f	o
Evolvulus alsinoides	G	Th	o	-	-	-	-	o	f	f	o
Fagonia cretica	FH	Th	o	-	-	-	-	o	f	f	o

1	2	3	4	5	6	7	8	9	10	11	12
Farrsetia hamiltonii	G	Th	f	o	o	f	f	f	r	r	f
Fimbristylis miliacea	FH	Cr	-	-	-	-	-	-	-	-	-
Gisekia pharnaciodes	G	Th	o	r	r	r	r	r	o	o	-
Heliotropium curassavicum	G	Th	-	-	-	-	-	-	r	r	-
H. marifolium	FH	Th	-	-	-	-	-	-	o	la	o
H. indicum	FH	Th	-	-	-	-	-	-	r	r	-
Heylandia latebrosa	G	Th	-	-	-	-	-	o	o	o	-
Hibiscus obtusifolius	G	Th	o	-	-	o	-	-	-	-	-
Indigofera cordifolia	G	Th	r	-	-	-	-	-	-	r	-
I. enneaphylla	G	Th	o	-	-	r	-	o	r	r	-
I. hochstetteri	G	Th	-	-	-	-	-	r	r	r	-
I. linifolia	G	Th	r	-	-	-	r	r	-	-	-
I. oblongifolia	G	Th	o	-	-	-	r	r	-	-	-
Ipomoea pes-tegridis	G	Th	r	-	-	-	o	o	o	r	-
Justicea simplex	G	Th	o	-	-	-	-	-	-	-	-
Kochia indica	H	Th	o	-	-	-	-	-	-	-	-
Launaea asplenifolia	G	Th	-	-	-	-	-	o	f	-	
L. chondrilloides	G	Th	r	-	-	-	-	o	o	o	o
L. nudicaulis	G	Th	r	-	-	-	-	-	-	-	-
Leucas urticaefolia	G	Th	o	-	-	-	-	-	o	f	-
Mariscus squarrosus	FH	Cr	r	-	-	-	-	o	o	o	o
Marsilea aegytiaca	FH	Cr	r	-	-	-	-	-		-	-
Melanocenchrus jacquimontii	FH	Th	o	-	-	-	-	-	o	f	-
Melilotus alba	G	Th	o	-	-	-	-	-	o	f	-
M. indica	G	Th	-	-	r	-	-	-	o	f	-
Momordica dioica	G	Cp	-	-	-	-	-	-	o	o	-
Oligochaeta ramosa	G	Th	o	-	-	-	-	-	o	-	-
Oxalis corniculata	G	Cr	-	-	-	-	-	-	o	o	-
Panicum antidotale	G	Ch	o	-	-	-	-	-	-	-	-
P. turgidum	G	Ch	o	-	-	-	-	-	r	f	-
Phyla nodiflora	G	Ch	o	o	r	-	r	r	o	f	o
Phyllanthus fraternus	G	Th	o	-	-	-	-	-	o	o	-

1	2	3	4	5	6	7	8	9	10	11	12
P. maderaspatensis	G	Th	r	-	-	-	-	-	r	r	-
Pergularia daemia	G	Cp	o	-	-	-	-	-	o	o	-
Polygala erioptera	G	Th	o	-	-	-	-	-	r	-	-
Portuleca oleracea	FH	Th	o	r	r	o	o	o	o	o	-
P. pilosa	FH	Th	-	-	-	-	-	-	o	l.f	-
P. quadrifida	FH	Th	-	-	-	-	-	-	o	l.f	-
Psoralea odorata	G	Th	o	-	-	r	-	o	-	-	-
Pulicario crispa	FH	Th	o	-	-	o	-	o	o	o	-
P. lanceolata	FH	Th	-	-	-	-	-	-	f	l.f	-
P. rajputanae	FH	Th	r	-	-	-	-	-	-	r	-
P. wightiana	FH	Th	o	-	-	r	o	o	-	-	-
Rhynnchosia minima	G	Th	-	-	-	-	-	-	o	o	-
Schoenefeldia gracilis	G	Th	o	-	-	-	-	-	-	r	-
Scirpus roylei	FH	G	o	-	-	-	-	-	o	-	-
S. tuberosus	FH	G	o	-	-	r	o	o	r	-	-
Sericostoma pauciflorum	G	Ch	o	-	-	-	-	o	-	-	
Sesbania portulacastrum	FH	Th	f	r	r	f	-	f	-	r	-
Setaria glauca	G	Th	-	r	-	-	r	-	f	f	-
S. verticillata	FH	Th	-	-	-	-	-	o	r	r	-
Sisymbrium irio	G,	Th	-	-	-	-	-	-	r	o	-
Sphaeranthus indicus	FH	Th	f	r	o	-	r	f	o	o	o
Sporobolu diander	FH	TH	-	r	o	-	r	o	f	f	o
S. helovolus	H	Th	f	-	-	f	f	o	o	r	-
S. marginatus	FH	Th	f	o	o	f	f	f	a	a	f
S. tremulus	H	Th	-	o	-	o	-	o	-	-	o
Suaedda fruticosa	H	Ch	f	f	f	f	f	f	f	l.a	a
S. maritima	H	Ch	-	-	-	-	-	-	-	-	-
Tephrosia purpurea	G	Ch	o	-	-	o	-	-	o	o	o
Tragus biflorus	FH	Th	o	-	-	o	r	o	-	-	-
Trianthema portulacastrum	G	Th	o	o	r	o	r	o	f	f	f
T. triquetra	FH	Th	f	o	o	f	o	o	f	f	f
Tribulus terestris	G	TH	o	-	-	o	-	-	o	f	-

1	2	3	4	5	6	7	8	9	10	11	12
Vernonia cinerea	FH	Th	o	r	r	o	o	r	o	o	-
Zaleya reedimita	G	Ch	o	-	-	-	-	-	r	r	-
Zygophyllum simplex	H	Th	o	o	o	r	o	o	r	r	-

The symbols depict as :

G - Glycophytes, H = Halophytes; FH = Facultative halophytes; M = Mega-micro-phanerophytes; N=Nano pha nereophytes; Ch = Chamaephytes; He = Hemicryptophytes; Cr = Creptyphytes (geophytes); Th = Therophytes; Cp = Climbing phanrophytes; a = abundent; f = frequent; o = occasional; r = rare; 1= local; - = absent, v = very, P = Pachpadra, L = Lunsarkar, Gh = Ghagar belt, S = San karala, T = Thob, Gc = Gujarat command areas, M = Mathura and nearby areas, H P = Haryana & Punjab areas. R = Rann of Kutchh.

(5) Littoral woodlands

These are woody formations which are either associated with backwater mangroves or are influenced by spring tides. They usually form the border of mangroves and separate them from inland forests. They are characterised by evergreen species but light foliaged species like *Casuarina equisetifolium* (which forms pure natural fringe in Casuraina Bay of Andamans) are also found. The important species which are found along most of the coastal areas include *Aglaia cucullata, Barringtonia sp., Bronlowia tersa, Clerodendron inerme, Caesalpinia crista, Cerbera manghas, Derris trifoliata, Dalbergia spinosa, Thespesia populnea, Hibiscus tiliaceus, Calophyllum inophyllum, Intsia bijuga, Pongamia pinnata and Terminalia catappa*. Besides above species, *Manilkara littoralis, Licuala spinosa, Pandanus sp., Premna corymbosa, Cordia subcordata, Scaevola taccada, Hernandia peltata, Guettarda speciose, Tournefortia ovata, Dolichandrone spathacea and Ochrosia oppositifolia* are common in Bay Islands. Along Gujarat coast *Salvadora oleiodes, S. persica and Tamarix troupii* are common. *Finlaysonia maritima, Sarcolobus carinatus, Derris spp., and Flagellaria indica* are prominent climbers on mangroves and littoral trees. *Dischidia nummularia, Hoya parasitica and Hydnophytum formicarum and ferns such as asplenium nidus and Drynaria quercifolia* are interesting epiphytes. Many authors have classified many of these species along with mangroves but these may be segregated as associate species (Table 11.5).

(6) Mangroves

The mangroves inhabit in the intertidal estuarine regions, sheltered coast lines, logoons and creeks. These dominated by submerged sclerophylous plant species which are taxonomically unrelated and constitute a dynamic and complex ecosystem represented by associated plant and animal species of varied life forms. The best mangrove formations are seen where the tidal regime is normal with a constant mixing of sea water and fresh water and where the temperature does not optimally go below 20 °C and usually the minimum temperature does not vary more than 5 °C. These plants have unique morphological and physiological adaptations for their growth and reproduction in saline habitats. In India, 37 species of exclusive mangrooves and many associate mangrove species are found in the mangal formations. These are inhabited by several interesting animals such as amphibians, salt water crocodiles, sea turtles, water monitor lizards, snakes, wild pigs, monkeys, even tigers in Sunderbans, deer and several indigenous and migratory birds.

In essence, the short term economic gains have received priority over the long term benefits bestowed by mangroves and these valuable resources are dwindling at fast rate.

An optimistic assessment (Government of India 1987) reveals that about 6740 km^2 area is left under mangrove forests which also includes other halophytic shrubs and patchy vegetation. The major causes of mangrove destruction are over-exploitation by traditional and non-traditional users; conversion of mangrove area to agriculture and aquaculture; domestic and industrial developments; construction of channels and harbours; pollution due to oil, waste disposal and hazardous chemicals; diversion of fresh water; and construction of salt ponds. As a result of the unsustainable exploitation of mangrove ecosystems, a number of plant and animal species which are largely or exclusively depend on them, are presently at the verge of extinction.

Loss of fringing mangroves exposes the shore line to accelerated erosion, which may be further increased under scenarios of global climate change and sea level rise. There is immediate need to develop a viable strategy to utilise mangrove resources on a sustainable basis and to declare mangrove forests as conservation zones. The undisturbed mangrove areas (in Andaman & Nicobar Islands and Sunderbans) are to be saved from destruction. A national policy for the conservation and sustainable use of mangrove ecosystems has to be evolved and put into operation with the involvement of all related interests.

The viability of an ecosystem can be judged from its diversity of species which indicates the survival value of the community. A rich gene pool means a higher adaptation potential. It leads to increased stability in an ecosystem. For example the mangrove ecosystem of Andaman & Nicobar Islands is a stable rich in both floral and faunal biodiversity as compared to other parts of the country. With maximum number of exclusive mangrove (35) and 132 associate vascular and non-vascular (including epiphytes) species and faunal diversity (Dagar 1982, 1995). In the mangrove ecosystems there is large accumulation of plant debris. To a wide range of animals, the litter is valuable feed either as such or after microbial degradation. Many of them make use of mangrove communities as habitat, nursery ground and source of food. The debris also serves as breeding ground for the juveniles of many types of fish, crustaceans and other fauna. Mangrove roots and branches serve as a good shelter for large number of organisms including many beautiful orchids, ferns, jungerminiales, mosses, lichens, algae, fungi, bacteria, ciliates, nematodes and amphipods which colonize and form food for many kinds of fish. Many of the fish species utilize the mangrove water as good nursery and breeding ground. Even the sea turtles use the sandy beaches bordered by littoral species for their breeding purposes. Thus the mangrove ecosystems not only provide fuelwood and charcoal, timber and construction material, pulp, tannin, food and beverages, honey, fodder, medicine, and stabilization of the coasts, but also sustain diverse communities of flora by providing them with habitat, energy and nutrition. In the mangrove ecosystems, phyto-, and zoo-planktons become supplementary source of nutrition to the small fish and other aquatic life and fish culture ponds may be created in cleared portions or behind the mangrove without further denudation of mangroves. Low lying marshy areas connected with sea offer good scope for brackish water fish and prawn culture. Mullets grow well in mangrove waters. The prominent commercial molluscs, the black lip pearl oyster and green mussel culture offer good scope in intact mangrove areas.

The Sunderban mangroves of Ganga delta form the largest belt covering about 4200 km^2 area intersected by criss-crossing rivers and estuaries. These mangroves present complex ecological conditions due to their vastness in extent and ramifications of the riverine system leading of the several islands. The diversity of habitats which are ideal for the growth of mangroves, results in zonation and development of different associations.

Physiognomically also these mangroves are highly variable, depending upon the dominant species which may be *Excoecaria agallocha, Avcennia officinalis, Sonneratia apetala* or *Ceriops decandra*. Each of these gregarious species may constitute a pure formation. Besides these, *Bruguiera gymnorrhiza, Xylocarpus granatum, X. moluccensis, Aegiceras cornifuculatum, Phoenix paludosa* and *Rhizophora mucronata* are common. In the eastern part of the delta *Heritiera fomes* formation attains up to 30 m height. From Port Canning to Gosaba, *Phoenix paludosa* grows as an elegant palm associated with *Excoecaria agallocha*. Further south, *Avicennia* formations are prominent. The mangroves of Sajnakhali are dense formations of *Xylocarpus moluccensis, Sonneratia apetala, Excoecaria agallocha* and *Bruguiera gymnorrhiza* forming the top storey and *Ceriops decandra* in the lower stratum. In Duttar range, *Aegiceras corniculatum, S. apetala* and *X. granatum* are abundant.

Table 11.5 : Distribution of prominent mangrove associates along Indian coasts.

Species	*(Family)*	*Life form*	*West Coast*					*East coast*					*A. N.*
1	*2*	*3*	*4*	*5*	*6*	*7*	*8*	*9*	*10*	*11*	*12*	*13*	*14*
Acrostichum aureum	(Pteridacaeae)	Fern, shrub	-	f	-	f	la	f	la	la	-	-	la
Aeluropus lagopoides	(Poaccae)	Grass	a	f	la	o	-	-	o	la	f	f	-
Aglaia cocullata	(Meliaceae)	Tree	-	-	-	-	-	-	o	o	-	-	f
Arthroonemum indicum	(Chenopodiaceae)	Undershrub	la	f	-	-	-	-	o	o	f	f	-
Atriplex stocksii	(Chenopodiaceae)	Undershrub	o	r	-	-	-	-	-	-	-	-	-
Barringtonia racemosa	(Barringtoniaceae)	Tree	-	o	f	o	o	o	f	f	f	f	f
Brownlowia tersa	(Tiliaceae)	Shrub	-	-	-	-	r	-	f	f	-	o	f
Caesalpinia bonduc	(Caesalpiniaceae)	Climber	-	o	o	f	f	o	o	f	o	o	f
C. crista	(Caesalpiniaceae)	Climber	-	o	o	f	f	r	o	f	o	o	f
Calophyllum inophyllum	(Clusiaceae)	Tree	r	o	-	o	r	o	o	o	o	r	f
Cerbera manghas	(Apocynaceae)	Tree	r	o	o	o	o	o	r	o	o	o	f
Clerodendrum inerme	(Verbenaceae)	Shrub	o	la	o	o	la	o	la	o	o	o	f
Crinum asiaticum	(Amarallidaceae)	Undershrub	o	o	o	r	o	-	-	o	o	o	-
Dalbergia spinosa	(Fabaceae)	Shrub	-	-	o	r	-	-	o	o	f	o	o
Derris trifoliata	(Fabaceae)	Climber	-	o	o	r	o	o	o	o	f	f	f
Dolichandrone spathacea	(Bignoniaceae)	Tree	-	-	-	-	o	-	o	o	-	-	f
Fimbistyllis ferruginea	(Cyperaceae)	Sedge	o	o	o	o	o	r	o	o	o	o	f
Finlaysonia obovata	(Asclepiadacea)	Climber	-	-	-	-	-	-	f	f	-	-	f
Fiagellaria indica	(Fiagellariaceae)	Climber	-	-	-	-	-	-	o	o	-	-	f

1	2	3	4	5	6	7	8	9	10	11	12	13	14
Heliotropium curassavicum	(Boraginaceae)	Herb	o	-	-	-	-	-	r	o	f	f	-
Hernandia peltata	(Hernandiaceae)	Tree	-	-	-	-	-	-	-	-	-	-	o
Hibiscus tiliaceous	(Malavaceae)	Tree	f	f	o	f	o	f	f	f	o	r	vf
Holophila beccarii	(Poaceae)	Grass	-	f	-	-	-	-	-	-	-	-	o
Intsia bijuga	(Fabaceae)	Tree	-	-	-	-	-	-	o	o	-	-	o
Ipomoea pes-capre	(Convolvulaceae)	Trailer	o	f	f	f	f	f	f	f	f	f	o
I. tuba	(Convolvulaceae)	Climber	-	-	-	-	-	-	-	-	o	-	la
Licuala spiosa	(Arecaceae)	Shrub	-	-	-	-	-	-	-	-	-	-	o
Manilkara littoralis	(Sapotaceae)	Tree	-	-	-	-	-	-	-	o	-	-	la,f
Mucuna gigantea	(Fabaceae)	Woodyclimber	-	-	-	-	o	-	f	o	-	-	f
Myriostachya wightianna	(Poaceae)	Grass	-	la	-	-	-	-	o	o	la	r	r
Ochrasia oppositifolia	(Apocynacceae)	Tree	-	-	-	-	-	-	-	-	-	-	+
Pandanus fascicularis	(Pandanaceae)	Tree	-	o	o	o	o	f	-	f	f	f	f
P. furcatus	(Pandanaceae)	Tree	r	o	-	-	-	-	-	f	f	o	-
Pongamia pinnata	(Fabaceae)	Tree	-	o	-	o	r	o	f	f	o	-	f
Porteresia coarctata	(Poaceae)	Grass	-	o	la	-	-	-	f	la	+	-	-
Premna corymbosa	(Verbenaceae)	Shrub	-	o	-	o	r	o	f	f	o	-	f
Salacia chinensis	(Hippocrataeaceae)	Climbing shrub	-	-	-	-	-	-	r	r	-	-	-
Salicornia brachiata	(Chenopodiaceae)	Herb	f	f	-	-	-	-	o	r	r	f	-
Salvadora oleoides	(Salvadoraceae)	Tree	o	r	r	-	-	-	-	-	-	-	-
S. persica	(Salvadoraceae)	f	o	r	o	-	-	-	o	o	r	-	
Sarccolobus carinctus/ globosus	(Asclepiadaceae)	Climber	-	-	-	-	-	-	o	f	o	-	o
Scaveola taccada	(Goodeniaceae)	Tree	-	-	-	-	-	-	-	-	-	-	f
Scirpus littoralis	(Cyperaceae)	Herb	f	o	-	-	-	-	o	o	o	f	la
Sesuvium portulacastrum	(Aizoaceae)	Trailer	f	f	o	-	r	-	o	o	f	f	-
Spinifex littoreus	(Poaceae)	Trailer	f	f	-	f	-	-	-	-	la	la	-
Stenochlaena palustrae	(Polypodiaceae)	Climber	-	-	-	-	o	-	-	r	-	-	r
Stictocardia tiliifolia	(Convolvulaceae)	Climber	-	o	-	-	-	-	f	o	f	r	r
Suaeda fruticosa	(Chenopodiaceae)	Herb	f	f	f	-	f	-	f	f	f	f	-
S. maritima	(Chenopodiaceae)	Herb	o	r	r	o	o	o	o	o	f	o	-

1	2	3	4	5	6	7	8	9	10	11	12	13	14
S. monoica	(Chenopodiaceae)	Herb	-	r	r	o	-	-	-	-	o	-	-
Tamarix troupii	(Tamaricaceae)	Shrub	o	r	-	-	-	-	o	I	-	-	-
Terminalia catappa	(Combretaceae)	Tree	f	o	o	-	-	-	o	o	o	o	f
Thespesia populnea	(Kalvaceae)	Tree	o	o	-	-	r	o	f	o	o	o	f
Thuarea involuta	(Poaceae)		-	-	-	o	-	-	-	o	-	-	la
Tournefortia ovata	(Boraginaceae)	Shrub	-	-	-	-	r	-	-	-	-	-	f
Tylophora tenuis	(Asclepiadaceae)	Climber	-	-	-	-	-	-	o	o	r	o	o
Urochondra setulosa	(Poaceae)	Grass	f	r	-	-	-	-	-	-	-	-	-
Vitex negundo	(Verbiaceae)	Shrub	-	-	-	-	-	-	-	-	-	o	o
Zoysiasatrella	(Poaceae)	Grass	f	o	o	o	o	-	o	o	o	o	r

Depictions as in Table 6.

The glory of Sunderban mangroves is however, disappearing rapidly due to increasing biotic pressure. Many species which were once dominating the mangrove stands have rare or are all together absent. Some of them have been listed as endangered (Naksar and Guha Bakshi, 1987). Among the important threatened mangroves in many localities *Heritiera fomes, Ceriopsn decandra, C. tagal, Rhizophora apiculata, R. mucronata, and Kandelia candal* are important species. *H. formes* had been a common dweller of the Sunderbans delta region is now very occasionally found in the tidal swamps, especiially in the western zone, where the salinity level is high up to 35 dS m^{-1} and human interference is much more. It has been exploited indiscriminately for its timber value. Only a few stunted plants are found in the eastern parts of the Sunderbans reserve forest. *Nypa fruticans* has also been over exploited by the native people for its leaves used for thatching. Many areas have been reclaimed for agricultural purposes. In many disturbed areas the growth of the plants is stunted. Species such as *Heritiera littoralis* which at the beginning of the century was quite famous for its wide spread occurrence in Sunderbans and now at many sites we find only very few plants here and there. Similarly *Aegialitis rotundifolia* has become rare. At many place the prominent raizophoraceae is ncw not abundant and *Nypa fruticans* is almost dying.

Silting and rapid trend of deforestation has resulted in new land reclamation for agriculture and fisheries in the Ganga delta. Which is also facing neotectonic changes due to geomorphologic, climatic and anthropogen causes (Naksar 1991). All these factors have resulted in a continuous change in the river systems and upheaval of lands. Several tidal rivers like Piyali, Bidyadhari, Matla, Ichhamati and Kultati of the Sunderbans regions of India have lost their connecting links with original flow of the river Ganga (Hooghly). All these geomnorphological changes have created tremendous pressures on the mangrove ecosystems of the Sunderbans. It has been documented by the fossil woods and pollen grain studies that only 5000 years back the mangrove *Heritiera fomes* was dominated in the western part of the Sunderbans and was in existennce *in situ* in the greater Calcutta (Banerjee et al. 1984).

The rular people of these Sunderbans mangrove forest areas are very poor and they have to depend on these mangrove forests for their livelihood. They are exploiting the forest products without giving little chance for natural regenerations. The gradual deforestation and inhabilitation have dwindled the entire mangrove ecosystem to an alarming state. For example, the Gosaba Island which at the beginning of twentieth century was a virgin land, covered with dense mangroves and inhabited by variety of wild life including tigers in the forest and crocodiles in the channels is now a highly populated area with practically very negligible halophytic vegetation. Similarly, the human population is gradually encroaching to the south and the mangrove belt is decreasing proportionately. The densely populated agricultural zone with mesophytic vegetation of the Sagar Island, Kaddwip, Namkhana, Frazergume, Banti, Sonakhali, Sandeshkhali, Pathankhali, Surjaberia, Satzelis, Mollakhali, Golabari, Deulbari, Ranigarh, Bharatgarh, Kachuberia and many others were under dense tidal mangrove forest only 150-200 years back (Naksar and Guha Bakshi, 1987).

The Mahanadi mangroves covering an area of about 200 km² are in a degraded state due to conversion for agriculture and development of port facilities at Paradweep. Dense forests are seen in Bhatarkanika estuarine mudflats and deltaic creeks between the rivers Devi and Dharma. *Phoeniix paludosa* and *Aegialitis rotundifolia* occur more towards estuarine connditions. *Rhizophora mucronata, Avicennia alba, A. officinalis, Ceriops decandra, C. tagal, Sonneratia alba, Bruguiera gymnorrhiza, B. parviflora, B. cylindrica, Xylocarpus granatum* and *Kandeliia candal* grow luxuriantly in the tidal zone. *Heritiera fomes, H. littoralis, Sonneratia apetala, Cynometra ramiflora, Lumnitzera racemosa* and *Excoecaria agallocha* are prominent under the freshwater of the tidal zone. *Acanthus ilicifolius* is found in elevated fringe areas. In tidal Burabhalanga estuary, species of *Avicennia, Rhizophora, Sonneratia, Phoenix* and Excoecaria are noteworthy elements. But most of the areas particularly away from the river remain highly degraded.

The mangroves of Godavari and Krishna estuaries depend on the frequency and amount of flooding of the river and configuration of the coast. During the hot season, salinity increases considerably and only certain taxa such as *Aviceennia, Excoecaria* and *Sonneratia* with wide ecological amplitudes exist. Only nearer to lagoons *Ceriops decandra, Bruguiera gymnorrhiza, Rhizophora mucronata* and *R. apiculata* can be found frequently but are rare elsewhere. *Scyphiphora hydrophyllacea* once reported to be widely distributed, is found in few patches. Much of these mangroves have been over-exploited.

The mangroves of the Cauvery deltic system are discontinuous. The mangals of Pichavaram show a marked zonation where *Avicennia marina* forms pure stands and *Rhizophora apiculata* and *R. mucronata* grow well along the channels and creeks. These are mixed with *Bruguiera cylindrica, Ceriops decandra* and *Sonneratia apetala*. Other species in this region include *Lumnitzera racemosa, Aegiceras corniculatum* and *Excoecaria agallocha*. The freshwater zone is generally dominated by *Acanthus Illicifolius, Avicennia marina* is found more towards the marine zone and *A. officinalis* towards freshwater zone influenced by Coleroon estuarine water.

Along the Kerala coast, there are left only about 1671 ha of mangrove forests (Chand Basha 1992). The distribution, of course, is discontinuous and patchy. Kannur district with a total mangrove area of 755 ha tops the list followed by Kozhikode (293 ha) and Ernakulam (260 ha). An interesting feature of the mangrove areas along the Kerala coast is that as much as 88 percent of them areunder private ownership. There has been extensive reclamation of wetlands in Kerala, but it is poorly documeted and, therefore, difficult to assess the extent of such activity. It is, however, believed that once there was a dense cover of mangrove along the coast. Although mangrove have been traditionally exploited as forests, they received little attention as a useful resource. Diversion of mangrove areas for agricultural purposes (paddy cultivation, establishment of coconut groves), fisheries (formation of fish ponds, prawn culture), forestry operations (use of shallow estuaries and backwater margins for timber stacking and haulage by wood based industries), retting of coconut husks and coir making, industrial development of the coastal areas, development of infrastructure facilities such as roads, rail line, ports etc. are important which led to destruction of mangroves along the Kerala coast.

Mangroves of Karnataka coast are highly degraded. These are formed on shallow coastal waters, estuaries or lagoons. *Avicennia officinalis, A. marina, Excoecaria agallocha, Acanthus ilicifolius, Rhizophora mucronata, Ceriops tagal, Sonneratia caseolaris, Kandelia candal* and *Heritiera littoralis* are frequent species forming pure or mixed associations.

The mangroves of Goa region cover about 200 km^2 consisting of seven estuaries of which Zuari, Mandovi, and the Cambarjua canal connecting them with harbour cover about 75% mangrove area. *Avicennia officinalis, Rhizophora mucronata* and *Sonneratia alba* are the prominent species which are associated with *R. apiculata, S. caseolaris, Kandelia candal, Bruguiera gymnorrhiza, B. parviflora, Aegiceras corniculatum, Excoecaria agallocha and Acanthus ilicifolius*. These stands are not rich in their biodiversity due to the excessive biotic prssure but during recent times good efforts have been made to regenerate mangrove stands by planting the trees manually.

The mangroves of Maharashtra coast are under tremendous biotic pressure at almost all the sites. Many species which were reported as common have become rare or all together absent on many of these sites. For example, *Lumnitzera racemosa* reported to be common in Bandra area is not seen now. Species of *Avicennia* which are tolerant to biotic pressure are found as stunted bushes. *Kandelia candal* has gone out of site at many places. Species of *Avicennia, Rhizophora Mucronata*, species of *Sonneratia, Kandelia candal, Bruguiera gymnorrhiza, Aegiceras corniculatum, Excoecaria agallocha* and *Acanthus ilicifolius* are the prominent species found in the region. At Elephanta Island *Avicennia alba* and *Sonneratia caseolaris* are the only surviving trees. *Acanthus ilicifolius* bushes are also seen. At many localities bushes of *Avicennia marina* are most prominent showing the capacity of tolerance to biotic pressure. All along Gujarat coast the Rhizophoraceae element is very rare. Species of *Avicennia* form gregarious but stunted stands. At some places species of *Aegiceras, Bruguiera, Rhizophora, Acanthus*, and *Sonneratia* are found as remanants.

The Andaman and Nicobar Islands harbour some of the best developed mangroves which are comperatively less disturbed. Their is distinct zonation pattern. *Rhizophora apiculata, R. stylosa*, and *R. mucronata* are the prominent towards sea where others can not stand. Species of *Bruguiera, Sonneratia, Xylocarpus, Aegiceras, Avicennia, Ceriops*, and *Scyphiphora* are common in middle zone while species of *Excoecaria, Lumnitzera, Cynomitra, and Heritiera* are more common towards land. *Nypa fruticans, Acanthus* spp., and *Phoenix paludosa* are common along creeks. The distribution of *Acanthus ebractatus, A. volubilis, Lumnitzera littorea, Bruguiera sexangula, Rhizophora stylosa and R. lamarckii* is interesting in these islands as these are not frequent elsewhere. In recent years due to developmental activities these stands also are under high biotic pressure and many species are becoming rare. For example, *Lumnitzera racemosa* was frequnt at Carbins' Cove and Wandoor is no more seen there.

Nypa, Heretiera, Phoenix and *Xylocarpus* are found nneither in Bombay region nor in South India. Based on a detailed survey a comprehensive list of 37 exclusive mangrove species along with their distribution in different coastal areas has been prepared (Table

11.6). It is interesting to observe that a general survey has has indicated that many species which were predominent sometimes back in different regions are either not found there or are very rare. In general all the stands including of the Andamans are highly disturbed and need immediate attention. Those stands which survive need legal protection and attempts should be made to regenerate them artificially. Where ever it is possible, the seedlings of important species like *Avicennia* may be planted. Besides the over exploitation of these resources there are other reasons also which are responsible for the dwindling of mangrove ecosystems. Khushoo (1986) has stated that role of pollutannts is very detrimental for the deterioration of these coastal or deltaic vegetations in Bombay, Madras, Cochin, Ernakulum to Trivandrum, Andhra Pradesh, Kranataka and Orissa. Yadav (1985) also mentioned that out of 397 tonnes of pollutants per day from Calcutta metropolitan city, about 23 tonnes pollutants are discharged through river Hooghly everyday and the majority of these pollutants are also discharged through the Kulti river (Sakha Bidyadhari) every day. All these pollutants are directly or indirectly coming to the estuarine water of the Sunderbans, which have also adverse effect on the mangrove ecosystems of the Sunderbans.

Associate Flora and Fauna

Marine Flora: Marine algae are found everywhere around the coast expect in very muddy and steep shores. The sandy beaches with shallow waters strewn with rocks and boulders and the coral reefs further into the sea, facilitate luxuriant growth of marine algae. We are concerned here with those floating algae which come on shores and are trapped in the roots of mangroves. These become a part of the food-web in mangrove ecosystem. The common genera collected from the coastal waters of Bay Islands include *Acetabularia, Caulerpa, Codium, Dictyosphaeria, Enteromorpha, Halimeda-(Chlorophyceae); Padina, Sargassum, Turbinaria*-(Phaeophyceae); and *Ceramium*-(Rhodophyceae). *Cymodocea, Enhalus, Halophila* and *Thalassia* are the common "sea grasses" (Angiosperms) which grow in sea water and are especially common on sandy soils in quite waters up to a metre or so beneath the ebb level.

In a study of mycoflora of soil around pneumatophores of *Sonneratia* in the Andamans Chauhan et al. (1980) isolated 60 fungal species out of which, species of *Aspergillus, Penicillium, Trichoderma, Fusarium* and *Rhizoctonia* were the most frequent fungi. Most of these were characteristic microfungi of acidic soils. In a separate study with mosses asociated with mangroves and littoral trees, *Bartramia madurensis, Bryum capillare, Calymperes andamense, C. manii, C. kurzianum, C. nicobarense, C.punctuatum, Ectropothecium manii, E. ramuligerum, Fissidens asperisetus, F. ceylonensis, F. robinsonii, Garckea phascoides, Hyophila involuta, Hypnum plumeforme, Isopterygium andamanicum, I. microplumosum, Leucobryum sp., Leucophanes nicobaricum, Philonotis turneriana, Thyvidium andamense, T. manii, T. nicobaricum and Vesicularea kurzii* are among prominent species widely distributed on the trunks and branches in these forests. Many of these species are endemic in islands and need more attention.

Table 11.6. Distribution of mangroves in different coastal regions of India

Species			*East coast*					*West coast*			*A&N ISL.*
1	*2*	*3*	*4*	*5*	*6*	*7*	*8*	*9*	*10*	*11*	*12*
Acanthus ebracteatus	-	-	-	-	-	-	-	-	-	-	o

1	*2*	*3*	*4*	*5*	*6*	*7*	*8*	*9*	*10*	*11*	*12*
A. ilicifolius	f	f	f	la	la	f	o	f	f	f	f,la
A. volubilis	-	-	-	-	-	-	r	-	r	-	-
Aegialitis rotundifolia	-	-	-	-	-	-	f	o	-	-	+?
Aegiceras corniculatum	o	f	o	o	r	-	f	-	o	f	f
Avicennia alba	r	f	-	f	r	-	o	f	o	la	f
A. marina	la	la	o	la	o	-	f	f	f	la	f
A. officinalis	f	f,la	la	a	la	-	f	f	f	la	f
Bruguiera cylindrica	-	o	r	-	-	-	o	f	o	o	r
B. gymnorrhiza	f	f	f	o	f	f	f	f	f	o	vf
B. parviflora	-	o	o	-	r	r	o	f	-	r	la
B. sexangula	-	-	-	-	-	-	o	r	r	-	o
Ceriops decandra	-	-	-	-	o	r	la	o	o	o	r
C. tagal	o	f	o	f	-	-	o	o	-	o	a
Cynomitra iripa	-	-	-	-	-	-	r	-	-	-	o
C. ramiflora	-	-	-	f	-	-	o	o	-	o	a
Excoecarcia agallocha	r	f	f	f	p	-	f	f	f	ff,	la
Heritiera fomes	-	-	o	-	-	-	o	-	-	-	-
H. kanikensis	-	o	-	-	-	o	-	-	-	-	-
H. littoralis	-	-	r	r	-	-	-	o	-	-	f
Kandelia candel	-	o	o	o	-	-	vr	f	-	-	+?
Lumnitzera littorea	-	-	-	-	-	-	-	-	-	-	la
L. racemosa	-	fo	-	-	r	o	f	o	f	o	f
Nypa fruticans	-	-	-	-	-	-	o	?	-	-	v,la
Phoenix paludosa	-	-	-	-	-	-	f	o	-	o	o
Rhizophora apiculata	f	f,la	f	f	f	-	f,la	a	a	la,f	a
R. lamarckii	-	-	-	-	-	-	-	-	-	-	vr
R. mucronata	la	o	a	la	la	la	la	la	la	la	f
R. stylosa	-	-	-	-	-	-	-	-	o	r	o
Scyphiphora hydrophyllacea											
Sonneratia alba	-	f	f	o	-	-	-	f	f	-	o
S. apetala	-	o	-	-	o	-	o	o	la	o	o
S. caseolaris	-	o	o	o	-	-	o	o	-	-	f

1	2	3	4	5	6	7	8	9	10	11	12
S. griffithii	-	r	-	-	-	-	r	o	-	-	r?
Xylocarpus granatum	-	o	-	-	-	-	f	o	r	o	f
X. mekongensis	-	-	-	-	-	-	f	o	-	f	r
X. moluccensis	-	-	-	-	-	-	f	r	r	-	o

* State/Union Territories of: (1) Gujarat, (2) Maharashtra, (3) Goa, (4). Karnataka, (5) Kerala, (6) Lakshadweep, (7) West Bengal, (8) Orissa, (9) Andhra Pradesh, (10) Tamil Nadu; (11). Andaman & Nicobar Islands.

Distribution letters depict : a = abundant/dominant, f = frequent, o = occasional, rr = rare, v = very, 1 = local, and -= absent. a > f > o > r

Fauna : Sunderban mangrove forests are the site of dangerous maneating "Royal Bengal Tiger" (*Panthera tigris*), the most ferocious crocodiles (*Crocodilas indigo, C. porosus*), dolphin (*Platensita ganfetica*), sharp footed hogs (*Sus salvinus*), spotted deer (*Axis axis*), migratory birds, poisonous snakes (*Najas najas, Viper russel*), and many insects along with enormous criss-cross rivers and creeks, Naskar and Chakraborty (1984) reported about 120 euryhaline fishes and crustaceans. It is interesting to find the sites where fishes (*Boleophthalamus*) creep up the trees, and the tides in the same creeks run in opposite direction twice a day.

The mangrove of Andaman and Nicobar islands are supporting a rich variety of animal life. Though it has not yet been possible to identify and catalogue all the species, from the studies so for, about 5100 species belonging to all major groups of animals have been recorded from the islands (Subba Rao, 1989). There are about 100 species of freshwater animals which include mostly introduced species. Terrestrial animal communities are represented by 2100 species and marine include 2900 species. The former include 1500 species of insects, while in the latter category molluscs and fishes together consist 1540 species in the Bay Islands. Out of the species mentioned above quite good number of them are associated with mangroves. So far, 8 species of polychaetes, 100 of molluscs, 59 of crustaceans, 6 of echinoderms, 2 of sipunculids, 35 of fishes, 3 of amphibians, 7 of reptiles, 53 of birds and 8 of mammals have been reported by Das and Dev Roy (1989) to be associated with mangroves of the Bay Islands. An overall analysis of the distribution of the diverse group of animals indicates quite a good number of endemics among them (Table 11.7). Over 300 new and interesting species of the lower invertebrate group of animals comprising the groups Nematoda, Copepoda, Annekida, Turbellaria, Gastrotricha, Kinorhyncha, Ostracoda, Isopoda, Amphipoda, Tardigrada, Mollusca and Echinodermata have been recorded and prescribed by Rao (1980).

There are certain borers on mangroves of these islands including molluscan borers (Shipworms or Teredinids) and crustacean borers (Sphaeroma) which thrive well within the tidal limits. Tikader et al. (1986) found that many of the animals of mangroves areas are common to mud flats including fiddler crabs, hermit crabs, cerithids littorinids on the leaves of the plants and mud skippers (*Periophthalmus and Boleophtalmus*) are common in the muddy substratum of mangroves and a frog *Ranna Cancrivora* is also commonly found in mangrove swamps. Gastropod molluscs are common as epifauna whereas bivalves constitute infaunal elements and bore into the wood and littorinids (*Littorina scabra*), netritids (*Nerita atropurpurea, N. articulata*) and edible oysters are common on the branches of prop roots of mangrove trees. *Sesarma longipes* and *Cardiopoma carnifex* are the common burrowing crabs.

An aggregation of the mangrove snail *Terebralia palustris* is quite widespread in the muddy substratum of many sites of the islands. Many genera of molluscs such as *Cassis, Chicotes, Haliotis, Lampis, Strombus, Tridaena, Trochus* and *Turbo* are usually collected from sea-shores. The commercially important icthyofauna found in mangrove areas are *Chanos, Mugilcephalus, M. dussumieri, Liza macrolepis, Lates calcarifer, Serranus* spp., etc. which are caught by hooks and nets for domestic consumption by the people. The edible prawns like *Penaeus monodon, P. semisulcatus, P. merguiensis, Metapenaeus dobsoni, M. ensis and M. affinis* are caught with the help of nets during the receding of high tide and the edible crab *Scylla serrata* which is a delicacy, is caught during the low tides (Dagar 1987, Balachandra, 1988).

Table 11.7 : Distribution of diverse groups of animals in aquatic, semi-aquatic, and terrestrial habitats of Bay Islands

Group	*Species report in Islands*	*No. of endemics*	*Associated with mangroves*
(A) Aquatic annd semi-aquatic components			
Sponges	70	-	-
Coelennterates	147	-	-
Corals	179	-	-
Polychaetes	161	-	8
Earth worms	21	7	2
Leeches	2	-	-
Molluscs	941	87	100
Crustaceans	407	-	59
Echinoderms	331	-	6
Fishes	750	24	35
(B) Terrestrial componnents			
Mammals	58	33	8
Birds	242	95	53
Reptiles	83	23	7
Amphibians	18	6	3
Insects-mites	1500	121	35
Termites	39	19	-
Spiders-scorpians	62	28	-
Pycnogonoides	8	-	-
Cenntipedes	17	-	3
Millipedes	5	-	-
Meiofauna	324	193	43

Source: Subba Rao (1989); Das and Dev Roy (1989).

About 80 species of reptiles and amphibians are known from these islands and of these at least 31 come to the sea shore at some part of their lives (Tikader et al., 1986). Amongst them salt water crocodiles and sea water turtles are of greater significance. The sea turtles are internationally threatened species and out of the existing seven species, five, namely, the Green turtles (*helonia mydas*), the Hawks bill (*Eretomochelys imbricata*), the Leatherback (*Dermochelys coriacea*), the Loggerhead (*Caretta caretta*) and the Olive Ridley (*Lepidochelyss olivacea*) are found in the Andaman waters (Khan, 1987). These turtles were hunted to a great extent in the past for food by settlers, tribals and other poachers as a result of which their population has decreased alarmingly. Salt water crocodile (*Crocodylus porosus*) has been enlisted as one of the most endangered species in the Red Data Book. It was once found in abundance in these islands but during the early period of settlement, they were hunted ruthlessly for their skin and as a result, their population has decreased abruptly and these are seen only in limited mangrove areas. Water monitor lizard (*Varanus salvator and V. amanensis*) inhabit mangroves and coastal littoral zones and have suffered badly because in past they were hunted mainly for flesh and their population has become exceedingly low. Sometimes the endemic wild boars (*Sus scrofa var. andamanensis and nicobarensis*) are also seen in mangrove swamps.

Sea shores particularly the mangroves also offer a veritable feeding and breeding ground to various kinds of wild life including various kinds of birds. Many of the birds are the seasonal visitors to these islands. The mangrove vegetation and associated communities form a food chain. The broad base of the food pyramid is formed by the mangrove plants. The ground dwelling surface animals such as hermit crabs, mud whelks and other snails; burrowing crabs, shrimps, worms, tree dwellings such as oysters, insects, snails and birds derive food from leaves, branches, flowers and fruits of mangroves. Some of the littorinids feed on debris which in turn form the diet of mud skippers. Some edible crabs become the diet of a part of fidder crabs which along with other crabs are the favourite food of frogs and snakes. Serpent eagle and some other birds are prey birds feeding on crabs and other animals. Thus a complete and complex food–web is established in mangrove swamps.

We may draw a conclusion that the saline habitats particularly the coastal (mangroves and other associate formations) are rich in biodiversity but these are under tremendous anthropogenic pressure and dindling at fast rate. These need immediate attention. We must also keep in mind that these are very important resources and a large coastal population of dependent upon these resources for their livelihood. They not only, get fuel and timber from these put also get directly or indirectly food, honey and other forest and sea produces for their livelihood. These also play a vital role for conserving coast line and are also life support system for coastal aquatic fishery and other wild life. In order to prevent global collapse we need a far-reaching environmental and cultural revolution which should lead us to a shift from consumption to sustainable utilization. Conservation and protection of the coastal and marine resources.

ACKNOWLEDGEMENTS

I am thankful to Director, Central Soil Salinity Research Institute for encouragement and necessary facilities. I am also thankful to Indian Council of Agricultural Research, New Delhi for providing addition grant in the form of A.P. Cess-Fund Scheme.

LITERATURE CITED

Aronson, J.A. (1989). Haloph: Salt Tolerant Plants for the World–A Computerized Global Data Base of Halophytes with Emphasis on Their Economic Uses. Univ. of Arizona Press, Tucsonn Az, USA.

Balachandra, L. (1988). A comprehensive account of the mangrove vegetation of Andaman and Nicobar Islands. Indian Forester 114 : 741-751.

Banerjee, M., Sen, P.K. and Dastidar, A.G. (1984). On the depositional condition of Halocene sediments of Bengal Basin with remarks on environmental condition of the soft grey deposition in Calcutta. Indian Geotechnical conference (Calcutta), Dec. 1984, pp. 63-69.

Chand Basha, S. (1992). Mangroves of Kerala-a fast disappearing asset. Indian Forester 118 : 175-190.

Chauhan, S.K., Tyagi, V.K. & Nagar, N.L. (1980). Mycroflora of soil around pneumatophores of *Sonneratia acida* L. in Andaman Islands. Journal of Indian Botanical Society 59 : 281-285.

Dagar, J. C. (1982). Some ecological aspects of mangrove vegetation of the Andaman and Nicobar Islands in India, Sylvatrop, Phillippine Forestry Research Journal 7 : 177-216.

Dagar, J.C. (1987). Mangrove vegetation, its structure, ecology, management and importance with special reference to Andaman and Nicobar Islands. Proceedings Symposium on Management of Coastal Ecosystems and Oceanic Resources of the Andamans. Andaman Science Association CARI, Port Blair pp. 8-23.

Dagar, J.C. (1995). Ecology of Halophytic vegetation in India : a review. Internation Journal of Ecology and Environmental Sciences 21 : 273-296.

Dagar, J.C., Mongia, A.D. and Bandyopadhyay, A.K. (1991). Mangroves of Andaman and Nicobar Islands. Oxford & IBH Publishing Co. Pvt. Ltd., New Delhi.

Das, A.K. and Dev Roy, M.K. (1989). A General Account of the Mangrove Fauna of Andaman and Nicobar Islands. Zoological Survey of India, Calcutta.

Government of India. (1987). Mangrove in India-Status Report. Ministry of Environment and Forestry, New Delhi.

Khan, I.H. (1987). Conservation of endangered marine species in the Andamans pp. 86-70. In : Proceedings Management of Coastal Ecosystems annd Oceanic Resources of Andamans. Andaman Sciennce Association, Port Blair.

Khushoo, T.N. (1986). Presidential address of 63rd session of Indian Science Congress Association, New Delhi.

Naskar, K.R. (1991). Mangrove ecosystemss in the Indian sub-continent with special reference to the Sunderbans, W.B. Journal Indian Society of Coastal Agricultural Reearch 9 (1-2) : 455-462.

Naskar, K.R. and Chakraborty, N.M. (1984). Studies on the economic fauna from the Sunderbans delta in West Bengal. Journal Indian Society of Coastal Agricultural Research 2(1) : 56.62.

Naskar, K.R. & Guha Bakshi, D.N. (1987). Mangrove Swamps of the Sunderbans. Naya Prokash, Calcutta.

Naveh, Z. (1995). From biodiversity to ecodiversity: new tools for holistic landscape conservation. International Journal of Ecology and Environmental Sciences 21 : 1-16.

Rao, G.C. 1980. On the zoogeography of the intertidal meiofauna of the Andaman and Nicobar Islands, Indian ocean. Records of Zoological Survey of India 77 : 153-178.

Sen, D.N. Rajpurohit, K.S. and Wissing, F.W. (1982). Survey and adaptive biology of halophytes in Western Rajasthan, India. pp 61-78. In : Sen, D.N. and Rajpurohit, K.S. (eds). Contribution to the Ecology of Halophytes. Dr. W. Junk Publishers, the Hague.

Singh, N.T. (1994). Land degradation and remedial measures with references to salinity, alkalinity, waterlogging and acidity. In : Deb, D.L. (ed.) Natural Resources Management for Sustainable Agriculture and Environment. Ankar Pub., New Delhi, pp. 442.

Szabolcs, I. (1989). Salt Affectd Soils. CRC Press, Inc. Boca Raton, Florida, USA.

Subba Rao, N.V. 1989. Fauna of Andaman and Nicobar Islands : diversity, endemism, endangered species and conservation strategies. pp. 74-82. In :

Saldanha, C.J. (Ed.) Andaman, Nicobar and Laksha Dweep-An Environmental Impact Assessment. Oxford & IBH Publishing Co. Pvt. Ltd. New Delhi.

Tikader, B.K., Daniel, A. and Subba Rao, N.V. (1986). Sea Shore Animals of Andaman & Nicobar Islands. Zoological Survey of India, Calcutta.

Yadav, J.S.P. (1987). Presidential address in National Seminor on Problem and Management of Coastal Agricultural Research, Indian Society of Coastal Agriculture, Canning Town.

Chapter - 12

Environment and Diversity of Mangrove Plant Communities in India

*L.K. Banerjee**

ABSTRACT

Mangrove Community in India with its 6740 Km^2 littoral area along the deltas, estuaries, backwaters and lagoons represent the third largest formation of the world, supporting more than 116 angiosperms, ferms, algaes and fungi. Mangrove plant Communities which are restricted up to the limit of saline tide water flow are found to be well developed along the East costal environment in India and Andaman & Nicobar Islands in comparison with that of the West coastal environment. To provide better knowledge regarding the different environmental factors and the mangrove plant diversities, major environmental parameters related to the change of plant diversity pattern have been discussed in this paper.

INTRODUCTION

Sheltered part of Indian coast lines in the tropical environmental conditions is well-known for greater diversity of flora and fauna as well as a greater degree of specialization in structure, organization and functioning of their constituent ecosystem. Among the latter, mangrove ecosystem commands the highest importance by virtue of their biological productivity, specialized adaptive diversity, complexity in the ecological processes and finally the importance of the biodiversity resources which are used directly or indirectly. Other amenities provided by these biodiversity resources for protection and conservation of the environment is so much valued that the cost could not be evaluated with the help of monetary measure. About 6740 Sq. Kms. littoral region including deltas, estuaries, backwaters, bay islands and lagoons in India is protected by extensive cover of mangals (mangrove communities), being the third largest formation of the world after Indonesia and Australia. In India, distribution of mangrove diversity is found in Andaman & Nicobar group of oceanic islands, Sundarbans in the Gangetic delta of West Bengal; Mahanadi deltaic complex of Orissa Coast; Coringa, Godavari and Krishna delta of Andhra Pradesh; Cauvery

*Botanical Survey of India, 1-Sudder Street, Calcutta-700016

delta and Tanjavur in Tamil Nadu; Narmada, Tapi estuaries, Gulf of Cambay, Saurashtra and Kutch in Gujarat; Bombay, Ratnagiri, Malvan, Devgad and Vijayadurg in Maharashtra; Mandovizuari estuary of Goa; Coondapur, Hannover and Malpe in Karnataka and Cochi, Vembanad, Quilon, Trivandrum, Kananore, Kozhikod and Kottayam in Kerala state. The species diversity is very low in this ecosystem due to dominance of large physical forces of saline tide water and lack of stable substrate. Total 116 species of higher plants including a number of economically important species have been recorded from Indian Mangrove Forests. (Banerjee *et al.* 1989).

The most abundant mangrove species are *Avicennia marina* (Forsk.) Vierh. *A. alba* Bl., *A. officinalis* L., *Rhizophora apiculata* Bl., *R. mucronata* Lamk., *Bruguiera Gymnorrhiza* (L.) Savingay, *B. Parviflora* (Roxb). Wt. & Arn. Ex Griff., *B. cylindrica* (L.) Bl., *Excoecaria agalocha* L., *Ceriops decandra* (Griff.) Ding Hou, C. tagal (Perr.) C.E. Robin., *Xylocarpus granatum* Koen., *X. mekongensis* Pierre., *Heritiera fomes* Buch.–Ham., *Sonneratia apetala* Buch.-Ham., *S. alba* J. Sm., *S. Crasiolaris* (L.) Engl., *S. griffithii* Kurz, *Lumnitzera racemosa* Wild., *Aglaia cucullata* (Roxb.) Pellegrin, *Xylocarpus molluccensis* (Lamk.) M. Roem., *Aegiceras corniculatum* (L.) Blanco, *Aegialitis rotundifolia* Roxb., *Acrostichum aureum* L., *Kandella candel* (L.) Druce, *Brownlowia tesa* (L.) Kosterm. The most extensively occurring palms and grasses are *Phoenix paludosa* Wurumb., *Nypa fruicans* Roxb., *Porteresia coarctata* (Roxb.) Tateoka and *Myriostachya wightiana* (Nees ex Steud.) Hook. F. 7 species of orchids, 3 species of ferns are found in the forest. There is also a large number of climbers, creepers and shrubs in addition to the numerous algal and fungal species that occur in the wet forest floor and the tree trunks. It has been reported that a total of 425 vertebrate (excluding fishes) in the mangrove regions including 49 species of mammals, 53 species of reptiles, 3156 species of birds and 8 species of amphibians, in addition to moths, butterflies, wasps, beetles and other invertebrates (Hussain 1996). Though total number of mangrove species diversity is more along the Mahanadi deltaic complex covering an area of only 251 Sq. Km. but species richness is largest in the Sundarbans of the Gangetic delta consisting of an area of 4266 Sq. Km. It is observed that mangrove diversity gradually tapers from the Sundarban-Mahanadi complex to the Andaman & Nicobar Islands and to the southwards of the Godavari, Krishna and Cauvery delta of the East Coast and is much less along the west coastal regions from the down south in Kerala towards Karnataka, Goa, Maharashtra and Gujarat coast lines. Resulting investigation reveals that different environmental factors which maintain ecological processes in the mangrove ecosystem are responsible for such type of differences. Ecological processes involved in the saline tidal areas where plants, animals, soil microbes and surrounding physical environment are interconnected by a simple or complex processes through which a continuous exchange and assimilation of energy occurs. In the mangrove ecosystem, the internal processes like energy fixation, biomass accumulation, mineral cycling, organic deposition etc. are closely related to some external environmental factors like adequate supply of water, pool of available nutrients, stability of substrates etc. Both the internal and external processes in the ecosystem are responsible for biological diversity, species richness, structural and functional activities and pattern of developments. River discharge is a prime source of nutrition from the fresh water flow to the mangrove system. It also serves balancing net evaporation loss from its surface. In India, different rivers such as, the Ganga, the Mahanadi, Godavari-Krishna and Cauvery along the east coast and Narmada, Tapi and other smaller ones along the West Coast maintain distinct changes in their quality of discharges. Among these, the river Ganga has maximum annual discharge (971 km^3 /year) with about 1130 t/ Km^2/yr sediment yield (Milliman & Meade 1983). The river Godavari with 84 km^3/yr and 310 t/km^2/yr. Other rivers have considerably lower discharges and sediment yields in comparison with the above rivers.

In most of the case inorganic mineral ions by the unimpended flow of fresh water and microbial decay of materials by larger benthic fauna, specially the crabs which transfer this organic materials into soluble inorganic form, are the main source of nutrient pool to the mangrove system. Seasonal fresh water runoff, tidal velocity and sediment load have very positive effect to the deposition of sediments in the intertidal zone which can provide new sites for colonisation of mangroves. Due to topographic diversity, variation of river discharges, differences in tidal amplitude and the amount of fresh water flow in different deltas, estuaries and backwater system differ from region to region and also from distance to distance even in the same estuary and delta. This is probably the major environmental factors accompanied with various unstable nature of the substrate on which mangal diversity differs.

On the basis of topographic diversity, degree of fresh water flow (salinity ranges_, river discharges, sediment load and tidal amplitude, mangals in India are divided into two major types, (1) Outer Estuarine Mangals and (2) Inner Estuarine or Riverine Mangals. Again with the help of small scale variation in environmental and ecological processes, the inner estuarine mangals which extend up to the limit of saline tidal water can be distinctly divided into three types, (1) True Mangals, (2) Semi Mangals and (3) Transitional Mangals. (Banerjee 1990. Rao, Sastry 1974).

METHODS

Field survey were conducted in different deltas and estuarine regions in India in different seasons. In all the mangrove forests, outer estuarine and inner estuarine sites were chosen for studying community structure with the help of line transact methods and sample plots were taken only at accessible areas having as much community variations as possible. Importance Value Index (I.V.I.) for each species was calculated following Misra (1968) and Philips (1959). Soil samples were analysed following Jackson (1958) and Piper (1942). Water salinity (ECE mmohs/cm.) was recorded with the help of conductivity metre. Seasonal tide levels were verified from the Indian Tide Table Data (Survey of India 1970-75).

RESULTS

Outer estuarine environment show less protected from the direct action of the bay and exhibits higher impact of web action sediment load, maximum siltation and high water salinity conditions in comparison with the inner estuarine areas. The mangrove species which are capable to excrete excess salt with the help of petiolar salt glands or increased water storage mechanism in the leaf surface are only found restricted for withstanding the situation of the environment. Composition of species diversity in this community according to the value of I.V.I. and corresponding environmental factors are shown in Table 12.1.

Table 12.1. Composition of plant diversity with I.V.I. (Estuarine mangals)
* Length of transact = 38 metre

Species	*F*	*D*	*R.F.*	*R.D.*	*R.A.*	*I.V.I.*
1	2	3	4	5	6	7
Avicennia marina	5	40	8.62	9.32	14.24	32.18
Sonneratia griiffithii	4	27	6.89	6.29	18.62	31.80
Ceriopos tagal	4	25	6.89	5.42	0.82	13.13

1	2	3	4	5	6	7
Lumnitzera racemosa	2	14	3.44	3.26	3.02	9.72
Bruguiera perviflora	3	9	5.17	2.9	2.91	9.17
Bruguiera cylindrica	2	19	3.44	4.42	0.06	7.92
Aegialitis rotundifolia	2	6	3.44	1.39	0.04	4.87

Impact of environmental factors

Soil particle size depth 0-20 cm

Clay (%)	*Silt (%)*	*Fine Sand (%)*	*Coarse sand (%)*	*pH (%)*	*Organic matter*	*EC-mm ohs/cm ranges*
31	44	19	7	6.8	2.5	30-12

Water Salinity EC				*Tide level in m*			
J-M	*A-J*	*J-S*	*O-D*	*J-M*	*A-J*	*J-S*	*O-D*
33	29	18	26	3.5	4.5	6.0	3.5

*The data from 10 line transacts were computed

Inner estuarine regions or riverine areas from the base of funnel-shaped estuarine mouth towards the 2-10 km distance are dissected by several creeks and cannals forming different fringes, islets and tidal floods. The vigour of the sea-surf, strong winds and continuous salt spray from the bay are not so effective over this regions due to sheltered conditions. This seems to be most favourite habitat of many mangrove species and floristic diversity becomes much more rich in this type. Periodic input of fresh water flow from the upstream regions, change of tidal levels, water salinity and sediment type due to distance from the sea probably the reasons for good growth and diversity of the mangroves along these region.

Mangrove species which are found more or less restricted in this regions with the adaptive characters of stilt roots and vivipary for withstanding this environmental conditions are usually named as True Mangrove Communities or "EU-Mangals". The composition of the species diversity and the different environmental factors which differ from the outer estuarine regions are shown in Table 12.2.

Table 12.2 : Composition of plant diversity with I.V.I. (Eu mangals)
Length of transact = 38 metre

Species	*F*	*D*	*R.F.*	*R.D.*	*R.A.*	*I.V.I.*
1	2	3	4	5	6	7
Rhizophora apiculata	5	33	8.62	7.69	19.42	35.73
Rhizophora mucronata	5	36	8.62	8.39	15.32	32.33
Excoecaria agallocha	5	40	8.62	9.32	14.24	32.18
Ceriops decandra	5	94	8.62	21.91	1.60	32.13
Avicennia officinalis	4	27	6.89	6.29	18.62	31.80
Bruguiera gymnorhiza	3	19	5.17	4.42	10.81	20.41

1	2	3	4	5	6	7
Kandelia candel	3	22	5.17	5.12	10.1	20.39
Sonneratia caseolaris	1	6	1.72	1.36	11.21	14.29
Xylocarpus granatum	2	19	3.44	4.42	8.36	14.22
Aegiceras corniculatum	2	14	3.44	3.26	3.02	9.72
Phoenix paludosa	1	6	1.72	1.39	5.9	9.01
Sonneratia apetala	3	3	5.17	0.69	2.21	8.07
Brownlowia tersa	2	19	3.44	4.42	0.06	7.92
Xylocarpus moluccensis	1	3	1.72	0.69	2.7	5.11
Avicennia alba	2.	6	3.44	1.39	0.04	4.87
Finlaysonia obovata	1	2	1.72	0.46	0.01	2.19

Impact of environmental factors:

Soil particle size depth 0-20 cm

Clay (%)	*Silt (%)*	*Fine Sand (%)*	*Coarse sand (%)*	*pH (%)*	*Organic matter*	*EC-mm ohs/cm ranges*
42	30	22	3	6.9	3.0	24-9

Water Salinity EC				*Tide level in m*			
J-M	*A-J*	*J-S*	*O-D*	*J-M*	*A-J*	*J-S*	*O-D*
25	21	11	22	3.0	3.5	5.5	2.5

*The data from 10 line transacts were computed.

Further, towards the middle part of the inner estuarine regions where influence of fresh water flow, water salinity and sediment type distinctly differ from the above type are found associated with different type of mangrove diversities and different adaptive capacities such as, development of woody breathing roots, knee roots, snake type lenticellate roots etc. for adjusting nonaerated soil condition of the habitat. This type is named as "Semi Mangals". Species composition and different environmental factors are shown in Table 12.3.

Table 12.3. Composition of plant diversity with I.V.I. (Semi mangals)

* Length of transact = 38 metre

Species	*F*	*D*	*R.F.*	*R.D.*	*R.A.*	*I.V.I.*
1	2	3	4	5	6	7
Heritiera fomes	4	40	9.09	13.12	16.4	38.6
Excoecaria agallocha	4	35	9.0	11.6	15.0	35.4
Bruguiera sexangula	4	22	9.0	7.0	10.2	26.2
Avicennia officinalis	3	12	6.8	3.8	12.4	23.0
Cynometra mimosoides	2	19	4.5	6.0	9.8	20.3
Aglaia cucullata	3	16	6.8	5.1	8.1	20.0

1	2	3	4	5	6	7
Phoenix paludosa	1	25	2.2	8.6	3.4	13.6
Intsia oiguja	2	16	4.5	5.3	3.8	13.6
Cerbera manghas	2	9	4.5	2.8	4.9	12.2
Xylocarpus mekongensis	1	5	2.2	1.6	4.5	8.3

Impact of environmental factors:

Soil particle size depth 0-20 cm

Clay (%)	*Silt (%)*	*Fine Sand (%)*	*Coarse sand (%)*	*pH (%)*	*Organic matter*	*EC-mm ohs/cm ranges*
50	29	17	2	7.1	3.5	15-6

Water Salinity EC				*Tide level in m*			
J-M	*A-J*	*J-S*	*O-D*	*J-M*	*A-J*	*J-S*	*O-D*
20	16	6	17	2.0	3.0	3.5	2.0

*The data from 10 line transacts were computed

Towards more inland part of the inner estuarine region or the hinter land areas where the topography is more or less elevated and devoid of regular tidal flow and the habitat is changed due to better consistency of soil particles, less salinity and more organic matter contains are associated with such mangrove diversities which do not show any morphological adaptive features and they can transit even to the fresh water conditions for certain period. This is named as "Transitional Mangals". The composition of species and its environmental factors are show in Table 12.4.

Table 12.4 : Composition of plant diversity with I.V.I. (Transitional mangals)
* Length of transact = 38 metre

Species	*F*	*D*	*R.F.*	*R.D.*	*R.A.*	*I.V.I.*
1	2	3	4	5	6	7
Dalbergia spinosa	4	25	6.89	5.42	0.82	13.13
Acanthus ilicifolius	2	20	4.5	6.4	1.5	12.4
Hibiscus tiliaceous	2	13	4.5	4.1	2.6	11.2
Derris trifoliata	1	9	2.2	2.8	0.4	5.4
Tamanix troupii	1	7	1.72	1.73	0.02	3.37

Impact of environmental factors:

Soil particle size depth 0-20 cm

Clay (%)	*Silt (%)*	*Fine Sand (%)*	*Coarse sand (%)*	*pH (%)*	*Organic matter*	*EC-mm ohs/cm ranges*
25	3	40	24	6.3	4.5	5-1

1		2	3	4	5	6	7
Water Salinity EC				*Tide level in m*			
J-M	*A-J*	*J-S*	*O-D*	*J-M*	*A-J*	*J-S*	*O-D*
18	14	5	15	1.0	2.0	2.5	1.0

*The data from 10 line transacts were computed

DISCUSSION

It is observed that considerable changes in the composition of plant diversity pattern as well as the zonation of mangals from the mouth of estuary up to the limit of salt water tide of all the mangrove regions in India can be recognised with the consequence of changing environmental parameters. Levels of plant distribution and species diversity vary in a predictable way along the gradients of primary environmental variables such as, tidal amplitudes, water salinity, stability of substrates etc. The easily measurable environmental factors should be very useful for rapid identification of species diversity and its zonation areas. The incidence of this environmental variability and species diversity pattern can be well discussed with the present observations. The exposed outer estuarine environment showing maximum percentage of silt, maximum tidal amplitudes and maximum range of salinity (Table 12.1) tend to be more pronounced for certain restricted mangrove species diversity which are adapted with the mechanism of excess salt excretion by salt glands of storing the excess salt in the leaf surface for withstanding this high tidal, salinity bound environment. Similarly, the inner riverine sheltered areas where velocity of tide becomes higher due to innumerable creeks and channels, tidal amplitude ranges from 5.5 to 2.5 m, salinity ranges from 21-25 Ecmmohms/cm and soil particle size dominated by 42% clay (Table 12.2) are providing best habitat of many mangrove diversities which are adapted by means of stilt roots and vivipary, probably to withstand the high velocity bound tidal environment. Further towards the middle riverine part where soil is dominated by 50% clay, salinity ranges from 15-20 Ec-mmohms/cm and tidal amplitude ranges from 3.5-2 m (Table 12.3) appearance of different mangrove diversity is very prominent. Due to more influence of fresh water and presence of stiff mud these plant communities are adapted with woody pneumotophores for withstanding the less aerated condition of the environment. Finally the hinter land areas awards more inland regions where soil is dominated by 40% fine sand, tidal amplitude ranges from 1-2.5 m and salinity ranges from 5-18 Ec-mmohms/cm (Table 12.4.) dominated by different mangrove communities which do not show any morphological adaptation to indicate lessed environmental conditions.

It may be concluded that the components of biodiversity develop some efficiency over its surrounding environment and the efficiency with which each mangrove species tolerate these environmental conditions largely determines its position along the inter-tidal zone and its nature of morphological adaption.

LITERATURE CITED

Banerjee, L.K. (1987). Ecological studies on the mangals of Mahanadi estuarine delta, Orissa. *Tropical Ecology* 82: 117-125.

Banerjee, L.K., A. R. K. Sastry & M.P. Nayar (1989). *Mangroves in India Identification Manual*. Botanical Survey of India, Calcutta.

Banerjee, L.K. & T.A. Rao (1990). *Mangroves of Orissa Coast and Their Ecology*. Bishen Mahendra Pal Singh, Dehra Dun.

Hussain, Z. (1996). *Wetlands, Biodiversity and Ramsar Convention*. Ed. A.J. Hails-Ramsar Convention Bureau, Ministry of Environment & Forest, New Delhi.

Jaction, M.L. (1958). *Soil Chemical Analysis*. Constabl & Co. Ltd. 498. London.

Milliman & Meade (1983). *In Nature and Resources*. UNESCO, 32 : NO. 2.

Misra, R. (1968). *Ecology Workbook*. Oxford & IBH Co. 244.

Phillips, E.A. (1959). *Methods of Vegetation Study*. Henry, Halt and Company Inc.

Piper, C.S. (1950). *Soil & Plant Analysis*. Interscience Publishers Inc. New York, 368.

Rao, T.A. & A.R.K. Sastry (1972). An Ecological Approach Towards Classification of Coastal Vegetation in India I. Strand Vegetation. Indian Forester 98 : 594-607.

SECTION - IV

DIVERSITY OF TERRESTRIAL ECOSYSTEMS : ANALYTIC APPROACH

Chapter - 13

Floristic Diversity in India: An Overview

*G.H. Dar**

ABSTRACT

India has a rich floristic diversity which is due to its characteristic geographical location and an enormous area of more than 3,000 million hectares with a wide range of physiographic and climatic conditions. It comprises 12 biogeographic zones with 16 major and 221 minor forest type, very rich and diverse aquatic and marshland vegetation, varied types of cultivated plants, and a rich assemblage of weeds. In all, over 49,000 species are reported in the Indian flora, which form 12.51% of all known floristic species on the earth. The angiosperm-flora contains about 17,000 species and is the most well known. Of these species, about 15% are trees, a significant proportion belong to aquatic and wetland category, a good representation is of medicinal plants, while about 33% are generally reported as endemics. Overall, about 10% of the flowering plant species are considered as threatened under various threat categories, especially in several hot spots of the country.

Key Words : aquatic and wetland plants, endemic and threatened plants. Floristic diversity, hot spots, species richness, tree flora.

INTRODUCTION

Biological diversity means the variety of life and its processes; including the variety of living organisms, the genetic differences among them, and the communities and ecosystems in which they occur (Keystone, 1991). Floristic diversity, therefore, refers to the variety and variability among constituent entities of flora in a given area.

India is megadiversity country with about 132,219 species of biota (49,219 species belonging to 'traditional' plant groups and 83,000 species to animal groups). The country ranks 12th in the world in its number of species and represents two of the major global realms. It is a signatory to the International Convention like Ramsar, World Heritage, CBD (Convention on Biological Diversity) and CITES (Convention on International Trade in

*Post-Graduate Department of Botany, The University of Kashmir, Srinagar - 190 006, India.

Endangered Species of Wild Fauna and Flora). The enormous diversity in the flora and fauna of this region is due to its large size and extreme variations in physiography and climate.

INDIA: PHYSIOGRAPHY AND CLIMATE

India, has a characteristic geographic location ($8^0.45$ to 37^0. 6N latitude and $68^0.70$ to $97^0.25$ E longitude) with a land frontier of about 15,200 km and a coastline of 7,516 km. With a total area of 3 287 263 sq. km (3 287 million hectares), it is the seventh largest country in the world and the second largest in Asia.

Physiographically, the Indian mainland may be divided into three distinct regions: the Himalaya, the Indo-Gangetic Plain, and the Peninsular India (Deccan Plateau). The Himalaya form a mountain chain along the extreme northern margins of India, extending east-west about 2,500 km in length, varying in width between 240 and 340 km, occupying an area of about 500,000 sq. km in India, and rising from the low-lying Indian plains to the highest peaks at Kanchenjunga (8,586 m) in Sikkim. The Indo-Gangetic Plain comprises alluvial low lands lying parallel to the south of the Himalayan mountain range. It stretches from Assam in the east to Punjab in the West (ca. 2,400 km), contains some of the major Indian rivers and is agriculturally high productive. The Peninsular India lies south to the Indo-Gangetic Plain, the two being separated roughly by the boundary of the Tropic of Cancer. It stretches from 22^0N to 8^0N and is flanked on east by the Eastern Ghats and west by the Western Ghats. A narrow coastal plain with a unique climate and biota lie between the western Ghats and the Arabian sea and between the Eastern Ghats and the Bay of Bengal.

Climatically, the Indian region is extremely diverse due to variations in topography, altitude and overall monsoon patterns. Whereas the country has some of the hottest places on earth, there are also some of the coldest plances; the rainfall ranges from almost zero in the dry desert of the west to the world's highest value in some parts of northeast India. The Himalayan range in the north greatly affects the climate of India. The extreme northern parts have a colder (montane temperature) climate, most of the Deccan Peninsula and northeastern parts are tropical, while the central and northern India reveal a nearly subtropical climate. The higher elevations in the Himalaya show alpine climate followed by arctic toward mountain summits.

FLORISTIC DIVERSITY

Measures of diversity should, ideally, include both the number and abundance of components of diversity over time (Langner and Flather, 1994). However, the data available on floristic diversity in India are of general type and not sufficient to allow a thorough discussion on definite trends over time. Thus, the present account is largely a summarized general survey of floristic diversity in India at vegetational and species richness levels.

VEGETATIONAL DIVERSITY

India is a vast country with varied edaphic conditions and characteristic geographical situation having latitudinal, longitudinal and attitudinal zonations. This results in a great variety of climates and immense types of habitats, supporting an array of vegetational patterns. For purposes of studying floristics, several floristic/biogeographic/ botanical regions have been recognised by various workers in the country (Clarke, 1898; Hooker,

1906; Razi, 1955; Puri, 1960; Chatterjee, 1962; Rao, 1974; Udvardy, 1975; Jain, 1983 1989; Takhtajan, 1986; Rodgers and Panwar, 1988; Sharma, 1993). Rao (1994) has followed a slightly modified version of the classification proposed by Rodger and Panwar, and recognised 12 biogeographic zones in India as given in Table 13.1.

Table 13.1 : Biogeographic zones of India

S. No.	*Biogeographiz zone*	*Political boundary*	*Vegetation types*
1	*2*	*3*	*4*
1.	Trans-Himalaya	Ladakh (Jammu & Kashmir), Lahul and Spiti (Himachal Pradesh).	Alpine scrubs; trees conspicuously absent except planted ones.
2.	West Himalaya	Jammu & Kashmir (except Ladakh), Himachal Pradesh (except Lahul & Spiti), Garhwal and Kumaon (Uttar Pradesh).	Tropical, Subtropical, Temperate, Subalpine, Alpine.
3.	East Himalaya	Sikkim, Darjeeling (West Bengal), Arunachal Pradesh.	Tropical, Subtropical Temperature, Subalpine.
4.	North-East India	Assam, Manipur, Meghalaya, Mizoram, Nagaland, Tripura.	Tropical, Subtropical, Temperate, Subalpine.
5.	The Indian Desert	Mainly western and north-western region of Rajasthan and a part of Kutch region of Gujarat.	Vegetation of sand-dunes, sandy and hammcky plains, gravely and rocky plains, saline tracts.
6.	Semi-arid zone	Punjab plains, Haryana Delhi; fringes of Jammu, Himachal Pradesh; western edge of Uttar Pradesh; eastern Rajasthan and Gujarat; north-west Madhya Pradesh.	Deiduous forests, Open scrub jungles.
7.	Gangetic Plain	From eastern Rajasthan through Uttar Pradesh to Bihar and West Bengal.	Secondary vegetation of xerophytes and open dry grasslands.
8.	Western Ghats	Narrow stretch from south of Tapati river to Kanyakumari along the West coast.	Dry scrubs; dry deciduous, moist deciduous, semi-evergreen and evergreen forests; Sholas', grasslands.
9.	Deccan Peninsula	Most of Maharashtra, Madhya Pradesh, Karnataka, Tamil Nadu, Andhra Pradesh, Orissa, Bihar.	Dry deciduous, moist deciduous, thorn scrub, and semi-evergreen forests.

1	2	3	4
10.	Indian coasts	Coast line of India from Gujarat to Cape Comorin (Kanyakumari) in the west & from Sunderbans to Cape Comorin in the east.	Submerged vegetation, Mangrove and Beach forests.
11.	Andaman and Nicobar Islands	Group of 348 islands in the Bay of Bengal.	Submerged and Strand vegetation; Mangrove, Tidal, evergreen, semi-evergreen and deciduous forests; grasslands.
12.	Lakshadweep islands	Archipelago of 27 small islands in the Arabian sea.	Virtually no natural vegetation left, coconut plantations common.

FOREST COVER AND TYPOLOGY

The figures given for forest cover of India vary. According to Central Forestry Commission (1980) total forest cover of India was 748 million hectares which constituted 22.75% of the total land mass. The Satellite imagery puts the figures at 642 million hectares, some reports maintain that forest cover of India has been depleted to 16-19% of its land mass. According to satellite data of NRSA, India is losing its natural forests at the rate of 1.3 million hectares every year (Kushalapa, 1992).

As per the Forest (Conservation) Act of 1988, 60% of the land area in the hills should be under forest cover, while for the country as a whole it should be 33%. The dense forest cover (More than 40% crown density) for whole of India is only 11.71%. However, dense forest cover for the Himalayan belt is 21.79%, although it ranges from 4.94% in Jammu and Kashmir to 65.13% in Arunachal Pradesh (Table 13.2).

Table 13.2: Dense forest cover in the Indian Himalayas (in sq. km.).

S. No.	*State*	*Geographic area*	*Dense forest cover (>40%)*	*Percentage*
1	2	3	4	5
1.	Jammu and Kashmir	222,236	10,986	4.94
2.	Himachal Pradesh	55,673	8,911	16.00
3.	Uttar Pradesh (8 hill districts)	51,082	17,448	34.15
4.	Sikkim	7,096	2,403	33.86
5.	West Bengal (hill district - Darjeeling)	3,075	1,109	36.06
6.	Assam	78,438	15,842	20.19
7.	Arunachal Pradesh	83,743	54,542	65.13
8.	Nagaland	16,579	3,531	21.29

1	2	3	4	5
9.	Meghalaya	22,429	335	14.73
10.	Manipur	22,327	5,309	23.77
11.	Tripura	10,486	1,825	17.40
12.	Mizoram	21,081	4,279	20.29
	Total	**59,4245**	**129,490**	**21.79**

Source: Anonymous (1993).

Almost all types of forests, ranging from scrub forest to the tropical evergreen rain forest, coastal mangroves to the temperate and alpine scrub occur in India. A comprehensive attempt has been made to classify the forests of India by Champion and Seth (1968). They have recognized 16 major forest types (Table 13.3). Which in turn comprise 221 minor type. Tropical moist and tropical dry deciduous forests constitute the bulk of total forest cover in India, 37% and 28.6% respectively; none of the remaining 14 types reaches 10%. Tropical moist forests, popularly known as the Rain Forests, are species rich habitats and support the maximum biodiversity (Wilson, 1989).

Table 13.3: Major forest type in India (After Champion and Sethi, 1968).

S. No.	*Forest type*	*Area in sq. km.*	*Percentage*
1.	Tropical Wet Evergreen Forest	51,249	8.0
2.	Tropical Semi-Evergreen Forest	26,424	4.1
3.	Tropical Moist Deciduous Forest	236,794	37.0
4.	Littoral and Swamp Forest	4,046	0.6
5.	Tropical Dry Deciduous Forest	186,620	28.6
6.	Tropical Thom Forests	16,491	2.6
7.	Tropical Dry Evergreen Forest	1,404	0.2
8.	Subtropical Broad-leaved Hill Forest	2,781	0.4
9.	Subtropical Pine Forest	42,377	6.6
10.	Subtropical Dry Evergreen Forest	12,538	2.5
11.	Montane Wet Temperate Forest	23,365	3.6
12.	Himalayan Moist Temperate Forest	22,012	3.4
13.	Himalayan Dry Temperate Forest	312	-
14.	Subalpine Forest	18,628	2.0
15.	Moist Alpine Forest	-	-
16.	Dry Alpine Scrub	-	-

Aquatic and Marshland Vegetation

India is quite rich in rivers, lakes and wetlands. These water bodies are distributed in all the regions including the cold and zone of Ladakh to warm and zone of Rajasthan. According to an estimate there are in India a total of 2,175 natural wetlands with an area of 14,60,107 hectares and 65,254 man-made wetlands covering an area of 28,89,266 hectares (Anonymous, 1989).

Aquatic Vegetation

The aquatic vegetation of India is very rich and diverse. All taxonomic groups from phytoplankton through filamentous algae, liverworts, mosses, ferns to woody angiosperms are represented in different wetland types. Many attempts have been made to classify the aquatic vegetation. In the present account it is described under the following four broad categories:–

Free-floating types

These species float freely on the surface of water and usually occur in pure formations, forming carpet-like stretches. They include algae such as *Chara carollina, Nitella hyalina;* pteridophytes such as *Azolla pinnata, Salvinia auriculata, S. natans* and angiosperms such as *Eichhornia crassipes, Lemna aequinoctialis, L. gibba, L. trisulca, Pistia stratiotes, Spirodela polyrrhiza, Wolffia globasa* and *W. microscopica.*

Rooted floating-leaf types

These include species which are rooted in the mud at the bottom and have usually long petioles (and scapes) carrying leaf blades on the surface of water and flower projecting above. The main representatives of this group are *Batrachium trichophyllum, Euryale ferox, Hydrocharis dubia, Marsilea minuta, Nelumbo nucifera, Nymphaea alba, N. nouchali, N. stellata, N. tetragona. Nymphoides indica, N. peltata, Potamogeton natans* and *Trpaa natans* var. *bispinosa.*

Submerged types

These are rooted, completely submerged hydrophytes. The common species in this group include *Blyxa spp., Ceratophyllum demersum, Hydrilla verticillata, Limnophila indica. Myriophyllum spicatum, Najas graminea, Nechamandra altemifolia; Ottelia alismoides, Potamogeton crispus, P. lucens, P. pectinatus, P. pusillus, Vallisneria spiralis* and many species of Podostemaceae.

Emergent types

These are hydrophytes which are rooted at the bottom of water body with major upper portion of the plant well emerging above the surface of water. Some conspicuous representatives of this group are *Acorus calamus, Alisma plantago-aquatica, Butomus umbellatus, Cyperus articulatus, Echinochloa crusgalli, Eleocharis palustris, Juncellus serotinus, Juncus articulatus, J. inflexus, Linoophyton obtusifolium, Monochoria vaginalis, Phragmites karka, P. australia, Polygonum spp., Ranunculus sceleratus, Saccharum spontaeum, Sagittaria trifolia, Scirpus articulatus, S. maritimus, S. triqueter, Sparganium ramosum, Typha angustata, T. elephantina and Veronica anagallis-aquatica.*

Marsh Vegetation

In addition to the a foregiven categories of aquatic plants, marsh vegetation deserves mention. This type of vegetation comprises plants which grow in moist marshy habitats, remaining saturated with water for a major portion of the year. Such plants can also withstand partial inundation and are then difficult to demarcate from other terrestrials which are capable of sustaining under periods of inundation. Marsh habitats can be of two types.

Fresh water marshes

These include natural marshes along the banks of rivers, lakes, streams, ponds, as well as the artificially created paddy fields which remain water-logged for some months of the year. These habitats support a rich variety of plant species belonging to different families. The following are some of the dominant species belongting to this category: *Acorus calamus, Aeschynomene aspera, A. indica, Alternanthera paronychioides, A. sessilis, Anmania baccifera, Ampelopteris prolifera, Bergia nnoides; Coldenia procumbens, Commelina longifolia, Cyperus spp., Dopatrium junceum, Eleocharis spp., Enhydra fluctuans, Equisetum spp., Eriocaulon spp, Grangea maderaspatana, hydrolea zeylanica, Hygrophila auriculata, Juncus spp., Limnophila spp., Ludwigia spp., Lythrum salicaria, Marsilea quadrifolia, Monochoria vaginalis, Neptunia oleracea. Paspalidum geminatum, Polygonum flaccidum, P. glabrum, P. minus, P. nepalense, Prunella vulgaris, Ranunculus sceleratus* and *Scirpus spp.*

Salt marshes

These comprise the saline marshes along the sea coast and support salt-tolerant species such as *Acanthus illicifolius, Arthrocnemum indicum, Atriplex stockii, Peplidium maritimum, Salicornia Brachiata, Sesuvium portulacastrum, Suaeda fruticosa, S. maritima and S. Nudiflora*. This kind of vegetation is markedly different from other types and is usually described under mangrove vegetation (cf. Littoral and Swamp Forest type in Champion and Seth, 1968).

Species Diversity

Species diversity is a function of both the number of species (referred to as richness) and the proportional number of individuals within each species (referred to as abundance or evenness) in a defined area or group of organisms. The following account is an overview of species richness in Indian flora.

According to the great British botanist Sir J. D. Hooker (1906). "The Flora of British India is more varied than that of any other country of equal area in the eastern hemisphere, if not in the globe". A spectrum of over 49,000 species has been reported in Indian flora, which account for 12.51% of the known floristic species of the world. These species are distributed under various taxa which can be classified into four kingdoms and compared to those of the world as shown in Table 13.4.

The Himalaya accounts for a considerable component of the floristic richness in India, harbouring 40.20 % of the total species of fungi and land plants reported from the country. The elements of floristic diversity from diverse ecological systems dispersed throughout the five biotic provinces recognised in the Himalaya by Rodgers (1985): Trans Himalaya, Northwest Himalaya, West Himalaya, Central Himalaya and Eastern Himalaya. About 50% (8,000 species) of the flowering plants recorded from India grow in the Himalaya, with highest richness (5,000 species, 62.5%) occurring in the Eastern ranges. Table 13.5 outlines the tentative status of various floristic groups in the Himalaya vis-a-vis India.

Table 13.4: Status of floristic richness in India in a global context.

Kingdoms with included taxa	*Approx. number of species described*		*%age of species in India with reference to world*
	India	*World*	
Monera			
Viruses*	Unknown	5,000	-
Bacteria and Cyanobacteria (=blue-green algae)	850	4,700	18.1
Protista			
Algae	2,500	40,000	6.25
Fungi			
Fungi	23,000	47,000	48.94
Lichens	1,940	17,000	11.41
Plantae			
Bryophytes	2,843	16,000	17.77
Pteridophytes	1,022	13,000	7.86
Gymnosperms	64	750	8.53
Angiosperms	17,000**	250,000	6.8
Total	**49,219**	**393,450**	**12.51**

*Some biologists do not classify viruses as living organisms.

** In addition, there are about 282 subspecies and 2,203 varieties (*Bot. Survey of India*, 1996, personal communication)

From Table 13.5 it is clear that the Himalayan region contains from 30% (fungi) to 68.75% (gymnosperms) of the known Indian flora. This territory also prossesses from 30.73% (lichens) to 87.50% (gymnosperms) of the total species endemic to India. The extent of this species richness is well appreciated considering the fact that the Indian Himalayan States/ Districts represent only 18% of the total land mass of India (see Table 13.2) and just 0.11% of that of the world (510, 100,500 sq. km.). The species richness in the Himalaya is greatest in the Eastern/Central ranges and declines a sharply through the West Northwest and Trans Himalaya.

Table 13.5: Status of floristic richness in Himalaya in Indian context.

Taxanomic Group	*Approx. number of species described*		*%age of species in Himalaya with reference to India*
	Himalaya Region	*India*	
1	2	3	4
Fungi	6 900 (1 890)*	23 000 (3,000)*	30.00(63.00)*

1	2	3	4
Lichens	1 159(130)	1,940(423)	59 74(30.73)
Bryophytes			
Liverworts	500(115)	843(166)	59.31(69.27)
Mosses	1 237(450)	2 000(820)	61.85(54.87)
Pteridophytes	600(150)	1,022(200)	58.70(75.00)
Gymnosperms	44(07)	65(08)	68.75(87.50)
Angiosperms	8 000 (3,200)	17 000 (5 400)	47.05 (59.25)
Total	**18 440(5,942)**	**45 869 (10 117)**	**40.20(59.31)**

*Figures in parentheses represent the number of taxa considered endemic to the respective region.

Source: Modified from Singh and Hajra (1996).

ANALYSIS OF SPECIES RICHNESS IN ANGIOSPERM-FLORA

The angiosperm-flora of India comprises about 17,000 species, representing 6.8% of the world's known flowering plants, the country ranking the 10th in the world in its number of species (Fig. 13.1). These species are distributed in 2,250 genera and about 315 families. The first ten large families with respect to number of species are presented in Table 13.6.

Table 13.6: First ten large families of angiosperms in India.

S. No.	*Name of the family*	*Number of species (genera)*
(i)	Poaceae	1,225 (225)
(ii)	Asteraceae	1,000 (161)
(iii)	Orchidaceae	990 (145)
(iv)	Fabaceae	775 (123)
(v)	Rubiaceae	495 (90)
(vi)	Cyperaceae	449 (24)
(vii)	Euphorbiaceae	419 (74)
(viii)	Lamiaceae	393 (68)
(ix)	Acanthaceae	379 (84)
(x)	Scrophulariaceae	356 (66)

In addition of this, several families possess more than 100 species each, e.g. Rosaceae (217 spp.), Asclepiadaceae (209 spp.), Apiaceae (202 spp.), Primulaceae (165 spp.), Brassicaceae (164 spp.), Lauraceae (163 spp.), Convolvulaceae (161 spp.), Ranunculaceae (160 spp.), Liliaceae (159 spp.), Ericaceae (150 spp.), and Gentianacae (147 spp.). On the other hand, there are more than 60 families in India which are represented by just a single

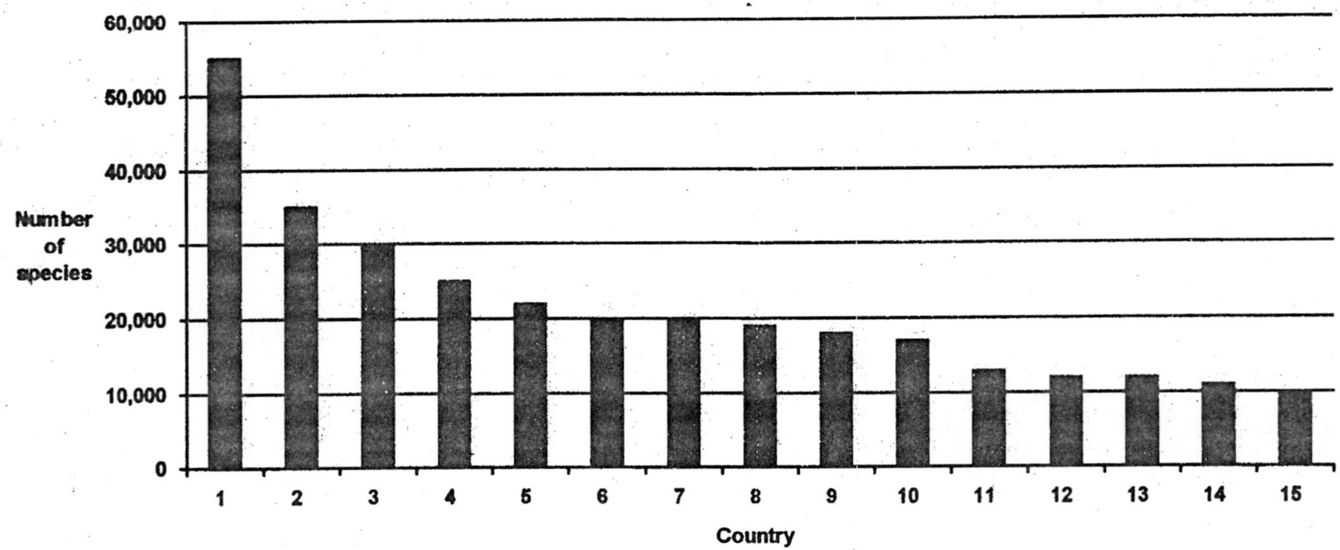

[1. Brazil 2. Colombia 3. China 4. Mexico 5. USSR (the former Union of Soviet Socialist Republics) 6. Indonesia 7. Venezuela 8. USA 9. Australia 10.India 11. Peru 12. Malaysia 13. Thailand 14. Zaire 15. Papua New Guinea]

Fig. 13.1 : The 15 most plant-rich countries.

species each. Some of these are Coriariaceae, Illiciaceae, Philydraceae, Ruppiaceae, Siphonodontaceae, Tetracentraceae and Turneraceae, etc.

The insectivorous families are represented by Lentibulariaceae (36 spp.), Droseraceae (3 spp.), and Nepenthaceae (1 sp.). The parasitic families are represented by Orobanchaceae, (54 spp.), Loranthaceae (46 spp.), Cuscutaceae (12 spp.), Santalaceae (10 spp.), Burmanniaceae (7 spp.), Balanophoraceae (6 spp.) Mitrastemonaceae (1 sp.), and Rafflesiaceae (1 sp.)

TREE-FLORA

Of the 17,000 species of flowering plants in India, about 2,560 (15%) species belong to trees (Nayar, 1989). The first eleven large families with respect to number of tree species are listed in Table 13.7.

Table 13.7: First eleven large families of angiosperms in India with respect to number of tree species.

S. No.	*Name of the family*	*Number of tree species*
(i)	Euphorbiaceae	200
(ii)	Lauraceae	150
(iii)	Annonaceae	100
(iv)	Fabaceae	100
(v)	Moraceae	100
(vi)	Rubiaceae	100
(vii)	Arecaceae	80
(viii)	Rutaceae	80
(ix)	Meliaceal	7
(x)	Caesalpiniaceae	60
(xi)	Mimosaceae	60

The Himalayan tree flora is also rich, representing 9% of the total angiosperm flora of the region and over 28% of the total estimated Indian tree species (Dhar, 1996). Some of the highly valuable timber species of the world grow in India. These include Teak (*Tectona grandis*), Sal (*Shorea rubusta*), Sheesham (*Delbergia sisoo*), Deodar/Himalayan cedar (*Cedrus deodara*), Himalayan silver-fir (*Abies pindrow*), Chir pine (*Pinus roxburghii*), Himalayan blue-pine (*Pinus wallichiana*), White siris (*Albizia procera*), Toon (*Toona ciliata*), Himalayan spruce (*Picea smithiana*), Lombardy poplar (*Populus nigra var. Italica*), and Himalayan Poplar (*Populus ciliata*), etc.

AQUATIC AND WETLAND PLANTS

The aquatic flowering plants are abundant in India and represent about half of the worlds total aquatic angiosperms (Rao, 1994). These are distributed in 64 families with Poaceae having the maximum number of genera (15) and Cyperaceae the maximum number of species (77) (see Table 13.8).

The larger aquatic and wetland angiosperm genera with respect to number of species are *Cyperus* (28 spp.); *Potamogeton* (14 spp.); *Eleocharis* (12 spp.); *Scirpus* (11 spp.); *Rotala* (10 spp.); *Carex* and *Limnophila* (9 spp. each); *Utricularia* and *Polygonum* (8 spp. each); *Nymphaea* (7 spp.); *Aponogeton* and *Myriophyllum* (6 spp. each); *Nymphoides* and *Lemna* (5 spp. each). Significantly all of the aquatic and wetland species of *Aponogeton* growing in India are endemic to the country. Species of Drosera, Aldrovandra and Utricularia comprise characteristic insectivorous plants.

MEDICINAL PLANTS

The Indian flora has a rich representation of medicinal plants. Within the country, the Himalaya, especially its western range, have the richest concentration of medicinal and aromatic plants (Arora, 1993). Some of the medicinal plants yield valuable medicines, while the majority of them are used in ethnomedicines as well as raw drugs in different parts of the country. Some of the important medicinal plant include *Aconitum deinorrhizum, A. heterophyllum, Actaea Acuminata, Amebia benthamii, Artemisia, spp., Asparagus racemosus, Atropa acuminata, Berberis aristata, B. Asiatica, Bergenia ciliata, B. Stracheyi, Centrella asiatica, Colchicum luteum, Coptis teeta, Costus speciosus, Dactylorhiza hatagirea, Digitalis purpurea, Dioscorea deltoido, Ephedra gerardiana, Gentiana kurroo, Habenaria commelinifolia, Hedychium spicatum, Hydnocarpus kurzii, Hyoscyamus niger, Illicium griffithii, Inula racemosa, Jurinea macrocephla, Mahonia nepalensis, Mucuna prurita, Myrica esculenta, Nardostachys grandiflora, Ocimum sanctum, Panax pseudoginseng, Picorhia kurrooa, Plumbago zeylanica, Podophyllum hexandrum, Rauvolfia serpentina, Rheum australe, R. nobile, Saussurea costus, Swertia chirayita, Taxus wallichiana, Valeriana hardwickii, V. jatamansi, Viola odorata, V. serpens, Zanthoxylum alatum,* etc. Because of over-exploitation and habitat destruction/modification, many of medicinal plants have become endangered.

Table 13.8. Distribution of aquatic and wetland flowering plants of India in families among dicots and monocots.

S. No.	*Name of the family*	*Number of species (genera)*
1	2	3
Dictoyledons		
1.	Acanthaceae	05(02)
2.	Aizoaceae	01(01)
3.	Amaranthaceae	05(03)
4.	Apiaceae	04(04)
5.	Asteraceae	11(09)
6.	Balsaminaceae	01(01)
7.	Boraginaceae	04(03)
8.	Brassicaceae	09(04)
9.	Cabombaceae	02(01)
10.	Callitrichaceae	04(01)

1	2	3
11.	Caryophyllaceae	03(03)
12.	Ceratophyllaceae	03(01)
13.	Convolvulaceae	02(01)
14.	Droseraceae	03(02)
15.	Elatinaceae	04(02)
16.	Fabaceae	08(05)
17.	Gentianaceae	03(03)
18.	Haloragaceae	08(03)
19.	Hippuridaceae	01(01)
20.	Hydrophyllaceae	01(01)
21.	Lamiaceae	06(03)
22.	Lentibulariaceae	08(01)
23.	Lythraceae	15(03)
24.	Menyanthaceae	06(02)
25.	Nelumbonaceae	01(01)
26.	Nymphaeaceae	10(03)
27.	Onagraceae	09(02)
28.	Podostemaceae	22(10)
29.	Polygonaceae	14(02)
30.	Rannuculaceae	06(03)
31.	Rubiaceae	02(02)
32.	Saururaceae	02(02)
33.	Scrophulariaceae	22(09)
34.	Sphenocleaceae	01(01)
35.	Sterculiaceae	01(01)
36.	Tamaricaceae	01(01)
37.	Tiliaceae	01(01)
38.	Trapaceae	01(01)
39.	Urticaceae	01(01)
40.	Vertebanceae	01(01)
Monocotyledons		
41.	Alismataceae	09(05)

1	2	3
42.	Amaryllidaceae	01(01)
43.	Aponogetonaceae	06(01)
44.	Araceae	10(06)
45.	Arecaceae	03(02)
46.	Butomaceae	01(01)
47.	Commelinaceae	06(04)
48.	Cyperaceae	77(13)
49.	Eriocaulaceae	01(01)
50.	Hydrocharitaceae	12(08)
51.	Juncaceal	05(01)
52.	Juncaginaceae	01(01)
53.	Lemnaceae	13(04)
54.	Limnocharitaceae	02(02)
55.	Najadaceae	04(01)
56.	Orchidaceae	01(01)
57.	Poaceae	24(15)
58.	Pontederiaceae	03(02)
59.	Potamogetonaceae	14(01)
60.	Ruppiaceae	01(01)
61.	Sparganiaceae	01(01)
62.	Typhaceae	03(01)
63.	Xyridaceal	07(01)
64.	Zannichelliaceae	01(01)

ENDEMIC PLANTS

The flora of India shows great affinity with the floras of Indo-Malayan and Indo-Chinese regions. It infact represents a confluence of floras from the surrounding countries/ regions like Afghanistan, Tibet, China, Myanmar, Indonesia, Philippines, Persia, Arabia, Sri Lanka, Malaysia, Japan, Europe and even from widely separated continents like Africa, America and Australia. Due to this fact, Hooker (1906) believed that India has no flora of its own but an admixture of floras from adjacent countries.

However, subsequent critical studies have revealed that, even so, the Indian flora depicts considerable endemism and uniqueness. Chatterjee (1939) reported 6,850 species of flowering plants as endemic to India. Nayar (1980) reported 141 (about 6.5%) genera as endemic to India, which are distributed over 47 families. Of these 111 genera are monotypic and 68 genera restricted to the Himalaya (40 in the eastern, 11 in western, and the rest 17 occur throughout the Himalayan range). According to Nayar (1989), the number of flowering plant species endemic to the present political boundaries of this country is 4,900 out of a total of 15,000, i.e. about 33%. Singh and Hajra (1996) report 5,400 endemics in 17,000 angiosperm species of India which comes to 31.76%. In fact, India, is said to harbour more endemic species of plants than any other region of the world except Australia (Rao, 1994).

The endemic taxa show high concentration in several areas of different biogeographical zones in India. Such areas where the endemics are mainly located have been called as endemic centres of the country or, more fashionably, "hot spots" of endemic diversity (Table 13.9).

Table 13.9 : List of major endemic centres of plants in India.

S. No.	*Endemic centres*
1	2
1.	Karakoram and Ladakh in Trans/Kashmir, Himalaya
2.	Kumaon in Garthwal Himalaya.
3.	Siwaliks in W. Himalaya.
4.	Terai
5.	Sikkim Himalaya in E. Himalaya.
6.	Arunachal Pradesh in E. Himalaya.
7.	Lushai hills in NE India.
8.	Tura-Khasi hills (Meghalaya) in NE India.
9.	Aravallis in Semiarid zone.
10.	Bundelkhand in Gangetic Plain.
11.	Chotanagpur plateau in Deccan.
12.	Panchamarhi-Satpura ranges in Deccan.
13.	Similipal and Jeypore hills of Orissa in Deccan (E Ghats).
14.	Bastar and Koraput hills in Deccan.
15.	Vishakhapatnam hills and Araku Valley in Deccan.
16.	Tirupati-Cuddappa hills in E. Ghats
17.	Marathwada hills in Deccan.
18.	Saurashtra-Kutch in Deccan.
19.	Mahabaleshwar-Khandala ranges in W. Ghats.
20.	Agumbe-Phonda ranges in W. Ghats.
21.	Ratnagiri and Kholaba ranges in W. Ghats.
22.	Nilgiris, Silent Valley and Wynaad in W. Ghats.

1	2
23.	Anamalais in W. Ghats.
24.	Idduki-Sabarigir in W. Ghats.
25.	Palni-Yercaud.
26.	Kalakad and Agastyamalai hills in W. Ghats.
27.	Andaman islands.
28.	Great Nicobar islands.

The areas ranking high in endemism are the Northeast India, followed by southern parts of Peninsular India and Northwestern Himalaya. While the Eastern Himalaya has about 1 500 endemics out of a total of 4,500 estimated species, Peninsular India has 2 000 out of 6,000 species, and the Northwest Himalaya has 1,000 endemics out of a total of 3 000 species (Rao, 1994). The Himalayan region as a whole contributes over 59% of all the endemic angiosperm species in India (see table 13.5). The Andaman and Nicobar Islands contribute 185 species to the endemic flora of India (Nayar, 1989). The family poaceae, one of the largest, has about 360 endemic taxa (Jain, 1988). According to Rao (1994), the largest endemic genera in India are *Nilgirianthus* and *Pteracanthus* of the family Acanthaceae, being represented by 20 species each.

THREATENED PLANTS

Meaningful studies on threatened plants of India started with the initiation of the Project on Study, Survey and Conservation of Endangered Flora (POSSCEF) in 1980 by the Botanical Survey of India (BSI). Jain and Sastry (1980) published short accounts of 134 threatened plant species. The seminar on "Threatened plants of India", organised by the BSI at Dehra Dun in 1981 and publication its proceedings (Jain and Rao, 1983), stimulated studied on threatened plants on a large scale. Since then lists of rare or threatened taxa have been published for several States/regions of India. Recently, the BSI has published a national Plant Red Data Book (Nayar and Sastry 1987, 1988, 1990). Based on this and other works, Rao (1994) reported 1,566 taxa (species, subspecies and varieties) of flowering plants as rare and endangered in India (Table 13.10).

Table 13.10: Flowering plant families with number of threatened taxa (species, sub-species and varieties) in India.

S. No.	*Family*	*Number of threatened taxa*
1	2	3
1.	Acanthaceae	32
2.	Aceraceae	08
3.	Amaranthaceae	03
4.	Anacardiaceae	09
5.	Annonaceae	35

1	2	3
6.	Apiaceae	43
7.	Apocynaceae	05
8.	Aquifoliaceae	04
9.	Araliaceae	05
10.	Aristolochiaceae	01
11.	Asclepiadaceae	53
12.	Asclepiadaceae	54
13.	Balanophoraceae	03
14.	Balsaminaceae	40
15.	Begoniaceae	19
16.	Berberidaceae	26
17.	Bynoniaceae	01
18.	Bixaceae	01
19.	Bombacaceae	02
20.	Boraginaceae	07
21.	Brassicaceae	06
22.	Burseraceae	02
23.	Caesalpiniaceae	16
24.	Campanulaceae	05
25.	Capparaceae	09
26.	Caryophyllaceae	12
27.	Celastraceae	07
28.	Chrysobalanaceae	01
29.	Circaeasteraceae	01
30.	Clusiaceae	09
31.	Combretaceae	01
32.	Connaraceae.	02
33.	Convovulaceae	10
34.	Crassulaceae	07
35.	Cucurbitaceae	09
36.	Dilleniaceae	01
37.	Dipterocarpaceae	04

1	2	3
38.	Droseraceae	01
39.	Ebenaceae	01
40.	Elaeagnaceae	01
41.	Elaeocarpaceae	08
42.	Elatinaceae	01
43.	Ericaceae	30
44.	Euphorbiaceae	31
45.	Fabaceae	73
46.	Flacourtiaceae	02
47.	Fumariaceae	01
48.	Gentianaceae	10
49.	Geraniaceae	03
50.	Gesneriaceae	08
51.	Hamamelidaceae	01
52.	Hippocrateacea	05
53.	Hypericaceae	03
54.	Icacinaceae	04
55.	Ixonanthaceae	01
56.	Juglandaceae	01
57.	Lamiaceae	28
58.	Lauraceae	19
59.	Lentibulariaceae	08
60.	Linaceae	01
61.	Loganiaceae	02
62.	Loranthaceae	02
63.	Lythraceae	07
64.	Magnoliaceae	12
65.	Malpighiaceae	03
66.	Malvaceae	10
67.	Melastomataceae	13
68.	Meliaceae	09
69.	Menispermaceae	08

1	2	3
70.	Mimosaceae	08
71.	Molluginaceae	01
72.	Moraceae	01
73.	Myristicaceae	02
74.	Myrsinaceae	08
75.	Myrtaceae	22
76.	Nepenthaceae	01
77.	Nyctaginaceae	02
78.	Nympheaceae	02
79.	Olacaceae	03
80.	Onagraceae	01
81.	Orobanchaceae	02
82.	Papavaraceae	03
83.	Periplocaceae	01
84.	Philadelphaceae	01
85.	Piperaceae	02
86.	Pittosporaceae	01
87.	Plumbaginaceae	01
88.	Podostemaceae	13
89.	Polygalaceae	02
90.	Primulaceae	08
91.	Proteaceae	02
92.	Rafflesiaceae	02
93.	Ranuncullaceae	17
94.	Rhamnaceae	03
95.	Rosaceae	16
96.	Rubiaceae	98
97.	Rutaceae	04
98.	L. Sabiaceae	03
99.	Santalaceae	01
100.	Sapindaceae	03
101.	Sapotaceae	09

1	2	3
102.	Saurauiaceae	01
103.	Saxifragaceae	02
104.	Scrophulariaceae	11
105.	Solanaceae	03
106.	Sterculiaceae	06
107.	Stylidiaceae	02
108.	Styracaceae	03
109.	Symplocaceae	03
110.	Ternstroemiaceae	01
111.	Tetracentraceae	01
112.	Theaceae	03
113.	Thymelaeaceae	02
114.	Tiliaceae	02
115.	Urticaceae	01
116.	Valerianaceae	02
117.	Verbenaceae	03
118.	Violaceae	03
119.	Vitaceae	09
120.	Zygophyllaceae	01
Monocotyledons		
121.	Alismataceae	01
122.	Amaryllidaceae	04
123.	Aponogetonaceae	02
124.	Araceae	38
125.	Arecaceae	21
126.	Burmnanniaceae	01
127.	Commelinaceae	19
128.	Cyperaceae	31
129.	Dioscoreaceae	10
130.	Eriocaulaceae	09
131.	Haemodoraceae	02
132.	Hydrocharilaceae	02

1	2	3
133.	Iridaceae	03
134.	Juncaceae	02
135.	Lemnaceae	02
136.	Liliaceae	38
137.	Marantaceae	02
138.	Musaceae	01
139.	Orchidaceae	152
140.	Pandanaceae	01
141.	Poaceae	133
142.	Smilacaceae	01
143.	Stemonaceae	01
144.	Triuridaceae	01
145.	Zingiberaceae	23

Source: Rao (1994).

Overall, about 10% (1,700) of the flowering plant species in India are presently evaluated as threatened under one or the other Red Data Category. According to Nayar (1989), 93 species could not be relocated in India during the past fifty years. Seventeen of angiosperm species in India are presumed to have already become extinct (WCMC, 1992; Table 13.11). Furthermore, unless the habitat loss and population pressure are contained, we may lose another 100 species by the turn of this century (Nayar, 1989).

Table 13.11 : Records of extinct angiosperm species in India

S. No.	*Name of species*	*Family*	*State/region wherefrom recorded*
1	*2*	*3*	*4*
Dictyledons			
1.	*Dicliptera abuensis* Blatter	Acanthaceae	Rajasthan.
2.	*Ophiorrhiza brunois Wight & Arn.*	Rubiaceae	Karnataka, Kerala, Tamil Nadu.
3.	*O. caudata* C. Fischer	"	Kerala.
4.	*O. radicans* Gardn.	"	Kerala.
5.	*Sterculia khasiana* Debberm	Sterculiaceae	Meghalaya
6.	*Wendiandia angustifolia* Wight ex Hook f.	Rubiaceae	Tamil Nadu

1	2	3	4
Monocotyledons			
7.	*Calanthe whiteana* King & Pantl.	Orchidaceae	Sikkim.
8.	*Carex repanda* C. B. Clarke	Cyperaceae	Meghalaya.
9.	*Corypha taliera* Roxb.	Palmae	India.
10.	*Dipcadi concanense* (Dalz.) Baker	Liliaceae	Western Ghats.
11.	*D. reidii* Deb & Dasgupta	"	Western Himalaya
12.	*Eragrostis rottleri* Stapf	Poaceae	Tamil Nadu.
13.	*Eriochrysis rangacharii* Fischer	"	Tamil Nadu.
14.	*Hedychium marginatum* C.B. Clarke	Zingiberaceae	Nagaland.
15.	*Hubbardia heptaneuron* Bor	Poaceae	Karnataka.
16.	*Pleione laginaria* Lindley	Orchidaceae	Meghalaya.
17.	*Urginea polyphylla* Hook.f.	Liliaceae	Deccan Peninsula.

"HOT SPOTS" OF FLORISTIC DIVERSITY

The three major "hot spots" of biodiversity in India are Andaman and Nicobar islands, Eastern Himalaya and eastern India, and Southern Western Ghats. These areas are predominantly covered by tropical wet evergreen forests (Rain Forests). These forests constitute only 8% of the Indian forests and just 1.5% of the country's total land surface. Even so, these forests support about 7,000 species which is 41% of the Indian angiosperm flora. Compared to this, the wet evergreen forests of the world occupy 6% of the earth's surface and hold about 90,000 (36%) of the global species of higher plants. The concentration of 41% of the total Indian higher plant species in only 1.5% of its land area covered by tropical wet evergreen forests, coupled with the fact that these forests are biologically and ecologically complex and more fragile, involves great risk, rendering such areas of utmost conservation importance and "hot spots" of biodiversity.

It is being increasingly realised that the endemics constitute the fundamental and the most important components of a particular flora or fauna. Since the endemic taxa in India (as also elsewhere are facing a great threat to their survival, mainly due to increasing anthropogenic effects, some workers hold that all the 28 endemic centres in the country (see Table 13.9) should be considered as hot spots and prioritized for conservation purposes. However, most of these endemic centres fall in the three major hot-spot zones mentioned above.

The Andaman and Nicobar islands form a major repository of diverse groups of plants. Of the estimated 1,500-1,800 plant species in Andaman, 150 species are supposed to be endemic. Nicobar islands are said to contain about 700 species of higher plants, of which some 70 are endemic. A number of species in the Andaman and Nicobar group of islands are reported to be extremely rare; 65 of the endemic species could not be collected even after repeated surveys (Nayar, 1989).

The Eastern Himalaya and Eastern India form the eastern gateway for migration of flora from adjacent countries like Myanmar, China Bhutan, Japan etc. This region is a crucible of floristic and genetic diversity. It supports about 7,500 species of plants and has the largest number of endemics and Schedule I species in the country MacKinnon and MacKinnon, 1986). This area, alongwith Yunnan, is considered as the "cradle of flowering plants" (Takhtajan, 1969).

The Western Ghats is known, according to several phytogeographers, as the "Malabar" province. It shows great plant diversity, harbouring as many as 4,000 species of flowering plants, of which 56 genera and 2,100 species are endemic (Nayar, 1982). This region is also a rich germplasm centre for a number of wild relatives of economically important crops like cereals and millets, legumes, tropical and subtropical fruits, vegetables, spices and condiments, etc.

LITERATURE CITED

Anonymous. (1989). *A Conservation of Wetlands in India*. Ministry of Environment and Forests, Govt. of India, New Delhi.

Anonymous. (1993). *The State of Forest Report 1991*. Forest Survey of India, Ministry of Environment and Forests, Govt. of India, New Delhi.

Champion, H.G. and Sethi, S.K. (1968). *A Revised Survey of Forest Types of India*. Manager of Publications, Govt. of India. New Delhi.

Chatterjee, D. (1939) Studies on the endemic flora of India and Burma. *Journal of Royal Asiatic Society*, Bengal, (N.S.), 5: 19-67.

Chatterjee, D. (1962). Floristic patterns of Indian Vegetation. pages. 32-42. In : Maheshwari, P., Johri, B. M. and Vasil, I.K. (Ed.) *Proceedings of Summer School in Botany*, Darjeeling. Sree Saraswati Press, Calcutta.

Clarke, C. B. (1898). Sub-areas of British India illustrated by the detailed distribution of the Cyperaceae in that empire. *Journal of Linnaean Society* (Botany), London 34 : 1-146.

Dhar, U. (1996). Overview of Himalayan Biodiversity. pages 3-20. In : Gujral, G.S. and Sharma, V. (Editors) *Changing Perspectives of Biodiversity Status in the Himalaya*. The British Council, New Delhi.

Hooker. J.D. (1906). *A Sketch of the Flora of British India*. Oxford. (Reprinted-Bishen Singh Mahendra Pal Singh, Dehra Dun.

Jain, S.K. (1983). *Flora and Vegetation of India–An Outline. Botanical Survey of India*, Calcutta

Jain, S.K. (1988). The grass flora of India–a synoptic account of uses and phytogeography. Bulletin Botanical Survey of India, 28: 229-240, 1986.

Jain, S.K. (1989). Botanical regions and flora of India. *Everyman's Science*: 213-223.

Jain, S.K. and Rao, R.R. (Editors) (1983). *An Assessment of Threatened Plants of India*. Proceedings of the seminar held at Dehra Dun, 14-17 Sept. 1981. Botanical Survey of India, Calcutta.

Jain, S.K. and Sastry, A.R.K. (1980). *Threatened Plants of India : A State-of-the-Art Report*. Botanical Survey of India, Calcutta.

Keystone, C. (1991). Final consensus report on the Keystone policy dialogue. Analysis of the species-area relationship. *Ecology*, 47 : 831-843.

Kushalapa, K.A. (1992). Saving natural forests. *Indian Forester*, 118 (5): 344-347.

Langner, L.L. and Flather, C.H. (1994). *Biological Diversity; Status and Trends in the United States.* United States Department of Agriculture, Forest Service General Technical Report Rm-244.

MacKinnon, J. and MacKinnon, K. (1986). *Review of the Protected Areas System in the Indo-Malayan Realm.* International Union of Conservation of Nature, Gland.

Nayar, M. P. (1980). Endemism and patterns of distribution of endemic genera (angiosperms). *Journal Economi; Taxonomic Botany* 1: 99-110.

Nayar, M.P. (1982). Endemism flora of Peninsular India and its significance. *Bulletin Botanical Survey of India.* 22: 12-33, 1980.

Nayar, M.P. (1989). In Situ conservation of wild flora rescurces. *Bulletin Botanical Survey of India,* 29(1-4): 319-333. 1987.

Nayar, M.P. and Sastry, A.R.K. (Editors) (1987, 1988, 1990). *Red Data Book of Indian Plants* Vols. 1-3 Botanical Survey of India, Calcutta.

Puri, G.S. (1960). *Indian Forest Ecology.* Vols. 1-2. Oxford Book and Stationery Company, Calcutta.

Puri, G. S., Meher-Homji, V. M. Gupta, R.K. and Puri, S. (1983). Forest Ecology, *Phytogeography and Forest Conservation.* Vol. 1. Oxford & IBH Publishing Company, New Delhi.

Rao, A.S. (1974). The vegetation and Phytogeography of Assam-Burma. pages 204-246. In : Mani, M.S. (Editor) *Ecology and Biogeography in India.* W. Junk. B.V. Publishers, the Hague.

Rao, R.R. (1994). *Biodiversity in India* (Florstic Aspects). Bishen Singh Mahendra Pal Singh, Dehra Dun.

Razi, B.A. (1955) Some observations on Plants of South Indian hill tops and their distribution. *Proceedings of National Institute of Science, India,* 21B (2): 79-89.

Rodgers, W.A. (1985). Biogeography and protected area planning in India. In : Thorsell, J. (Editor). *Conserving Asia's Natural Heritage.* International Union of Conservation of Nature, Gland.

Rodgers, W.A. and Panwar, H.S. (1988). *Planning a Wildlife Protected Area Network in India.* Vols. 1-2 Wildlife Institute of India, Dehra Dun (Mimeographed).

Sharma, P.D. (1993). *Ecology and Environment.* Rastogi Publishers, Meerut, India.

Singh, D. K. and Hajra, P. K. (1996). Floristic diversity. pages 23-38. In : Gujrat, G.S. and Sharma, V. (Editors). *Changing Perspective of Biodiversity Status in the Himalaya.* The British Council, New Delhi.

Takhtajan, A. (1969) *Flowering Plants, Origin and Dispersal* (Translated by Jeffrey C.). Oliver Boyd, Edinburgh.

Takhtajan, A. (1986). *Floristic Regions of the World.* The University of California Press. (Indian Edition, 1988, Bishen Singh Mahendra Pal Singh, Dehra Dun).

Udvardy, M. D.F. (1975). *A Classification of the Biogeographical Provinces of the World.* International Union of Conservation of Nature, Occasional paper 18. Morges, Switzerland.

Wilson, E.O. (1989). Threats to Biodiversity. *Scientific American* 261: 108-116.

World Conservation Monitoring Centre. (1992). *Global Biodiversity: Status of the Earth's Living Resources.* Chapman and Hall, London.

Chapter - 14

Vegetation Structure and Composition under Different Landscape Elements at Kodayar in Western Ghats of Tamilnadu

S.M. Sudarapandian, S. Chandrashekran and P.S. Swamy

ABSTRACT

Vegetation structure and composition of various landscape elements such as natural forests (evergreen, moist deciduous), man-modified ecosystems (teak and rubber plantations, agroecosystems, wastelands), disturbance regimes (annual fire, occasional fire, mechanical disturbed, mechanical + fire affected) and fallow lands after shifting cultivation at Kodayar were studied. *Terminalia paniculata* and *Pterocarpus marsupium* dominant species in moist deciduous forest. Whereas *Hopea parviflora* and *Syzygium laetum* were dominated in the evergreen forest. Forest at higher elevation was dominated by *Agrostistachys meeboldii, Cullenia excelsa* and *Drypetes oblongifolia.* Stem density and basal area of the evergreen forest were two times greater than the moist deciduous forest. *Eupatorium* and *Ageratum* are two dominant exotic weeds forms part of the ecosystem in open and disturbed lands here. *Syndrella nodiflora* and *Ageratum conyzoides* are the most dominating weeds in agroecosystems. A sharp decline of exotic weeds in dominance and their contribution to the community beyond 5 years of fallow age was observed. The changes in species composition are largely due to transition in vegetation types influenced by anthropogenic perturbations and other abiotic factors.

Keywords : agrosystems, disturbed regimes, evergreen forests, fallow lands, forestry plantation, moist deciduous forests, wastelands,

INTRODUCTION

Tropical forests are amongst the richest and complex biological communities on earth, which has got tremendous power of self maintenance. But, most of our forests have lost

*Department of Plant Sciences School of Biological Sciences, Madurai Kamraj University, Madurai - 625 021

these qualities due to increasing biotic interference. The disappearance of these forests, at an estimated rate of 1-2% per year comes at a time when our knowledge of their structure and functional dynamics is woefully inadequate (Hubbell and Foster, 1992).

Deforestation, forest farming, change in land use pattern and development activities have done considerable harm to the environment in the recent past (Dibson *et al.* 1989). Apart from this imperial colonization and quest for food brought about many exotic introductions in to native environment (Ramakrishnan, 1991). These mindful and accidental introduction of exotic plants crossed their boundaries and became menance in neighbouring ecosystems. Introduced species are those that have been intentionally or accidentally transported by human agency to an area outside of their natural geographical distributional range. Invasive species are those introduced species that expand their population and distribution range in the new geographical location without further human intervention.

During the last few decades Western Ghats of Tamil nadu has been subjected to unscientific exploitation particularly for agriculture, construction of hydro-electric project, raising monoculture plantations (*Hevea braziliensis* M. Arg.- Rubber, *Acacia mangium* - Acacia and *Tectona grandis* L. - Tectona) and other developmental activities. The ever increasing demand for forest products and forest land, together with the alarming rate of population growth has put the remaining patches of forests on the verge of extinction (Tamart Bekele, 1994). A detailed quantitative and qualitative description and regeneration status of the remaining forests are necessary and timely as it will form the basis for future plans to manage and restore these vanishing resources. Therefore this study aims at: (1) analysis of vegetation structure and species composition of the natural forest ecosystems, man-modified ecosystems, disturbed regimes and early succesional communities developed subsequent to shifting cultivation at Kodayar and (2) an attempt was also made to understand the regeneration status of tree species in natural forests.

STUDY AREA

The study area at Kodayar (village), located 400 Km south of Madurai (77°15′ E, 8°29′ N) is at 250-1150 m elevation (Fig. 14.1). Kodayar is part of Agasthyamalai hill range and it falls with in the Veerapuli forest reserve in South India. This forest area has 30 kani (local tribals) settlements which occupies an area of 6.85 km^2. The mean annual rainfall recorded in the study sites were 2338 mm, and 3000 mm at lower elevation and higher elevation, respectively, of which 81% occurred from June to November (Fig. 14.2). December to March represents a brief dry period. Average monthly maximum and minimum temperatures were 30°C and 26°C in summer and 28°C and 24°C in winter, respectively, in lower elevation (250-650 m). Temperature does not vary much over the seasons at higher elevation. Mean maximum temperature was 24°C and the minimum was about 16°C.

METHODS

Phytosociological studies were done in different landscape elements such as natural forests (Forest at higher elevation (HEF), Evergreen forest (EF) and Moist Deciduous forest (MDF)), man-modified ecosystems (rubber and Teak plantations, Agroecosystems and Wastelands), disturbed regimes (Annual fire, occasional fire, mechanical disturbance, mechanical + fire disturbance) and fallow lands (1, 3, 5, 8, 20 years fallow land shifting cultivated land) at Kodayar (Table 14.3). The density, frequency, basal area and importance value index (IVI) were estimated by using 50 randomly placed quadrats (10 x 10 m^2) for trees (individuals with GBH more than 30 cm), saplings (>10 cm -<30 cm GBH), and shrubs

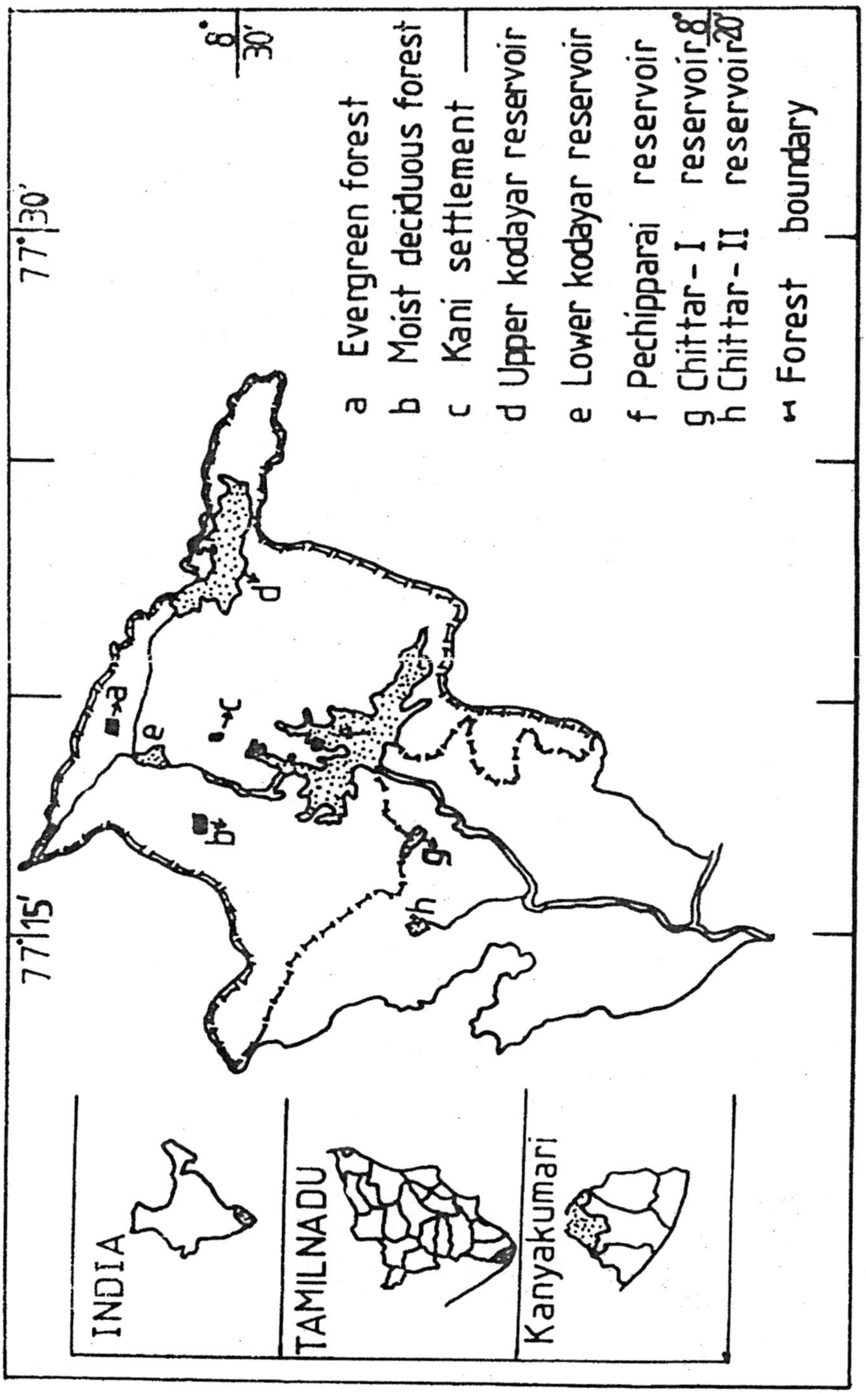

Fig. 14.1 : Map of the study area showing location of the study sites.

at each study site of natural forest ecosystems. Similarly lianas (all climbers of all sizes) whose base fell inside the plot (10 × 10 m^2) were enumerated. Similar number of quadrats (1 x 1 m^2) were studied for herbs at each study site (Kershaw, 1973; Misra, 1968). All epiphytes were not sampled in-this study. The plant samples were identified in the field with the help of Gamble (1925) and Pascal and Ramesh (1987) and later counter-checked with the reference material available at Botanical Survey of India, Coimbatore. A reference collection of specimens were preserved in the herbarium at MKU. Species diversity was calculated using:

$$H' = - (n_i/N) \log \text{In} (n_i/N)$$

Margalef (1968) Where H' = Shannon index of general diversity, n_i = importance value index of species i, N = importance value index of the community. The index of dominance of the community was calculated by Simpson's index (Simpson, 1949) as

$$c = (n_1/N)^2$$

Where c = index of dominance; n_i and N being the same as in the Shannon index of general diversity.

The family importance value was calculated as mentioned by Keel *et al.* (1993) (Cited in Ganesh *et al.*, 1996).

FIV = relative density + relative diversity + relative dominance. (for all numbers of a given plant family combined)

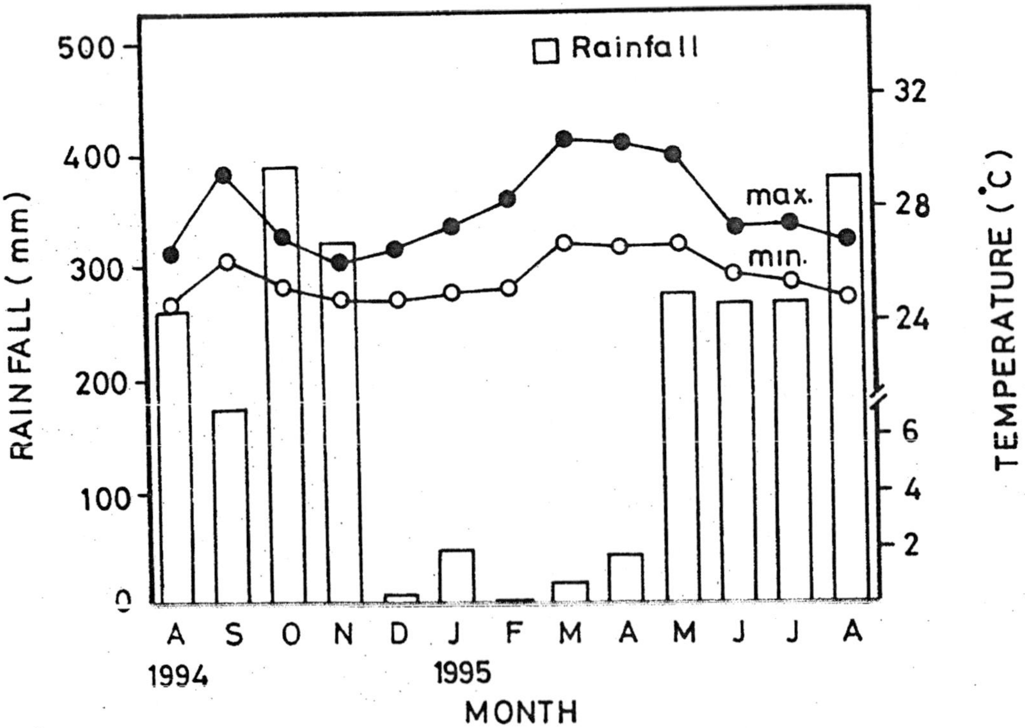

Fig. 14.2 : Mean monthly rainfall recorded at the study area at Kodayar in the Western Ghats

Relative density = (number of individuals of the species x 100)/total number of individuals in the sample.

Relative diversity = (number of species in the family x 100)/total number of species in the sample.

Relative dominance = (basal area of the family x 100)/total basal area in the sample.

RESULTS

A total of 170 plant species were recorded from three study sites at Kodayar (Table 14.1). Fifty eight plant species were recorded from the moist deciduous forest (MDF) representing in 30 families. Whereas in the evergreen forest (EF) 158 plant species belonging to 52 families were recorded. Forest at higher elevation (HEF) is represented by about 102 plant species belonging to 42 families. Fifteen tree species from 8 families in MDF, 45 from 22 families in EF and 32 from 18 families in EF; shrub community were 6 species from 5 families in MDF, 18 from 11 families in EF and 10 from 6 families in HEF; herbaceous community were 12 species from 6 families in MDF, 16 from 11 families in EF and 15 from 9 families in HEF; and Lianas were 5 species from 4 families in MDF, 20 from 14 families in EF and 14 from 10 families in HEF (Table 14.1). The MDF was dominated by members of Combretaceae family *Terminalia paniculata*, followed by *Pterocarpus marsupium*, *Buchanania lanzan* and *Careya arborea*. Whereas the EF was dominated by *Hopea parviflora* followed by *Syzygium laetum*, *I. brachiata* and *Xanthophyllum flavescens*. The HEF was dominated by *Agrostistachys meeboldii* and *Cullenia excelsa* followed by *Drypetes oblongifolia*, *Myristica dactyloides* and *Aglaia barberi*. For large trees like *T. paniculata*, *H. parviflora* and *C. excelsa*, relative dominance accounted for 41%, 44% and 55% respectively, of the IVI values in these study sites. Whereas for *B. lanzan*, *S. laetum* and *A. meeboldii* although being abundant, accounted for only 22, 21 and 22%, respectively.

Table 14. 1 : Importance value indices of the natural forest ecosystems (Moist deciduous MDF; Evergreen EF; Forest at higher elevation HEF) at Kodayar in Western Ghats of Tamilnadu

Name of the species		*Importance Value Indices*		
		MDF	*EF*	*HEF*
1	2	3	4	5
Trees				
Aglaia barberi Gamble	(Meliaceae)	-	0.694	24.140
Agrostistachys meeboldii Pax.& K.Hoffm.	(Euphorbiaceae)	-	-	61.540
Aporosa lind1eyana Baill	(Euphorbiaceae)	-	-	
Bridelia crenulata Roxb.	(Euphorbiaceae)	7.963	-	-
Buchanam'a lanzan Spr.	(Anacardiaceae)	40.437	-	-
Calophynum polyanthum Wall. ex Choisy	(Clusiaceae)	-	-	15.150
Careya arborea Roxb.	(Lecythidaceae)	31.558	-	-

1	2	3	4	5
Carallia brachi'ata (Lour.) Merr.	(Rhizophoraceae)	-	18.872	-
Cullem'a excelsa Wt.	(Bombacaceae)	-	-	61.370
Dillenia'a pentagyna Roxb.	(Dillemaceae)	26.302	-	-
Dimocarpus longan Lour.	(Sapotaceae)	-	4.724	-
Diospyros bourdilloni Brandis	(Ebenaceae)	-	4.413	-
D. panicutata Dalz.	(Ebenaceae)	-	6.063	-
Drypetes oblongifolia (Bedd.) A. Shaw	(Euphorbiaceae)	-	-	37.640
Emblica officinalis Gaertn.	(Euphorbiaceae)	21.350	-	-
Gluta travancorica Bedd.	(Anacardiaceae)	-	-	-
Gomphandra tetrandra Sleumer	(Icacinaceae)	-	2.761	13.430
Holigarna arnottiana J. Hk.	(Anacardiaceae)	-	-	7.840
Hopea parviflora Bedd.	(Dipterocarpaceae)	-	103.584	-
Ixora brachiata Roxb.	(Rubiaceae)	-	22.169	
Kingiodendron pinnatum Harms	(Fabaceae)	-	3.136	
Mastixia arborea (Wt.) Bedd.	(Cornaceae)	-		6.620
Mesua ferrea L.	(Clusiaceae)	-	0.716	-
Myristica dactytoides Gaertner	(Myristicaceae)	-	-	25.710
Polyalthia wightii	(Annonaceae)	1.135		
Pterocarpus marsupium Roxb.	(Leguminoseae)	40.234	-	-
Pterospermum rubiginosum Wt. & Arn.	(Sterculiaceae)	-	6.043	-
Scolopia crenata (Wt. & Arn.) Clos.	(Flacourtiaceae)	-	10.061	-
Stereospermum personatum (Hassk). Chatterjee	(Bignoniaceae)	-	8.001	-
Syzygium laetum Gandhi	(Myrtaceae)	-	39.195	-
S. mundagam (Bourd.) Chithra	(Myrtaceae)	-	0.865	1.220
Terminalia arjuna Wight & Arn.	(Combretaceae)	19.488	-	-
T. *chebula* Retz.	(Combretaceae)	14.319	-	-
T. crenulata Roth.	(Combretaceae)	6.137	-	-
T. paniculata Roth.	(Combretaceae)	82.036	-	-
Vitex altissima L.	(Verbinaceae)	-	10.980	-
Xanthophynum ftavescens Roxb.	(Xanthophyllaceae)	-	6.185	-
Others		9.041 (4)	51.538 (28)	44.030 (22)

1	2	3	4	5
Shrubs				
Boehmeria c1oineru1ifera Miq.	(Urtucaceae)	-	5.250	-
Calamus brand isii Becc.	(Arecaceae)	-	138.010	65.150
C. *travanconcus* Bedd.	(Arecaceae)	-	-	73.820
Diotacanthus grandis (Bedd.) Benth ex. Clarke	(Acanthaceae)	-	-	12.960
Euonymus crenulatus Wall.	(Celastraceae)	-	-	14.820
Eupatorium odoratum L.	(Arteraceae)	135.750	-	-
Heticteres isora L.	(Sterculiaceae)	89.031	-	-
Ixora elongata Heyne	(Rubiaceae)	-	15.480	-
I. nigricans Br.	(Rubiaceae)	-	9.940	
Leea indica (Burm. f.) Merr.	(Vitaceae)	-	0.570	-
Hemecylon heyneanum Benth.	(Melastomaceae)	-	26.100	-
Hemecylon sp.	(Melastomaceae)	-	15.560	-
Nilgirianthus foliosus (Wight) Bremek	(Acanthaceae)	-	-	46.640
Ochlandra travancorica Gamb.	(Gramineae)	-	12.260	-
Psychotria sp.	(Rubiaceae)	-	21.970	-
Rauwolfia beddomei Hk. F.	(Apocynaceae)	-	10.500	-
Saprosma corymbosum (Bedd.) Bedd.	(Rubiaceae)	-	-	82.180
Symplocos macrophylla Wan. ex DC.	(Symplocaceae)	-	1.990	-
Thespesia tampas D. & G.	(Malvaceae)	8.960	-	-
Zizypus oenoptia (L.) Miller	(Rhamnaceae)	53.770	-	-
Others		12.489 (2)	42.37 (7)	4.43 (4)
Herbs				
Apama siliquosa Lam.	(Aristolochiaceae)	-	21.250	-
Bolbitis prolifera	(Pteridophytes)	-	-	18.960
Chasalia ophioxloides (Wall.) Cralb.	(Rubiaceae)	-	37.090	-
Chtoranthes brachystachys L.	(Chloranthaceae)	-	51.070	-
Curculigo orchiodes Gaertn.	(Hypoxidaceae)	21.130	-	31.870
C. *trichocarpa* (Wight) Bennet & Raizada	(Hypoxidaceae)	-	11.400	6.930
Digitaria ciliaris Koel	(Gramineae)	13.200	-	-
Geodorum densiftorum Schlechter	(Orchidaceae)	-	10.310	3.230
Globba orixensis L.	(Zingiberaceae)	7.140	-	-

1	2	3	4	5
Grass		-	88.950	-
Hemidesmus indicus R. Br.	(Asclepiadaceae)	5.430	-	-
Kalanchoe sp.	(Crassulaceae)	-	-	17.910
Leptochilus lanceolatus	(Pteridophytes)	-	26.280	90.530
Memecylon sp.	(Melastomaceae)	-	18.800	43.790
Neurocalyx calycinus (Benn.) Robins	(Rubiaceae)	-	5.990	-
Ophiorrhiza sp.	(Rubiaceae)	-	4.050	-
Panicum sp.	(Gramineae)	3.190	-	-
Pteris argyraea T. Moore	(Pteridophytes)	-	-	18.140
P. confusa T.G.Walker	(Pteridophytes)	-	-	26.510
Pteris sp.	(Pteridophytes)	-	15.560	-
Selaginella sp.	(Pteridophytes)	-	-	27.520
Themeda cymbania Hackel	(Gramineae)	238.210	-	-
Zingiber reseum (Roxb.) Roscoe	(Zingiberaceae)	-	2.170	8.040
Others		11.700 (6)	7.08 (3)	6.57 (4)
Lianas				
Ancistrodadus heyneanus Grab.	(Ancistrocladaceae)	-	132.010	-
Asparagus sp.	(Liliaceae)	33.21	-	-
Calycopteris floribunda Lam.	(Combretaceae)	127.410	-	-
Canthium angustifolium Roxb.	(Rubiaceae)	-	8.860	3.250
Canthium sp.	(Rubiaceae)	-	20.140	-
Connarus wightii J.Hk.	(Connaraceae)	-	13.510	104.650
Derris benthamii Thw.	(Papilionaceae)	-	7.130	3.410
Elaeagnus kologa Schlecht	(Elaeagnaceae)	-	-	23.670
Entada scandens Benth.	(Luguminaceae)	-	20.570	-
Hiptage benghalensis (L.) Kurz	(Malpighiaceae)	-	-	23.250
Jasminnum azoricum L.	(Oleaceae)	-	5.820	3.230
Luvunga sarmentosa (B1.) Kurz.	(Rutaceae)	-	7.890	42.270
Piper sp.	(Piperaceae)	6.660	8.670	37.750
Randia rugulosa Thw.	(Rubiaceae)	-	6.300	-
Smilax zeytanica L.	(Liliaceae)	-	13.780	9.590
Strychnos wallichiana DC.	(Loganiaceae)	-	14.010	3.230

1	2	3	4	5
Toddalia asiatica Lam.	(Rutaceae)	-	1.900	16.910
Vipris sp.	(Rutaceae)	-	6.870	4.000
Vitaceae	(Vitacea)	-	7.240	-
Zizyphus rugosa Lam.	(Rhamnaceae)	115.420	7.210	-
Others		17.300 (1)	18.670 (4)	24.800 (2)

Values in parenthesis represent number of species which have less than 5 IVI values and unidentified species.

The shrub community of the MDF was dominated by *Zizypus* sp. and *Thespesia lampas.* Whereas the EF was dominated by *Calamus* followed by *Memecylon* and *Symplocos macrophylla.* The HEF was dominated by *Saprosma corymbosum, Calamus travancoricus, Calamus brandiss* and *Nilgrianthus folious.* Rubiaceae is the dominant family in terms of species richness for shrub community in the EF and HEF sites.

The herbaceous community was dominated by *Themeda cymbaria* in MDF. Whereas the EF was dominated by *Curculigo, Strobilanthus, Chasalia ophioxyloides* and *Memecylon* sp. HEF was dominated by *Leptochllus lanceolatus* and *Rungia wightiana* followed by *Ophiorrhiza, Memecylon* and *Curcullgo orchioides.* The MDF was poorly (3 species only) represented by large lianas. While 20 species of lianas were recorded in the EF. Similarly, HEF was represented by 14 species. However, lianas (>10 cm DBH) occurred in low densities in these forests. *Calycapteris florlbunda* was the dominant species in the MDF. While *Ancistrocladus hyneanus, Entada, Connarus wightii, Canthium angustifolium, Randia rugulosa* and *Luvunga sarmentosa* were dominant species in the EF. HEF sites were dominated by C. *wightii* and -L. *sarmentosa.*

Combretaceae and Euphorbiaceae were dominant families in terms of species richness in the MDF (Table 14.2). Whereas in the EF, Myrtaceae, Euphorbiaceae, and Moraceae were dominant families in terms of species richness, while, HEF was dominated by Lauraceae and Euphorbiaceae. However, the familial importance value indices of trees showed that the Combretaceae was the most important family in the MDF. Whereas in the EF and HEF, Dipterocarpaceae and Euphorbiaceae contributes most to the forest tree community structure.

Table 14.2 : Family Importance value (FIV) of tree species in the tropical forest ecosystems at Kodayar in the Western Ghats

Family	*MDF*	*EF*	*HEF*
1	2	3	4
Acanthaceae	-	-	3.579 (1)
Anacardiaceae	32.138 (1)	6.684 (2)	11.300 (2)
Annonaceae	7.202 (1)	-	3.493 (1)
Apocynaceae	-	6.466(2)	-

1	2	3	4
Bignoniaceae	-	7.232 (1)	-
Bombacaceae	-	-	51.670 (1)
Bridiliaceae	-	3.368 (1)	-
Clusiaceae	-	5.586 (2)	15.068 (1)
Combretaceae	121.725 (5)	-	-
Cornaceae	-	-	6.543 (1)
Cycadaceae	7.343(1)	-	-
Dilleniaceae	22.785 (1)	-	-
Dipterocarpaceae	-	88.633 (2)	-
Ebenaceae	-	9.792 (2)	6.182 (1)
Euphorbiaceae	28.873 (2)	9.139 (3)	85.856 (4)
Fabaceae	37.320 (1)	3.649 (1)	-
Flacourticeae	-	11.786 (2)	3.483 (1)
Icacinaceae	-	3.701 (1)	9.501 (1)
Lauraceae	-	5.901 (2)	26.157 (7)
Lecythidaceae	26.848 (1)	-	-
Meliaceae	-	5.109 (2)	21.893 (2)
Moraceae	-	11.133 (3)	4.862 (1)
Myristicaceae	-	-	18.620 (1)
Myrtaceae	-	39.722 (5)	9.058 (2)
Podocarpaceae	-	-	3.599 (1)
Rhizophoraceae	-	16.393 (1)	-
Rosaceae	-	3.080 (1)	-
Rubiaceae	-	17.500 (1)	9.549 (2)
Sapindaceae	-	4.809 (1)	-
Staphyleaceae	-	-	3.886 (1)
Sterculiaceae	-	8.553 (2)	-
Verbenaceae	-	8.501(1)	
Xanthophyllaceae	-	5.416 (1)	-
Unidentified	15.753 (2)	17.937 (5)	5.734 (1)

Values in parentheses represent number of species.

Table 14.3: Stem density, basal area, species diversity and dominance Indices of different life-forms in tropical forest ecosystems at Kodayar in the Western Ghats.

Forest type	*Altitude (m)*	*No. of species**	*Stem density trees/ha⁻*	*Basal area m^2/ha*	*Species diversity index*				*Dominance index*			
					Tree	*Shrub*	*Herb*	*Liana*	*Tree*	*Shrub*	*Herb*	*Liana*
MDF	300	58	450	.28.05	2.196	1.063	0.910	1.224	0.143	0.424	0.639	0.345
EF	500	158	862	90.44	2.612	1.916	2.380	2.249	0.153	0.240	0.113	0.216
HEF	1150	102	1256	79.508	2.276	1.709	2.749	2.071	0.122	0.211	0.151	0.180

*includes trees, shrubs, herbs, tree seedlings and saplings.

Stem density and basal area are shown in Table 14.3. Species richness in terms of number of species was greatest in the EF with 158, and lowest in the MDF with 58 species. Stem density and basal area of the EF were twice greater than the MDF. While, HEF sites showed greater density and low basal area when compared to the EF. Similarly, tree sapling and seedling density and basal area of the EF was greater than the MDF. Diversity indices were greater in the EF when compared to the MDF. Dominance index value was greater in MDF compared to other study sites except for tree species.

Similarity indices of tree species are presented in Table 14.4. No similarity was found between MDF and other study sites. Low similarity was observed among the EF and HEF sites.

Table 14.4: Similarity indices of different life forms in tropical forests at Kodayar in the Western Ghats.

	MDF	*EF*	*HEF*
MDF	-	0 T	0 T
		0 S	0.074 S
		0 H	OH
		0.16 L	0.105 L
EF	-	0.130 T	0.323 S
		0.071 H	0.667 L
HEF			-

T - Trees; S - Shrubs; H - Herbs; L - Lianas

Regeneration potential

Tree saplings: (>3 cm - <10 cm DBH)

Most of the dominant tree species had good representation in the saplings (Table 14.5). In MDF, dominant saplings were *B. lanzan* and *Emblica officinalis.* Whereas the EF was dominated by *I. brachiata, S. laetum, H. parviflora O. erythrocarpa* and *Psychotria* sp. HEF was dominated by *Octotropis travancorica* followed by *A. meeboldii D. oblongifolia* and Jsonandra *lanceolata.* Eighteen, 35 and 40 species of saplings were recorded from the MDF, EF and HEF, respectively.

Table 14.5: Tree seedlings and saplings recorded in the tropical forest ecosystems at Kodayar in the Western Ghats.

Name of the species		*Density of Species/ha*		
		MDF	*EF*	*HEF*
1	2	3	4	5
Acrocarpus fraxinifoliu Wt.	(Leguminosae)	-	4	-

1	2	3	4	5
Actinodaphone campanulata J. Hk.l	(Lauraceae)	- -	- -	68 (4)
Aglaia barberi Gamble	(Meliaceae)	- -	- (8)	520 (56)
A. Jainii Viswa & Ramachan	(Meliaceae)	-	76	-
Agrostistachys indic Dalz.	(Euphorbiaceae)	- -	188 (26)	904 (48)
A. meeboldii Pax. & K. Hoffm.	(Euphorbiaceae)	-	-	694 (204)
Albizia sp.	(Leguminosae)	4	-	-
Alstonia scholaris R. Br.	(Apocynaceae)	2	6	-
Antiaris toxicaria Lesch.	(Moraceae)	-	2	-
Antidesma menasu Miq. ex Tul.	(Euphorbiaceae)	- -	6 (4)	8 (20)
A. zeylanicum lam.	(Euphorbiaceae)	- -	40 (8)	- -
A. fusiformis Thw.	(Euphorbiaceae)	34	-	-
Arenga wightii Griff.	(Arecaceae)	-	22	-
Artocarpus heterophyllus Lam.	(Moraceae)	- -	30 -	16 (4)
A. hirsutus Lam.	(Moraceae)	2	-	-
Atalantia wightii Tanaka	(Rutaceae)	-	2	-
Baccaurea courtallensis M. Arg.	(Euphorbiaceae)	- -	20 (2)	- -
Beilschmiedia gemmiflora Kosterm	(Lauraceae)	- -	48 -	76 (8)
Bridelia crenulata Roxb.	(Euphorbiaceae)	52 (8)	- -	- -
Buchanania lanzan Spr.	(Anacardiaceae)	636 (82)	- -	- -
Calophyllum polyanthum Wall. ex Choisy	(Clusiaceae)	- -	56 -	148 (24)
Careya arborea Roxb.	(Lecythidaceae)	534 (26)	- -	- -
Caryota urens L.	(Arecaceae)	-	312	-
Casearia bourdillonii Mukh.	(Flacourtiaceae)	-	-	60

1	2	3	4	5
Cassia fistula L.	(Caesal pinaceae)	12 (18)	- -	- -
Chionanthus teprocarpa Thw.	(Oleaceae)	- -	108 (10)	- -
Chrysophyllum lanceolatum (Bl)DC.	(*Sapotaceae*)	- -	4 (2)	- -
Cinnamomum malabathrum (*Burm.* f.)Bl.	(Lauraceae)	-	14	256
C. *verum J.S. Presl.*	(Lauraceae)	- -	- -	1212 (56)
Clerodendrum infortunatum L.	(Verbenaceae)	-	-	16
Cryptocarya bourdillonii Gamble	(Lauraceae)	- -	- -	76 (8)
Cullenia excelsa Wt.	(Bombacaceae)	- -	- -	436 (12)
Cycas sp.	(Cycadaceae)	22	-	-
Dalbergia latifolia Roxb.	(Leguminosae)	40	-	-
Dillenia pentagyna Roxb.	(Dilleniaceae)	20 (16)	- -	- -
Dimocarpus longan Lour.	(Sapotaceae)	- -	1304 (8)	- -
Diospyros bourdilloni Brandis	(Ebenaceae)	-	(2)	-
D. paniculata Dalz.	(Ebenaceae)	-	(6)	-
D. pruriens Dalz.	(Ebenaceae)	- -	166 (46)	- -
Diospyros sp.	(Ebenaceae)	- -	12 -	16 (12)
Diotacanthus sp.	(Acanthaceae)	-	-	(84)
Drypetes oblongifolia (Bedd.) A. Shaw	(Euphorbiaceae)	- -	- -	1400 (112)
Emblica officinalis Gaertn.	(Euphorbiaceae)	196 (102)	- -	- -
Epiprinus mallotiformis (Mueller) Croizat	(Euphorbiaceae)	- -	- -	(4) (36)
Erythroxylum obtusifolium J. Hk.	(Erythroxylaceae)	-	(2)	-
Eugenia thwaitesii Duthie	(Myrtaceae)	-	(8)	-
Euonymus sp.	(Clesteraceae)	-	-	(4)

1	2	3	4	5
Excoecaria crenulata Wt. Ic. t.	(Euphorbiaceae)	-	-	6
Fahrenheitia zeylanica (Thw.) A. Shaw	(Euphorbiaceae)	-	2	-
Ficus hispida L.f.	(Moraceae)	-	(2)	-
Ficus sp.	(Moraceae)	-	6	-
		-	(2)	-
Garcinia travancorica Bedd.	(Clusiaceae)	-	116	-
		-	(6)	-
Glochidion ellipticum W.	(Euphorbiaceae)	-	2	-
Gluta travancorica Bedd.	(Anacardiaceae)	-	4	-
Glycosmis mauritiana (Lam.) Thw.	(Rutaceae)	-	(4)	-
Gomphandra tetrandra Sleumer	(Icacinaceae)	-	52	52
		-	(4)	(44)
Grewia tiliaefolia Vahl.	(Tiliaceae)	14	-	-
Holigarna arnottiana J. Hk.	(Anacardiaceae)	-	-	4
		-	-	(8)
Hopea parviflora Bedd.	(Dipterocarpaceae)	-	82	-
		-	(94)	-
Hopea sp.	(Dipterocarpaceae)	-	(6)	-
Hunteria corymbosa Roxb.	(Apocynaceae)	-	12	-
Hydnocarpus alpinia Wt.	(Flacourtiaceae)	-	298	-
		-	(32)	-
Isonandra lanceolata Wt.	(Sapotaceae)	-	302	804
		-	(20)	(84)
Ixora brachiata Roxb.	(Rubiaceae)	8	2 166	-
		-	(134)	-
I. lawsonii J. Hk.	(Rubiaceae)	-	18	-
I. notoniana Wall.	(Rubiaceae)	-	10	80
		-	-	(4)
Ixora sp.	(Rubiaceae)	-	156	-
		-	(38)	-
Ixora sp.	(Rubiaceae)	-	6	-
Kingiodendron pinnatum Harms	(Fabaceae)	-	22	-
		-	(6)	-
Lepisanthes decipiens (Wt. & Arn.) Thw.	(Sapotaceae)	-	20	-
		-	(2)	-
Litsea laevigata (Nees) Gamble	(Lauraceae)	-	(2)	-

1	2	3	4	5
L. mysorensis Gamble	(Lauraceae)	- -	- -	360 (36)
Macaranga peltata (Roxb.) Mueller	(Euphorbiaceae)	6	-	-
Mallotus beddomei J. Hk.	(Euphorbiaceae)	- -	- -	23 (28)
M. muricatus Wt. Bedd.	(Euphorbiaceae)	-	-	(8)
M. philippensis M. Arg.	(Euphorbiaceae)	- -	24 (4)	- -
M. stenanthus M. Arg.	(Euphorbiaceae)	- -	- (2)	312 (4)
Maesa inica D.C.	(Myrsinaceae)	-	(2)	-
Mangifera indica L.	(Anacardiaceae)	-	14	-
Mastixia arborea (Wt.) Bedd.	(Cornaceae)	- -	- -	12 (4)
Meiogyne ramarowii (Dunn) Gandhi	(Annonaceae)	- -	14 (2)	- -
Memecylon sp.	(Melastomaceae)	-	68	-
Mesua ferrea L.	(Clusiaceae)	- -	78 (16)	- -
Microtropis wallichiana Wt.	(Celastraceae)	-	(2)	-
Miliusa wightiana J. Hk. & Thw.	(Annonaceae)	-	340	16
Murraya paniculata (L.) Jacq.	(Rutaceae)	- -	8 -	20 (4)
Myristica dactyloides gaertner	(Myristicaceae)	- -	32 (4)	208 (44)
M. malabarica Lam.	(Myristicaceae)	-	(4)	-
Nageia wallichiana (Presl.) Kuntze	(Podocarpaceae)	-	-	16
Neolitsea zeylanica (Nees) Merr.	(Lauraceae)	- -	446 -	816 (4)
Nothopedytes foetida (Wt.) Sleumer	(Icacinaceae)	-	(2)	-
N. racemosa (Dalz.) Ramam.	(Anacardiaceae)	- -	- -	328 (76)
Octotropis travancorica Bedd.	(Rubiaceae)	- -	- -	636 (324)
Odina wodier Roxb.	(Anacardiaceae)	4	-	-

1	2	3	4	5
Olea dioica Roxb.	(Oleaceae)	192 (8)	- -	4 -
Orophea erythrocarpa Bedd.	(Annonaceae)	- -	74 (8)	- -
O. ramarowii (Dun) Gandhi	(Annonaceae)	- -	- (6)	476 (60)
Pavetta thomsonii Berm.	(Rubiaceae)	-	(10)	-
Persea macrantha (Nees) Kosterm.	(Lauraceae)	- -	- -	136 (8)
Phaeanthus malabaricus Bedd.	(Annonaceae)	- -	476 (64)	- -
Phoenix loureirii Kunth.	(Arecaceae)	4 (2)	- -	- -
Polyalthia wightii	(Annonaceae)	72 (12)	- -	- -
Prunus ceylanica (Wt.) Miq.	(Rosaceae)	- (4)	90 (14)	16 -
Psychotria anamallayana Bedd.	(Rubiaceae)	- -	- -	1612 (56)
P. globicephala Gamble	(Rubiaceae)	-	(2)	-
P. nigra (Gaertn.) Alston	(Rubiaceae)	- -	398 -	1456 (24)
Psychotria sp.	(Rubiaceae)	- -	3088 (74)	- -
Pterocarpus marsupium Roxb.	(Leguminoseae)	470 (2)	- -	- -
Pterospermum Diversifolium Bl.	(Sterculiaceae)	- -	76 (2)	- -
P. rubigionosum Wt. & Arn.	(Sterculiaceae)	2 -	- (2)	- -
Sapindus emerginatus Vahl.	(Sapindaceae)	-	(2)	-
Schefflera racemosa Harms.	(Araliaceae)	-	-	8
Scolopia crenata (Wt. & Arn.) Clos.	(Flacourtiaceae)	- -	698 (22)	20 -
Stereospermum personatum (Hassk). Chatterjee	(Bignoniaceae)	- -	2 (10)	- -
Symplocos racemosa Roxb.	(Symplocaceae)	-	-	168

1	2	3	4	5
Syzygium laetum Gandhi	(Myrtaceae)	-	642	120
		-	(168)	-
S. gardneri Thw.	(Myrtaceae	-	110	-
			(4)	(4)
S. mundagam (Bourd.) Chithra	(Myrtaceae)	-	12	168
		-	(4)	(12)
Syzygium sp.	(Myrtaceae)	-	(14)	-
T. bellirica Roxb.	(Combretaceae)	6	-	-
		(4)	-	-
T. chebula Retz.	(Combretaceae)	190	-	-
		(2)	-	-
T. paniculata Roth.	(Combretaceae)	1026	-	-
		(46)	-	-
T. tomentosa W. & A.	(Combretaceae)	92	-	-
Tricalysia apiocarpa Gamble.	(Rubiaceae)	-	-	88
		-	(2)	-
Turpinia malabarica Gamble.	(Staphyleaceae)	-	-	8
Vateria indica L.	(Dipterocarpaceae)	-	2	-
Vitex altissima L.	(Verbinaceae)	8	-	-
		-	(4)	-
Walsura trifolia Harm.	(Meliaceae)	-	16	-
Wendlandia notonaiana W. ex Wt. & Arn.	(Rubiaceae)	-	(2)	-
Xanthophyllum flavescens Roxb.	(Xanthophyllaceae)	2	230	-
		-	(62)	-
Others		10	474	130
		2*	7*	4*
		(10)	(24)	(64)
		(3)*	(7)*	(3)*

Values in the parenthesis represent number of saplings in different study sites; * refers to the number of unidentified species in the study sites.

Tree seedlings: (<3 cm DBH)

Most of the dominant species had good representation in seedlings except *H. parviflora* in EF (Table 14.5). In the MDF, dominant species of seedlings were *T. paniculata*, *B. lanzan*, *C. arborea*, *E. officinalis*, *O. dioica* and *T. chehula*. Whereas the EF was dominated by *Psychotria* sp. followed by I. *brachiata*, *D. longan*, *O. erythrocarpa*, *P. nigra* and *S. laetum*. However, HEF was dominated by *P. anamallayana*, *P. nigra*, *D. oblongifolia*, *C. verum* and A. *indica*. Twenty nine, 56, and 50 species of tree seedlings were recorded in MDF, EF and HEF, respectively.

Importance value indices of common herbaceous species in man-modified ecosystems in Western Ghats are presented in Table 14.6. In rubber plantations *E. odoratum, L. camera* and other grasses were dominant. *P. polystachyom, H. isora* and *Hyptis suaveolens* were dominant species under teak plantations. *S. nodiflora, A. conyzoides, H. suaveolens* and *E. odoratum* were some of the exotic weeds recorded in agroecosystems. In man-modified ecosystems and wastelands, exotic plants contributed about 37% of the total importance value indices, whereas under natural forest ecosystems their contribution nil except in disturbed regions (c.f.Table 14.1). Diversity index of the herbacoeous community showed that the greater value in the wastelands followed by agroecosystems, rubber plantations and teak plantations. Whereas dominance index showed reverse trend.

Phytosociological analysis of herbaceous community under various disturbance regimes are presented in Table 14.7. Recurrence of annual fires inhibited the exotic plant invasions compared to occational fire regime and mechanical disturbance regime. Under the sites where recurrence of annual fires are common, the grasses such as *T. cymbaria, P. polystachyom, C. travancorensis* and *I. cylindrica* dominated the herbaceous community. *E. odoratum* could invade here but without much success. Greater infestation of exotic weeds found under mechanical disturbance regime combined with occational fire disturbance regime. Diversity index was greater under mechanical disturbance regime compared to others, while, dominance index was greater under annual fire regime compared to others.

Table 14.6 : Importance value indices of herbaceous species during the peak growth period (November) in man-modified ecosystems at Kodayar in Western Ghats of Tamilnadu.

Name of the species	*Rubber*	*Teak*	*Agroeco-systems*	*Wasteland*
1	2	3	4	5
Exotic species				
Ageratum conyzoides L.	8.5	-	21	4
Eupatorium odoratum L.	54.5	53.5	7	39
Hyptis suaveolens (L.) Poit	8.5	12.5	6	23
Imperata cylindrica L.	9.0	-	3	-
Lantana camara L.	29.0	4.0	-	60
Mimosa pudica L.	-	-	5	5
Syndrella nodiflora (L.) Gaertn	4.0	11.0	64	28
Native species				
Barleria sp.	1.0	-	25	16
Borreria hispida (L.) K. Sch.	11.5	-	11	7
Cassia tora L.	-	-	-	21
Clitoria ternatea L.	1.0	8.5	-	-
Digitaria ciliaris Koel.	56.0	10.5	20	25
Helicteres isora L.	-	37.0	11	10
Isachne sp.	47.5	-	22	-

1	2	3	4	5
Pennisetum polystachyon (L.) Sch.	32.5	65.0	-	-
Scleria lithosperma L.	3.5	11.0	-	-
Themeda cymbaria L.	5.0	43.0	-	-
Thespesia lampas Dalz & Gibs	-	14.0	-	-
Urena lobata L.	3.5	-	44	5
Tree seedlings and others	25.0 (5)	29.5 (6)	60 (7)	55 (10)
Diversity index	2.38	2.30	2.43	2.66
Dominance index	0.10	0.13	0.11	0.09

Herbaceous community under the fallow lands are shown in Table 14.8 . One year old fallow was dominated by the herbaceous exotic weeds like *A. conyzoides*, *H. suaveolens* and *S. nodiflora*, in the subsequent fallows they are replaced by shrub community exotic weeds like *E. odoratum* and *L. camera*. After 5 years fallow regrowth of total exotic plants IVI reduced sharply and in a 20 year old fallow, they were almost replaced by shrub and tree community.

Table 14.7 : Importance value indices of herbaceous community under various disturbance regimes at Kodayar in Western Ghats of Tamilnadu.

Name of the Species	*Annual fire*	*Occational fire*	*Mechanical disturbance*	*Mechanical + fire disturbance*
Exotic species				
Ageratum conyzoides L.	-	-	6.6	4.1
Eupatorium odoratum L.	19.3	6.6	98.7	33.8
Hyptis suaveolens (L.) Poit	-	-	12.7	23.0
Imperata cylindrica L.	17.7	-	6.7	-
Lantana camara L.	-	8.2	5.2	60.2
Mimosa pudica L.	-	-	-	5.4
Syndrella nodiflora (L.) Gaertn	-	-	2.0	28.4
Native species				
Cymbopogan travancorensis Bor.	42.3	-	-	-
Pennisetum polystachyon (L.) Sch.	44.1	-	64.7	-
Themeda cymbaria Hackel.	159.0	-	9.9	-
Other grasses	10.3	102.5	27.0	26.1
Dicots	7.3	128.2	66.5	114.0
Diversity index	**1.46**	**2.30**	**2.96**	**2.43**
Dominance index	**0.33**	**0.13**	**0.18**	**0.10**

Table 14.8 : Importance value indicies of plant species in different successional communities following shifting cultivation at Kodayar in Western Ghats of Tamil Nadu.

Name of the Species	*Age of the fallow*				
	1	*3*	*5*	*8*	*20*
Exotic species					
Ageratum conyzoides L.	26.43	8.2	1.1	-	-
Eupatorium odoratum L.	62.40	112.03	142.8	77.9	2.1
Hyptis suaveolens (L.) Poit	44.16	35.29	22.1	-	-
Imperata cylindrica L.	-	1.82	-	4.6	-
Lantana camara L.	18.52	28.12	24.21	19.4	5.2
Syndrella nodiflora (L. (Geartn)	32.58	10.82	1.82	-	-
Native species					
Barleria sp.	2.17	-	-	-	-
Borreria hispida (L.) K. Sch.	4.9	4.2	1.92	-	-
Clitoria ternatea L.	2.02	0.02	2.42	5.75	-
Digitaria ciliaris Koel.	-	2.09	4.28	7.04	22.8
Hemedesmus indicus L.	4.4	2.95	3.32	1.81	-
Isachne sp.	-	4.17	2.18	12.42	3.6
Leucus biflora L.	-	1.82	-	-	-
Penicum sp.	-	1.93	4.46	-	12.2
Pennisetum polystachyon (L.) Sch.	77.7	49.99	52.1	72.87	-
Scleria lithosperma L.	-	-	1.82	2.86	-
Urena lobata L.	12.68	4.24	-	-	-
Tree seedling and others	12.64	24.31	35.47	95.41	254.1

Similarity indices of herbaceous community under natural and man-modified ecosystems are presented inTable 14.9. No similarity was occurred between the Evergreen forest and man-modified ecosystems. Low similarity indicies (0.12-0.35) was recorded in between the MDF and man-modified ecosystems. Greater similarity was observed among the rubber plantations, agroecosystems and wastelands. The common noxious plants in natural and man-modified ecosystems and their local uses were presented in Table 14.10.

DISCUSSION

High species richness is one of the characteristic features of humid tropical forest ecosystem (Parsons and Cameron, 1974). The Western Ghats harbour over 3500 species of flowering plants out of 17,000 species described in India (Khoshoo, 1995; Meher-Homji and Pascal, 1996). The Agasthyamalai range in the southern Western Ghats includes more than 2000 (57%) of 3500 species. Of these, 170 plant species occurred in the st..dy sites at Kodayar.

Table 14.9 : Similarity indices of herbaceous community under natural and man-modified ecosystems at Kodayar in the Western Ghats.

	MDF	*EF*	*Rubber*	*Teak*	*Agroecosystem*	*Wasteland*
MDF	-	-	0.25	0.33	0.12	0.19
EF		-	0	0	0.06	0
Rubber			-	0.42	0.67	0.67
Teak				-	0.43	0.45
Agroecosystem					-	0.68
Wasteland						-

The species of trees >30 cm GBH, over 140 species have been recorded in the Malayan forest (Whitmore, 1984). The number of tree species estimated in the Western Ghats varied; Silent Valley was 84 per ha (Singh*et al.*, 1981), Nelliampathy was 30 per ha (Chandrashekara and Ramakrishnan, 1994) and Kakachi was 45 per ha (Ganesh *et al.*, 1996). According to Whitmore (1984) in tropical rainforest, the range of tree species count per ha is about 20 to a maximum of 223. The number of tree species >30 cm GBH in Kodayar forest sites ranged from 15-45 per 0.5 ha and this number is at the lower side of the range given in wet evergreen forest. Similar observations are made by others (Gentry, 1988). While studying tree species richness of upper Amazonian forests, Gentry (1988) stated that pronounced dry season and relatively low annual precipitation (2000 mm) factors expected to correlate with low species richness. This could also be partly attributed to (a) smaller sampling area (Sukumar *et al.*, 1992; Condit, 1995) (b) enumerations were restricted to only adult tree species (> 10 cm DBH (Gentry, 1988). In the process juveniles (especially pioneer species) which comes up in the canopy gaps were ignored.

The tree density (>10 cm DBH; ranged 450 -1256 trees/ha) recorded in the present study is at the higher end values reported by Gentry (1988) for Amazonian forests (Range 205-858 trees/ha). The tree density values obtained in the present study was comparable with available reports in southern Western Ghats (Chandrashekara and Ramakrishnan, 1994; Ganesh*et al.*, 1996). However, the value was lower than the values reported by Pascal (1988) and Jose *et al.* (1994) in the Western Ghats at Karnataka and Eravikulam (Kerala), respectively. Burgess (1961) recorded a basal area of 73.6 m /ha for trees (>30 cm GBH) over a small area (0.08 ha) at Gum Gum Sabah. Basal area of trees in the Kodayar forest sites were within the range recorded by others (Ganesh *et al.*, 1996). The differences might be attributed to altitude, species composition, age and degree of disturbances.

The tree species (>10 cm DBH) diversity index in the study sites at Kodayar was in the range of 2.2 - 2.6, which was lower than those recorded in tropical rainforest of Barro Colarado Island (4.8; Knight, 1975), Silent Valley (4.89; Singh *et al.*, 1981), Nelliampathy (2.95; Chandrashekara and Ramakrishnan, 1994), and Kakachi (3.37; Ganesh *et al.*, 1996). Differences in the computations of Shannon index could make comparison difficult, because of the differences in the area sampled, lack of uniform plot dimension and standard girth or diameter classes. On the other hand, the value obtained for the concentration of

Table 14.10 : Common noxious plants in natural and man-modified ecosystems and their local uses

Plant name	*Family*	*Origin*	*Occurence*	*Uses*
Ageratum conyoides L.	Asteraceae	Mexico	Abundent weed in agro-ecosystem also spread over in rubber plantation, teak plantation, fallow fields, forest, edges. and wasteland at Kodayar forest[1]	Green manure
Eupatorium adenophorum Sprangel	Asteraceae	Mexico	Above 1200 m in Western Ghats of Kodayar[1] Nilgris, Yercad.[2]	Green manure Medicine
Eupatorium odoratum L.	Asteraceae	America	Upto 1000 m in Western Ghats of Kodayar It is abuntent weed in rubber and teak plantation, grassland, agroecosystems, disturbed forest and wasteland[1].	Green manure Medicine
			It spreads over from Nagarcol to kudal along the coastal areaf Maharastra[2]	Package material
Lantana camara L.	Verbenaceae	Tropical America	Upto 1250 m in Kodayar forest abundent weed in rubber plantation, wasteland and less in disturbed forests.	Household materials fuelwood
Parthenium hysterophorus L.	Asteraceae	America	Wasteland, arid to semi-arid area roadsites and agrocultural fields[1]	Green manure
Prosopis juliflora DC.	Mimosaceae	America	Plains to foot hills, arid to semi-arid area waste-lands and abundant in irrigation tanks of Rama-nad district villages[1,2].	Charcoal making, fuel wood and fodder
Synidrella nodiflora (L.) Gaertner	Asteraceae	W. Indies	Weed in agro-ecosystems, rubber and teak plantation, wastelands and forest edges at Kodayar of Western Ghats.[1]	Green manure
Hyptis suaveolens (L.) Poit	Lamiaceae	Central	Forest edges, rubber and teak plantation agcroecosystems, grasslands at Kodayar[1]	Green manure
Imperata cylindrica (L.) Rausch	Poaceae	Old	Rubber plantation and disturbed forest at Kodayar of Western Ghats[1].	Thatchgrass

1. Present study
2. From Muniappan and Viraktamath (1993)

dominance for the tree layer (0.122 - 0.153) in the present study was greater than the value recorded in Nelliampathy (0.085; Chandrashekara and Ramakrishnan, 1994) and Silent Valley (0.06 -0.14; Singh *et al.*, 1981). The higher dominance value in the present study indicates single species dominance (around 30% of IVI value; c.f.Table 14.1) by *T. paniculata* and *H. parviflora* in both the moist deciduous and evergreen forest, respectively. However, the low value observed in the forest sites at higher elevation might be due to the dominance of more than two (A. *meeboldii*, *C. excelsa* and *D. oblongifollus)* species.

Anthropogenic perturbation to the larger extent can alter the structure and composition of forest ecosystem at Kodayar forests (Chandrasekaran and Swamy, 1995; Sundarapandian and Swamy 1996). Studies are available on the impact of perturbations on nutrient cycling and the relative fragility of forests (Ramakrishnan, 1992; Ram and Ramakrishnan, 1992; Horn and Sanford, 1992; Hunter and Parker, 1993; Turner *et al.*, 1994; Walker and Boneta, 1995; Carcaillet *et al.*, 1997). Fire affects the regeneration both directly through burning of seeds, seedlings and saplings and indirectly through its action on the soil. The later is often superficial and brief duration, through surface temperature may be high, reducing the organic matter, modifying soil texture, facilitating erosion and reducing nitrogen fixation (Schnell, 1971). Recurrence of annual fire remove the ground vegetal cover and litter fall and increased evaporation, which results to loss of soil moisture (Christenson and Muller, 1975), loss of clay content (Dawkins, 1939) and soil nutrients (Kouffman *et al.* 1993). Greater loss of nutrients during fire due to volatilization are reported by others (Alien, 1964; Lloyd, 1971; Walker and Boneta, 1995). In the present study MDF has been subjected to annual fire. This may be the reason for *Themeda* dominated grassland with *T. paniculata* and *P. marsuplum* in the present study. Repeated burning and other factors (disturbance) alter the course of succession by giving an advantage to certain species over others such as *T. paniculata* and *T. cymbaria* in the present study. This often leads to the replacement of forests by grasslands closely simulating an edaphic and climatic climax (Richards, 1952). It may also lead to the formation of scrub with scattered medium sized trees over large area. On the related question of the forest savannah boundary, most authors concur that most savannahs have an anthropogenic origin (Budowski, 1965; Schnell, 1971). Recurrence of annual fire leads to decline of diversity.

Though many forests in the Western Ghats are reported to have poor regeneration (Sukumar *et al.* 1992), the dominant species at Kodayar forest site showed adequate regeneration. However, the dominant species *H. parviflora* in evergreen forest showed poor regeneration (82 seedlings/ha). Poor regeneration of this species might be due to shorter viability of seeds and also could be greater seed predation as observed in the present study. Several interacting factors contribute to successful colonization by exotic plant species. Obviously climatic and edaphic similarities between the original habitat of the exotics and the new habitat are the important factors (Holdgatel, 1986). Thus, it is reasonable to expect high degree of colonization of invaders under open habitats with reduced competition created by a high degree of disturbance in man modified ecosystems, in the present study compared to that of natural forest ecosystems. *E. odoratum* and *L. camera* are of latin American origin (Ramakrishnan, 1992) produces abundant seeds dispersed by wind and birds respectively also regenerate through vegetative means. Variation in the degree of infestation of these weeds observed in the present investigation may be because of initial population structure, propagule size, site quality and the degree of disturbance. This high degree of infestation of *E. odoratum* under rubber plantation may be attributed the above reasons and also may be because of improper mechanical weeding.

Under regular annual fire regimes the stress tolerant grasses dominated the community. Bharucha and Shankaranarayan (1958) reported that fire stimulated the lush growth of grasses such as *Arlstida, Dichanthium, Themeda* etc. *E. odoratum* failed to colonize successfully under these envii onments may be because of competition for resources with stress tolerant grasses. If they succeed their growth to maturity was much slower here. Mecdonald (1993) reported that *E. odoratum* enabled savannah fires to have a grater impact on the forest/savannah marginal zone. In many cases, *L. camera, E. odoratum,* and *S. nodi flora co* existed. Where two or more exotic species co-exist in the same area, they may possibly do through mutual avoidance (Ramakrishnan, 1991). An important attribute of exotic invaders in the wide ecological amplitude, either through extreme phenotypic plasticity or due to ecological races adopted to specific habitats (Ramakrishnan 1991) and also allelopathic properties of exotic invaders that confer up on them a distinct edge over the existing native species. *E. odoratum* and L. *camera* being not normally consumed by cattle, these species have a selective advantage over palatable native species (Lal and Karia, 1969; Sastry).

Initial phase of secondary succession mostly determined by available propagule source in the soil, previous land use pattern and kind of disturbance (Ramakrishnan, 1992). It is reasonable to expect that the dominance of exotic weeds in abundance fallow lands because of their mode of regeneration like production of large number and highly mobile seeds to disturbed ecosystems such as fallowlands. The importance of these strategies were stressed by others (Hkayashi and Numuta, 1968; Salisbury, 1942). The reduction of exotic plants dominance in 8, 20 year fallow lands confirms with Toky and Ramakrishnan (1983) and Swamy and Ramakrishnan (1987).

It is concluded that the forest ecosystem at Kodayar is superior to the mid and low elevation forests in Western Ghats in terms of structural features of vegetation density, diversity and floristic composition. Recurrence of annual fires and anthropogenic perturbation leads to form a monotypic tree dominated grassland and allows alien plant invasion. Therefore, the forest ecosystems at Kodayar should be given due importance for conservation of its rich biodiversity. Effective management of exotic plant invasion should be given immediate attention in order to avoid further shrinkage of effective grassing land area and also be erosion of biodiversity. A wide range of options can be considered while designing management strategies. Considering the socio-economic problems of the country large scale exotic plant biomass utilization may be considered as one of the viable option in addition to chemical and biological control.

ACKNOWLEDGEMENTS

We thank the Ministry of environment and forests. Government of India for financial assistance through a research grant and the forest officials for extending help in the field. First author thanks to CSIR for the award of Senior Research Fellowship. We also extend our thanks to Deputy Director, BSI (SC), Coimbatore and Dr. B.R. Ramesh, French Institute of Pondicherry for their help in identification of plants. We also thank UGC for granting special assistance (DSA III) to School of Biological Sciences.

LITERATURE CITED

Alien, S.E. (1964). Chemical aspects of heather burning. *Journal of Applied Ecology*, 1:347-367.

Bharucha, F.R. and Shankaranaryanan, K.A. (1958). Studies on the grasslands of Western Ghats, India. *Journal of Ecology*, 46:681-705.

Budowski, G. (1965). Distribution of tropical American rain forest species in the light of successional processes. Turrialba (Costa Rica), 15: 40-52.

Burgess, P.F. (1961). The structure and composition of lowland tropical rain forest in North Borneo. Malayan Forester, 24:66-80.

Carcaillet, C., Barakat, H.N., Panaiotis, C. and Loisel, R.(1997). Fire and late-Holocene expansion of *Quercus ilex* and *Pinus pinaster* on Corsica. Journal of Vegetation Science, 8:85-94.

Chandrashekara.U.M. and Ramakrishnan, P.S. (1994). Vegetation and gap dynamics of a tropical wet evergreen forest in the Western Ghats of Kerala, Indian Journal of Tropical Ecology, 10:337-354.

Chandrasekaran, S. and Swamy, P.S. (1995). Changes in herbaceous vegetation following disturbance due to biotic interference in natural and man-made ecosystems in Westernghats. Tropical Ecology, 36:213-220.

Christensen.N.L. and Muller, C.M. (1975). Effect of fire in factors controlling plant growth in *Adenostoma chaparral*. Ecological Monograph, 45:29-45.

Condit, R. (1995). Research in large, longterm tropical forest plots. Trends in Ecology and Environment, 10:18-22.

Dawkins, C.J. (1939). Tussock formation by *Schoenus nigricans:* The action of fire and water erosion. Journal of Ecology, 27:78-82.

Dobson, A., Jolly, A. and Rubenstein, D. 1989. The green house effect and biological diversity. Trends in Ecology and Evolution, 4: 64-68.

Gamble, J.S.-(1925). Flora of the Presidency of Madras. Adiard and Son, London. Volume 1-3, pp.2017.

Ganesh, T., Ganesan.R., Soubadra Devy, M., Davidar.P. and Bawa.K.S. (1996). Assessment of plant biodiversity at a mid-elevation evergreen forest of Kalakad-Mundanthurai Tiger reserve. Western Ghats, India. Current Science, 71: 379-392;

Gentry, A.H., (1988). Tree species richness of upper Amazonian forests. Procedings of National Academic Science(USA), 85:156-159.

Hayashi, I. and Numuta, M. (1968). Ecology of pioneer species of early stages in secondary succession II. The seed production. The Botanical Magazine, Tokyo, 80:55-66.

Holdgatel, M.W.(1986). Summary and conclusions: Characteristics and consequence of biological invasions. Philosophical Transactions of the Royal Society London.

Horn, S.P. and Sanford, Jr. R.L. (1992). Holocene fires in Costarica. Biotropica, 24:354-361.

Hubbell, S.P. and Foster, R.B. (1192). Short-term dynamics of neotropical forest: Why ecological research matters to tropical conservation and management. Oikos, 63:48-61.

Hunter, J.C. and Parker, V.T. (1993). The disturbance regime of an old-growth forest in coastal California. Journal of Vegetation Science, 4:19-24.

Jose.S., Sreepathy.A., Mohan Kumar, B. and Venugopal.V.K., (1994). Structural, floristic and edaphic attributes of the grassland-shola forests of Eravikulam in Peninsular India. Forest Ecology Management. 65:279-291.

Kauffman, J.B., Cummings, R.L., Salcedo, D.L. and Sampai, I.H. (1993). Biomass and nutrient dynamics associated with slash fires in neotropical dry forest. Ecology, 74:140-151.

Kershaw, K.A. (1973). Quantitative and dynamic plant ecology (2nd Ed.). Edward Arnold, London. 308 pp.

Khoshoo, T.N., (1995). Census of Indians biodiversity Tasks ahead. Current Science, 69: 14-17.

Knight, D.H. (1975). A phytosociological analysis of species rich tropical forest on Barro Colorado Island, Panama. Ecological Monograph, 45: 259-284.

Lal, M. and Keira, D.B. (1960). *Lantana* poisioning in domesticatide animals. Indian.Veterinary Journal, 37:263.

Lloyd, P.S; (1971). Effects of fire on the chemical status of herbaceous communities of the Derbyshire Dales. Journal of Ecology, 59:261-273.

Margalef, R. (1968). Perspective in Ecological theory. University of Chicago Press, Chicago.

MacDonald, I.A.W. (1983). Alien trees, shrubs and creepers invading indigenous vegetation in the Hluhluwe-Umfolozie Game Reserves complex in Natal. Bothalia, 14:949-959.

Meher-Homji, V.M. and Pascal, J.P. (1996). Ecodiversity in the Western Ghats. In: P.S.Ramakrishnan, A.N. Purohit, K.G. Saxena, K.S. Rao and R.K. Maikhuri (Editors) Conservation and Management of Biological Resources in Himalaya. Oxford and IBH Publishing Co.Pvt.Ltd. New Delhi, pp.259-289.

Misra, R. (1968). Ecology work book. Oxford and IBH Publication, New Delhi. 244 pp.

Muniappan, R. and Viraktamath, C.A.(1993). Invasive alien weeds in Western ghats. Current Science, 64:555-557.

Parsons, R.F. and Cameron.D.S. (1974). Maximum plant species diversity in terrestrial communities. Biotropica, 6:202-203.

Pascal, J.P. and Ramesh, B.R. (1987). A field key to the trees and lianas of the evergreen forests of the Western Ghats (India). Institut Francais de Pondicherry, Pondicherry, India. 236 pp.

Pascal, J.P. (1988). Wet evergreen forests of Western Ghats of India: Ecology and Structure floristic composition and anthropoligical impact. Institut Francais de Pondicherry, Pondicherry. pp.26-27.

Ram, S.C. and Ramakrishnan, P.S. (1992). Fire and nutrient cycling in serai grasslands of Cherrapunji in North Eastern India. International Journal of Wildland Fire, 2:131-138.

Ramakrishnan, P.S. (1991). Ecology of Biological invasion: an overview, p.1-19. In: Ramakrishnan, P.S. (Editor) Ecology of biological invasion in the tropics. International Scientific Publications, New Delhi.

Ramakrishnan, P.S. (1992). Shifting agriculture and sustainable development. Unesco, MAB, Paris. 424 pp.

Ramakrishnan, P.S. (1992). Tropical forest: exploitation, conservation and management. Impact. UNESCO., 42:149-162.

Richards, P.W., (1952). The tropical rain forest an ecological study Cambridge Univ. Press. 4th ed. with correction 1972. 450 pp.

Salisbury, E.J. (1942). The reproductive capacity of plants. Bell, London. 244 pp.

Sastry, M.S. (1963). Investigation of toxicity of *Lantana camera* L. Indian Veterinary Journal, 40: 79-82.

Schnell, R. (1971). Introduction of la Phytogeographie Compares des pays tropicaus 2 vol. Paris. Gauthier - Villars, 452 pp.

Simpson, E.H. (1949). Measurement of diversity. Nature (London), 163:688.

Singh, J.S., Singh, S.P., Saxena, A.K. and Rawat, Y.S. (1981). The Silent Valley forest ecosystem and possible impact of proposed hydroelectric project: reports on the Silent Valley study; Ecology Research Circle, Kumaun University, Nainital. 69 p.

Sukumar.R., Dattaraja.H.S., Suresh, H.S., Radakrishnan, J. Vasudeva.R., Nirmala.S. and Joshi.N.V. (1992). Long term monitoring of vegetation in a tropical deciduous forest in Mudumalai, Southern India. Current Science, 62:608-616.

Sundarapandian, SM. and Swamy, P.S. (1996). Influence of herbaceous species composition on fine root biomass production in disturbed and undisturbed deciduous forests of Western Ghats in India. Acta Oecologica, 17:163-176.

Swamy, P.S. and Ramakrishnan, P.S. (1987) Weed potential of Mikania micrantha H.B.K. and its control in fallows after shifting agriculture (Jhum) in North-east India. Agriculture Ecosystems and Environment, 18:195-204.

Tamrat Bekele. (1994). Phytosociology and ecology of a humid Aformontane forest on the central Plateu of Ethiopia. Journal of Vegetation Science, 5:87-98.

Toky, O.P. and Ramakrishnan, P.S. (1983). Secondary succession following slash and burn agriculture in north eastern India. I. Biomass, litterfall and productivity. Journal of Ecology, 72:735-745.

Turner, M.G., Hargrove, W.W., Gardner, R.H. and Romme, W.H. (1994). Effects of fire on landscape heterogeneity in Yellowstone National park, Wyoming. Journal of Vegetation Science, 5:731-742.

Walker, L.R. and Boneta, W. (1995). Plant and soil responses to fire on a fern-covered landslide in Puerto Rico. Journal of Tropical Ecology, 11:473-479.

Whitmore.T.C. (1984). Tropical rainforest of the Far East (2nd Edition). Oxford University Press, Oxford.

Chapter - 15

Phytodiversity and Biological Spectrum of Patnitop and Adjoining Hills in J&K, India

Shashi Kant and Koshal Kr. Sharma**

ABSTRACT

By dint of a range of landscapes and environmental regimes, ranging from perpetual snow cover to equatorial tropical conditions and from mangroves and humid tropics to hot and cold deserts, India nurtures a rainbow of flora and fauna and is recognised as one the world's top "Twelve Megadiversity" nations. The vastness of the plant genetic resource available to us can be gauged from the fact that 238 families comprising of 15,000 angiospermic plant species (16% of global wealth of flowering plants of 425 families recognise to be existing in the world) are known to exist in this sub-continent. The overall plant wealth of the country is assessed to be 45,000 species i.e., 12% of the global plant wealth of which 33% is endemic. During the course of study, extensive exploration of phytodiversity from the study area which falls in north-western Himalaya has been done. More than 375 tracheophyte species (Angiosperms, Gymnosperms and Pteriodophytes) belonging to 105 families have been recorded. The dominant families in the area are Asteraceae (47 sps.) Lamiaceae (25 sps.) Fabaceae (21 sps.). The phytodiversity of the region has been classified into various life forms, which are shown in descending order : Therophytes 29.8%, Hemicryptophytes 26.4%, Chamaehytes 15.2%, Macrophanerophytes 10.4%, Nanophanerophytes 8.5%, Geophytes 3.05%, Hydrophytes 3.2%, Lianas 2.6% and Epiphytes 0.2%. When compared to Raunkiaer's normal forms, it shows thero-hemicryptophytic phytoclimate of the region.

Key words: biological spectrum, life-forms, monotypic, phytoclimate, phytodiversity, Raunkiaer's normal forms, tracheophytes.

*Department of Biosciences, University of Jammu, Jammu - 180006

INTRODUCTION

Planet Earth is endowed with a rich variety of life forms and the teeming millions of living organisms have been well-knit by the laws of nature. The interdependence of the various life forms-starting from the unicellular primary producers to the complexly built higher plants and animals is a unique feature of this green planet.

Biodiversity, as this assemblage of life forms is referred to, has now been acknowledged as the foundation for sustainable livelihood and food security. Scientists have estimated that more than 50 million species of plants and animals, including invertebrates and micro-organisms occur on earth and hardly 2 million of them have been described by man so far.

Biological Diversity refers to the variety and variability among living organisms and ecological complexes in which they occur. Diversity can be defined as the number of different items and their relative frequency. For biological diversity, these items are organized at many levels ranging from complete ecosystems to the biochemical structures that are the molecular basis of heredity (genes). Thus, the term encompasses different ecosystems, species, genes and their relative abundance.

The Indian region alone has given to the World nearly 167 economical plants whose centres of origin/diversity are located in India, along with their–320 species–of wild relatives and land–races.

Ehrlich and Ehrlich (1992) have given four crucial reasons for the preservation of biodiversity. These are:

(a) Ethics

Homo sapiens has a responsibility to act as a conscientious steward towards the other organisms that share the earth with humanity. This is not necessarily altruistic but moral environmental existentialism as it contributes to sustainable development.

(B) Aesthetic Values

Landscape variation influences cultural developments and provides component and/ or situations for gardening, recreation, the visual arts (including nature films), and ecotourism.

(c) Economic Values (Goods)

Humanity has advanced through the development of technologies to utilize many components of biodiversity. These components give rise to marketable products which have an economic value. Agriculture; for example, depends largely on ecosystem components which have been domesticated to produce food and fibre. Many industries are also based on agricultural or silvicultural produce. Modern biotechnology is opening up even more opportunities for–the utilization of ecosystem products.

(d) Ecosystem Services

The persistence of humanity is dependent on a variety of factors amongst the most vital of which are the maintenance of a favourable atmosphere composition (the Gaia hypothesis of Lovelock, 1979), biogeochemical cycling, soil genesis and maintenance, water cycling and pest control.

In relation to ethics, the responsibility of society to preserve biodiversity is not altruistic in the short or the long term. This is because its preservation provides gene pools and gene components which represent opportunities for the future. The benefits of modern biotechnology alone furnish substantive reasons why biodiversity should be preserved. Undoubtedly, future biotechnologists will develop methods that will facilitate greater exploitation of living organisms than currently in use. Scientific advances will, therefore, enhance the value of living organisms. On this basis alone, i.e., that valuable pharmaceutical, industrial, and agrochemical products and environmental improvements, can be generated from or through it, the preservation of biodiversity should be a major component of "sustainable development" policies.

The largely unexplored potential of biodiversity to generate goods with an economic value is another reason for preserving biodiversity. As per Lovejoy (1994) there are many such products already available and others are in the development stage. Some are the familiar products of agriculture and forestry, whilst others are–less familiar and highly specialized.

EROSION OF BIODIVERSITY

Since life began, extinction has always been a part of evolution. Mass extinctions occurred during a few periods, apparently due to the results of relatively abrupt geological or climatic changes. But in most periods the rate of species formation has been greater than the rate of species extinctions, and biological diversity has gradually increased. Recently, in evolutionary history, the human species has derived great economic value from ecosystem, species and genetic diversity and recognised the intrinsic values of diversity. But now that values are being recognised, there is evidence that the world may be entering into another period of massive reduction in diversity. This time, humans are the cause, and it appears that the consequences will be loss of a substantial share of Earth's valuable resources (Biodiversity).

Drastic reductions in populations of wild animals and plants are not new and have long been recognised as consequences of intensive hunting, fishing and gathering. For example, great bison herds of North America were depleted in 19th Century, as were stocks of various whales and birds species. The now barren hills of southern China's coasts and islands were deforested 1,000 years ago. Erosion from the deforested and overgrazed, Armenian hills, which eventually led to the demise of productive agriculture in Mesopotamia, apparently began over 2,000 years ago. These kinds of changes undoubtedly caused–local and regional losses of diversity.

What is new today is the scale on which resource degradation is occurring and thus the rate at which diversity is apparently being reduced. Fishing, hunting, grazing, and extractions beyond the carrying capacity of ecosystems continue today but the effects are being greatly exacerbated by degradation of ecosystems and significant reduction in natural areas.

Acceleration of resource degradation and diversity loss is partly a consequence of population growth especially in rural areas of developing countries where compound growth rates are often more than one percent per year. It is also a consequence of technologies developed over the past century that have enabled human to devastate natural ecosystems even where population densities or growth–rates are moderate.

Biologists and agriculturists have thus become alarmed about the scale of plant and animal resource degradation during last two to three decades. The alarm stems from observations of extensive reduction in habitat, coupled with a growing understanding about how such changes adversely affect diversity.

During the next 20 to 30 years, the World could lose more than a million species of plants and animals–primarily because of environmental changes due to humans. At 100 species per day, this extinction rate will be more than 1,000 times the estimated "normal" rate of extinction. The list of lost, endangered and threatened species includes both plants and animals. About 10% of temperate region plant species and 11% of the world's 90,000 bird species are at some risk of extinction. In the tropics, the destruction of forests threatens 1,30,000 species–which live nowhere else.

CONSERVATION OF BIODIVERSITY

Biological diversity can be preserved for posterity in two ways–*in situ* and *ex situ*. *Ex situ* maintenance of species is provided by botanical gardens, zoos, and aquaria; and of gene pools by germplasm banks (seed stores, *in vitro* collections and field gene banks) and gross-root collections of plant cultivars and animal breeds. The World's zoos have capacity to sustain captive breeding programmes for no more than 900 vertebrate species-perhaps even fewer. Botanical gardens, probably, have a grater capacity with respect to plant species. But clearly it is possible to maintain *ex situ* conservation of only a tiny fraction of World's species.

In situ Conservation is done by protecting areas rich in biodiversity. These include biosphere reserves, natural parks, sanctuaries, etc. The primary objective of this concept is to save for the present and future use, the diversity and the integrity of biotic communities of plants and animals within natural ecosystem and to safeguard the genetic diversity of species on which their continuing evolution depends. Such reserves are to comprise terrestrial and marine ecosystems and coincide with National Parks and Sancturies.

INTRODUCTION TO THE STUDY AREA:

For the present study Patnitop and Adjoining forest areas (J&K State-India) were selected. Patnitop is 109 kms towards North of Jammu city. It is located on Jammu-Srinagar National Highway. The adjoining areas include Chenani, Lapri, Nathatop, Kud, Karla, Shabgarh, Sanasar, Nachhar, Batote, and Devi Ka Padar Forest. The whole study area is in latitude 33°01'23" N to 33°08'37' N and longitude of 75°15'0" E to 75°23'07" E, and lies on both East and West of Jammu-Srinagar National Highway. The study area is the inter zonal region of two Districts of J&K State, from Chenani to Patnitop ridge, (Southern aspect) the area lies in Udhampur District; while from Patnitop ridge towards Batote-Nacchar, (Northern aspect), the area lies in Doda District. The study area is a part of other Western Himalayas (Fig. 15.1 & 15.2).

Physiography of the Study area

The area is highly mountainous terrain, marked with rugged topography comprising of high ranges, steep hills and deep valleys, cliffs and sloping meadows. The flat areas are almost absent. The variation in the altitude is from 1,000 m.a.s.1 (Nachhar) to 2,800 m.a.s.l Shahgarh and Nathatop. The area lies in the outer/lesser Western Himalaya (Southern flanks of the Pir Panchal) just adjacent to Shivaliks. Broadly speaking there are two major watersheds which in turn comprise of many small sub-water-sheds. These are Chenani

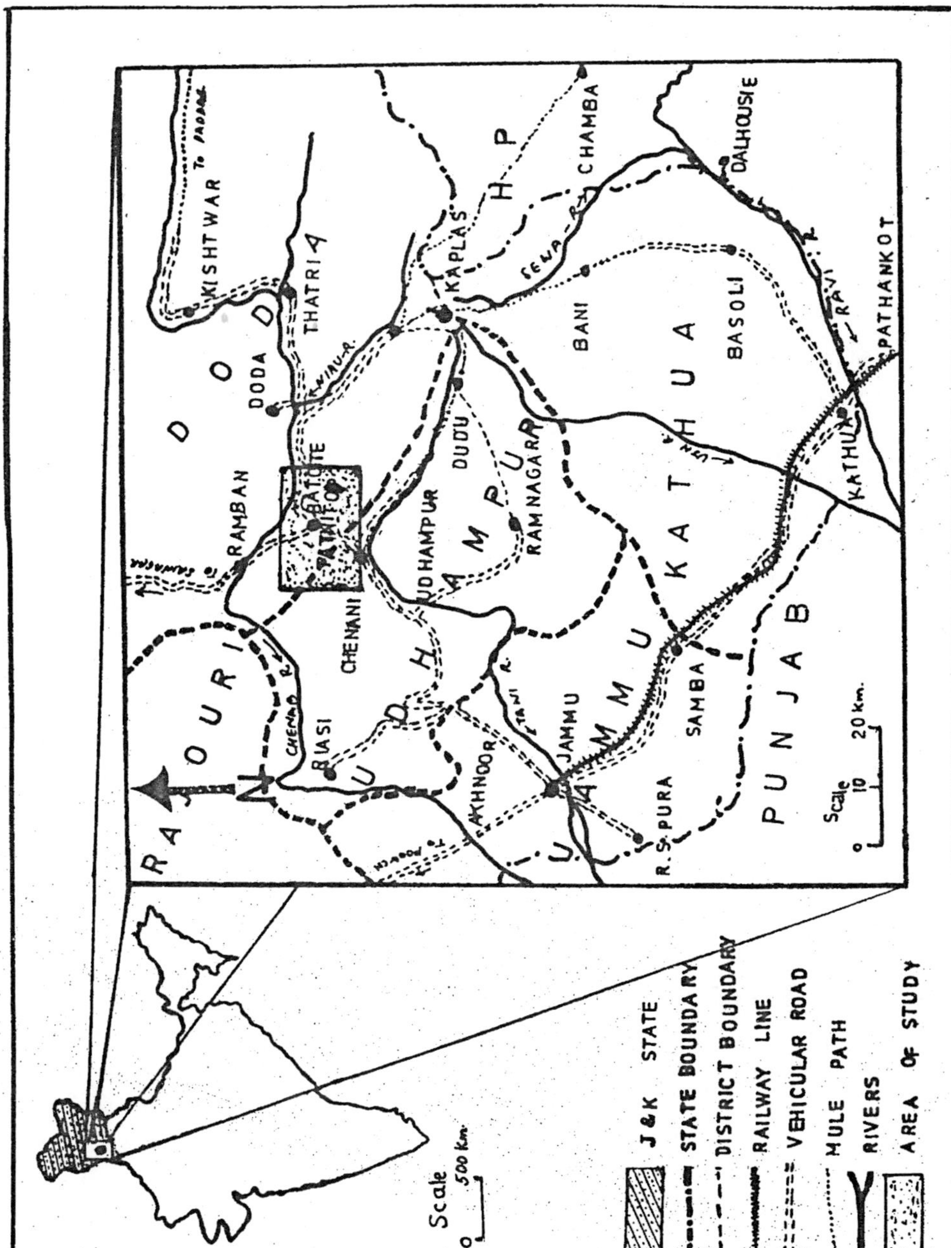

Fig. 15.1 : Location Map Showing the Area of Study

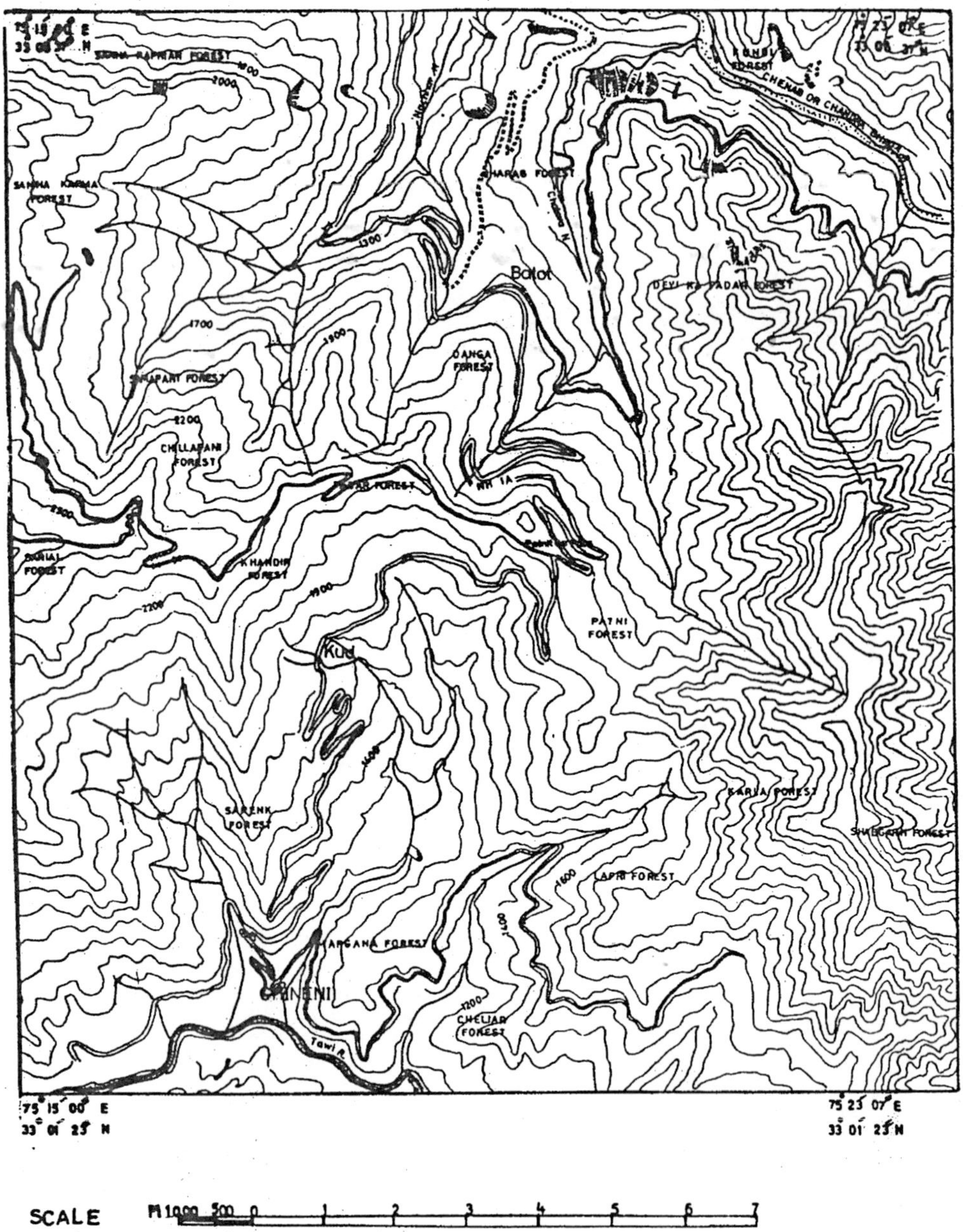

Fig. 15.2 : Detailed Map of The Study Area

watershed which comprises of southern aspect of Patnitop ridge. River Tawi flows through Chenani watershed from Patnitop to Udhampur along the national highway. There are many small streams, and tributaries on the southern aspect of the study area, which merge with Tawi river. Similarly on the northern aspect, there are two big Nallahs–Nachhar and Chakwa flanking the Western and Eastern sides of Batote town, respectively. Many small tributaries of these nallahs join them and ultimately they merge with Chenab river (Chanderbagha) at dugga forests near Pira and Baglior, respectively, in the extreme north of the study area. The two watersheds are connected through two passes crossing the main ridge which are Patni-pass–and Nathatop-pass.

The altitude of various study sites range from 1,000 m.a.s.l to 2,800 m.a.s.l. e.g. Nachhar (1,000 m.a.s.l); Batote (1,600 m.a.s.l.); Sanasar (2,076 m.a.s.l); Chenani (1,200 m.a.s.l); Samroli (1,126 m.a.s.l); Nathatop (2,741 m.a.s.l); Shabgarh (2,800 m.a.s.l); Patnitop (2,060 m.a.s.l); Kud (1,650) m.a.s.l); and Karla (2,051 m.a.s.l). The altitude of different sites have an important role in the climatic change and consequently shaping the type of vegetation at different altitudes.

MATERIALS AND METHODS

For the present study, mountainous terrain "Patnitop and Adjoining Hills" (J&K) India was selected. For the phytodiversity studies, monthly collections of tracheophytes (Trees, shrubs, ferns and grasses) at their blooming stages were made from all the possible habitats of the area such as road sides, along the perennial and seasonal streams and mostly within the forest areas towards the core of the forest upto several hundred meters towards the centre and in different directions of terrain. Monthly collections of the plants were done in the entire area of study to collect the baseline data. The plants collected were pressed, dried and preserved in herbarium sheets by routine methods. In addition to herbarium, plants have also been photographed. For biological spectrum the species of the area have been grouped into various life form classes. The percentage of the various life form classes put together constitutes the biological spectrum. The biological spectrum of the area has been draw and compared with Raunkiaer's normal spectrum as with the spectra available for other localities.

OBSERVATIONS AND DISCUSSION

Floristic composition

There is a great floral diversity in this region. The region is known for not only some endemic species but also is very rich in the distribution of various families of the angiospermic flora. There are as many as 52 families which are monotypic. This is very significant from the point of view of the species diversity under–great danger–or threat of extinction.

Table 15.1 : Pteridophytic diversity with number of genera and species and the status of the families.

S. No.	*Family*	*Genera*	*Species*	
1	*2*	*3*	*4*	
1.	Selaginellaceae	1	1	Monotypic

1	*2*	*3*	*4*	
2.	Equisetaceae	1	1	Monotypic
3.	Ophioglassaceae	1	1	Monotypic
4.	Polypodiaceae	1	2	
5.	Sinopteridaceae	1	1	Monotypic
6.	Cryptogammaceae	1	1	Monotypic
7.	Pteridaceae	1	2	
8.	Adiantaceae	1	3	
9.	Heminitidaceae	1	1	Monotypic
10.	Hypolepidaceae	1	1	Monotypic
11.	Thelypteridaceae	1	2	
12.	Aspleniaceae	1	4	
13.	Athyriaceae	1	3	
14.	Dryopteridaceae	3	9	
15.	Bleachnaceae	1	1	Monotypic
	Total	**17**	**33**	

Table 15.2 : Gymnospermic diversity with number of genera and species and the status of the families.

S. No.	*Family*	*Genera*	*Species*	
1.	Pinaceae	4	5	
2.	Taxaceae	1	1	Mono-typic
		5	6	

Table 15.3 : List of the angiospermic families with number of genera and species along with their status

Dictos

Sl. No.	*Family*	*Genera Species*		
1	*2*	*3*	*4*	*5*
1.	Ranunculaceae	4	6	
2.	Berberidaceae	1	1	Monotypic
3.	Podophyllaceae	1	1	Monotypic
4.	Fumariaceae	2	3	

1	*2*	*3*	*4*	*5*
5.	Brassicaceae	9	13	
6.	Violaceae	1	2	
7.	Polygalaceae	1	1	Monotypic
8.	Caryophyllaceae	3	5	
9.	Hypericaceae	1	2	
10.	Malvaceae	1	1	Monotypic
11.	Tiliaceae	1	1	Monotypic
12.	Geraniaceae	1	2	
13.	Oxalidaceae	1	3	
14.	Balsaminaceae	1	1	Monotypic
15.	Ruttaceae	1	1	Monotypic
16.	Simaroubaceae	1	1	Monotypic
17.	Meliaceae	1	1	Monotypic
18.	Celastraceae	1	1	Monotypic
19.	Rhammaceae	1	1	Monotypic
20.	Sapindaceae	1	1	Monotypic
21.	Hippocastanaceae	1	1	Monotypic
22.	Anacardiaceae	3	3	
23.	Coriariaceae	1	1	Monotypic
24.	Fabaceae	13	21	
25.	Caesalpiniaceae	2	2	
26.	Rosaceae	12	21	
27.	Saxifragaceae	1	1	Monotypic
28.	Crassulaceae	1	1	Monotypic
29.	Hamamelidaceae	1	1	Monotypic
30.	Lythraceae	1	1	Monotypic
31.	Punicaceae	1	1	Monotypic
32.	Onagraceae	2	3	
33.	Cucurbitaceae	1	1	Monotypic
34.	Cactaceae	1	1	Monotypic
35.	Datiscaceae	1	1	Monotypic
36.	Apiaceae	9	11	
37.	Araliaceae	1	1	Monotypic

1	*2*	*3*	*4*	*5*
38.	Caprifoliaceae	2	4	
39.	Rubiaceae	3	5	
40.	Valerianaceae	1	2	
41.	Asteraceae	30	47	
42.	Campanulaceae	1	1	Monotypic
43.	Primulaceae	2	3	
44.	Myrsinaceae	2	2	
45.	Oleaceae	2	3	
46.	Apocynaceae	2	2	
47.	Asclepiadaceae	1	1	Monotypic
48.	Loganiaceae	1	1	Monotypic
49.	Gentiannaceae	2	5	
50.	Boraginaceae	3	5	
51.	Convolvulaceae	2	3	
52.	Cuscutaceae	1	1	Monotypic
53.	Solanaceae	5	6	
54.	Scrophulariaceae	5	9	
55.	Acanthaceae	3	3	
56.	Verbenaceae	2	2	
57.	Lamiaceae	16	25	
58.	Plantaginaceae	1	2	
59.	Amaranthaceae	2	3	
60.	Chenopodiaceae	2	3	
61.	Polygoniaceae	2	8	
62.	Lauraceae	1	1	Monotypic
63.	Thymelliaceae	2	3	
64.	Elaeagnaceae	1	1	Monotypic
65.	Euphorbiaceae	4	6	
66.	Buxaceae	1	1	Monotypic
67.	Ulmaceae	1	1	Monotypic
68.	Cannabinaceae	1	1	Monotypic
69.	Urticaceae	4	4	
70.	Moraceae	2	3	

CONSTITUTIONAL PROVISION FOR ENVIRONMENT IN INDIA

The Government of India within five years of the Stockholm Conference (1972), took the following steps to meet the need for environment protection.

1. The 42nd Constitutional Amendment was made (1976) to include **"Environmental Protection"** as a Constitutional Obligation. **"Article 48A",** it lays down *"The State shall endeavour to protect and improve the environment and to safeguard the fores' and wildlife of the country"*

2. **Article 51A,** relates to fundamental duty. This article runs:

   "It shall be the duty of every citizen of India to protect and improve the natural environment including forests, lakes, rivers and wildlife and to have compassion "for living creatures".

NATIONAL COMMITTEE ON ENVIRONMENT PLANNING AND COORDINATION (NCEPC)

The Committee was also soon set up after Stockholm Conference (1972), and was concerned with issues relating to:

Appraisal of Development Projects

Human Settlement Planning

Survey of Ecosystems (e.g. wetlands, etc.)

Spread of Environmental Education

"TIWARI COMMITTEE" ON ENVIRONMENT

This Committee set up in 1980 by Government of India to make recommendations on environmental issues. It recommended the following:

1. A comprehensive review of the State and Central Acts, e.g. The Water (Prevention and Control) Act, 1974, Indian Forests Act, 1927, Insecticides Act, 1968.
2. New legislation on the areas not covered by the present laws (e.g. toxic substances etc.)
3. The introduction **"Environmental Protection"** in the Concurrent List of the Seventh Schedule.
4. Later the Committee recommended Establishment of a separate **'Department of Environment' (DOE).**
   - DOE was established on **Nov. 1, 1980** by the Government of India.

The following functions were assigned to DOE:

- To act as nodal agency for environmental protection and eco-development in the county.
- To carry environmental appraisal of development projects through other Ministers/Agencies as well as directly.
- To have administrative responsibility for:

  (i) Pollution monitoring and regulations

  (ii) Conservation of critical ecosystems designated as Biosphere Resources

  (iii) Conservation of Marine Ecosystem.

This was followed by creating National Committee on Environment Planning.

NATIONAL COMMITTEE ON ENVIRONMENT PLANNING (NCEP)

However, the NCEP was soon replaced by a National Committee on Environment Planning (NCEP), with the following functions:

- Preparation of an annual **"State of Environment Report"** for the country.
- Establishing an **Environmental Information and Communication System** to propagate environmental awareness through the mass media.
- To sponsor **Environmental Research**
- The arrange **public hearings** or **conferences** on issues of environmental concerns.

Here the word "ENVIRONMENT" is used in its widest sense. It means all external environmental conditions and factors affecting **Human, Animal and Plants.**

The external factors include **climate/water/noise/temperature, soil** etc.

And, the State was also empowered to take effective steps to prevent the damage to the environment and ecology.

ENVIRONMENT PROTECTIONS ENACTMENTS

At the Stockholm Conference, the them Prime Minister, **Smt. Indira Gandhi** said,

".........extreme forms in which questions of **population or environmental pollution** are posed, obscure the total view of **political economic** and **social** situations .... It is said that in country after country, progress should become synonymous with an assault on nature .... Among the rest mankind, we in India – in spite of Ashoka – have been guilty of wanton disregard for the curses of our sustenance".

To meet these challenges, India also embarked on several **legislative measures** for the **protection of environment,** and **for maintaining ecological balance.** These were:

- The water (Prevention and Control) Pollution Act, 1974
- The Forest (Conservation) Act, 1980
- Air (Prevention and Control) Act, 1981.

These Acts were amended from time to time to make them more effective.

- Another feature was to establish **independent Ministry of Environment and Forest** by the Central Government in 1985.

ENVIRONMENT PROTECTION ACT, 1986

Till 1980's, the emphasis was chiefly to prevent and **control pollution.**

- In 1986, Government of India passed a comprehensive **Environment Protection Act (1986)** as our aftermath of **Bhopal tragedy of 1984 to cover** many aspects **other than Prevention and control of Pollution.**
- The Act changed the whole concept and gave enough **scope** to the Government and even an individual to act and go **to Court** for **Pollution prevention.**

The Act was passed for:

(i) the protection of environment,

(ii) regulation of discharge of pollutants,

(iii) handling of hazardous substances,

(iv) speedy response in the event of accidents threatening environment and

(v) deterrent punishment to those who endanger human environment, safety and health.

SEVENTH (VII) AND EIGHTH (VIII) PLAN STRATEGIES

These plans laid down well-defined strategy for environmental protection.

VII Plan – Strategy was to achieve substantial development in harmony with (natural) environment.

VII Plan's main components in regard to environment protection were:

- Pollution monitoring and control
- Environmental impact assessment
- Resource conservation
- Eco-development
- Environmental research promotions
- Environmental education, training and awareness
- Environment Information
- Coordination and liaison with State Governments and Union Territories
- Environment Policy and Law
- Strengthening the organisations.

VIII Plan's contribution was to further strengthen the Environment Policies.

A Cell was also set up by the Government of India, to ensure effective **implementation of anti-pollution measures** for the protection of environment by.

- Launching of new programmes [e.g. cleaning of rivers (like Ganga and Yumuna)]; Project of Afforestation etc.
- India also became signatory to Earth Summit in 1992.

NATIONAL CONSERVATION STRATEGY AND POLICY STATEMENT OF ENVIRONMENT AND DEVELOPMENT

Policy Statement

Immediately after the Earth Summit (1992) India has taken several steps in the direction of **Conservation of Environment and Development** in harmony with Environment. The National Conservation Strategy and Policy Statement on Environment and Development was a step in that direction.

- New Acts, **Amendments/Regulations** were made, and existing ones modified e.g.

  Wild Life/Water/Forest/Air/Environment Protection Act, 1986/Motor Vehicles Act. Public Liability/Insurance Act, 1991/Coastal Zone, 1991 Amended in 1994, 1997/Env. Audit Notification/Eco Mark Notification 1994/Public hearing Notification 1978 etc.

National Environmental Appellate Authority, 1997

Coastal Zone Management Authority, 1998

The following new Institutions/department/organisations were created by the Government of India from time to time:

Departments created (S&T, Biotech, Ocean Development, Space, Energy, Central Pollution Control Board (CPCB) and State Pollution Control Board (SPCB)

- Ministry of Environment and Forests
- Forest Research Institute (FRI), Dehradun
- Forest Survey of India (FSI)
- Botanical Survey of India (BSI)
- Zoological Survey of India (ZSI)
- National Environmental Engineering Institute (NEERI) (CSIR)
- National Wasteland Board
- Centre for Environmental Education
- Wadia Institute of Himalayan Geology
- National Museum of National History

The four-key corner stones of sustainability for industry are:

- Environment
- Low cost, and energy efficient extraction techniques
- Safety and health and
- Effective consultations with stakeholders. (*Ref.- Man and Child for Sustainable Development)*

PATH OF ENVIRONMENTAL SCIENCES

A study of environments in Kashmir and elsewhere calls for a new approach to environmental sciences. In combines the study of laws of nature and ecological systems in Kashmir Valley. It also aims to produce sustainable development which is a positive concept and depends upon the application of sciences in harmony and collaboration with nature. We are reminded of Einstein's address in honour of Max Planck in 1918 when he defined the motivation for the study of science. Einstein said: "The Temple of science is a multi-faceted building. In it, many engage in science is a multi-faceted building. In it, many engage in science out of joy in flexing their intellectual muscles, or for utilitarian ends..." Modern concept of sustainable development no doubt is a joy of scientific research as were are dealing with the nature and ecosystem. The whole of biosphere indeed is open for a new study of relatedness of various laws of nature. An ecological and biological approach to science and law has produced a new paradigm in science. And we are aware of this new scientific path which has been opened by many international covenants and declarations made by the United Nations, such as the World Charter for Nature (1982) etc.

SPECIAL FEATURES OF KASHMIR'S

Ecology and Environments

In our concern of environment protection and sustainable development in Kashmir, we have to highlight some special features of this valley. Indeed a leading scientist of India and a colleague in the School of Environmental scientist in Jawaharlal Nehru University calls the entire Kashmir Valley as an important biosphere reserve. The United Nations has held a conference on man and biosphere programme. Professor P.S. Ramakrishna advises on the importance of these reserves for the protection of biosphere. There are 301 reserve listed b the UNESCO world over which are also testing areas for sustainable approach for development. As a Man and Biosphere (MAB) study of IUCN points out, biosphere reserves help in conservation of ecosystems and genetic resources, monitoring and research, and providing facilities for education and training. Biosphere reserve include scientific nature reserve, parks, natural monuments, wildlife sanctuary, cultural heritage landscapes, multi-use management areas, natural biotic areas, historical archaeological areas, and world heritage sites. There are 14 biosphere reserve in India. Beyond some studies made by scholars from forest departments, no overall ecological study seems to have been made of Kashmir environments. There are good number of research works on 'the Himalaya mountain ecology by India and by the United Nations. At Kathmandu in Nepal and in Almora in Uttaranchal, Ecological Institutes have been set up for study and research of emerging problems of regions in Himalayas. However, so far we do not have one interdisciplinary research centre in Kashmir. During my academic visits to discuss environmental issues in Kashmir in June 2003, I came across a lot of growing awareness and concern for environment protection. Many scholars would like to chart out the path and plan for sustainable development keeping in view various ecosystems in the valley. Prominent academic places where Department of Environment, where Professor A.R. Yusuf has a large number of projects in hand; Post-graduate Department of Environment in S.P. College, and the State Pollution Control Board located in Raj Bagh area which has some research scientists employed. S.P. College is a historic institution. It is celebrating its centenary year in 2004. The Principal, Dr. S.G. Sarwar is a distinguished environmentalist. The college has an excellent laboratory for environmental research. I had occasion to discuss with Dr. Sarwar regarding setting up of a new centre of excellence for environment education in S.P. College with assistance of the Ministry of Environment and Forests, Government of India. This Ministry has already established such centres in Ahmedabad, Chennai etc. A proposal to this effect from the Department of Environment, Jammu and Kashmir Government to MOEF, It will save and protect Kashmir environment for future, and provide guidance for sustainable development. To nominate the whole of valley as a biosphere reserve will be an important consideration for the overall scientific and economic purposes.

Some Problems of immediate Environmental Concern

Kashmir is located in 32° to 37° N latitude and 72° to 80° East longitude. It has an area of 222236 sq. km. including 78932 forms part of Western Himalayas. The valley areas is 150 km. long work on agriculture, 85% live in villagers, and 23% of land is used for cultivation. 61% of land is under forests (See, S.K. Chadda, *Ecology of Kashmir*, 1990). Seven wetlands are being developed in Jammu and Kashmir state which include Wular lake in the valley. Kashmir has a unique biodiversity reserves of forests, wet lands, flora and fauna.

Most of the urgent environmental problems are due to air increase, water pollution, land degradation, forest depletion and increase in population. Air pollution due to vehicles has become an urgent health hazard.

In Srinagar, private vehicles have increased and roads are choked. Enforcement of air pollution standards is an urgent necessity. There are increasing cancer cases. Courts have an important role in this respect. Now eco-clubs have been established for conservation and environment protection. Public interest litigation is needed to clean the valley like the Dehradun Valley. The new project on Dal lake was a result of Court order based on public interest litigation.

Regarding water resources: a new study is being drafted to provide clean water to people and arrange for drainage systems. Along with clean water supply, the disposal of sewage by public bodies needs a citizens ' movement. Recycling of sewage will provide new sources of manure and energy. Removal of biomedical waste has become important for which purpose, the State Pollution Control Board has prepared research works.

The natural reserves of Dal lake, Wular lake, Anchar lake are some unique features of Kashir's s natural heritage. A new policy is being evolved to protect these lakes from erosion and human encroachment. There are estimated one and a half lakh people in the catchment area of Dal lake alone. There are 18 more lakes situated in the upper areas of Kashmir valley. About them not much is known regarding their environmental status.

Eminent Writers

To understand the issues of environment of Kashmir we have to refer to great writers, poets and historians. In modern times poets like Mehjoor, Zinda Kaul have inspired us on the harmony of Kshmiri people with the natural environments. Mehjoor's following poems is a reminder to the global environment movement: "Wulu ha bagbano nav baharuk shan pada kar, khilan gul ghat karan bulbul, the theee Saman, pada kar". Kalhanas Rajtarangni is also full of references to "shrines, temples and monuments of Kashmiri". He was not only a historian, writes R.S. Pandit, but a great artist and poet. Professor Agha Ashraf Ali has also in our times contributed substantially his thoughts on environment protection of Kashmir.

In recent past, Samsar Chand Kaul wrote a book "Beautiful Valley of Kashmir and Ladakh", 1942. Kaul was a teacher of natural history, writes C.E. Tyndale-Biscoe in his foreword. This book is worth any price being a true reflection on the environments of Kashmir valley. Sone of the chapters are on Gulmurg valley, valleys of Pantsal range, Konsar Nag lake, Liddar and Sind valleys, Wardown valley, Pahalgam valley, Gangbal lakes, Erin valley, Gurais valley, Wular lake, wild animals monasteries etc. If we are to map the ecological systems of the valleys mentioned and protect historical monuments, Samsar Chand Kaul is the guide. Similarly Ram Chand Kak wrote an excellent book on "Ancient Monuments of Kashmir", 1933. Some of the monuments he describes are Shankaracharya temple, Khangah of Shah Hamdan, Patter Majid, Tomb of Zain-ul-abidin, Jama Masjid, Hari Parbat Fort, Mughal Gardens, Harwan excavations, Pandrethan, Pampur, Avantipur, Martand, Vernag, Patan Manasbal, Gurur, Wular lake etc. A Foucher writers in this foreword to Kak's book that the charm of Kashmir environment is due to combination of a mysterious life of nature and an artistic landscape. Foucher writers: "Personally, I have often thought that three there remains more classic antiquity still living in the shadow heat of Kashmir jugle than the theatrical scenes of the Mediterranean shores". On Wular lake, a writer in this seminar namely, Nahida Assad Basu has made a study of Wular lake as her thesis for Master's degree in ecology and environment. She has a good description of the geography, geo-economic aspect and historical monuments situated in and around this lake. New avenues for tourism, fisheries, power generation are discussed. Her work was guided by Mr. M. Sayeed, an outstanding engineer, and calls for study by experts.

No wonder that the Government in Kashmir is keen to protect the heritage of Kashmir Valley as we cannot let it go down the history. The temples and mosques are being repaired. Khir Bhawani and Mattan Temples along with some shrines have been given grants recently for repairs. Besides, eco-tourism to places like Amarnathji, Char-e-Sharif etc. will help economy in the State. Among writers Mr. Jag Mohan, Honourable Minister of Culture, has made a formidable attempt in his book "My Frozen Turbulence in Kashmir" to describe the links of glorious past with the present, and drawn a hopeful view of future of Kashmir based on harmony. Finally, Walter R. Lawrence published his master piece book in 1895: "The Valley of Kashmir" that provides us an objective and creative account of Kashmir as a people, their multi-cultural heritage, flora, fauna, archaeology and social life. Lawrence's is one of the best-published works on Kashmir.

Floristically, the State of Jammu and Kashmir lies outside the Indian Region. The Indian floristic region comprises Peninsular India, Gangetic Plain, flanks of the Himalayas, and the island of Sri Lanka (Good, 1964). The north-west Indian desert is a part of the African sub-kingdom. Eastern Assam forms part of the Indo-Malayan sub-kingdom. The North-West Himalayas, including Jammu and Kashmir form the southern most part of the Boreal (north temperate) Kingdom (Plates I-IV).

Within the Indian Region, the study area is phytogeographically the most complex region. The flora has passed through various stages during the geomorpholocal evolution of this region. This region has been colonised at different times by humid tropical Malayan forms, tropical African forms, temperate and alpine north Asiatic-European forms, sclrophyllous Mediterranean forms, temperate EastAsian (Sino-Japanese) forms, and semi-arid Central Asian forms. The rigorous environment has further acted upon this mosaic of geographical forms leading to the extinction of species, breaking up distribution ranges, or inducing genetic variation with or without speciation.

The demarcation of the flora of the State into mainly two regions is affected in the different phytogeographic systems. Clarke (1898) and Jain (1990) have considered this region to be homogenous on the basis of distribution of family Cyperaceae, and vegetation types, respectively. Chatterjee (1962) and Hooker (1907) have demarcated this region into two subregions on the basis of endemism and largest families of plants, respectively. Good (1964)) has placed Ladakh and the Karakorum range in the Western and Central Asiatic region, and the Pir Panjal range into Sino-Japanese region. Rodgers and Panwar (1988) have divided the State into three biogeographic provinces viz., Ladakh, NW Himalaya, and Punjab Plains. In contrast, the whole State is treated as a single agroclimatic zone on the basis of physiographic, climatic and agricultural features by the Planning Commission (Sehgal et al., 1990).

A comparison of the largest families and genera of plants of the selected areas in the different natural regions gives clues about the phytogeographical variation with the State.

A comparison of the largest families reveals temperate taxa predominate in Ladakh and Srinagar viz., families *Asteraceae, Brassicaceae, Rosaceae,* and *Caryophyllaceae.* However, the present study area which from Patnitop–Nathatop-Sanasar falls in the temperate region shows a different hierarchy of largest families as compared to Ladakh annd Srinagar.

LARGEST FAMILIES

LADAKH	*SRINAGAR*	*JAMMU*	*PATNITOP AND ADJOINING AREA*
1	2	3	4
(Kachroo et al.,)	(Singh & Kachroo, 1976)	(Sharma & Kachroo, 1981)	(Author, 1997)
Asteraceae	Asteraceae	Leguminosae	Asteraceae
Brassicaceae	Poaceae	Poaceae	Laminaceae
Leguminosae	Leguminosae	Asteraceae	Poaceae
Caryophyllaceae	Cyperaceae	Cyperaceae	Fabaceae
Polygonaceae	Brassicaceae	Scrophulariaceae	Rosaceae
Lamiaceae	Rosaceae	Lamiaceae	Brassicaceae
Gentianceae	Lamiaceae	Euphorbiaceae	Apiaceae

1	2	3	4
Rosaceae	Scrophulariaceae	Acanthaceae	Scrophulariaceae
Scorphulariaceae	Apiaceae	Amaranthaceae	Solanaceae
Chenopodiaceae	Caryophyllaceae	Brassicaceae	Cyperaceae

LARGEST GENERA

LADAKH	SRINAGAR	JAMMU	PATNITOP AND ADJOINING AREA
Astragalus	Carex	Cyperus	Nepeta
Polygonum	Polygonum	Ficus	Sonchus
Artemisia	Ranaculus	Euphorbia	Artemisia
Potentilla	Nepeta	Ipomoea	Rumex
Nepeta	Taraxacum	Indigofera	Eri geron
Gentiana	Potentilla	Fimbristylis	Leucas
Pedicularis	Cyperus	Scirpus	Veronica
Corydalis	Astragalus	Cassia	Poa
Saussurea	Artemisia	Desmodium	Lactuca
Ranaculus	Veronica	Crotalaria	Cardamine

The sequence of the largest families in this area: *Asteraceae, Lamiaceae, Poaceae and Fabaceae*. At the same time the perusal of the Table as above, makes it clear that the ten largest families in the flora of Ladakh, Srinagar and the present study are almost the same except that in the flora of Ladakh families *Polygonaceae, Gentianaceae* and *Chenopodiaceae* are well represented which are not so dominant in the temperate flora of Srinagar and the present study area. The only family dominant in the flora of Srinagar which is not as dominant in the flora of the study area is *Caryophyllaceae*. However, the dominance sequence of the top ten families in all the three zones is different and this is mainly attributed to the trampling the grazing in the area to a large extent. Considering the largest genera also, both Ladakh and Srinagar show dominance of northern extratropical elements like the boreal and mediterranean. *Cartex* and *Polygonum* have the largest number of species in Srinagar, this may be due to the presence of a large number of aquatic habitats in this region. The dominance of boreal taxa in Ladakh is misleading, since this element is composed of mostly weeds of cultivation which do not form a part of the natural landscape. The largest genera of Jammu are, however, the tropical genera. The largest genera in the study area *Nepeta, Erigeron, Artemisia, Sonchus* etc. This is largely because these genera are very hardy and can withstand the biotic pressures, though the temperate species should dominate the area.

The flora of the Jammu region is very different from the rest of the State. Whole range of new families are added to the flora like *Annonaceae, Menispermaceae, Resedaceae, Flacourtiaceae, Elatinaceae, Bombaceae, Sterculiaceae, Malpighiaceae, Leeaceae, Moringaceae, Combrataceae, Myrtaceae, Molluginaceae, Aizoaceae, Ebenaceae, Ehretiaceae, Martyniaceae,*

Zingiberaceae, Areaceae. This region has lesser representation of the temperate families like *Brassicaceae, Caryophyllaceae* and *Rosaceae,* and the predominantly tropical ***Leguminosae, Euphorbiaceae, Acanthaceae*** and ***Convolvulaceae*** have greater representation this area.

The temperate forest trees like *Pinus walliciana, P. roxburghii, Cedrus deodara, Abies pindraw, Neolitsea consimiilis* and *Qurcus incana* are present in the study area, restricted to the higher mountain slopes, whereas the rest of the study area which falls in the sub-temperate region has the plants like *Lannea coromandelica, Punica granatum, Cassia fistula, pistacia integrrima, Olea cuspidata, Dodonaea viscosa* etc. The rest of the region in Jammu has semi-arid to tropical vegetation with trees like *Acacia nilotica, Lannea coromandelica, Butea monosperma, Ficus species, Diospyros montana, cordifolia, Bombax ceiba, Terminalia bellerica, T. chebula* and *Syzgium cuminii.*

In Jammu and Kashmir the figures for edemism are available for the alpine flora (Dhar and Kachroo, 1983). Endemism in Kashmir Himalaya amounts to 58% in flowering plants (39% in dicots and 19% in monocots). The 15 dominant families of flowering plants in the area show range of endemisms from 17.19% **(Poaceae)** to 54.90% **(Primulaceae).** Considering the widespread nature of these families, the high endemisms sports the hypothesis that the Himalaya is a secondary centre of speciation.

Endemism in legumes of Himalayas has been studied by Rao and Husain (1993). As many as 35 taxa belonging to 17 genera are endemic to this region. However, there are more genera endemic to the Himalayan region.

According to Rao and Husain (loc cit) the west Himalayan region is for richer in endemic legumes than east Himalaya, indicating that there has been a greater diversification of legumes in the West Himalaya.

Endemisms in Himalayan pteriodophytes has been described by Bir (1993). He has reported 19 taxa of ferns and fern allies to be endemic to Western Himalayas, and 14 taxa endemic to Eastern and Western Himalayas.

Figures for endemisms in the Kashmir Himalayas are higher than the figures for the whole Himalayan chain, and also higher than the adjacent regions in the Himalayas. The extent of endemism in relation to some adjacent phytochoria according to Dhar et al. (1987) is given below :

Kashmir Himalaya	36.5%
Pamir	4.0
Himalaya	28.8
W. Tienshan	13.0
Hindukush	20.0
Alps	62.0

The extremes of climate and higher elevation of the lofty mountains have been greatly responsible for a rich concentration of endemic plants in the Himalaya.

The information on plant distribution patterns in general, and endemism in particular, is insufficient to estimate the actual magnitude of floristic influence of the surrounding regions on the flora of this region, and to locate centres of speciation or endemism within the boundaries of the State. There is need for intensive studies on endemism, cytogeography, phytogeography and ecology for a better understanding of evolutionary processes in the region.

In the study area *Pinus wallichianna* is dominant along with *Cedrus deodara* as co-dominant. The other trees which are occasionally seen include *Picea smithiana, Taxus wallichiana* and the deciduous trees *Rhus succedanea, Aesculus indica, Juglans regia, Quercus incana* etc. The shrub species of these forests are : *Cotoneaster bacillaris, Sarcococca pruniformis, Vibernum grandiflorum, Berbris lycum* etc. The ground flora of the region includes *Potentilla nepalensis, Polygonum amplexicaule, Rumex nepalensis, Saussurea heteromella, Traxacum officinale, Pediculris pecitinata, Digitalis purpurea, Galium elegans, Gentiana argentea, Fragaria vesca, Androsace rotundifolia, Erigeron alpinus, Sonchus oleraceous* etc.

The deodar excels best from 1846 m to 2615 m. Though *Pinus wallichiana* is found upto elevations of 2800m, above 2350 m the blue pine begins to be replaced by the silver fir *Abies pindrow* which becomes dominant. *Pinus roxburghii, Olea cuspidata, Ficus palmata, Pistacia integerrima* are restricted upto the elevation of 1500 m.

In Pir Panjal, the blue pine zone extends from 1602 m to 2000 m. In this Pinus *wallichiana* dominated forest type, the other trees which are occasionally seen include *Cedrus deodara, Picea smithiana, Taxus wallichiana,* and the deciduous trees *Sorbus, Aesculus, Acer, Juglans* etc. The shrub species of these forests are : *Cotoneaster affinis, Rosa macrophylla, Parrotiopsis jacquemontiana* etc.

In the silver fir zone the shrub component is the same as in the blue pine zone upto 2750 m but higher up *Salix denticulata* is frequently found. The other common shrubs in this region are: *Skimmia laureola, Lonicera myrtilis, Berberis jaeschkeana* and *Rubus* sp. The herbaceous component in this forest comprises *Androsace rotundifolia, Bisterta amplexicaule, Campanula colorata, Cucubalus baccifer, Cynoglossum sp., Fragaria nubicola, Galium asperuloides, Geranium wallichianum, Geum urbanum, Laminum album, Lotus corniculatus, Myriastiis wallichii, Nepata linearis, Oxalis corniculata, Stellaria monosperma, Strobilanthes attenuatus, Taraxacum officinale, Tricholepis elongata, Valeriana hardwickii, Veranica melissaefolia, Viola pilosa, Trillium govanianum. Stipa sibirica* is found in forest clearings, *Sedum ewersii* in rock crevices; moist shelters are inhabited by Aconogonum alpinu and *Persicaria thapsus* grow along forest borders. Along the river banks, common plants are : *Carduus edelbergi, Cardamine macrophylla, Impatients brachycenta, I. glandulifera, Ligusticum thomsonnii, Myosoton aquaticum, Senecio ligularis* etc., Swampy stretches are colonized by *Acorus calamus, Alisma plantago-aquatica, Nasturtium officinale, juncus spp., Veronica beccabunga* and the marsh orchid *Dactylorhiza hatagirea.*

The alpine zone begins at about 3000m elevation where the trees start disappearing. An alpine scrub form of vegetation exists at altitudes above 3350m. The dominant plants of this landscape are the shrubs *Juniperus squamata, Rhododendron hypenanthum, Cotoneaster microphyllus,* and dwarf specimens of *Salix flabellaris,* and *Rosa macrophylla. Cassiope fastigata* forms a health-like growth. Among herbs common species : *Corydalis govaniana, Meconopsis aculeata, Aconogonium rumicifolium, Actaea spicata, Aquilegia moorcroftiana, Dispsacus mitis, Inula royleana, Rhodiola heterodanta, Saussurea candolleana, Thalictrum cultratum* etc.

There is great floral diversity in the study area. The region is known for not only some endemic species but also is very rich in the distribution of various families of the angiospermic flora. There are as many as 105 families belonging to Angiosperms, Gymnosperms and Pteridophytes. In all 375 species of the 260 genera belonging to these families have been represented.

The phytodiversity of the study area depicts that 75 families of the Dicots harbour the major flora of the area with 202 genera and 295 species. Monocots are meagerly represented by 36 genera and 41 species belonging to 13 families. There are only 2 families of the Gymnosprems represented by 5 genera and 6 species, while as the Pteriodphytes have 17 genera and 33 species spread in 15 families. There are a large number of monotypic families and genera in this region.

There is a great floral diversity in this region. The region is known for not only some endemic species but also is very rich in the distribution of various families of the angiospermic flora. There are as many as 52 families which are monotypic. This is very significant from the point of view of the species diversity under great danger or threat of extinction. Rao and Murti (1990) found that north east India is a major centre for orchids, ferns, bamboos, species of *Musa, Rhododendron* and *Hedychium*. In contrast to this, the study area which falls in the North-Western Himalaya does not show any preponderance of orchids, bamboos or *Rhododenron* and *Musa*. Instead we have the dominant species of *Sarcococca, Vibernum, Rubus, Prinsepia* and *Daphne*.

Appanah et al. (1993), in the Malaysian rain forests found a very high species richness of lianas and confirmed the existence of intercontinental variation in liana floristics, with Annonaceae being the dominant liana family in Asia. In the present study it has been observed that lianas constituted a very insignificant proportion of the total floristic diversity of the area and *Hedra nepelansis* is the dominant liana species. The other species belong to families *Ranunculaceae, Rosaceae, Liliaceae, Convovulaceae*, etc. in contrast to the family *Annonaceae* which is dominant in Asia. This is expected because Malaysian rain forests are in the tropical zone while as the study area is temperate conifer forest. George and Varghese (1993) reported three major vegetation communities viz., *Anogeissus–Albizia, Albizia Gyrocarpus* and *Euphorbia* scrub from tropical dry deciduous forest ecosystem of Western Ghat. In contrast to this in our study area with temperate conifer forest the main vegetation communities consist of *Cedrus–Pinus–Wallichiana–Vibernum; Pinus roxburghii–Prinsepia–Dodonaea; Abies–Picea–Cotoneaster and Quercus–Pinus wallichiana–Berberis* species.

Dhar et al. (1997) identified mid-elevation oak forests (*Quercus leucotrichophora* and *Q. lanuginosa*) as the most critical communities from the Kumaun Himalaya. They also stressed that the expansion of a native canopy dominant, *P. roxburghii*, threatens other native broad leaved forest and facilitates invasion of non-natives in herb and shrub layers. The present study also finds the same results and suggests the proper management interventions for developing priorities for conservation.

The floristic studies of the study area showed that out of 105 families of Tracheophytes, **Asteraceae** is the most dominant family having 30 genera and 47 species. The other dominant families are **Lamiaceae** (16 genera, 25 species); **Poaceae** (14 genera, 17 species); **Fabaceae** (13 genera, 21 species); **Rosaceae** (12 genera and 21 species). The overall Tracheophytic diversity represented in the study area has been 260 genera annd 375 species.

Life forms and biological spectrum:

A life form is **"the sum of the adaptation of the plant to climate".** After the complete enumeration of the flora of Patnitop and adjoining area, life form classification and construction of biological spectra was carried out. For this purpose all the Tracheophytes were taken into consideration. A total of 375 tracheophytes listed from the study area have been classified into 9 life form classes. The units of classification used are modified after Raunkiaer (1918). On the basis of the position of perennating buds on the plants and the degree of their protection during adverse conditions the various classes used in the present study are as under

Therophytes (TH): all annuals; perennating buds are those of the embryo in the seeds.

Hydrophytes and Helophytes (HH): water, marshy and swampy plants with perennating organs submerged in water during unfavourable season.

Geophytes (G): perennating buds below ground.

Hemi-cryptophytes (H): renewal buds situated at ground level; including biennials.

Chamaephytes (CH): woody or semi-woody perennials with renewal buds situated at 25 cm height or less above ground surface.

Nanophanerophytes (N): shrubs with vegetative organs less than 2 m above soil.

Phanerophytes (PH): trees with vegetative organs more than 2 m above soil.

Lianas (L): climbing plants: and

Epiphytes (E): which settles upon the trunk and branches of the trees.

Life-form classes of Patnitop and adjoining areas (Raunkiaer's classification as modified by Braun Blanquet, 1932) are as under:

Therophytes (TH)

1. *Adiantum caudatum*
2. *Aidantum capillus-veneris*
3. *Adiantum venustum*
4. *Agrimonia eupatoria*
5. Alliariaofficinalis
6. *Amaranthus hybridus*
7. *Anagallis arvensis*
8. *Arabidopsis thaliana*
9. *Aster molliusculus*
10. *Aster tibetieus*
11. *Atriplex tatarica*

12. *Bellis perennis*
13. *Aster alpinis*
14. *Bidens tripartita*
15. *Brassica campestris Napus*
16. *Brassica juncea*
17. *Brachyactis umbrosa*
18. *Bromus pectinatus*
19. *Bromus tectarum*
20. *Brunella vulgaris (Prunella Vulgaris)*
21. *Buplerum tenue*
22. *Calamintha umbrosa*
23. *Campanula colorata*
24. *Capsella bursa-pastori*
25. *Cardamine flexuosa*
26. *Cardamine hirsuta*
27. *Cardamine impatiense*
28. *Chenopodium album*
29. *Chenopodium murale*
30. *Coleus forskohlii Syn. Coleus barbatus.*
31. *Commelina bengalensis*
32. *Cosmos sulphureus*
33. *Cousinia alpina*
34. *Crepis multicaulis*
35. *Datura stramonium*
36. *Erigeron alpinus*
37. *Erigeron bonariensis*
38. *Erigeronn canadensis*
39. *Erigeron linifolius*
40. *Eriophorum comosum*
41. *Euphorbia indica*
42. *Euphorbia helioscopia*
43. *Filago germanica*

44. *Fumaria parviflora*
45. *Galisoga parviflora*
46. *Galium elegans*
47. *Galium mollugo*
48. *Galium rotundifolium*
49. *Gerbera kunzeana*
50. *Hordeum murinum*
51. *Impatiens brachycentra*
52. *Koeleria cristata*
53. *Lactuca brunoniana Syn. Prennanthus brunniana*
54. *Lactuca dissecta*
55. *Lactuca scariola*
56. *Lamium album*
57. *Lens esculenta*
58. *Lolium tomulantium*
59. *Matricaria disciodia*
60. *Mazus surculosus*
61. *Medicago denticulata*
62. *Medicago lupulinna*
63. *Melilotus indica*
64. *Micromeria biflora*
65. *Myriactis nepalensis*
66. *Myriactis wallichi*
67. *Nasturtium officinale*
68. *Ocimum canum*
69. *Oenothera drummondii*
70. *Oenothera rosea*
71. *Onosma thomsonii*
72. *Ophioglossum reticulatum*
73. *Oplismenus burmannii*
74. *Origanum vulgare*

75. *Petunia alba*
76. *Pilea umbrosa*
77. *Physalis minima*
78. *Poa annua*
79. *Poa bulbosa*
80. *Poa rigida*
81. *Polygonum nepalensis*
82. *Polypogon littoralis*
83. *Rumex dentatus*
84. *Rumex maritimus*
85. *Rumex nepalensis*
86. *Scandix pecten-veneris*
87. *Scirpus mucronatus*
88. *Setaria viridis*
89. *Silene vulgaris*
90. *Sisymbrium trio*
91. *Solanum nigrum*
92. *Sonchus arvensis*
93. *Sonchus asper*
94. *Sonchus brachyotus*
95. *Sonchus oleraceous*
96. *Stachys floccosa*
97. *Stellaria media*
98. *Strobilanthus atropurppurens*
99. *Tagetus minuta*
100. *Taraxacum officinale*
101. *Torilis japonicus*
102. *Torilis stocksiana*
103. *Trichodesma indicum*
104. *Trigonella emodi*
105. *Trigonella pubescens*

106. *Tusilago farfana*
107. *Verbena officinalis*
108. *Veronica capitata*
110. *Veronica persica*
111. *Youngia japonica*
112. *Zea mays*

Epiphytes (E)

1. *Cuscuta reflexa*

Lianas (L)

1. *Clematis grata*
2. *Clematis montana*
3. *Convolvulus arvensis*
4. *Hedera nepalensis*
5. *Ipomoea purpurea*
6. *Momordica dioica*
7. *Rosa brunoniana*
8. *Smilax parvifolia*
9. *Smilax vaginata*
10. *Vicia sativa*

Geophytes (G)

1. Arisaema jacguemontii
2. Colchicum luteum
3. Cyperus iria
4. Dioscorea deltoidea
5. Gagea dshungarica
6. Goodyera repens
7. Narcissustazetus
8. Ophiopogon intermedius Syn. Mondo intermedius
9. Podophyllum haxandrum
10. Sauromatum guttatum
11. Solanum tuberosum
12. Trillium govanianum

13. Tulipa stellata

Hydrophytes and Helophytes (HH)

1. *Acorus calamus*
2. *Caltha palustris*
3. *Cyperus distans*
4. *Eleocharis palustris*
5. *Equisetum arvense*
6. *Juncus glaucus*
7. *Lemna minor*
8. *Mentha longifolia*
9. *Mentha sylvestris*
10. *Selaginella pennata*
11. *Slum latijugum*
12. *Stellaria aquatica*

Hemicrotophytes (H)

1. *Achillea millefolium*
2. *Ajuga parviflora*
3. *Achyranthes aspera*
4. *Achyranthes polystachya*
5. *Anaphalis hiplinenis*
6. *Anaphalis royaleana*
7. *Androsace rotundifolia*
8. *Androsace umbillata*
9. *Anemone oblusiloba*
10. *Arabis amplexicaulis*
11. *Asplenium adiantum-nigrum*
12. *Asplenium dalhousiae*
13. *Asplenium oxyphylum*
14. *Asplenium viride*
15. *Athyrium drepanopterum*
16. *Athyrium otophorum*

17. *Athyrium thelypteroides*
18. *Bergenia ligulata (Saxifraga ligulata)*
19. *Buplerum falcatum Marginatum*
20. *Calanthe tricarinata*
21. *Carex condensata*
22. *Centella asiatica*
23. *Cerastium glomeratum*
24. *Chaerophyllum villosum*
25. *Chrysopogon montinus*
26. *Cichorim intybus*
27. *Cyclosorus spp.*
28. *Cyrtomium caryostideum*
29. *Dactylis glomerata*
30. *Epilobium royleana*
31. *Foeniculum vulgare*
32. *Gentiana argentea*
33. *Gentiana kurroo*
34. *Geum urbanum*
35. *Girardinia heterophylla*
36. *Gnaphalium luteo-album*
37. *Gymnogramme javanica Syn. Conigramme fraximia*
38. *Gypsophilla crastiodes*
39. *Heracleum cachemirianum Syn. Peucedomum dhana*
40. *Launaea nudicaulis*
41. *Lecaus capitata*
42. *Lecuas cephalotes*
43. *Lecuas lanata*
44. *Lecuas martinicensis*
45. *Leontopodium alpinnum*
46. *Nephrodium filix-mas*
47. *Nephrodium molle*

48. *Nephrodium moulmeinense*
49. *Nephrodium truncatum*
50. *Onychium japonicum*
51. *Pedicularis pectinata*
52. *Pellea nitidula*
53. *Pennisetum orientale*
54. *Plantago lanceolata*
55. *Plantago major*
56. *Polygala.abyssinica*
57. *Polygonum amplexicaule*
58. *Polygonum hydropiper*
59. *Polygonum longisetum*
60. *Polypodium microrhizom*
61. *Polypodium polypoides*
62. *Polystichum aculeatum (Aspidium)*
63. *Polystichum dentatum*
64. *Polystichum diaphanum*
65. *Polystichum thomsoni*
66. *Potentilla leschenanltiana*
67. *Potentilla neplensis*
68. *Pteridium aquilinum*
69. *Pteris cretica*
70. *Pteris vittata*
71. *Pterotheca falconeri*
72. *Ranunculus laetus*
73. *Rubia cordifolia*
74. *Salvia lanata*
75. *Saussurea candicans*
76. *Saussurea heteromella*
77. *Sedum adenotrichum*
78. *Senecio chrysanthemoides*

79. *Senecio nudicaulis*
80. *Senecio graciliflor*
81. *Solidago virga-aurea*
82. *Wertia ciliata Syn. S. speciosa*
83. *Swertia cordata*
84. *Swertia odorata*
85. *Thalaspi alpestre*
86. *Thelypteris erruvescence*
87. *Thymus serphyllum*
88. *Trifolium arvense*
89. *Trifolium pratens*
90. *Trifolium repens*
91. *Urtica dioica*
92. *Valriana himalayana*
93. *Valeriana pyrolaefolia*
94. *Verbascum thapsus*
95. *Veronica anaglis-arvensis*
96. *Veronica capitata*
97. *Vicatia coniifolia*
98. *Violapatrinii*
100. *Woodwardia radicans*

Macrophanerophytes (M)

1. *Abies pindrow*
2. *Aesculus indica*
3. *Atlanthus altissima*
4. *Alnus nitida*
5. *Cassia fistula*
6. *Cedrus deodara*
7. *Celtis australis*
8. *Dalbergia sisso*
9. *Ficus palmata*

10. *Ficus roxburghii*
11. *Grewia oppositifolia*
12. *Juglans regia*
13. *Lannea coromandelica*
14. *Mallotus philippienensis*
15. *Melia azedarach*
16. *Morus nigra*
17. *Olea cuspidata*
18. *Parrotipsis jcquemontiana*
19. *Picea smithiana*
20. *Pinus roxburghii*
21. *Pinus wallichiana*
22. *Pistacia integerrima*
23. *Platanus orientalis*
24. *Populus ciliata*
25. *Prunus armeniaca*
26. *Prunus persica*
27. *Punica granatum*
28. *Pyrus communis*
29. *Pyrus pashia*
30. *Pyrus malus*
31. *Quercus floribunda (Q. dilatata)*
32. *Quercus incana (Q. Leucotrichophora)*
33. *Rhus succedanea*
34. *Robinia psendo-acacia*
35. *Salix serioscaspa*
36. *Salix oxycarpa*
37. *Sapium sebiferum*
38. *Taxus wallichiana*
39. *Zanthoxylum alatum*

Nanophanerophytes (N)

1. *Adhatoda vasica*
2. *Berberis lycium*
3. *Buddleia paniculata*
4. *Carissa spinarum*
5. *Colebrookia oppositifolia*
6. *Daphne oleoides*
7. *Daphne patyracea*
8. *Datisca cannabina*
9. *Debregeasia salisifolia*
10. *Dodonaea viscosa*
11. *Elaeagnus umbellata*
12. *Euphorbia royleana*
13. *Gymnosporia royleana*
14. *Indigofera gerardiana*
15. *Indigofera heterantha*
16. *Jasminum humile*
17. *Jasminum officinale*
18. *Maesa indica*
19. *Myrsine africana*
20. *Neolitsea consimilis*
21. *Nerium indicum*
22. *Opuntia vulgaris*
23. *Pandanus tectorius*
24. *Prinsepia utilis*
25. *Randia tetrasperma*
26. *Rhamnus virgatus*
27. *Ricinus comminis*
28. *Rosa indica*
29. *Vibernum grandiflorium*
30. *Vitex negundo*
31. *Wikstroemia canescens*

32. *Woodfordia flouribunda*

Chamaephytes (CH)

1. *Arabidopsis wallichii*
2. *Argyrolobium flaccidum*
3. *Artemista glauca*
4. *Artemisia maritiana*
5. *Artemista parviflora*
6. *Artemisia vulgaris*
7. *Astragalus grahimanus*
8. *Astragalus Kashmiriensis*
9. *Cannabis sativa*
10. *Coriaria nepalensis*
11. *Corydalis documbens*
12. *Corydalis diphylla*
13. *Cotoneaster bacillaris*
14. *Cotoneaster microphylla*
15. *Cynanchum vincetoricum*
16. *Cynodon dactylon*
17. *Cynoglossum lanceolatum*
18. *Cynoglossum microglochin*
19. *Cynologssum wallichii*
20. *Desmodium tiliaefolium*
21. *Desmodium wallichii*
22. *Dicliptera bupleruroides*
23. *Digitalis lanata*
24. *Digitalis purpurea*
25. *Duchesneindica (Fragaria indica)*
26. *Erianthus versicolor*
27. *Fragaria vesca*
28. *Geranium nepalensis*
29. *Geranium wallichianum*

30. *Hypericum patulum*
31. *Hypericum perforatum*
32. *Ipomoa cornea*
33. *Lespedeza cuneata*
34. *Lespedeza gerardiana*
35. *Lonicera alpigena*
36. *Lonicera quinquelocularis*
37. *Lonicera rupicola*
38. *Lotus corniculatus*
39. *Malva neglecta*
40. *Nepeta clarkai*
41. *Nepeta elliptica*
42. *Nepeta erecta*
43. *Nepeta govaniana*
44. *Nepeta spicata*
45. *Oxalis acetosella*
46. *Oxalis corniculatus*
47. *Oxalismaritiana*
48. *Plectranthus japanicus*
49. *Plectranthus rugosus*
50. *Rubus ellipticus*
51. *Rubus niveus*
52. *Rubus ulmifolius*
53. *Rumex hastatus*
54. *Sarcococca pruniformis*
55. *Solanum psudo-capsicum*
56. *Spiraea bella*

Table 15.5: Total number of species and percentage of different life form classes.

Life-form class	*No. of species*	*percentage*
1	2	3
Therophytes (TH)	111	29.8

1	2	3
Hemicryptophytes (H)	100	26.4
Chamaephytes (CH)	57	15.2
Macrophanerophytes (M)	39	10.4
Nanophanerophytes (N)	32	8.5
Geophytes (G)	13	3.5
Hydrophytes and Helophytes (HH)	12	3.2
Lianas (L)	10	2.6
Epiphytes (E)	1.	0.2
Total	375	100.00

Biological Spectrum/ Phytoclimatic Spectrum:

Biological spectrum represents (%age of the total species in the community) distribution of species among the various life forms. Raunkiaer (1918) proposed four main phytoclimate series of the earth:

Phanerophytic climate–in the tropics

Therohytic climate–in the desert

Hemicryptophytic climate–in the greater part of the cold temperate zone Chamaephytic with a fair–in cool climate. Proportion of geophytes.

Thus, through calculation of percentage value of each life form, biological spectrum of the study area has been obtained and has been compared with Raunkiaer' Normal Biological spectrum of Hemicryptophytic climate (Table 15.6; Fig 15.4).

A detailed floristic survey was made including observations on habit, habitat, height, form and nature of perennating organs. The 375 species of tracheophytes collected from the study area fall under the following life form classes : Therophytes (TH) 111, Hemicryptophytes (H) 100, Chamaephytes (CH) 57, Macrophanerophytes (M) 39, Nanophanerophtes (N) 32, Hydrophytes and Helophytes (HH) 12, Geophytes (Cryptophytes (G) 13, Lianas (L) 10 and Epiphytes/Parasite (E) 1. The biological spectrum indicates the phytoclimate to be that of Thero-Hemicryptophytic type. Compared to the normal spectrum of Raunkiaer, Therophytes show maximum deviation (+16.8%), Chamaephytes show (+6.2%) deviation. On the contrary there is corresponding decrease in the life forms percentage of the order of for Macrophanerophytes (-17.6%) and Nanophanerophytes (-6.5%). In the rest of the life forms there is minor deviation from the normal spectrum.

There are 52 families of the tracheophytes which are represented by only single genera and species (monotypic) Table - 15.1; 15.2; and 15.3).

A total of 375 Traceophytes, in the present study, have been classified into 9 life form classes according to Raunkiaer's (1918) classification as modified by Braun-Blanquet (1932).

Table 15.6 : Comparison of biological spectrum of the study area with Raunkiaer's (1918) normal biological spectrum.

Life forms										
	TH	*HH*	*G*	*H*	*CH*	*N*	*M*	*L*	*E*	*TOTAL*
Total Number of Species	111	12	13	100	57	32	39	10	01	375
% age of life form	29.8	3.2	3.5	26.4	15.2	8.5	10.4	2.6	0.2	100.0
% age of life form in normal spectrum	13.0	2.0	4.0	26.0	9.0	15.0	28.0	-	3.0	100.0
	+	+	-	+	+	-	-		-	
% age Deviation	16.8	1.2	0.5	0.4	6.2	6.5	17.6	-	2.8	

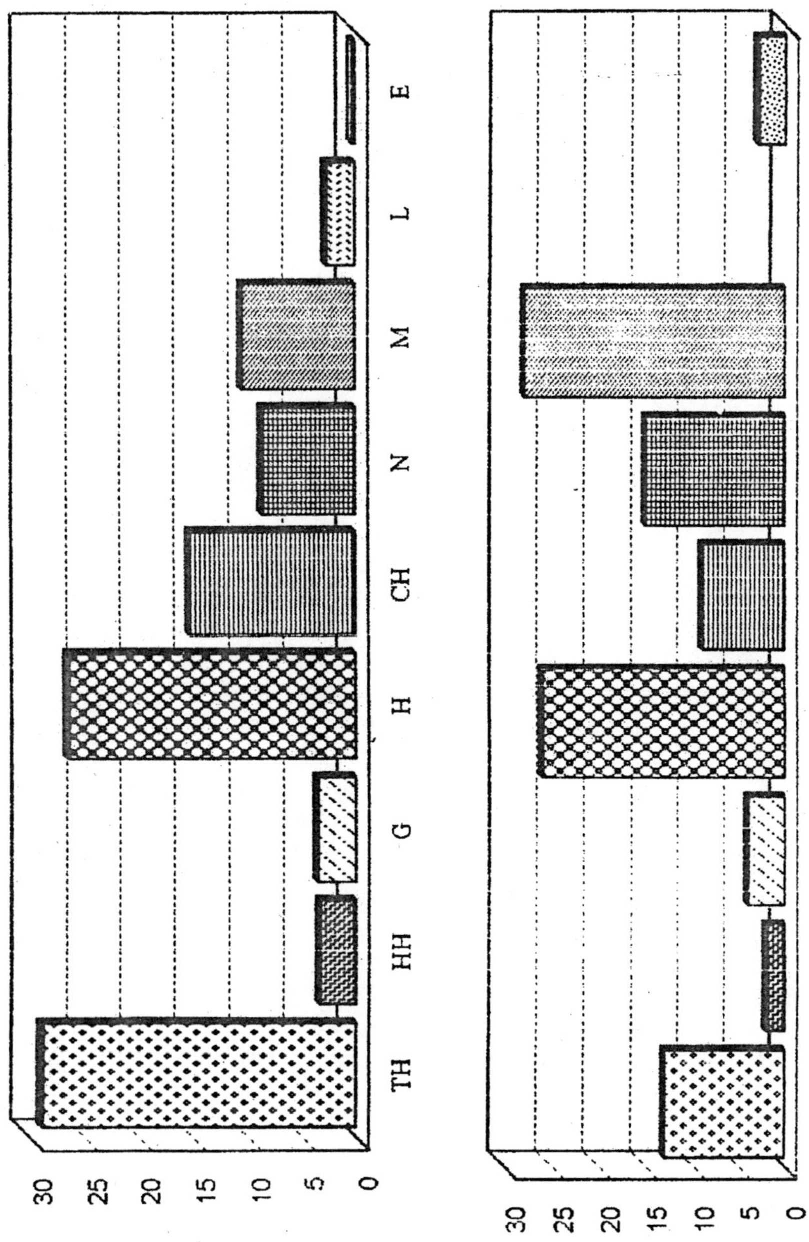

Fig. 15.4 : (a) Biological Spectrum of Patnitop and Adjoining Area.
(b) Raunkiaer's Normal Biological spctrum

Raunkiaer (1918) proposed Phanerophytic climate–in the tropics; Therophytic climate–in the deserts; Hemicryptophytic climate–in the greater part of the cold temperature zone and Chamaephytic with a fair proportion of Geophytes–in cool climate. Ansari and Singh (1979) observed a therophytic phyto-climate in Madhaulia forest of Gorakhpur rather than the phanaerophytic one that would be predicted by the Raunkiaer's hypothesis. They concluded that this was probably due to biotic interferences. Kapoor and Singh (1990) also found a hemicryptophytic-therophytic association in temperate grasslands of Shimla Hills. This was also slightly different and the reasons were attributed to heavy grazing in the area. Sharma annd Dhakre (1993) presented a thero-nanophanerophytic biological spectrum of the flora of Shahjahanpur District, U.P. The dominance of therophytes was again because of the grazing on the area.

In the present study a thero-hemicryptophytic phytoclimate has been worked out. The normal expected of this area is hemicryptophytic phyto-climate. The dominance of therophytes (29.8%) is because of greater biotic interferences in the area coupled with heavy grazing pressure of livestock of the nomad and trampling effect due to touristic activities (Table 15.5 and 15.6).

According to Raunkiaer (1934), the phyto-climate of all given regions may be characterized by the life forms which is the biological spectrum of that region and which exceeds of the percentage of the life forms in the normal spectrum.

Raunkiaer formulated the system with the definite purpose of utilizing the flora of a given track as an indicator of its climate. He firmly believed that the phyto-climate is characterized by the statistics of the life forms least adapted to certain climate which dominate the percentage of the flora than the others. The life forms of the present study area show a unique feature. There is a dominance of therophytes in lower altitudes indicating a warm dry climate (Dansereau, 1951, 1957). According to Pandeya (1961, 1964), and Pandeya et al. (1968) their higher percentage indicates a pronounced biotic or anthropogenic influence which maintains the vegetation open for further invasion of annuals with the result that good phanerophytic flora may be very much reduced in course of time. Beadle (1951) and Agrawal (1975) concluded that in such infected area the phytoclimate is not likely to give a correct indication of the vegetation. A similar conclusion is also drawn by Meher-Homji and Misra (1974) and Shah et al. (1981). There have been a large number of studies biological spectra of different regions of India (Bharucha & Ferreira, 1941; Srivastava, 1944; Lakshmanan, 1962; Wali, 1966; Srivastava, 1969; Singh and Yadav, 1974; Sapru, 1975; Kaul and Sarin, 1976; Rajwar and Gupta, 1981; Srivastava et al., 1981; Kapur, 1982; Singh and Joshi, 1983; Dhar and Koul, 1986).

When compared with the Raunkiaer's normal spectrum, the spectrum of the region showed hemicryptophytes and therophytes having a greater divergence from those in the normal, reflecting a thero-hemicryptophytic plant climate of the region. The biological spectrum of the area compared with the Raunkiaer's normal and those of the adjoining regions in the Siwalik ranges as well as in the Kashmir Himalayan ranges are given in Fig. 15.5.

These diagrams are self explanatory and indicative of the fact that these areas though adjacent to each other represent different phytoclimate because of varied amounts of biotic interferences. As per Raunkiaer's classification, the phytoclimate of the area should be phanero-hemicryptophytic, but the results of the present investigations do not correspond to normal because of many factors, at macro-meso-and micro-climatic level forming the biochores.

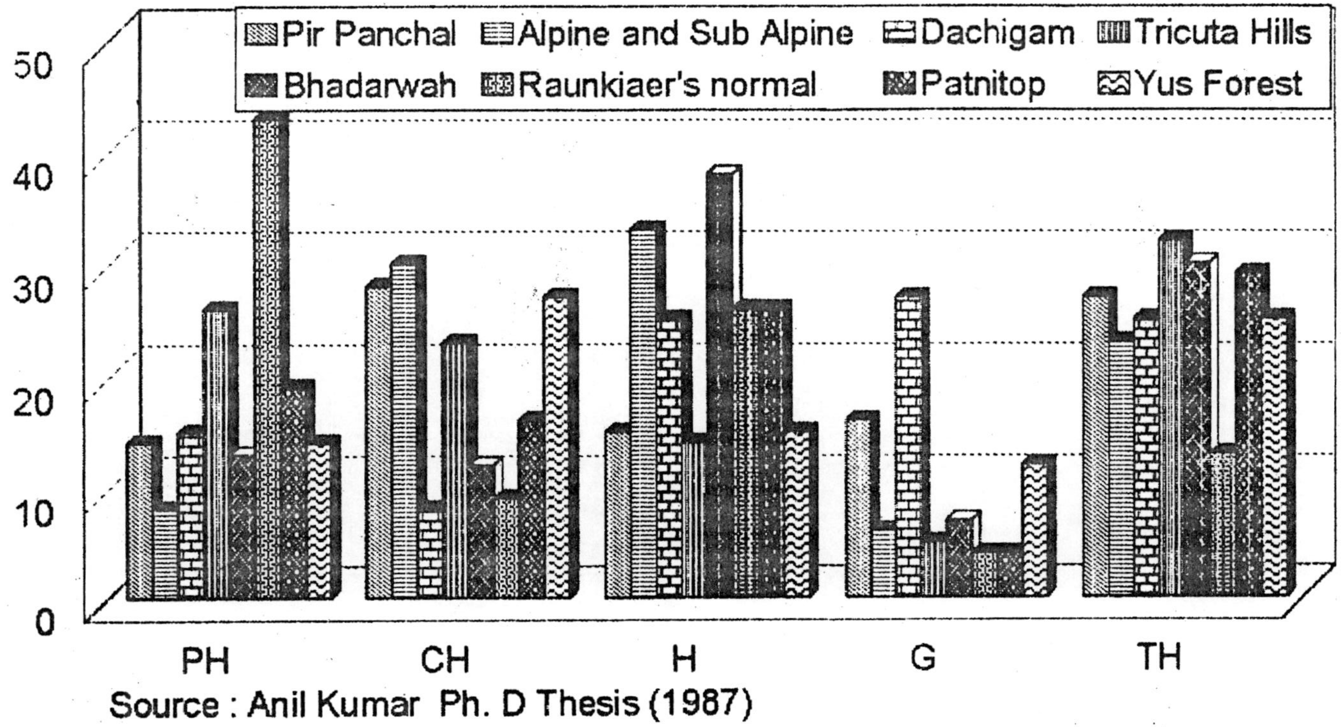

Fig. 15.5 : Biological Spectrum compared with Raunkiaer's Normal and Biological Spectra of Adjoining Areas.

Table 15.7: Biological spectra of some areas in north-West Himalayas.

Life Forms	*TH*	*HH*	*G*	*H*	*CH*	*N*	*M*	*L*
Patnitop forests	29.8	3.2	3.5	26.4	15.2	8.5	10.4	2.6
Raunkiaer's Normal	13.0	2.0	4.0	26.0	9.0	15.0	28.0	-
Yus forests	25.3	5.7	11.7	15.4	27.7	5.7	7.5	0.6
Trikuta forests	31.6	1.7	4.6	13.8	22.2	14.7	9.3	1.5
Bhadarwah forests	29.3	3.3	4.5	37.1	11.5	6.1	6.3	1.4

The fact that the therophytes are commonly along the disturbed grounds, the trempled paths within the forests, suggested their biotic dispersal. The reduction of phanerophytes is because of cutting and lopping, geophytes through successive grazing and trampling. The present findings find support in the statement that the flora of a region is not merely the product of the climate but also due to its history. The thero-hemicryptophytic plant climate may also be attribute to the irregular topography within the area. The irregularity might also have caused suitable niches for hemicryptophytic and therophytic type of life form and not for others. Infact, the present study area exhibits a divergent phytoclimate in departure with Raunkiaer's proposed classes. This thero-hemicryptophytic phyto-climate is attributed to the biotic disturbances, biochores in these area, longitudinal and latitudinal differences and the prevalent microclimate. All these operate either singly or in unison in one or the other way to determine the phytoclimate of the area and not the climate of the region alone. Kaul and Sarin (1976) working on another belt and aspect of the Bhadarwah forests, reported chamaephytes and hemicryptophytes as the largest life form classes. Our studies though showed a considerable (more than 70%) increase in the chamaephytes and geophytes do not record them as the major life form classes which could influence the phytoclimate of the area. The marshy plants also registered more than double the values when compared to Raukiaer's normal.

The higher percentage of hemicryptophytes with the increase in the altitutde is perhaps due to the abundance of moisture and precipitation in the form of snow for the major part of the area which is more congenial to the perennial growth of the species. Cain (1950) also reported that drier stepps have smaller number of hemicryptophytes than the steppes with more rainfall. On the other hand high percentage of therophytes in the present area may be due to biotic influence, viz., grazing and scrapping by animals and men respectively. The high value for therophytes is an indicator of the amount of influence of man and animal (Bharucha and Dave, 1944). According to Cain (1950) over-grazing tends to increase the percentage of therophytes through introduction and spread of weedy species. The present investigation falls in line with these findings.

LITERATURE CITED

Agarwal, S.K. (1975). Biological spectrum of the flora of Western Rajasthan. *Journal of Biological Sciences* 18 : 43-44.

Ansari, A.A. and Singh, S.K. (1979). Biological spectrum of the Madhaulia forest of Gorakhpur. *Indian Journal of Forestry*, 2(2): 153-157.

Appanah, S. ; Gentry, a.H. and LaFrankie, J.V. (1993). Liana diversity and speciesrichness of Malaysian rain forests. *Journal of Tropical Forest Science*, 6(2): 116-123.

Beadle, N.C.W. (1951). The misuse of climate as an indicator of vegetation and soils. *Ecologia*, 32 : 343-345.

Bharucha, F.R. and Ferreira, B.D. (1941). The biological spectra of the Matheran and Mahableshwar flora. *The Journal of Indian Botanical Society*, 20 : 195-198.

Bharucha, F.R. and Dave, R.N. (1944). The biological spectrum of grassland association. *Journal University of Bombay*, 13: 15-16.

Bir, S.S. (1993). Uniqueness of the Pteridophytic flora of the Himalayas and Conservation of threatened elements, in : Himalayan Biodiversity : Conservation Strategies, U. Dhar (ed.), G.B. Pant Institute of Himalayan Environment and Development, Almora, pp. 65-66.

Braun-Blanquet, J. (1932). Plant sociology (Translated by G.D. Fuller and H.S. Conard), McGraw Hill, New York.

Cain, S.A. (1950). Life form and phytoclimate, *Botanical Review*, 16 : 1-32.

Chatterjee, D. (1962). Floristic patterns of Indian vegetation. In : Proceedings of Summer School Botany, Darjeeling, P. Maheshwari, (ed.) pp. 32-42.

Clarke, C.B. (1898). On the sub-areas of British India, illustrated by the detailed distribution of the Cyperaceae in that Empire. *Journal Linnean Society (Bot.) Land on*, 34 : 1-146.

Dansereau, P. (1951). Description and recording of vegetation on structural basis. *Ecology*, 32 : 172-229.

Dansereau, P. (1957). Biogeography–An Ecological Perspective, Vol. I, New York.

Dhar, U. (1996). Overview of Himalayan Biodiversity. In : Changing Perspectives of Biodiversity Status in the Himalaya, The British Council, New Delhi, pp. 3-20.

Dhar, U. and Koul, A. (1986). Life form classification and Biological spectrum of alpine-subalpine flora of Kashmir Himalayas. *Research and Development Reporter*, 3(1): 25-28.

Dhar, U. and Kachroo, P. (1983)). Alpine flora of Kashmir Himalaya, Scientific Publishers, India. Dhar, U.; Virjee and Kachroo, P. 1987. Endemisms in Kashmir–a retrospect, in : Western Himalaya, Environment, Problems and Development, Vol. 1 (Pangtey, Y.P.S. and Joshi, S.C.) (eds.), Gyanodaya, Prakashan, Nanital, India, pp. 413-420.

Ehrlich, P.R. & Ehrlich, A.H (1992). The value of biodiversity. *Ambio.*, 21 : 219-226.

George, M. and Varghese, G. (1993). Species Diversity and structural variation of a tropical dry deciduous forest ecosystem of Western Ghat. *Indian Journal Tropical Biodiversity*, 1 : 30-36.

Good, R. (1964). The Geography of the Flowering Plants. London, 518 pp.

Hooker, J.D. (1907). A sketch of the flora of British India, Oxford University Press, U.K.

Jain, S.K. (1990). Botanical Regions, and flora of India. Everyman's Science, 24 (6) : 213-223.

Kachroo, P.; Sapru, B.L. and Dhar, U. (1977). Flora of Ladakh. Bishen Singh Mahendra Pal Singh, Dehra Dun.

Kapoor, K.S. and Singh, R. (1990). Life forms and biological spectrum of the temperate grasslands with varying biotic pressures in Shimla Hills, Himachal Pradesh. *Range Management and Agroforestry*, 11 (2) : 109-114.

Kapur, S.K. (1982). Phytoecological studies in Trikuta hills with special reference to the distribution pattern of economic plant products. PH.D. thesis, Univ. of Jammu.

Kaul, V. and Sarin, Y.K. (1976). Life form classification and the biological spectrum of the flora of Bhadarwah. *Tropical Ecology*, 17 (2) : 132-138.

Kumar, Anil (1987). Phytosociological and Productivity studies of Bhadarwah Forests, Jammu (J&K). Ph.D. Thesis, University of Jammu.

Kumar, Koshal (1997). Studies on plant diversity of Patnitop and adjoining areas and impact of biotic activities. Ph.D. Thesis, University of Jammu.

Lakshmanan, N.K. (1962). The application of Raunkiaer's life forms. *The Journal of Indian Botanical Society*, 41 : 585-589.

Lovejoy, T.E. (1994). The quantification of biodiversity : an esoteric quest or a vital component of sustainable development ? Phil. *Transactions of Royal Society London*, B345 : 81-87.

Lovelock, J.E. (1979). Gaia: A new look at life on Earth. Oxford University Press, UK. 192 pp.

Meher-Homji, V.M. and Misra, K.C. (1973). Phytogeography of the Indian sub-continent Progress of plant Ecology (R. Misra et al. ed.), 1 : 67-71.

Mehrotra, A. and Kushalaya, K.A. (1994). An approach for sustaining biodiversity in India. *Indian Forester*, 120 (9) : 759-772.

Pandeya, S.C. (1961). Ecology of grasslands of Sagar, M.P. I Grassland map of the area on physiongomic basis. The *Journal of Indian Botanical Society*, 40 : 592-600.

Pandeya, S.C. (1964). Ecology of grasslands of Sagar, M.P. II A–composition of the fenced grassland association IIB–composition of the association open to grazing or occupying special habitat. *The Journal of Indian Botanical Society*, 43 : 577-605; 606-639.

Pandeya, M.T. and Oza, G.M. (1994). Biodiversity for the masses. *Indian Forester*, 120 (9) : 824-826.

Rao, R.R. and Husain, T. (1993). Himalayan Legumes : diversity and strategies, in : Himalayan Biodiversity : Conservation Strategies, U. Dhar (ed.), G.B. Pant Institute of Himalayan Environment and Development, almora, pp. 253-266.

Rao, R.R. and Murti, S.K. (1990). North-East India–a major centre for plant diversity in India.*Indian Journal of Forestry*, 13 (3) : 214-222.

Rajwar, G.S. and Gupta, S.K. (1981). Flora and biological spectrum of Khoh Valley. District Pauri (Garhwal). *Indian Journal Forestry*, 4 (1) : 49-55.

Raunkiaer, C. (1918). Uber das biologische Normal spectrum, Kgl. Danksk. Vidensk. Selsk. Biol. Meddle, 1 : 4.

Raunkiaer, C. (1934). The life forms of plant and statistical plant geography (collected papers of C. Raunkiaer). Clarendon Press, Oxford. 639 pp.

Rodgers, W.A. and Panwar, H.S. (1988). Planning a Wildlife Protected Area Network in India, Vols. 1-2, WII, Dehra Dun.

Sapru, B.L. (1975). Vegetational studies in Jhelum Valley, Botanique, VI (2 & 3) : 151-164.

Sarin, S. (1981). Management of minor forestry produce : perspective and alternative frameworks for research and analysis. *Indian Forester*, 107 (7): 397-411.

Sehgal, J.C.; Mondal, D.K.; Mondal, C. and Vadi, S. (1990). Agro-ecological regions of India, NBSS Tech. Bull. No. 24 NBSS & LUP, Nagpur, 73 pp.

Shah, G.L.: Yadav, S.S. and Menon, A.R. (1981). Biological spectrum of the flora of Gujrat State. *Bulletin of Botanical Survey of India*, 23 (1-4) : 25-29.

Sharma, B.M. and Kachroo, P. (1981). Flora of Jammu and plants of neighbourhood, Vols. 1 & 2. Bishen Singh Mahendra Pal Singh, DehraDun.

Sharma, M. and Dhakre, J.S. (1993). Life form classification and Biological spectrum of the flora of Shahjahanpur District, Uttar Pradesh (India). Indian Journal of Forestry, 16 (4) : 366-371.

Singh, G. and Kachroo, P. (1976). Forest flora of Srinagarand plants of neighbourhood. Bishen Singh Mahendra Pal Singh, Dehra Dun, pp. 210.

Singh, R. and Joshi, M.C. (1983). Biological spectrum of the sand dune herbaceous vegetation at Narhar near Pilani (Rajasthan). *Indian Journal of Ecology*, 10(2) : 344-346.

Singh, J.S. and Yadav, P.S. (1974). Seasonal variation in composition, plant biomass and net primary productivity of a tropical grassland at Kurukshetra, India. Ecological Monographs, 44 : 351-376.

Srivastava, G.D. (1944). The biological spectrum of Allahabad Flora. *The Journal of Indian Botanical Society*, 23 : 1.

Srivastava, M.M. (1969). Plant Ecology of Alkaline soils of Varanasi. Ph.D. thesis, Banaras Hindu University.

Srivastava, M.M.; Srivastava, P.K. and Kukreti, U.D. (1981). Biological spectrum of marshy grassland at Rajput, Dehra Dun (U.P.), India. Indian Journal of Forestry, 4 (3) : 220-223.

Wali, M.K. (1966). Life forms and biological spectrum of Lolab Valley, Kashmir, in relation to climate. Journal Bombay Natural History Society. 63 (1) : 115-122.

Chapter - 16

Host Range and diversity of Aphidophagous Insects (Predators) of Kashmir Himalaya, with New Host Aphid / Plant Records

R.C. Bhagat and T.A. Matta**

ABSTRACT

This work incorporates data pertaining to host range and diversity of 26 species of aphidophagous predators, belonging to four insect orders viz., Coleoptera, Diptera, Hemiptera and Neuroptera. These predatory species prey upon more than 45 aphid species, damaging 56 species of plants, distributed over 22 families. A total of about 73 different host/ prey aphid – predator couples are being recorded for the first time from Kashmir, besides, 8 species of predators feeding on some unidentified aphid species on 8 plant species, showing 13 predator-aphid associations. With the new host record of predators, a total of about 116 different host aphid/predator species couples are recognized from Kashmir. Apart from this, an up-to-date catalogue of predator species – prey aphid species and host plant species is added.

Key-words: Aphidophagous insect predators, host range, new host/aphid plant record.

INTRODUCTION

The aphids or 'plant lice' or Kashmir are found to damage different types of vegetation, including important species by sucking sap. Aphids belong to the family Aphididae and order Homoptera of class Insecta, are characterized by small soft, more or less pear-shaped bodies, size usually 1-5 mm, relatively slow moving and live in colonies. The adults are both apterous (wingless) as well as alate (winged). They have a great biotic potential as reflected by their parthenogenicity, viviparity and fast development. The control

*P.G. Departmet of Zoology, The University of Kashmir Hazrat-bal, Srinagar-190006, Kashmir (J & K), India

measures of aphids are diverse, incorporating both chemical and biological methods. In the recent years, biological control of aphids by utilizing natural enemies–parasites and predatory insects, has shown promising results in different parts of the v. orld.

In Kashmir, various insect predators of aphids on diverse plant species and families, in different localities and regions, belong to various orders and families viz., Coleoptera (Bruchidae, Coccinellidae), Diptera (Cecidomyiidae, Chamaemyiidae, Drosophilidae and Syrphidae), Hemiptera (Anthocoridae) and Neuroptera (Chrysopidae). The larvae and adults of Lacewings (Chrysopids), Ladybird beetles (Coccinellids), are commonly occurring predators of this region. The aphido–phagous coccinellids, syrphids and cecidomyiids play a important role in natural biological control of aphids of Kashmir.

The earliest records of syrphids of Kashmir Himalayas have been given by Hull (1950) and Kapur (1942). Later, Zaka-ur-Rab recorded *Syrpbus confrator* Wied. preyed on apple woolly aphid, *Eriosoma lanigerum* and Ahmad and Vaishnnavi (1982) recorded *Aphidoletes aphidimyza* Rondani for the first time from Kashmir, predaceous on 7 aphid species. Bhagat and Lone (1984) recorded 17 speciies of Aphidophagous predators, belonging to different insect orders. Also, from this region, Bhagat and Masoodi (1986) reported for the first time *Chrysopa orestes* Bank, preying on *aphid gossypii* and *A. pomi.*

RESULTS AND DISCUSSION

The present communication provides valuable information on 26 species of aphidophagous predators of Kashmir, belonging to four orders, with respect to their host/prey aphid and host plant diversity, together with new prey aphid and host plant species records.

26 species of predators are found to exist in Kashmir, covering 13 species of Coleoprerans, 9 species of Dipterans, 3 species of Hamipterans and 1 species of Neuropteran. These predatory species prey upon more than 45 aphid species, distributed over 30 genera, infesting 56 species of plants, spread over 22 plant families, including economically important ones (Table 16.1 & 16.2). Now, a total of 116 different host aphid/predatory couples are recognized in this region. Out of these couples, about 73 host/prey aphid predator couples are being recorded for the first time from Kashmir by the present authors. In addition to these couples, the present authors have also encountered for the first time from this region, 8 different species of predators, preying on some unidentified aphid species, indicating 13 different predator-prey aphid and plant associations (Table : 16.1 & 16.2).

Table 16.1 : Number of different Predator spcies preying on different aphid species, infesting number of plant species belonging to different families in Kashmir.

Sl. No.	*Prey aphid species*	*Number of species of predator*				*Number of Host plant species*	*Name of Plant family*
		Col.	*Din.*	*Hem.*	*Neu.*		
1	*2*	*3*	*4*	*5*	*6*	*7*	*8*
1.	*Acyrthosiphon rubi*	1	3	-	-	1	Ros.
2.	*Amphicercidus tuberculatus*	1	-	1	-	1	Cap.
3.	*Anuraphis persicae niger*	1	1	1	-	2	Ros.

1	*2*	*3*	*4*	*5*	*6*	*7*	*8*
4.	*Aphis craccivora*	6	4	1	-	3	Pap*, Fab.
5.	*Aphis fabae*	1	-	-	-	Unidentified	-
6.	*Aphis fabae solanella*	1	4	-	-	2	Pol;, Pap.
7.	*Aphis gossypii*	1	-	-	1	1	Ast.
8.	*Aphis krosawai*	1	-	-	-	1	Thy.
9.	*Aphis nerii*	2	1	-	-	1	Sor.
10.	*Aphis pomi*	2	1	-	1	2	Ros.
11.	*Aphis spiracola*	-	1	-	-	1	Ros.
12.	*Aphis* sp.	3	4	-	-	3	Ast., Ros.
13.	*Brevicoryne brassicae*	-	1	-	-	1	Bra.
14.	*Brevicoryne* sp.	-	1	-	-	1	Bra.
15.	*Brachycaudus cardui*	-	-	1	-	1	Ast.
16.	*Callaphis juglandis*	2	1	-	-	1	Jug.
17.	*Capitophorus hippohaeus*	-	1	-	-	1	Ast.
18.	*Cavariella* sp.	-	1	-	-	1	Api.
19.	*Chaetosiphon chaetosiphon*	-	1	-	-	1	Ros.
20.	*Chaetosiphon* sp.	-	1	-	-	1	Ros.
21.	*Chaitophorus* nr. *niger*	1	3	-	-	1	Sal.
22.	*Chaitophorus pakistanicus*	1	1	-	-	1	Sal.
23.	*Chaitophorus populeti*	1	-	-	-	1	Sal.
24.	*Chromaphis juglandicola*	1	1	-	-	1	Jug.
25.	*Eriosoma lanigerum*	1	1	-	-	2	Ros.
26.	*Hyalopterus pruni*	4	4	-	-	2	Ros., Poa.
27.	*Indiochaitophorus furcatus*	1	2	1	-	2	Ulm.
28.	*Liosomaphis bimalayensis*	2	1	-	-	1	Ber.
29.	*Lipaphis erysimi*	2	2	-	-	1	Bra.
30.	*Macrosiphoniella artemisie*	-	-	1	-	1	Ast.
31.	*Macrosiphoniella sanborni*	2	-	-	-	1	Ast.
32.	*Macrosiphum rosae*	3	3	-	-	1	Ros.
33.	*Myzaphis rosarum*	1	-	-	-	1	Ros.
34.	*Myzaphis persicae*	1	-	-	-	1	Ros.
35:	*Neothoracaphis* sp.	1	-	-	-	1	Ham.

1	2	3	4	5	6	7	8
36.	*Pemphigus immunis*	-	-	1	-	1	Sal
37.	*Pemphigus mordvilkooi*	-	1	-	-	1	Sal.
38.	*Prociphilus* sp.	-	1	-	-	1	Cap.
39.	*Rhopalosiphum* sp.	-	1	-	-	1	Ros.
40.	*Schizoeurella indica*	-	1	-	-	1	Ulm.
41.	*Shivaphis celti*	1	2	-	-	1	Cel.
42.	*Tetraneura* sp.	1	-	-	-	1	Sim.
43.	*Tricaudatus polygoni*	2	-	-	-	1	Ros.
44.	*Uroleucon kashmiricus*	-	1	-	-	1	Urt.
45.	*Uroleuco* sp.	2	2	-	-	1	Ros.
46.	Unidentified aphid species	4	4	-	-	8	Ast., Big., Cap., Hip., Sol.

Abbreviations used

Col.–Coleoptera, Dip.–Diptera, Hem.–Hemiptera and Neu.–Neuroptera
Api.–Apiacae, Ast.–Asteraceae, Ber.–Berberidaceae, Big.–Bignoniaceae,
Bra.–Brassicaceae, Cap.–Caprifoliaceae, Cel.–Celtaceae, Fab.–Fabaceae,
Ham.–Hamamelidaceae, Hip.–Hippocastanaceae, Jug.–Juglandaceae,
Pap.–Paplionaceae, Poa.–Poaceae, Pol.–Polygonaceae, Ros.–Rosaceae,
Sal.–Salicaceae, Scr.–Sorophularaceae, Sim.–Simiarubiaceae,
Sol.–Solanaceae, Thy.–Thymeliaceae, Ulm.–Ulmaceae, Urt.–urticaceae.

The number of new record of prey aphid species observed during the present investigation are : 6 aphid species predated upon by *Adalia tetraspilota*, 4 species by *Chilocorus nigrita*, 1 species by *G. rubidus*, 6 species by *Coccinella undecipunctata*, 5 species by *Hippodamia variegata*, 6 species by *Oenopia conglobata*, 6 species by *Platynaspis saundersii*, 4 species by *Aphidoletes aphidomyza*, 14 species by *Leucopis* sp. 13 species by *Episyrphus balteatus*, 6 species by *Metasyrphus (M.) confractor* 1 species by *Sphaerophoria scripta* and 1 species by *Anthocoris* sp. (Table 16.2).

In addition to the aforementioned newly recorded prey/hosts of predators, the present authors have also witnessed 8 different species of predators feeding on some unidentified aphid species. These pedators of aphids, infesting various plant species are: *A. tetraspilota* feeding aphids on host plant, *Catalpa bignonoides* and *Aesculus indicus; Aiolocaria* sp. on *Viburum foetans: C. undecipunctata*, predaceous on aphid on *Artemisia turnfortiana; Platynaspis saundersii*, predated upon aphids on *Solanum miniatum* and *A. turnfortiana; A. aphidomyza* feeding on aphids on *A. turnfortiana* and *Aster multicanlis; Leucopis* sp. feeding on aphids on *A. multicanlis, Breeea arvense* and *C. bignonoides; E. balteatus* preying on aphids, infesting *Sonchus* sp. and *Sphaerophoria scripta* preyed on aphid, infesting *A. turnfortiana*, as given in Table 16.2.

Table 16.2 : A Catalogue of AphidophagousInsects (Predators) of Kashmir

	Predator/ Aphid Prey species	*Host Plant species*	*Author/s reporting & New host Aphid/ Plant Record (= NR)*
	1	2	3
	Order : Coleoptera		
	Family : Bruchidae		
1.	*Brouchidius* sp.		
	Amphicercidus tuberculatus	*Lonicera quinqueloris*	Bhagat & Lone (1984)
	Family 2. Chrysomelidae		
	Subfamily : Cassidae		
2.	*Aspidomorpha indica* Boheman		
	Acyrthosiphon rubi Narzikulov	*Rubus fruticosus*	-Do-
	Family 3 : Coccinellidae		
3.	*Adalia tetraspilota* Hope		
	Aphis oraccivora Koch	*Robinia pseudoacacia* *Medicao sativa*	-Do- NR
	Aphis gossypii Glover	*Zinnia elegans*	Bhagat and Lone (1984)
	Aphis sp.	Unidentified plant species	-Do-
	Callaphis juglandis Goeze	*Juglans regia*	NR
	Hyalopterus pruni Geoffroy	*Prunus domestica*	-Do-
	Indiochaitophorus furcatus Verma	*Ulmus villosa*	-Do-
	Liosomaphis himalayensis Basu	*Berberis zibbliana*	-Do-
	Myzaphis rosarum Kaltenbach	*Rosa webbiana*	Bhagat and Lone (1984)
	Neothoracaphis sp.	*Parrotiopsis jacquemontiana*	NR
	Unidentified aphid species	*Catalpa bibnonoides*	-Do-

1	2	3
Unidentified aphid species	*Aesoulus indicus*	-Do-
4. *Aiolocaria hexaspilota* Hope		
Tetraneura sp.	*Ailantus altissma*	Bhagat and Lone (1984)
5. *Aiolocaria* sp. Unidentified aphid species	*Viburnum foetans*	NR
6. *Chilocorus nigrita* (Fabricius)		
Callaphis juglandis (Goeze)	*J. regia*	-Do-
H. pruni	*P. domestica*	-Do-
Tricaudatus polygoni Narzikulov	*Spirea cantoniensis*	-Do-
Uroleucon sp.	*Chaenomeles japonica*	-Do-
7. *Chilocorus rubidus* (Hope)		
H. pruni	*P. domestica*	-Do-
8. *Coccinella undecipunctata*	*Linnaeus*	
A. craccivora	*R. pseudoacacia*	-Do-
Aphis nerii	*Verbascum thapsus*	-Do-
Aphis pomi de Geer	*Pyrus malus*	Bhagat and Lone (1984)
Aphis sp.	*Z. elegans*	NR
Lipaphis erysimi (Kaltenbach)	*Brassica campestris*	-Do-
Macrosiphum rosae Linnaeus	*Rosa indica*	-Do-
Unidentified aphid species	*Artemisia turfortiana*	-Do-
9. *Coccinella* sp.		
A. craccivora	*Medicago polymorpha*	Bhagat and Lone (1984)
10. *Hippodamia* (*Adonia*) *Variegata* (Goeze) (*Adonia variegata*)		
Aphis craccivora Koch	*Robinia psudoacacia*	NR
Chaitophorus nr. niger Mordvilko	*Populus caspica*	-Do-

1	2	3
Eriosoma lanigerum Hausmann	*Pyrus malus*	Kapur (1942)
Lipaphis erysimi Kaltenbach	*Brassica campestris*	NR
Macrosiphum rosae Linnaeus	*Rosa indica*	-Do-
Tricaudatus polygoni Narzikulov	*Spirea cantoniensis*	-Do-
11. *Hippodamia convergens* Guerin		
Aphis pomi de Geer	*Pyrus malus*	Rishi (1966)
Anuraphis persicae niger Smith	*prunus persica*	-Do-
Chromaphis juglandicola Kaltenbach	*Juglans regia*	-Do-
Hyalopterus arandinidis	*Prunus domestica*	-Do-
Myzus persicae (Sulzer)	Pyrus communis	-Do-
12. *Oenopia conglobata* Linnaeus		
(Synharmonia conglobata)		
Aphis craccivora	*Robinia pseudoacacia*	NR
Aphis fabae Scopoli	Unidentified plant species	Bhagat and Lone (1984)
Ghaitophorus pakhistanicus HRL	*Salix alba*	NR
Chaitophorus populeti	*Populus caspica*	-Do-
Liosomaphis himalayensis Basu	Berberis zibbliana	-Do-
Macrosiphum rosae Linnaeus	*Rosa indica*	-Do-
Shivaphis celti Das	*Celtis australis*	-Do-
13. *Platynaspis saundersi* Grotch		
A. crraccivora	*R. pseudoacacia*	-Do-
Aphis kurosawai Takahashi	*Daphne oleoides*	Bhagat and Lone (1984)

	1	2	3
	Aphis fabae solanella	*Astragalis polycanthus*	NR
	Aphis nerii Boyer de Fonsolm.	*Verbascum thapsus*	-Do-
	Aphis sp.	*Inula racemose*	-Do-
	Uroleucon kashmiricus Verma	*Urtica dioica*	-Do-
	Uroleucon sp.	*Chaenomeles japonicus*	-Do-
	Unidentified aphid species	*Solanum miniatum*	-Do-
	-Do-	*Artemisia turnfortiana*	-Do-
	Order : Diptera		
	Family 1 : Cesidomyiidae		
14.	*Aphidoletes aphidomyza* Rondani		
	A. craccivora	*R. pseudoacacia*	Ahmad and Vaishnavi (1982)
	A. nerii	*Verbascum thapsus*	NR
	A. pomi	*Prunus domestica*	Ahmad and Vaishnavi (1980)
	Aphis Fabae solanella Theobald	*Astragalis polycanthus*	NR
		Romex nepalense	Ahmed & Vaishnavi (1982)
	Aphis spiraecola	*Rubus antenifer*	-Do-
	Aphis sp.	*Inula racemose*	NR
	Capitophorus bippohaeus Walker	*Garduus eddelbergi*	Ahmad and Vaishnavi (1962)
	Chaetosiphon sp.	*Rosa macrophylla*	-Do-
	Macrosiphum ronae Linnaeus	*Rosa* sp.	Ahmad and Vaishnavi (1982)
	Rhopalosiphum sp.	*Prunus* sp.	-Do-
	Uroleucon sp.	*Chanomeles japonica*	NR
	Unidentified aphid species	*Attemisia turnfortiana*	-Do-
	-Do-	*Aster multicanlis*	-Do-
	Family 2 : Chamaemyiidae		
15.	*Leucopis* (*Leucopis*) sp.	*Pyrus malus*	Bhagat and Lone (1984)

1	2	3
Aphis sp.		
Hyalopterus pruni Geoffroy	*Phragmites australis*	-Do-
16. *Leucopi* sp.		
Acyrthosiphon rubi Narzilulov	*Rubus fruticosus*	NR
Aphiscraccivora Koch	*Robinia pseudoacacia*	-Do-
Aphis nerii BDF	*Verbascum thapsus*	-Do-
Aphis fabae solanella Theobald	*Astragalis polycanthus*	-Do-
	Rumex nepalense	-Do-
Aphis sp.	*Inula racemose*	-Do-
Brevicoryne sp.	*Capsella bursa pastoris*	-Do-
Chaitophorus nr. *niger Mordvilko*	*Populus caspica*	-Do-
Eriosoma langigeerum Hausmann	*Arunius* sp.	-Do-
H. pruni	*Prunu domestica*	-Do-
Macrosiphoniella sanborni Gillette	*Chrysanthemum morifolium*	Bhagat and Lone (1984)
Macrosiphum rosae Linnaeus	*Rosa indica*	NR
Prociphilus sp.	*Lonicerra* sp.	Bhagat and Lone (1984)
Shivaphia celti Das	*Celtis australis*	NR
Uroleucon kashmiricus Verma	*Urtica dioica*	-Do-
Uroleucon sp.	*Chaenomeles japonica*	-Do-
Unidentified aphid species	*Aster multicanlis*	-Do-
-Do-	*Breea arvense*	-Do-
-Do-	*Catalpa bignonoides*	-Do-
Family 3 : Drosophilidae		
17. *Scaptomyza* (*Parascaptomyza*) *pallida* (Zetterstedt)		

1	2	3
Aphis fabaesolanella Theobald	*Rumex nepalense*	Bhagat and Lone (1984)
Family 4 :Syrphidae		
18. *Episyrphus balteatus* De Geer		
A; Rubi Narzikulov	*Rubus frutucosus*	NR
Anuraphis persicae niger Smith	*Prunus amygdalus*	-Do-
A. craccivora Koch	*R. pseudoacacia*	Bhagat and Lone (1984)
A.f. solanella	*R. nepalense*	NR
Aphis sp.	*Inula racemose*	-Do-
Brevicorynae brassicae Linnaeus	*Brassica oleraceae*	-Do-
Callaphis juglandis Goeze	*Juglans regia*	-Do-
Chaitophorus nr *niger*	*Populus caspica*	-Do-
Chaitophorus pakistanicus HRL	*Saliex alba*	-Do-
H. pruni	*P. domestica*	-Do-
Indiochaitophorus furcatus Verma	*Ulmus wallichiana*	Bhagat and Lone (1984)
Liosomaphis himalayensis Basu	*Brberis zibbliana*	NR
Lipaphis erysimi Kaltenbach	*Brassica campestris*	-Do-
Macrosiphoniella sanborni Gillette	*Chrysanthemum morifolium*	-Do-
Macrosiphum rosea Linnaeus	*Rose indica*	-Do-
Schizoneurella indica Hills Ris Lambers	*Ulmus* sp.	Bhagat and Lone (1984)
Unidentified aphid species	*Sonchus* sp.	NR
19. *Mtasyrphus (M.) confrator* (= *Syrphus confractor*) Wiedmann		

1	2	3
Acrythosiphon rubi Narzikulov	*Rubus fruticosus*	NR
Aphis craccivora Koch	*Robinia pseudoacacia*	-Do-
Chaitophorus nr. *niger* Mordvilko	*Populus caspica*	
Eriosoma lanigerum Hausmann	*Pyrus malus*	Zaka-ur-Rab (1972)
	Arunius sp.	NR
Hyalopterus pruni Geoffroy	*Prunus domestica*	-Do-
Indiochaitophorus furcatus Verma	*Ulmus villosa*	-Do-
Shivaphis celti Das	*Celtis australis*	-Do-
20. *Sphaerophoria inndianna* Bigot		
H. pruni	*P. domestica*	Bhagat and Lone (1984)
21. *Sphaerophoria scripta* Linnaeus		
Cavariella sp.	*Hieraclium candicans*	NR
Unidentified aphid species	*Artemisiaturnfortiana*	-Do-
22. *Syrphus* sp.		
Chaetosiphon chaetosiphon Nevsky	*Rosa* sp.	Bhagat and Lone (1984)
Eriosoma lanigerum Hausmann	*Pyrus malus*	-Do-
Lipaphis erysimi Kaltenbach	*Brassica campestris*	-Do-
Order Hemiptera		
Family Anthocoridae		
23. *Anthocoris* sp.		
I. furcatus Verma	*Ulmus villosa*	NR
Pemphigus immunis Buckton	*Populus ciliata*	Bhagat and Lone (1984)

1	2	3
24. *Orius Heterorius minutus* Linnaeus		
Amphicerocidus tuberculatus, David, Narayanan & Rajsingh	*Lonicera quinquelocularis*	-Do-
Brachycaudus (Acaudus) Cardui L (= Brachycaudus cardui)	Crduus edelbergi	-Do-
25. *Orius (Orius) lindbergi* Wagner		
Anuraphis persicae niger Smith (= *Anuraphis persicae*)	*Prunus amygdalus*	-Do-
A. craccivora Koch	*Medicago sativa*	-Do-
Macrosiphoniella artemisae BDF	*Artemisia* sp.	-Do-
Pemphigus mordvilkoi Cholodkvosky	*Populus caspica*	-Do-
Order Neuroptera		
Family Chrysopidae		
26. *Chrysopa orestes* Banks		
Aphis gossypii Glover	*Solanum melongena*	Bhagat and Masoodi (1986)
Aphis pomi de Geer	*Pyrus malus*	-Do-

Abbreviation : NR = New host Aphid/Plant species record by the present authors.

The predator species feeding on aphid pests of economically important crops and plants are highlighted as under:–

(a) Aphid predators of fruit-trees

Among the aphid pests which damage the fruit trees of Kashmir are: *Anuraphis persicae niger* on *Prunus amygdalus* and *P. persica; Aphis pomi* and *Eriosoma lanigerum* on *Pyrus malus, A. pomi* also damage *Prunus domestica; Callaphis juglandis* and *Chromaphis juglandicola* on *Juglans regia; Hyalopterus pruni* on *Pruus domestica* and *Myzus persicae* on *Pyrus communis.*

A. P. niger is predated upon by *Hippodamia convergens* (coccinellid), *Episyrphus belteatus* (syrphid), *Orius* (*O.*) *lindbergi* (anthocorid). *A. pomi* is devoured in the fields by predatory species *viz., Aphidoletes aphidomyza, Chrysopa oresta, Coccinella undecipunctata* and *H. convergens.* The apple woolly aphid, *E. lanigerum,* as a serious pest of apple trees in Kashmir, is found to be predated upon by *H. variegata* (Coccinellid) and *Syrphus* sp.

Two widely distributed aphid pests of *J. regia viz.*, *C. juglandis* and *C. juglandicola* damage the leaves of walnut trees greatly in this region. *Adalia tetraspilota*, *Chilocorus nigrita* and *E. balteatus* feed on *C. juglandis*, whereas *H. convergens* has been found to suppress *C. juglandicola* in the fields. Great damage is rendered by aphid pest, H. pruni to the leaves of *P. domestica* and this pest is checked in the orchards by four coccinellid beetles (*A. tetraspilota*, *C. nigrita*, *C. rubidus* and *H. convergens*), one chamaemyiid (*Leucopis* sp.), three syrphids (*E. belteatus*, *Metasyrphus confractor* and *Sphaerophoria indiana*). *Mysus persicae* has been found to damage fruit tree, *P. communis* and is predated upon by H. convergens.

(b) Aphid predators of vegetables, oilseeds and fodder crop

The widely cultivated vegetable of this region viz., *Brassica oleraceae* is is damaged by aphid pest, *Brevicoryne brassicae*. Syrphidfly, *E. balteatus has* been reported to feed on *B. brassicae*. The main oilseed crop (*Brassica campestris*) of this region is damaged by aphid pest, *Lipaphis erysimi*. Four predators have been found to devour *L. erysimi* in Kashmir and these include *C. undecipuctata*, *H. variegata*, *E. balteatus* and *Syrphus* sp. Fodder crop–*Medicago polymorpha* and *M. sativa* is attacked by aphid pest *viz.*, *Aphis craccivora*, predated upon by *A. tetraspilota*, *Coccinella* sp. and *Orius* (*O.*)*lindbergi*. Brinjal (vegetable) (*Solanum melongena*) is damaged by *A. gossypii*, which is predated upon by a *Neuropteran viz.*, Chrysopa orestes (Lacewingfly), see Table (16.1 & 16.2).

(c) Aphid predators of ornamental plants

The aphid pests damaging ornamental plants of Kashmir include *Aphis gossypii* on *Zinnia elegans*, *Chaetosiphon* spp. on *Rosa* spp. *Macrosiphoniella sanborni* on *Chrysanthemum morifolium*, *Macrosiphum rosae* on *Rosa* spp. and *Tricaudatus polygoni* on *Spirea cantoniensis*. The number of different predator species preying on aphid species infesting ornamental plants are given in the Table 16.2. The populations of aphids of these economically important plants are suppressed in the fields by various predators such as *A. tetraspilota*, *Syrphus* sp., *A. aphidomyza*, *E. balteatus*, *C. undecipunctata*, *H. variegata*, *Oenopia conglobata* and *C. nigrita* and *Leucopis* sp. (See Table 16.1 & 16.2).

(d) Aphid predators of medicinal plants

The predators of various aphid species damaging medicinal plants observed in this region are: *A. aphidomyza*, feeding aphids on medicinal plants *viz.*, *Inula racemose* and *Rumex nepalense*; *C.undecipunctata* preyed on aphid on *Artemisia turnfortiana* and *Verbascum thapsus*; *E. balteatus* feeding on aphid species on *I. racemosa*; *Leucopis* sp., devouring four aphid species on medicinal plants *viz.*, *V. thapsus*, *R. nepalnse*, *I. racemosa* and *Capsella bursa pastoris*; *Scaptomyza pallida* feeding on one aphid species on *R. nepalense*; *Sphaerophoria scripta* devouring one aphid species on Hieraclium candicans (see Tables 16.1 & 16.2).

(e) Aphid predators of forest trees

The major aphid pests of forest trees and of silvicultural importance of Kashmir are: *Aphis craccivora* on *Robinia pseudoacacia*; *Chaitophorus* spp. on *Populus* spp. and *Salix alba*; *Indiochaitophorus furcatus* and *Schizonneurella indica* on *Pulmus* spp. ; *Neothoracaphis* sp. on *Parrotiopsis jacquemontiana*; *Shivaphis celti* on *Celtis australis* and *Tetraneura* sp. on *Ailantus altissima*. These aphids of forestry importance are suppressed in different areas and localities of Kashmir by predatory species such as *A. tetraspilota*, *Aiolocaria* sp. *Annthocoris* sp., *E. balteatus*, *H. variegate*, *Lucopis* sp., *M. confrator*, *O. conglobta and Orius* (*O.*) *lindbergi* and *Platynaspis saundersii*. The number of different predator species preying on different aphid species, infesting different forest tree species are preseented in Table : 16.1 & 16.2.

LITERATURE CITED

Ahmad, D. and Vaishnavi, Y.R. (1982): *Aphidoletesa phidomyza* Rondani (Diptera:.Cacidomyiidae) A new record from Kashmir. *Kashmir University Research Journal*, 2 : 46-47.

Bhagat, R.C. and Lone, M.A. (1984): New Records and host range of predators of aphids (aphididae: Homoptera) in Kashmir, India. *Science and Culture*, 50: 364-366.

Bhagat, K.C. and Kasoodi, M.A. (1986): Record of green lacewing fly, *Chrysopacrestes* Banks (Chrysopidae: Neuroptera) as a predator of the brinjal aphid, *Aphis gossypii* Glover (Aphididae: Homoptera). *Indian Journal of Plant Protection*, 13: 132

Hull, F.M. (1950): Studies upon Syrphid flies in the British Museum (Natural History). *Annual Magazine of Natural History*, 12 : 603-624.

Kapur, A.P. (1942): Bionomics of some coccinellidae, predaceous on aphids and coccids in North-India. Indian Journal of Entomology, 4: 49-66.

Zaka-ur-Rab (1972): Record of *Syrphus confractor* Wed., as predaceous on woolly aphid, *Eriosoma lanigerum* (Hausman), *Indian Journal of Entomology*, 34: 348-349.

Chapter - 17

Diversity of Aphidocolous Ants in Kashmir, with reference to anti-aphid Associations*

*T.A. Matta*** and *R.C. Bhagat****

ABSTRACT

The present investigation deals with 9 species of Aphidocolous ants, distributed over 8 genera, belonging to sub-families *viz.*, Formicinae, Myrmicinae and Dolichoderinae of the family Formicidae, have been encountered from Kashmir. 23 Species of aphids have been found to be associated with these aphidocolous ants, infesting 27 species of plants, belonging to 15 families from different habitats and localities of Kashmir. Studies have been made on ant-aphid association range and diversity. Field observations have been made on nature and extent of damage caused by aphid pests of economically important crops and plants. A catalogue on anti-aphid-host plant is also given.

Key Words : *Anti-aphid association range, Aphidocolous ants, diversity.*

INTRODUCTION

Some groups of ants (Hymenoptera : Formicidae) are known to attend aphids for collection of honey-dew, are called as aphidocolous ants and the aphids attended by ants as myrmecophilous aphids. The relationship between the two i.e., aphids and ant is mutualistic. The honey-dew excreted by aphids is an important part of their diet. Ants protect aphids by driving away the enemies or carrying them to the protected places. Aphidocolous ants, like other ants are social insects and most colonies contain three castes: Queen, males and workers. The queen usually starts a colony and does most of the egg

*Part of M.Phil dissertation, University of Kashmir

**P.G. Department of Zoology, The University of Kashmir, Srinagar-190006, Kashmir (J & K), India.

***Address for correspondence : P.O. Box No. 1250, G.P.O. Srinagar - 190 001, Kashmir (J&K), India.

laying in the colony. The males are winged, short lived, and die soon after mating. The workers are sterile wingless females and make up the bulk of the colony.

In India, in the recent years, investigations on aphidocolous ants, encompassing their records, diversity and interaction between aphids and their attendant from different regions, have been carried out by number of workers. The important publications pertaining to surveys/records on aphidocoluous ants, with ant-aphid association range in different regions of this sub-continent are given by : Roy and Behura (1980); Datta et al. (1982, 83), Datta and Raychaudhuri (1983) (on aphidocolous ants from West Bengal and Sikkim); Kurl and Chauhan (1986) (from Himachal Pradesh); Datta and Raychaudhuri (1985, 86) Devi and Singh (1987) (from North-east India).

As far as Kashmir region is concerned, no investigation has so far been accomplished on aphidocolous ants except the present work carried out under the M.Phil programme (q.v. Matta, 1996).

METHODS

Extensive surveys for various aphidocolous ants were performed in three districts of Kashmir viz., Anantnag, Baramulla and Srinagar, from April 1994–September 1995. The present survey was based on random sampling method, covering different ecological habitats, with special emphasis on agricultural fields, gardens, parks, orchards and other cultivated avenues.

The minimum equipment required for field collecting of relevant material (host plants, aphids, ants, *etc.*) included a pair of stout scissor, forceps, soft camel brush, collection tubes, polythene bags, specimen tubes, containing 70% alcohol as preservative, a field lens and camera, etc.

The aphid specimen, harbouring ants were collected in the field, from infested host plant by removing them, carefully from the infested part/s with the help of camel hair brush and preserved in 70% alcohol. Later, aphids were mounted on slides for identification. The aphidocolous ants were collected directly from aphid colonies with the help of forceps or by drowning/preserving them in alcohol, and later brought to the laboratory for identification. Host-plant specimens were also brought to the laboratory in separate polythene bags and preserved in herbarium for identification.

A number of a abbreviations have been used in the text, the keys to these are given as under: A.–Anantnag District; B.–Baramulla District; BG.–Botanical garden; FP.–Fodder plant; FST–Forest tree; MP.–Medicinal plant; ONP.–Ornamental Plant; ONT. Ornamental tree; KU.–Kashmir University Campus; S.–Srinagar District

RESULTS AND DISCUSSION

During the course of present investigation, a number of aphidocolous ants have been encountered for the first time from different habitats and localities of Kashmir. Extensive field studies were made to assess the distribution, ant-aphid association range and diversity, besides field observations were made on damage caused by aphid pests of economically important crops of the Valley. Aphidocolous ant-wise field studies, with associated aphids on various plants in different habitats and localities are discussed under the following heads:–

1. Camponotus oblongus Smith (Plate III)

(a) Material examined

5 ☿ tending *Acyrthosiphon rubi* Nazikulov on *Rubus fruticosus* (Rosaceae), MP, neglected place in garden, Shalimar, S, 26 June 1994; 3 ☿ tending *Aphis* sp. on Zennia *eleganse* (Asteraceae), ONP, BG, S, 20 October, 1994; 2 ☿ tending *Myzus persicae* (Sulzer) on *Prunus persicae* (Rosaceae), fruit tree, Nehru BG, S, 20 Oct. 94; Many ☿ tending *Uroleucon* sp. on *Chaenomeles japonica* (Rosaceae), ONP, Harwan Garden, S, 26-vi-95; Many ☿ tending *Mirolophium* sp. on *Urtica dioica* (Urticaceae), waste place in Park, Harwan, S 26 June 95; Many ☿ tending *Aphis* sp. on *Viburnum* sp. (Caprifoliceae), Shalimar, S, 25 June, 1995

(b) Field observations

This ant was widely prevalent on the foot hills of Zobarwan range, actively engaged with aphids, from June to November. The higher number of this ant was only found with aphid colonies on host plants viz., *Chaenomeles japonica* and *Urtica dioica*. The *Uroleucon* spp. are probably the most favourite aphids of this species of ants in the Valley. The aphid colonies attended by this ant, showed fewer number of alate forms and apterous forms. Some minor forms of thier species were found on host plant, *C. japonica*.

2. Componotus paria Emergy

(a) Material examined

2 ☿ tending *Uroleucon kashmiricus* Verma, on Urtica dioica (Urticaceae), waste place in garden, Shalimar, S 2 May 1995; One ☿ tending unidentified aphid on *Catalpa bignonoides*(Bignonicaceae), ONP, BG, S, 20 October 94.

(b) Field observations

This antspecies, like C. oblongus Smith was also found in the foot hills of Zabarwan range of Srinagar District. It was found engaged with aphids from May to July. No aphid predators were found in the aphid colony in the presence of *C. paria*. One minor form of this species was seen on plant *Catalpa bignonoides* in Nehru Memorial BG. This species also showed liking for aphid viz., *Uroleucon* sp. *C. parja* is being recorded for the first time for Kashmir Valley.

3. *Formica rufibarbis* Fabricius (Plate IV)

(a) Material examined

3 ☿ per colony tending *Cinariaujafilina* (del Guerico) on *Thuja orientalis* (Cupressaceae), ornamental plant, BG, KU, S, 16 July 1994, Many ☿ tending *Aphis fabae solanella* Theobald on *Astragalis polycanthus* (Paplionaceae), BG, KU, S, 22 June 1995/6 July 95; 3 ☿ tending *Chaitophorus* nr. niger Mordivilko on *Populus caspica* (Salicaceae), FST, Hazratbal, S, 1 June 1995; Many ☿ teding *Tricaudatus polygoni* (Nazkikulov) on *spirea cantoniensis*, (Rosaceae), ONP, BG, KU, S, 21 July 1994, Nehru BG, S, 6 May 1995; Many ☿ tending *Hyalopterus pruni* (Geoff). on *Prunus domestica* (Rosaceae), fruit tree, BG, KU, S, 27 May 1994; 10 ☿ tending *Aphis nerii* Boyer de Fonscolombe on *Verbascum thapsus* (Scrophulariaceae), MP, wild, Hazratbal, S, 8 July 1995, Many ☿ tending *Uroleucon* sp. on *Charenomeles japonica* (Rosaceae), ONP, BG, KU, S. 8 June 1994; 7 ☿ tending *Anuraphis persicae niger* Smith on *Prunus amygdalus* (Rosaceae), fruit tree, BG, KU, S, 3 May 1994; Many ☿ tending *Indiochaitophorus furcatus* Verma, on *Ulmus villosa* (Ulmaceae), FST, BG, KU, S, 7 July 1994; 5 ☿ tending *Aphis* sp. on

Inula racemosa (Asteraceae), MP, cultivated, BG, KU, S, 4 June, 1994. Many ☿ tending *aphis craccivora* Koch. on *Rubica pseudoacacia* (Paplionaceae), FST, BG, KU, S, 4 July 1994; *Shivaphis celti* Das on *Celtis australis* (Ulmaceae). FST,. BG, KU, S, 12 Augus, 1995; 10 ☿ tending *Uroleucon kashmiricus* Virma on *Urtica dioica* (Urticaceae), MP, waste place, Shalimar, S. waste place, Pahalgam, A, 25 June, 1994/27 July 95 respectively; 4 ☿ tending *Acyrthiosiphon rubi* Narzikulov on *Rubus fruticosus* (Rosaceae), waste place, BG, KU, S, 30 May, 94; 10 ☿ tending *Aphis fabae solanella* Theobald on *Rumex nepalense* (Polygonaceae), MP, wild, Pahalgam, A, 2 July 1995; Many ☿ tending *Liosomaphis himalayensis* Basu on *Berberis zibbliana* (Berbridaceae), thorny shrub, garden avenue, BG, KU, S, 8 May 1995. Many ☿ tending unidentified aphid, on *Hypericum perforatum* (Hypericaceae), MP, cultivated area, University Campus, Hazratbal, S, 8 July 1995, *Artemisia turnfortiana* (Asteraceae), MP, BG, KU, S, 14 June 95; *Solanum miniatum* (Solanaceae), MP garden avenue, BG, KU, S. 12 August. 1995; *Catalpa bignonoides* (Bignoniaceae), ONT, University Campus, Hazratbal, S, 20 May, 1994.

(b) Field observations

During the present detailed survey, *F. rufibarbis* was found in dominance as compared to other aphidocolous ants, also it showed a wide range of associations with the aphids, attacking variety of host plants, including economically important ones in Kashmir Valley. This ant was noticed in different types of ecological habitats of valley, ranging from low-land or high-altitude areas and localities.

Various valuable plants are attacked by myrmecophilous aphids, in association with *F. rufibarbis*, causing damage of different nature and extent. Ornamental plant viz., *Spirea cantoniensis* was observed in the field with damage by aphid, viz., *T. polygoni* though sucking sap from tender shoots and inflorescence, which resulted in stunted growth of the host plant. This aphid pest was active in the field during spring. High population densities of *myrmecophilous aphids* viz., *H. pruni* and *A. p. niger*, have ben found on fruit trees, *P. domestica* and *P. amygdalus* respectively. These aphids pests caused considerable damage to foliage and tender shoots of these fruit trees.

Various aphid species associated with *F. rufibarbis* have been observed to damage a number of medicinal plant species in different districts and localities of Kashmir. This ant greatly discouraged the predators, by attacking them with mandible.

4. *Lasius himalayanus*, Forel

(a) Material examined

3 ☿ leaf tending *Chaitophorus populti* (Panzer)on *Populus casela* (Salicaceae), forest tree, Social forest avenue, sopore, B, 14 June 1994 15 May 1995.

(b) Field observations

This ant was found in association with aphid pest, *C. populeti*, damaging leaves of poplar tree, *P. casale*. The aphids remained active from April to July. The field observation revealed the presence of *L. himalayanus* in association with aphid till the latter population vanished completely from the host plants. This aphidocolous ant is being recorded for the first time from Kashmir valley.

MYRMECOPHILE

A **myrmecophile** is an organism that lives in association with ants.

Myrmecophiles may have various roles in their host ant colony. Many consume waste materials in the nests, such as dead ants, dead larvae, or fungi growing in the nest. Some myrmecophiles, however, feed on the stored food supplies of ants, and a few are predatory on ant eggs, larvae, or pupae. Others benefit the ants by providing a food source for them. Many myrmecophilous relationships are obligate, meaning one or the other participant requires the relationship for survival. Some associations are facultative, benefiting one or both participants but not being necessary to their survival.

Myrmecophilous associations are best known in butterflies of the family Lycaenidae. Many lycaenid caterpillars produce nectar by specialized organs and communicate with the ants through sound and vibrations. The association with ants is believed to reduce the parasitisation of the butterfly caterpillars.

There are myrmecophilous beetles in the families Coccinellidae (e.g. the ladybird *Thalassa saginata*), Aphodiidae, Scarabaeidae, Lucanidae, Cholevidae, Pselaphidae, Staphylinidae, Histeridae and Ptiliidae. Myrmecophilous associations are also seen in various other insects such as aphids and treehoppers, as well as the hoverfly genus *Microdon* and several other groups of flies.

Some mites and spiders are also myrmecophilous, particularly some oribatid mites, which have been found to be obligate myrmecophiles.

Others myrmecophile groups include

- Orthoptera, like the cricket *Myrmecophilus acervorum*
- Diptera, like the stratiomyid fly *Clitellaria obtusa*
- Molluscs, like *Allopeas myrmekophilos*

The first major work in cataloguing British myrmecophiles was done by Horace Donisthorpe in his 1927 book *The Guests of British Ants.*

ANT-PLANT INTERACTIONS

Ant-plant interactions are geographically widespread, with hundreds of species of myrmecophytic plants in several families including the Leguminosae, Euphorbiaceae, and Orchidaceae. In general, myrmecophytes (or ant plants) usually provide some form of shelter and food in exchange for ant "tending," which may include protection, seed dispersal (see myrmecochory), reduced competition from other plants, hygienic services, and/or nutrient supplementation.

Three of the most common and important structural adaptations of ant plants are the domatia, Beltian bodies and extrafloral nectaries. Plant domatia are pre-formed nesting sites provided by the plant in the form of hollowed out stems, petioles, thorns, or curled leaves. The production of ant-specialized domatia has been documented in over 100 genera of tropical plants. Beltian bodies provide a high-energy food source to ants in the form of nutritive corpuscles produced on leaflet tips, and they have been described in at least 20 plant families. Extrafloral nectaries (EFN's) are known to occur in at least 66 families of angiosperm plants in both temperate and tropical regions, as well as some ferns, but are absent in all gymnosperms and are most abundant in the topics. EFN's being outside of the plant flowers are not employed in pollination; their primary purpose is to attract and sustain tending ants. Many plants can control the flow of nectar from the EFN's so that the availability of nectar varies according to daily and seasonal cycles. Because ants can respond quickly to changes in flow rate from EFN's, this may be possible mechanism by which plants can induce greater ant activity during times of peak herbivory and minimize overall costs of nectar production. The combined nutritional output of EFN's and Beltian bodies is a significant food source for tending ants, and in some cases can provide the total nutritive needs for an ant colony.

In exchange for nesting sites and food resources, ants protect plants from herbivores.

Perhaps the best-known example of ant-plant mutualism is in bullhorn acacias (*Acacia cornigera*) and their tending *Pseudomyrmex* ants in African savannas.This system was studied extensively by Dan Janzen in the late 1960's, who provided some of the first experimental evidence that ants significantly reduce herbivory rates of myrmecophytes, and since then may other studies have demonstrated similar results in other systems as well. In exchange for protection, the acacias provide domatia, Beltian bodies, and EFN's, and there is evidence that the *Pseudomyrmex* ants can survive exclusively on these food resources without having to forage elsewhere. For many plants, including the bullhorn acacias, ants can significantly reduce herbivory from both phytophagous insects as well as larger organisms, such as large grazing mammals. Obligately associated ant species are some of most aggressive ants in the world, and can defend a plant against herbivory by large mammals by repeatedly biting their attacker and spraying formic acid into the wound.

Myrmecophily is considered a form of indirect plant defense against herbivory, though ants often provide other services in addition to protection. Some ants provide hygienic services to keep leaf surfaces clean and deter disease, and defense against fungal pathogens has also been demonstrated. It is common for ants to prune epiphytes, vines, and parasitic plants from their host plant, and they sometimes thin the shoots of neighboring plants as well. In doing so, ants reduce plant-plant competition for space, light, nutrients, and water. Finally, current work focusing on ants' role in nutrient supplementation for plants has shown that in many ant-plant relationships nutrient flow is bidirectional. One study has estimated that while 80% of the carbon in the bodies of *Azteca* spp. workers is supplied by the host tree (*Cecropia* spp.), 90% of the Cecropia tree's nitrogen was supplied by ant debris carried to the tree as a result of external foraging. In light of these services, myrmecophily can be very advantageous in ensuring a plant's survival and ecological success.

ANT-INSECT INTERACTIONS

Ants tend a wide variety of insect species, most notably lycaenid butterfly caterpillars and hemipterans. Forty-one percent of all ant genera include species that associate with insects. In all ant-insect associations the ants provide some service in exchange for nutrients in the form of honeydew, a sugary fluid excreted by many phytophagous insects. Interactions between honeydew-producing insects and ants is often called trophobiosis, a term which merges notions of trophic relationships with symbioses between ants and insects. This term has been criticized, however, on the basis that myrmecophilous interactions are often more complex than simple trophic interactions, and the use of symbiosis is inappropriate for describing interactions among free-living organisms.

Hemiptera

Some of the most well-studied myrmecophilous interactions involve ants and hemipterans (earlier grouped in the order Homoptera which included the Auchenorrhyncha and Sternorrhyncha), especially aphids. There are around 4000 described species of aphids, and they are the most abundant myrmecophilous organisms in the northern temperate zones. Aphids feed on the phloem sap of plants, and as they feed they excrete honeydew droplets from their anus. The tending ants ingest these honeydew droplets then return to their nest to regurgitate the fluid for their nestmates (see trophallaxis). Between 90-95% of the dry weight of aphid honeydew is various sugars, while the remaining matter includes vitamins, minerals, and amino acids. Aphid honeydew can provide an abundant food source for ants (aphids in the genus *Tuberolachnus* can secrete more honeydew droplets per hour than their body weight) and for some ants aphids may be their only source of food. In these circumstances, ants may supplement their honeydew intake by preying on the aphid's once the aphid populations have reached certain densities. In this way ants can gain extra protein and ensure efficient resource extraction by maintaining honeydew flow rates that do not exceed the ants' collection capabilities. Even with some predation by ants, aphid colonies can reach larger densities with tending ants than colonies than without. Ants have been observed to tend large "herds" of aphids, protecting them from predators and parasitoids. Aphid species that are associated with ants often have reduced structural and behavioral defense mechanisms, and are less able to defend themselves from attack than aphid species that are not associated with ants.

Ants engage in associations with other honeydew-producing hemipterans such as scale insects (Coccidae), mealybugs (Pseudococcidae), and treehoppers (Membracidae), and most of these interaction are facultative and opportunistic with some cases of obligate associations, such as hemipterans that are inquiline, meaning they can only survive inside ant nests. In addition to protection, ants may provide other services in exchange for hemipteran honeydew. Some ants bring hemipteran larvae into the ant nests and rear them along with their own ant brood. Additionally, ants may actively aid in hemipteran dispersal: queen ants have been observed transporting aphids during their dispersive flights to establish a new colony, and worker ants will often carry aphids to a new nesting site if the previous ant nest has been disturbed. Ants may also carry hemipterans to different parts of a plant or to different plants in order to ensure a fresh food source and/or adequate protection for the herd.

Lycaenid butterflies

Myrmecophily among lycaenid caterpillars differs from the associations of hemipterans because caterpillars feed on plant tissues, not phloem sap, and therefore do not continually excrete honeydew. Caterpillars of lycaenid butterflies have therefore evolved specialized organs that secrete chemicals to feed and appease ants. Because caterpillars do not automatically pass honeydew, they must be stimulated to secrete droplets and do so in response to ant antennation, which is the drumming or stroking of the caterpillar's body by the ants' antennae. Some caterpillars possess specialized receptors that allow them to distinguish between ant antennation and contact from predators and parasites, and others produce acoustic signals that agitate ants, making them more active and likely better defenders of the larvae. As with homopteran myrmecophiles, ants protect Lycaenid larvae from predatory insects (including other ants) and parasitoid wasps, which lay their eggs in the bodies of many species of Lepidoptera larvae. The enemy-free space that ants can provide for lycaenids is significant: one study conducted by Pierce and colleagues in Colorado experimentally demonstrated that survival rates of *G. lygdymus* larvae declined 85-90% when ant partners were excluded. These interactions do not come without an energetic cost to the butterfly, however, and it has been shown that ant-tended individuals reach smaller adult sizes than non-tended individuals due to the costs of appeasing ants during the larval stage. Interactions with ants are not limited to the butterfly's larval stage, and in fact ants can be important partners for butterflies at all stages of their life cycle. For example, adult females of many lycaenid butterflies preferentially oviposit on plants where ant partners are present, possibly by using ants' own chemical cues in order to locate sites where juvenile butterflies will likely be tended by ants. Finally, while ant attendance has been most widely documented in Lycaenid butterflies (and to some extent riodinid butterflies), many other lepidopteran species are known to associate with ants, including many moths.

Multiple levels of myrmecophily

Many trophobiotic ants can simultaneously maintain associations with multiple species. Ants that interact with myrmecophilous insects and myrmecophytes are highly associated: species that are adapted to interact with one of these myrmecophiles may switch among them depending on resource availability and quality. Of the ant genera that include species that associate with ant plants, 94% also include species that associate with trophobionts. In contrast, ants that are adapted to cultivate fungus (leaf cutter ants, tribe Attini) do not possess the morphological or behavioral adaptations to switch to trophobiotic partners.Many ant mutualists can exploit these multi-species interactions to maximize the benefits of myrmecophily. For example, some plants will host aphids instead of investing in EFN's, which may be more energetically costly depending on local food availability. The presence of multiple interactors can strongly influence the outcomes of myrmecophily, often in unexpected ways.

SIGNIFICANCE OF MYRMECOPHILY IN ECOLOGY

Mutualisms are geographically ubiquitous, found in all organismic kingdoms, and play a major role in all ecosystems. Combined with the fact that ants are one of the most dominant lifeforms on earth, it is clear that myrmecophily plays a significant role in the evolution and ecology of diverse organisms, and in the community structure of many terrestrial ecosystems.

Evolution of positive interactions

Questions of how and why species coevolve are of great interest and significance. In many myrmecophilous organisms it is clear that ant associations have been influential in the ecological success, diversity, and persistence of species. Analyses of phylogenetic information for myrmecophilous organisms as well as ant lineages have demonstrated that myrmecophily has arisen independently in most groups multiple times. Because there have been multiple gains (and perhaps losses) of myrmecophilous adaptations, the evolutionary sequence of events in most lineages is unknown. Exactly how these associations evolve also remains unclear.

In studying the coevolution of myrmecophilous organisms, many researchers have addressed the relative costs and benefits of mutualistic interactions, which can vary drastically according to local species composition and abundance, variation in nutrient requirements and availability, host plant quality, presence of alternative food sources, abundance and composition of predator and parasitoid species, and abiotic conditions. Because of the large amounts of variation in some of these factors, the mechanisms that support the stable persistence of myrmecophily are still unknown. In many cases, variation in external factors can result in interactions that shift along a continuum of mutualism, commensalism, and even parasitism. In almost all mutualisms, the relative costs and benefits of interactions are asymmetrical; that is, one partner experiences greater benefits and/or fewer costs than the other partner. This asymmetry leads to "cheating," in which one partner evolves strategies to receive benefits without providing services in return. As with many other mutualisms, cheating has evolved in interactions between ants and their partners. For example, some lycaenid larvae are taken into ant nests where they predate on ant brood and offer no services to the ants. Other lycaenids may parasitize ant-plant relationships by feeding on plants that are tended by ants, apparently immune to ant attack because of their own appeasing secretions. Hemipterophagous lycaenids engage in a similar form of parasitism in ant-hemipteran associations. In light of the variability in outcomes of mutualistic interactions, and also the evolution of cheating in many systems, much remains to be learned about the mechanisms that maintain mutualism as an evolutionarily stable interaction.

Species coexistence

In addition to leading to coevolution, mutualisms also play an important role in structuring communities. One of the most obvious ways in which myrmecophily influences community structure is by allowing for the coexistence of species which might otherwise be antagonists or competitors. For many myrmecophiles, engaging in ant associations is first and foremost a method of avoiding predation by ants. For example, the caterpillars of lycaenid butterflies are an ideal source of food for ants: they are slow-moving, soft-bodied, and highly nutritious, yet they have evolved complex structures to not only appease ant aggression but to elicit protective services from the ants. In order to explain why ants cooperate with other species as opposed to predating on them, two related hypotheses have been proposed: cooperation either provides ants with resources that are otherwise difficult to find, or it ensures the long-term availability of those resources.

Community structure

At both small and large spatiotemporal scales, mutualistic interactions influence patterns of species richness, distribution, and abundance. Myrmecophilous interactions play an important role in determining community structure by influencing inter- and intraspecific competition; regulating population densities of arthropods, fungi, and plants; determining arthropod species assemblages; and influencing trophic dynamics. Recent work in tropical forests has shown that ant mutualisms may play key roles in structuring food webs, as ants can control entire communities of arthropods in forest canopies. Myrmecophily has also been key in the ecological success of ants. Ant biomass and abundance in many ecosystems exceeds that of their potential prey, suggesting a strong role of myrmecophily in supporting larger populations of ants than would otherwise be possible. Furthermore, by providing associational refugia and habitat amelioration for many species, ants are considered dominant ecosystem engineers.

Myrmecophily as a model system

Myrmecophilous interactions provide an important model system for exploring ecological and evolutionary questions regarding coevolution, plant defense theory, food web structure, species coexistence, and evolutionarily stable strategies. Because many myrmecophilous relationships are easily manipulable and tractable, they allow for testing and experimentation that may not be possible in other interactions. Therefore they provide ideal model systems in which to explore the magnitude, dynamics, and frequency of mutualism in nature.

aphidocolous ants belonged to genus *Monomorium*, tending (8) species of aphids. Ant *viz., Camponotus oblongus* followed *Monomorium* in dominancy, collected honey dew from 5 species of aphids. *Technomyrmex brunneus* attended only one species of aphid *Uroleucon* sp. These ant-aphid associations are being reported for the first time from their region of J&K State.

Table 17.1 : Ant-Aphid-Host Plat Catalogue

S. No.	*Ant/Aphid Species*	*Host Plant*
1	*2*	*3*
Family : Formicidae		
Sub-Family 1 : Formicinae		
1.	***Camponotus oblongus*** **Smith**	
	Acyrthosiphon rubi	*Rubus fruticosus*
	Aphis sp.	*Zennia eleganse*
	Aphis sp.	*Viburnum sp.*
	Microlophium sp.	*Urtica dioica*
	Myzus pericae	*Prunus persica*
	Uroleucon sp.	*Chaenomeles japonica*
2.	***Camponotus paria*** **Emery**	
	Uroleucon kashmiricus	*Urtica dioica*
	Unidentified aphid	*Catalpa bignonoides*
3.	***Formica rufibarbis*** **fabricius**	
	Acyrthosiphon rubi	*Rubusfruticosus*
	Anuraphis persicae niger	*Prunus amygdalus*
	Aphis craccivora	*Robinia pseudoacacia*
	Aphis fabae solanella	*Astragalis polycanthus*
	Aphis nerii	*Verbascum thapsus*
	Aphis sp.	*Inula racemosa*
	Chaetosiphon (Pentatrichopus)-glabrum	*Rubus fruticosus*
	Chaitophorus nr. niger	*Populus caspica*
	Cinara tujafilina	*Thuja orientalis*
	Hyalopterus pruni	*Prunus domestica*
	Indiochaitophorus furcatus	*Ulmus villosa*
	Liosomaphis himalayansis	*Berberis zibbliana*

1	2	3
	Shivaphis celti	*Celtisaustralis*
	Tricaudatus polygoni	*Spirea cantoniensis*
	Uroleucon kashmiricus	*Urtica dioica*
	Unidentified aphid species	*Artemisia turnfortiana*
4.	***Lasius himalayanus* Forel**	
	Chaitophorus populeti	*Populus casale*
	Uroleucon sp.	*Chaenomeles japonica*
Sub Family 2 : Myrmicinae		
5.	***Crematogaster politula* Forel**	
	Callaphis juglandis	*Juglans regia*
	Uroleucon kashmiricus	*Urtica dioica*
	Uroleucon (Uromelan) similis	*Urtica dioica*
6.	***Monomorium minutum* Mayr**	
	Acyrthosiphon rubi	*Rubus fruticosus*
	Aphis sp.	*Inula racemosa*
	Semiaphis nr. hieraclei	*Foeniculum vulgare*
	Shivaphis celti	*Celtis australis*
	Uroleucon kashmiricus	*Urtica dioica*
	Uroleucon sp.	*Chaenomeles japonica*
	Unidentified aphid species	*Artemisia turnfortiana*
	Unidentified aphi species	*Breea arvense*
7.	***Pheidole bhavanae* Bingham**	
	Aphis fabae solanella	*Rumex nepalense*
	Callaphis juglandis	*Juglans regia*
	Chaitophorus nr. niger	*Populus caspica*
	Uroleucon kashmiricus	*Urtica dioica*
8.	***Tetramonium christei* Forel**	
	Aphis sp.	*Inula racemosa*
	Brevicoryne sp.	*Capsella bursa pastoris*
	Melanaphis sacchari	*Sorghum vulgare*
	Semiaphis nr. *heraclei*	*Foeniculam vulgare*
	Uroleucon sp.	*Chaenomeles japonica*

1	2	3
Sub Family 3 : Dolichoderinae		
9.	*Technomyrmex brunneus* **Forel**	
	Uroleucon sp.	*Chaenomeles japonica*

The ants associated with aphids in the field help thriving of aphid population. Aphidocolous ants, therefore, are nuisance, leading to alate (winged aphid) production in ant attended–aphid populations. There is deterioration of host plants and also dispersal of aphids thereby attacking other host plants. The voracity of aphid predators and parasites is also affected in aphid colony, as aphidophagous insects are discouraged by attendant ants. During the present study, it was found that larvae of many predators (Diptera, Coleoptera Plate : III & IV) remained inactive in the presence of ants in the aphid colonies. In addition to these, limitations are put by some aphidocolous ants in biological control of aphids as same ants protect aphids from adverse weather conditions by transportation and providing shelter in their nest.

ACKNOWLEDGEMENTS

The authors are highly thankful to Head, P.G. Department of Zoology, Kashmir University, for providing necessary laboratory facilities. They also expressed their gratefulness to Dr. A.R. Naqshi ad Dr. G.Ḥ. Dar, P.G. Department of Botany, for their help in the identification of plant species.

LITERATURE CITED

Datta, S.K. and Raychaudhari, D. (1983). Taxonomy of the aphidocolous ants (Hymenoptera: Formicidae) of Sikkim and billy areas of West Bengal (I) Sub-family: Myrmicinae. *Akitu*, 56:14.

Datta, S.K. and Raychaudhari, D. (1985). One new species of ant, *Componotus horshoetus* (Hymenoptera: Formicidae) from Nagaland, North-East India. *Science and Culture*, 15 : 271-273.

Datta, S.K. and Raychaudhuri, D. (1986). Ant-aphid association of North-East India. *Aphidology in India*. 81-83.

Datta, S.K. Raychaudhuri, D. and Agarwala, B.K. (1982) Study on aphid tending ants in India I., New recordsof aphids and ant species in their association. *Entomon*, 7: 327-328.

Datta, S.K., Raychaudhuri, D. and Agarwala, B.K. (1983); Study on aphid tending ants in India. II. New records of aphid and ant species in their association. *Entomon*, 8: 23-25.

Devi, Manmohini, C. and Singh, T.K. (1987). Aphidocolous ants (Hymenoptera : Formicidae) in Manipur. *Entomon*, 12 : 309-313.

Matta, T.A. (1996). *Survey and Studies on some aphidocolous ants and predators of Aphids in Kashmir Valley*. M. Phil dissertation. P.G. Department of Zoology, Faculty of Science, University of Kashmir, Srinagar, Kashmir (J & K), India, 191 pp.

Kurl, S.P. and Chauhan, R.S. (1986). Mutualism between aphids and ants pages 251-257, In : Kurl, S.P. (Editor), *Proc. 2nd National Symposium on Recent Trends in Aphidological Studies*, Modinagar, U.P. India.

Roy, D.K. and Behura, B.K. (1980). There more species of ants attending aphids on Aphis gossypii Glover. *Newsletter, Aphidological Society of India* : 1:4.

SECTION - V

ARID ENVIRONMENTS : BIODIVERSITY

Chapter - 18

Floristic Diversity in Ladakh Region of J & K State, India

Irshad A. Nawchoo and G.M. Buth**

ABSTRACT

Ladakh is a harsh mountainous desert supporting a scanty vegetation. The alpine, the desert and the oastic elements are the representative features of the flora of Ladakh. The vegetation is characterised by stunted, tufted desert plants particularly prominent from Drass onwards. The floristic surveys and vegetational analysis reveal that distinct floristic links exist between Ladakh and Central Asia while only about 14 percent are endemic. The present studies list up the detailed floristic and crop diversity with a view to establish phytogeographic links with adjoining region of Central Asia.

Key Words : alpine, crops, desert, diversity, endemic, floristic, oastic, scrub, vegetation.

INTRODUCTION

Ladakh is a harsh and mountainous desert of wild beauty, set among the jagged peaks of north western Himalaya. Lying north of the main Himalayan watershed, beyond the reach of the summer moonsoon it is an exceptionally cold and dry area with an annual rainfall of less than 12 cm and the temperatures dropping to as low as -50°C. Its long winters are as severe as those of almost any inhabited region of the world.

Ladakh is one of the most elevated regions of the world. The lowest valleys of the region are located between 3000 m - 3500 m. It is situated between 30°15' to 80°15 'E longitude occyping an area of about 56,000 sq. km (including the area usurped by China). It is a vast mountainous tree-less tract and the general aspect is that of extreme barren topography. The habitation is concentrated along the headwaters of the important rivers.

The Karakoram and the Kuenlum ranges from the northern boundary of Ladakh region and separate it from Yarkand and Kotan provinces of Central Asia. On the east and south lie the Chinese districts of rudok and Chumurti. Southern boundaries are formed by

* Department of Botany University of Kashmir Srinagar - 190 006 - India

the Zanskar range of northwestern Himalaya which separates it from Lahul and Spiti valleys of Himachal Pradesh.

GEOGRAPHICAL BACKGROUND AND CLIMATE

The striking features of the region is the parallelism of its mountain ranges which stretch from southeast to northwest. The three distinct ranges (Kachroo *et al.* 1977) are the Trans-Himalaya, the Chushuk and the Trans-Tibetan. The Trans-Himalayan ranges divides the headwaters of the Sutluj from those of the Indus, the Trans-Tibetan ranges forms the natural boundary of Ladakh. This general direction of the mountain ranges determines the course of rivers as well as the boundaries of different regions in Ladakh.

The climate of the region is characteristically of cold desert type. During summer, days are usually very hot and dry while in winter, days are extremely cold and chilly.

The excessive dryness of the climate is chiefly due to elevation by which the air is so hot and rarified as to be incapable of holding much moisture in suspension. It is also partly due to the great radiations of heat from the bare sand by which any moisture is rapidly evaporated. The dryness of the climate increases with height (Cunningham, 1854).

Days and nights in both summer and winter are windy. During the nights, the wind flows in northeasternly direction and during the day it blows in southwesternly direction. In summer, the day breeze starts before midday and continues to rise with frequent strong quests till 4 p.m. When it reaches its maximum. Towards sunset, it changes to westnorthwest direction and gradually start abating. The night winds usually start after 10 p.m. and rise till midnight when it becomes fainter and gradually abates in morning. In winter the winds start rising at about 2 p.m. and reach the maximum by mid night when it gradually abates (Thomson 1852).

In Leh, Nubra and Changthang, it usually drizzles at intervals during the moonsoon season (July-August). Very rarely there is a heavy downpour lasting for more than 12 hours when all the rivers rise rapidly. Snowfall in Leh proper and Nubra valley is very meagre but at higher altitude of Changthang and on mountain tops in Leh and Nubra, it is very heavy and lasts for quite sometime.

The temperature and rainfall (Fig. 18.1) shows extreme variations during different months of the year. The summer lasts from June to September recording temperature as high as 30°C - 35°C. The winter is long lasting from October to May with temperature dropping to as low as - 20°C (or even as - 30°C) in Leh and Nubra. In Changthang area it drops to as low as - 45°C. The variation in day and night temperatures during summer is as great as 10°C - 15°C while in winter it is only 5°C - 10°C.

MATERIALS AND METHODS

For the purpose of study the Ladakh region was divided into 3 main areas Leh and Nubra, Drass and Kargil and Zanaskar. Periodic surveys were conducted in these three regions and extensive collections made. The plants were collected in triplicate and dried and processed using standard techniques. The identification were carried out in KASH (Kashmir University Herbarium) using standard floras; Hookar (1872-1897), Stewart (1917, 1972), Kachroo et al. (1977), Naqshi (1980), Dhar (1980), Dhar and Kachroo (1983) and Seybold and Kull (1985). Wherever necessary comparison with identified specimens were also made.

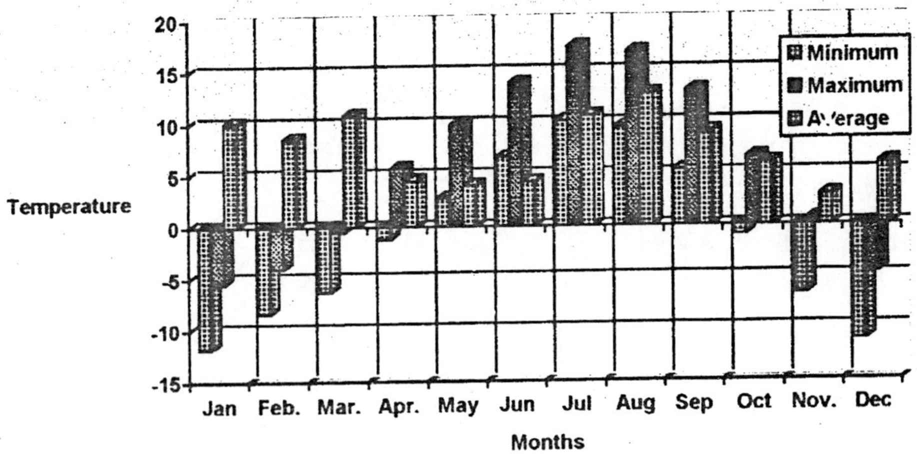

Fig. 18.1 : Graphic representation of Mean monthly temperature (Max. & Min.) and average rainfall in Ladakh.

The data were compiled on the basis of all available information in the form of floras, Monographs and other literature and computed using Stewart's (1972) work as the standard for flora of Ladakh because of its comprehensive and exhaustive details.

The information regarding climate was obtained from the Meteorological station at Leh and Srinagar.

RESULTS

Botany

Ladakh supports a scanty vegetation due to various odd factors operative in the region. The entire area from Zojila onwards has been classified as "alpine scrub "forest zone"1 (Kachroo *et al.* 1977). Trees are few in number (mostly introduced) and are cultivated in some specific pockets along streams and other water bodies. The herbaceous element found growing in the area is a representative mixture of Tibetan, Afganistan, Siberian and European alpine or temperate elements. These usually comprise some annual weeds and a few perennial scrubs commonly encountered within the places that have escaped erosion, or are located within places that have moisture retaining areas. The plants usually grow upto an altitude of 5600 m. Natural forests are absent in the whole region from Zojila to Chushool, in West and Nubra in North.

The Vegetation of Ladakh can be grouped into three representative elements, the alpine, the desert and the oastic elements. The alpine element is mostly confined to narrow belts of vegetation which receives part of the molten snow water and the upperbeds of the mountain streams. The desert flora which is very prominent in the upper reaches and the valleys is most closely related to the Tibetan and Siberian floristic elements. The Oasitic element comprises a variety of exotics as well as some indigenous plants.

Cultivated trees include *Salix elegans, S. alba, S. fragilis, S. Sclerophylla, Populus alba, P. euphratica, P. Candicans, P. nigra, P. ciliata, P. angustifolia, Prunus armeniaca, Juglans regia* and *Malus pumila.*

Common shrub by elements along streams, rivers and other water bodies are; *Hippophea rhamnnoides, Rosa webbiana, Lonicera spinosa, Myricaria elegans, M. germanica, Berberis ulicina, B. pachyclada, Ephedra prezewalsky, Caragana moorcroftiana* etc.

Common herbs include species of *Nepeta, Chenopodium, Artemesia, Astragalus, Lactuca, Senecio, Sonchus, Tanacetum, Waldheimia, Carex, Cyperus, Elymus, Plantago, Corydalis, Anaphalis, Potentilla, Ranunculus, Polygonum, Euphorbia, Veronica, Euphrasia*, etc.

Floristics and Diversity

Ladakh has been exposed to plant collecters since the time of Victor Jacquemont (1802-1832). The rich collections made during this period were sent to different herbaria in India and outside. The collectors in Ladakh have been Aitchinson (1856-58); Bellew (1948-50); Boshard and Trinkler (1927-28); Bower and Thoroid (1932); Doughlas (1950); Duthie (1890); Falconer (1834-38) Heyde (1860-70) Huegal (1935-37); Jacquemont (1801); Jaeschke (1860-70); Lacoste (1906); Meinertzhagen (1955); Moorcroft (1820-22); Picot (1892); Schmidt (1912-13); Stewart (1916-17) Thomson (1884); Vigne (1935), Webly (1896) and Young (1895) (After Naqshi 1980). These are followed by detailed reports of Hartman (1983) and Seybold and Kull (1985).

Stewart (1917) was the first to compile flora of Ladakh and Western Tibet. He also described physical features of the region. From 1912 to 1952 Stewart did extensive plant collection in Kashmir and Ladakh and brought out the Catalogue of Vascular Plants of west Pakistan and Kashmir in 1972. In this catalogue 1174 taxa from Ladakh, comprising 957 dicots, 210 monocots and 7 gymnospermes were listed.

The other major contribution on flora was made by Kachroo *et al.* (1977). Who on the basis of plant collection deposited in Kashmir University Herbarium (Kash) compiled the Flora of Ladakh. In this flora 611 species were described including 540 dicots, 65 monocots and 2 gymosperms. The recent attempt to compile a flora of Ladakh has been made by Naqshi (1980) who on the basis of his collections from 1975-1980 enumerated 472 species including 418 dicots and 54 monocots. (Fig. 18.2).

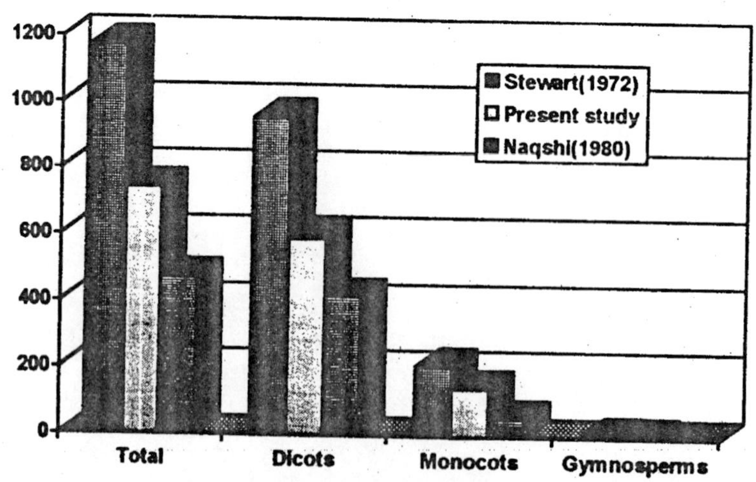

Fig. 18.2: Various compilations of flora of Ladakh.

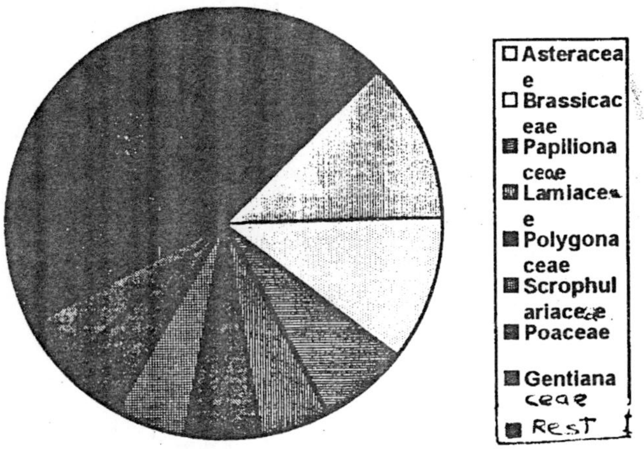

Fig. 18.3: Main represented families.

During the course of present investigation 740 species, of which 590 are dicots, 144 nonocots and 6 gymnosperms were collected. Asteracae was the largest represented family with 90 species followed by Brassicaceae (80 species), Papilionaceae (50 species), lamiaceae (40 species), Polygonaceae (40 species), Scrophulariaceae (39 species) Poaceae (38 species), and Gentianaceae (30 species), (Fig. 18.3) *Astragalus* was the largest genus with 20 species followed by *Potentilla* (15 species) *Nepeta* (14 species) *Artemiesia* (14 species) *Polygonum* (10 species), and *Sedum, Ranunculus* and *Pedicularis* (9 species each) (Fig. 18.4).

The floristic surveys and vegetational analysis reveal that distinct floristic links exist between Ladakh and Central Asia. About 50 per cent of the total flora is spread through diverse regions of Central Asia. The number of plants common between different areas and Ladakh is given in Table 18.1.

Table 18.1 : The adjoining floristic regions and taxa common with Ladakh.

S. No.	*Area*	*No. of Taxa Common with Ladakh*
1.	China and Tibet	485
2.	USSR	279
3.	Afghanistan	225
4.	North & Central Europe	117
5.	Kashmir	510
	Total No. of Taxa in Ladakh	**1174**

From the table it will be clear that 43.4% of total flora is common with Kashmir, 41.3% with China and Tibet, 23.7% with USSR, 19.1% with Afghanistan, 9.9% with north and Central Europe. (Fig. 18.5.)

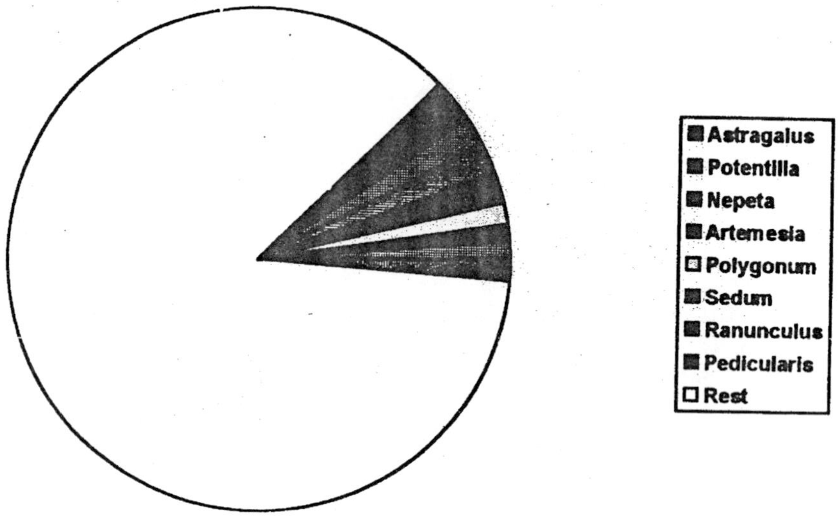

Fig. 18.4: Main represented genera

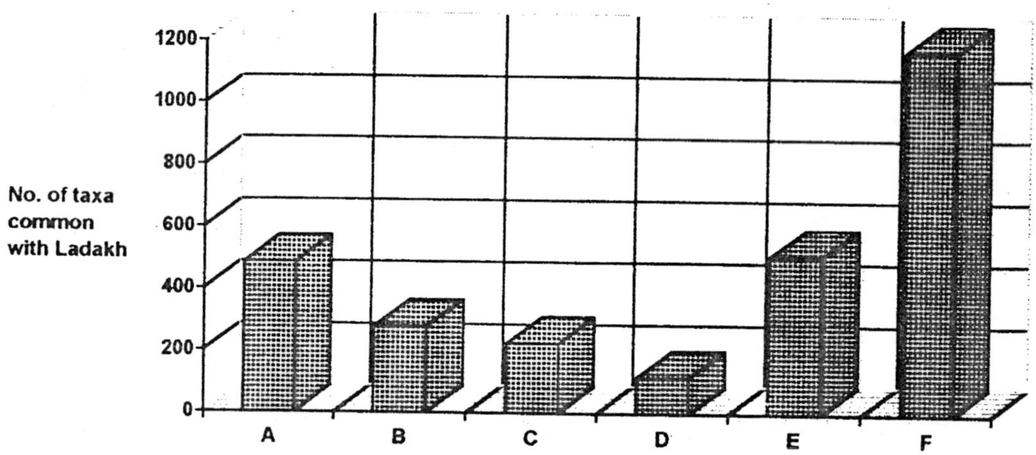

A	**China & Tibet**	**B**	**USSR**
C	**Afghanistan**	**D**	**North & Central Europe**
E	**Kashmir**	**F**	**Total No. of taxa in Ladakh**

Fig. 18.5: Graphic representation of taxa from Ladakh common with adjacent areas.

Only 164 taxa (13.6%) are endemic to the region. Of these 62 taxa be classified as narrow endemics as these are restricted to Ladakh only. The wide endemics fall under three categories i.e. endemic to Ladakh and Kashmir, endemic to Ladakh and Pakistan and endemic to Ladakh, Kashmir and Pakistan. (Fig. 18.6 & 18.7).

The taxa endemic to Ladakh are: *Actinocarya acaulis, A. tibetica, Allium loratom, Arenaria poltrichoides, A. straechyi, Artemesia salsoloides var. Wellbyi, Bassia divaricatata, Berberis ulicina,*

Braya trinkleri, Carex haematostma var. submacrogyne, C. stenophylla var. longipe dicellata, Chenopodium prostratum, Corydalis crassifolia, C. hendersonii, Dontostemon glandulosus, Draba iudlowiana, D. tennerrima, Elsholtzia eriostachya var. pusilla, Ermania albiflora, E. languinosa, Euphorbia tibetica, Gentiana aquatica var. karelinii, G. heugelli, G. moorcroftiana. var. maddenii, G. thomsonii, Hackelia meeboldii, Hamada kashmiriana, Heteropappus semiprostratus, impatiens thomsonii, Kobresia koelzii, Lasciocaryum densiflorrm, Leonotopodium monocephalum, Lychnis apetala, Microula tibetica, Nepeta ceorulascens, N. nivalis, N. paucifolia, N. tibetica, Pedicularis heydei, P. rhinanthoides ssp. speciosa, P. sevennedinii, Peucedanum malcolmii, Polygonum sibirum, Potentilla breviscissa, P. sericea var. polyschista, P. thomsonii, Ranunculus palifolius, R. involucratus, Saussurea dainellii, S. glanduligera, S. subulata, S. thomsonii, Sedum ewersii, S. fischeeri, Stellaria pusilla, Strachya tibetica, Thalictrum rutaefolium, Thermopsis inflata, Veronica koelzii, V. nana, V. uncinata and Waldhemia vestita.

The 33 taxa endemic to Ladakh and Kashmir are: *Anaphalis royleana var. concolor, Androsace muscoidea, Aster asteroides, Caragana bravifolia, Carex digocarpa, Cousinia flaconeri, Cynoglossum thomsonii, Epilobium latifolium ssp speciosum, Atricum fruticosum, Ermania stewartii, Euphrasia laxa, E. paucifolia, Gentiana Carinata, Gnaphalium thomsonii, Hedysarum microcalyx, Mattiastrum thomsonii, Megacarpaea polyandra, Nepeta eriostachya, Pedicularis brevirostris, P. kashmiriana var. Oranta, P. longiflora ssp. tubiformis, P. longiflora, Polygonum hydropiper var. eglandulosa, P. tortousum, Primula heydei, Pycnopimthus uniflorus, Salix alba var. coerulea, Saussurea graminifolia, S. leontodontondes, S. sacra, Saxifraga stella-areues, Scrophularia dentata and Tanacetum tibeticum.*

The 12 taxa endemic to Ladakh and Pakistan are: *Allium stoliczkai, Artemesia campbellii, Chesneya cuneata, Chrysanthemun tibeticun, Eriocyclia undua, Gentiana burkillii, Saussurea bracteata, S. nana. S. stoliczkai, Stellaria tibetica, Tanacetum artemisioides, and T. fruticulosum.*

The 57 taxa endemic to Ladakh, Kashmir and Pakistan are: *Anaphalis royleana, Androsace rotundifolia var. thomsonii, Arabis saxicola, Arthrophytom thomsonii, Bupleurum thomsonii, Carex lutobrumea, C. Plectobasis, Companula colorata, Corydalis tibetica, Cynanctum arnottianum, Delphinum cachemirianum, Draba cachemiraica, Eleocharis mitracarpa, E. Quinquelora, Euphorbia kanaorica, Gallium paucifloraum, Geranium aconitifolium, Heracleum thomsonii var. glabior, Inula clarkei, I. royleana, Jurinea ceratocarpa, J. himalaica, Kobresia pamiroalaica, Krascheninikovii ccerastioides, Lactuca lessertiaa ssp. lyrata, Lagotis globosa, L. kunawurensis, Nepeta clarkei, N. multicaulis, N. royleana, Oxytropis densa, O. nivalis, O. cachemirica, O. parviflora, O. pusilla, Pedicularis perpurea, P. pycnantha, P. seminovii, Physochclaena praelta, Pleurosperumum stellatum var. lindleyana, Polygonum rottboelloides, Potentilla atrosanguina var, cataclines, P. curvisata, P. monanathes, Primula farinosa var. armena, Pseudomertensia echioides, Rheum webbianum, Saussurea falconeri, Senecio tibeticus, Stellaria subumbellata Swertia coerulea, Urtica hyperborea, Valeriana clarkei, Veronica salina and Vicatia wolffiana.*

CROPS AND CROP DIVERSITY

Varieties of various crops adopted to the peculiar topo-environmental conditions prevailing in the region are given in Table 18.2. The cultivated cereals include varieties of barley and wheat, both showing diversity of form and yield. Although new varieties introduced by agricultural agencies are claimed high yielding and better in nutritive value, yet local forms are better adopted to the conditions of the region and could play a significant role in the crop improvement programme.

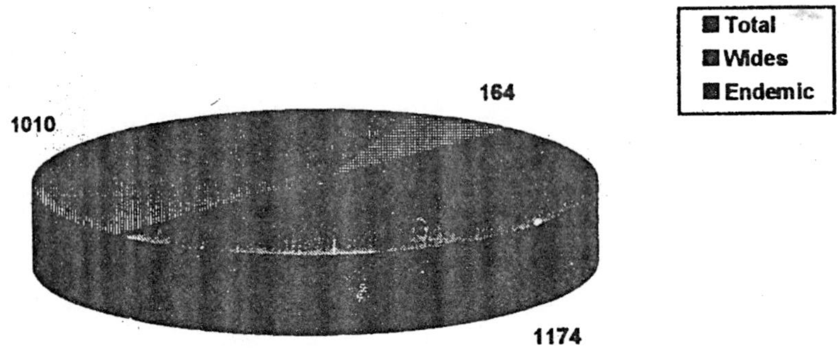

Fig. 18.6: Representation of wides and endemics

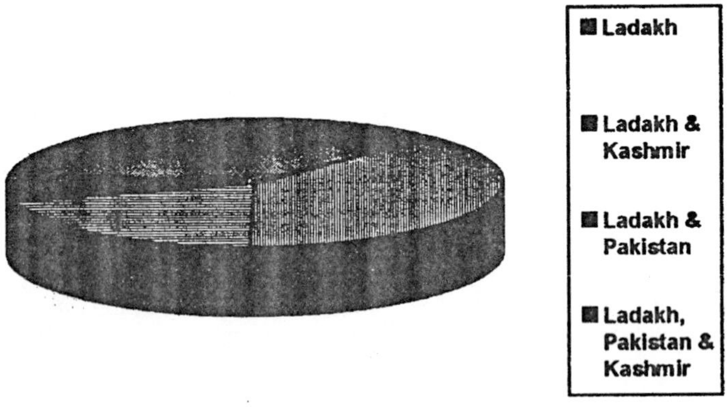

Fig. 18.7: Percentage representation of various types of endemics

Barley has several varieties *e.g.* "*naked barley*" with grains free from glume, "*Hulled barley*" (Swah) with adhering glumes. Tibetan awnless barley (*Grim*) is used mainly for preparing the local beer "*chhang*" and for making fried flour "*sattu*" a useful and convenient food. An awnless variety of wheat, "*toril*" is also grown.

Table 18.2 : Crops Grown in Ladakh

Crop	*Local Name*	*Botanical Name*
(a) *Cereals*		
(i) Wheat	Toril	*Triticum sp.*
(ii) Barley	Swah	*Hordeum vulgare*
(iii) Naked Barley	Grim	*H. Vulqare var. nudum*
(iv) Buckwheat	Tau	*Faqopyrum sp.*
(b) *Millets*		
(i) Common millet	Cheena	*Panicum milliaceum*
(ii) Italian millet	Kangni	*Setaria italica*
(c) *Pulses*		
(i) Pea	Shama	*Pisum sativum*
(ii) Sweet pea	Karaz	*Lathyrus* sp.
(iii) Broad bean	Nagsaran	*Vicia faba*
(iv) Lentil	Karze	*Lens esculenta*
(d) *Vegetables*		
(i) Turnip	Numa	*Brassica rapa*
(ii) Raddish	Labuh	*Raaphanus sativa*
(iii) Potato	Beeshaq	*Solanum tuberosum*
(iv) Tomato	Mirjoon	*Lycopersicum esculentum*
(v) Carrot	Sarakum	*Daucus carrota*
(vi) Knolkhol	Kushotam	*Brassica Oleracea*
(vii) Cauliflower	Phooleat	*B. Oleracea* var. *botrytis*
(viii) Cabbage	Bundi	*B. Oleracea* var. *capitata.*
(ix) Onion	Skotse	*Allium cepa.*
(x) Lettuce	Dums	*Lactuca sativa*
(xi) Spinach	Palak	*Spinacea oleracea*
(e) Oil Seeds		
Mustard	Neuskar	*Brassica compestris*
(f) Fodder		
Alfalfa lucerene	Ole	*Medicago sativa M. falcata*
(g) Fruits		
(i) Apple	Kushu	*Malus pumila*
(ii) Apricot	Phating/Chuli	*Prunus armeniaca*

Buckwheat is cultivated and used as a supplement to other food plants. The crop is suitable for growing in areas with a short growing season and deficient soils. It is rich in protein content and these qualities render it as a valuable nutritious subsidiary food for all sections of people in rural Ladakh.

A comparison of the grain characteristics of buckwheat has revealed a great diversity between different species in their chemical composition *Fagopyrun tataricum* possesses the highest hull groat ratio compared to other species of the genus, though the species has a relatively lower protein content. However, *F. sagitatum* and *F. Kashmirianum* by and large resemble each other in their protein content. Owing to its tall growth habit, higher percentage of hull in the grains and a high incidence of grain shattering *F. tataricum* stands apart as a distinct taxon. In respect of growth characteristics and grain composition *F. sagitatum* and *F. kashmirianum* show a good deal of resemblance to each other.

A notable feature observed in *Faqopyrum* is that the species breed true in monocultures. Though they are grown in mixed cultures in the fields of Ladakh, still the species breed true. On account of its high percentage of protein content, buckwheat offers great potential as a food subsidiary for the people of Ladakh.

Vegetables

The seeds of the vegetables showing extraordinary performance in Leh are not any special genetic varieties but are obtained from ordinary sources, except the seeds of raddish which in the olden days were obtained from Lhasa (Tibet) (Khan 1980). Turnips weigh from 50-100 g to even 2-4 k and a raddish weighs from 10 g to 100 g to even 1 kg.

Fruits

Among fruits apricots and apples show a considerable variation in shape, size, juice content and degree of taste. Both apples and apricots are a source of major income of about 50% of rural population of the Leh and Kargil districts.

Apples and apricots are grown in lower Indus Valley, Nubra Valley and Kargil areas. Popular varieties of apricots are "*chuli*" and "*phating*". Chuli is juicy and sweet and is eaten fresh. Phating is sweeter and but less juicy. It is dried and exported as a dry fruit. Apricot kernal is a source of a non drying oil which is used as a hair oil, for illumination and cooking purpose. The popular varieties of apple are: "*Mongol Kushu*" and "*tha kushu*". Former is larger, green at maturity, sweet but less juicy. The latter is smaller, bright red at maturity, sweeter, fragrant and more juicy than the former.

Alfalfa or "*Oal*" is the only cultivated fodder crop. A single sowing lasts for 10-15 years. Besides many species of *Melilotus* and *Astragalus* are also used as a fodder from wild.

DISCUSSION

Ladakh is botanically and geographically difficult to be separated from Tibet. Both belong to the phytogeographical region of Central Asia. Ladakh like Tibet supports scanty vegetation due to the various odd factors operative in the region. The alpine, the desert and oastic elements are the representative feature of the flora of Ladakh. The alpine element is mostly confined to narrow belts of vegetation which receive part of the molten snow water and the water from upper beds of mountain streams. The desert flora, which is very prominent in the upper reaches and valleys is most closely related to Tibetan and Siberian floristic elements. The oastic element comprises a variety of exotics as well as some indigenous plants (Kachroo et al. 1977).

The vegetation of Ladakh is characterised by stunned plants particularly prominent from Drass onwards. The vegetational cover gradually decreases from Zojila to Baltal, Baltal to Gumri, Gumri to Drass and Drass to Leh. At Fotula (4100 m) there is slightly greater vegetational cover as compared to adjoining areas largely due to local precipitation mainly during night (Hartman, 1983).

The transition in vegetation is more prominent from Drass to Leh. After crossing Zojila, the natural barrier between Ladakh and Kashmir, there is a gradual decrease in vegetation upto Drass. From Drass onwards one meets the barren and highly eroded mountain slopes with scattered green patches near moisture retaining areas to break the monotonous look. The vegetation in this transition zone is chiefly composed of alpine mesophytes. Characteristically the vegetation on Ladakh side of Zojila upto Drass is akin to alpine vegetation on Kashmir side of the pass, though natural forests are completely absent on Ladakh side (Kachroo *et al.* 1977). The common alpine mesophytes include, *Aster diplostephoides A. tibeticus, Astragalus rhizanthus, Delphinium cachemirianum, Podophyllum emodi, Lavatera kashmiriana, Impatiens balsamina, Pedicularis pectinata, P. siphonantha, P. albida, Corydalis crassifolia, C. thomsonii, Polygonum nepalens, Adonis chrysocyathus, Picororhiza kurooa, Poa annua, Bupleurum longicaule, Euphrasia vulgaris, Swertia thomsonii, Pleurogynae thomsonii and Trachydium roylae* etc.

From Drass, onwards these plants are gradually replaced by plants like *Echinops cornigreous, E. niveus, Astragalus cicerífolius, A. melanostachys, cousinia falconeri, C. thomsonii, Jurina ceretocarpa, Chesneya cuneata, Leontopodium nanum, Nepeta coerlascens, N. discolor, N. tibetica, Dracocephalum stamineum and Stachys tibetica* etc.

The vegetation at Leh chiefly comprises species of *Chenopodium, Kochia, Artemesia and other elements as Capparis spinosa, Morina longifolia, Halogeton glomeratus, Salsola collina, S. Kali, Perovskia abrotanoides, Tanacetum fruticuloum, Lepidium latifolium, Senecio coronopifolius, Lactuca orientalis, Chesneya cuneata, Astragalus zanaskariensis, A. rhizanthus, Nepeta coerulascense, N. discolor, N. tibetica, Arundo donax, Pennisetum orientale* etc.

One of the interesting features of the flora of Ladakh is that the alpine vegetation at Baltal and Gumri hardly share any common taxa with alpine vegetation of Leh (Hartman 1983). This is largely due to the fact that vegetation at various sites is mostly controlled by precipitation and temperature. The precipitation at Baltaj and Gumri is the chief causative factor for the vegetation abundance the gradual decrease in precipitation towards Leh results in general reduction in vegetation. The vegetation of Drass shows an intermediate condition sharing mesophytic and xeric relations with Kashmir and Leh, respectively (Kachroo *et al.* 1977).

There are no forests in Ladakh expecting plantations of willows and poplars along the banks of rivers and other water bodies. The woody shrubs and trees, both indigenous and introduced usually occur along the water bodies. In addition to many species of *Populus and Salix*, extensive plants of various shrubs and bushes form the bulk of man made forests. These include *Myriaria germanica, M. elegans, Rosa webbiana, Hippophae rhamnoides, Tamarix indica, Elaegnus hortensis, Caragana pygmae and C. moorcroftina.* To overcome the scarcity of fuel and timber, the state department of forests has launched a massive afforestation programme. In addition to planting indigenous trees, plantation of some fast growing exotic species which are being procured from homo climatic regions of the world which have a growth period as short as that in Ladakh are also done (Kachroo et. al. 1977). Planting of species like *Paspalum ambigum, Panicum millacium, Alopecurus himaliacus, dactylis glomeratus, Poa persica, Festuca sibirica* etc. has largely been encouraged by the state forest department. This will not only improve soil substratum and fodder development but will also help in checking wind erosion hazards in the region.

The marked similarity in geography and climate of Ladakh and adjoining regions of Central Asia is largely reflected in the distribution pattern of the flora. Out of a total of 1177 taxa almost 50% is widely distributed in adjoining areas. The percentage of endemic taxa in Ladakh is only 5.5% through further 8.4% are endemic to Kashmir, Ladakh and Pakistan.

The vast majority of Ladakhi's are farmers and live by subsistence farming in small village community working plots of between two and four acres. The cultivable land is commonly known as *'Zhing'*, good fertile land as *'Zhing-Zhang*, stony land as *'Ri-Zhing'* and meadoland as *'Thang-Zhing'*. Although land is used to the utmost, no attempt is made to work it beyond its capacity. The Ladakhi's are scrupulously careful to ensure that what is taken from the soil is returned. Cow dung and night soil are the chief manures so that even without following or crop rotation good yield are rendered possible, year after year.

Altitute and topography, chiefly determine the choice of crops. The varieties specially adopted to peculiar conditions prevailing in the land are cultivated. Wheat, locally called "Toril', barley called 'Swah' and awnless barley or 'grim' are the major crops. The flour from these cereals forms the major staple food and is used in a variety of forms as 'Kholaq', 'Khambir' or 'mok mok'. The cultivation of buckwheat and other millets, is limited yet these form important supplementary crops. Among vegetables turnip has long been in cultivation. European winter vegetables, properly introduced by missionaries are still under cultivation. Besides, cabbage, beet, turnip, Knolkhol, raddish, carrot, lettuce, peas etc. are introduced by the state agriculture department.

Since the implementation of various agriculture development projects, manifold improvement has been recorded in the average yield of field crops and vegetables. Many new varieties of crops and vegetables have been introduced which have greatly increased the yield. The HBI I variety of barley introduced by the department of Agriculture is reported to give a yield of 25 quintals/hectare while wheat ford, yields about 20 qts/hectare. Further it has been found that local barley varieties yield 10-12% less than HBI I while that of wheat gives 4-5 less yield.

The Central Defence Ministry has established a Field Research Laboratory at Leh where extensive experimental work is being done regarding the cultivation of vegetables under natural and controlled conditions. Besides, underground polyhouses have been established by the State Agriculture Department where seedlings are raised under controlled conditions and later transplanted in the open fields during short growing season. This technique has boosted up yield, especially that of vegetables by over 33%.

The grain crops are usually stored for winter in underground store rooms usually in rectangular boxes or baskets that are externally coated with a mixture of dung and red clay to keep away the rats. Since no vegetables can be cultivated during winter large quantity of vegetables are stored for winter in deep pits or underground basements. A raddish or a cabbage of large size cannot be consumed in one day thus slices cut from the vegetables in storage are cooked as an when desired.

Despite all the improvements and development programmes, the crop production in Ladakh is not good enough to meet the demands of the population scattered over such a vast tract. Every year large scale imports are to be made from Kashmir to meet the demands. The Agriculture Department and FRL have to perform in a way to increase the production of crops wherein they are able to fulfill the demands of the population, if not to produce surplus.

LITERATURE CITED

Cunningham, A. (1854). Ladakh-Physical, Statistical and Historical, Straeger Publishers, London.

Dhar, U. (1980). Plants of Zanaskar. Pages 48-52. In Kachroo, P. (Editor.) Ladakh - An integrated survey, University of Kashmir, Srinagar.

Dhar, U. and Kachroo, P. (1983). Alpine Flora of Kashmir Himalaya. Scientific Publishers, Jodhpur.

Hartman, Hans. (1983). Pflanzengesllschaften entlag der Kashmir route in Ladakh. Jahrab verein zum Schutz der Bergwelt 48 : 131-173. Munchen.

Hooker, J.D. (1872-1897). Flora of British India Vol. 1-7. L. Reeve and Co., London.

Kachroo, P., Sapru, B.L. and Dhar, U. (1977). Flora of Ladakh. Bishensingh Mahendrapalsingh Dehra Drun.

Khan, R. (1980). Ethno-botany and social fabric in Ladakh. pages 4-21. In : Kacharoo, P. (Editor). Ladakh-An integrated Survey, University of Kashmir, Srinagar.

Naqshi, A.R. (1980). Ladakh-Floristics. pages 29-47. In Kachroo, P. (Editor). Ladakh-An integrated Survey, University of Kshmir, Srinagar.

Seybold, S. And Kull, U. (1985). A contribution to the floristis and vegetation of Zanaskar. Botanischer Jahrab Systematic. 105:263-277. Switzerland.

Stewart, R.R. (1972). Flora of West Pakistan and Kashmir, Fakhri Press, Karachi.

Thomson, T. (1952). Western Himalaya and Tibet a narrative of Journey through mountains of northern India in 1847-48, Praeger Publishing Company, London.

Chapter - 19

Crop Diversity in Cold-Arid Ecoregion of Jammu and Kashmir, India

Gh. Hassan Dar, M.H. Wani, M.Y. Zargar and M.S. Mir***

ABSTRACT

Cold arid region covers about 43% of geographic area of Jammu and Kashmir state. The region is known for difficult terrain and harsh climate. Due to the concerted efforts by various agencies about three fold increase in crop diversity has been achieved during last three decades. Improved package of practice, adoption of controlled environment technology, introduction of short duration crops and soil and water conservation measures have been instrumental in greater diversification achieved in crop cultivations.

Key words : cold arid, crop diversity, Jammu and Kashmir, Ladakh, plant survival, protected production.

INTRODUCTION

In Jammu and Kashmir cold arid region comprises of 96,701 km² area covering about 43% of geographic area of the state. The cold arid zone has deserts of naturally degraded eco-system formed due to radical fluctuations in temperature, increased salinity, drought, meager precipitation coupled with poor nutrient status of the soil. The terrain is mostly mountainous and devoid of vegetation. The eco-physiological, physical and biological factors are responsible for sparse plant density and reduced plant growth. The region has essentially a single cropping season which extends from March-April to October. The single cropping season which extends from March-April to October. The cropping season in the region shows great variation from 2 months in areas above 4,000 m, 3 months in areas between 3,000-4,000 m to 6 months in areas below 3,000 m elevation (SKUAST, 1993a). Not

Division of Environmental Sciences, S.K. University of Agricultural Sciences and Technology, Shalimar, Srinagar, J&K 191 121

* Section of Agricultural Economics, Faulty of Agriculture, SKUAST, Wadhura, Sopore 193 201

** Regional Agriculture Research Station, SKUAST, Kargil 194 103

much crop diversity is seen in the region compared to temperate or subtropical regions of Jammu and Kashmir state. The present paper gives an account of diversification in cultivated crops grown in the region.

AGRO-CLIMATIC CONDITIONS AND PLANT SURVIVAL

The cold-arid region is classified as cryosic arid with very cold thermal and hydric index and has characteristic optimum summer thermal regime when weather remains almost stable. The wider diurnal variation in temperature during cropping season is due to very clear skies and quick loss of heat by sandy rich soils. The higher day temperature increases evapo-transpiration losses and salinity. The area exhibits large variation in seasonal temperatures (35^{0}C to - 50^{0}C) with long freezing winters for about 5-7 months. Precipitations are mostly in the form of snow during winters thus forcing farmers to adopt cultivation in the areas where irrigation facilities are assured. The region experiences average rainfall 83 mm in Leh and 120 mm in Kargil districts with relative humidity of 15-37% during cropping season.

Most of the soils in the region are coarse, porous and immature sandy soils (sand content 54-89%), prone to wind and water erosions (Bhatt, 1966; SKUAST, 1993b). The soils are alkaline, usually deficient in N and P and have low humus content. The region has only 0.17% area (approximately 20,000 ha) irrigated which forms 98% of the total net sown area (Sharma, 1987).

The physical characteristics of the region such as very low temperature, low atmospheric pressure, high solar radiations, low partial pressure of metabolic gases (O_2 and CO_2), udulating slopes, turbulent winds and light soils have resulted in dramatic and drastic changes in the eco-system (Chhaurasia and Singh, 1996 a,b). Plants in the region are mostly xerophytic and mesophytic in nature. During growth period continuous and successive changes in floristic pattern is observed. In order to survive the harsh ago-climate conditions, plants have adopted to the environment with certain modifications in their root system, vegetative growth and reproduction coupled with physiological specialization to overcome evapo-transpiration losses and frost damage. Most of the plant species growing in the region have deep and extensive root system to extract water and nutrients from less vulnerable portions of soil. The perenial plants have high osmotic concentration of cell sap which offers resistance against freezing injury.

CROP DIVERSITY

Agriculture in cold-arid eco-region, though primitive and traditional, is based on the principle of sustained organic farming. The cultivable land in the region in 1847 was 26,700 ha (Cunningham, 1854) whereas in 1983-84 it was reduced to about 19,000 ha revealing 27% loss in net sown area. During last decade due to the creation of more irrigation facilities and soil and water conservation efforts the net sown area has increased to 20,002 ha (Directorate of Economies and Statistics, 1996).

The major crops grown in the area prior to 1960 were barely confined to barley, wheat, mustard, apricot, grapes, millets, buckwheat, oats, lucerne, lentil, pea and a few vegetables like, turnip, raddish, onion, chinese cabbage, carrot, potato, lettuce and spinach etc. So very little diversification was observed in crop cultivation. However, with the introduction of improved cultivation technology, short duration crops, new and high yielding varieties and water shed management more diversification in crop raising has been achieved (Table 19.1). More than 116 crops belonging to food, oilseeds, fruits, vegetables, fodder, pulses legumes and forests are presently cultivated in comparison to only 42 cultivated prior to sixties. Besides, more than 58 ornamental plants are raised in the region. Amongst the major food crops millets (9,530 ha) and wheat (3,970 ha) cover about 62% of the net sown area in the region. The fodder, fruits and vegetables, pulses, and oilseed crops over approximately an area 4940, 920, 890 and 100 ha, respectively (Directorate of Economics and Statistics, 1996).

The crops listed in Table 19.1 are grown in cold arid Ladakh successfully, mostly on commercial scale. No doubt some crops in the initial stages of nursery growth (prior to transplantation) require polyhouse, trench or mulched condition just to overcome short season periodicity and initial harsh and wide climatic variations in March-April. Later on, the seedlings are transplanted in open fields. In oil seeds, linseed, safflower and sunflower, introduced in Ladakh, have survived the harsh conditions. Similarly, clover, sorghum, gram and pringos have been cultivated, though on smaller scale, for fodder purpose. Clover cultivation was introduced in Ladakh by the Department of Agriculture in late eighties and its commercial cultivation can be visualized in the Khurbathang area of Kargil. Pringos has been grown in Suru valley pasteurs, in and around Drass, while as sorghum and maize are cultivated in Leh and Nubra valleys. The area under fruit crops is mostly covered by apricot followed by apple. So far only 10 strains of apricot stand described from the region (Bhatia *et al.* 1977). In a recent general survey undertaken by SKUAST about 25 local strains have been observed but these are still underscript. Similarly, in addition to 6 local underscript strains, each in apple and grapes, many more varieties from Kashmir valley and Himachal Pradesh state have been introduced in Ladakh by the Department of Horticulture and SKUAST. More than 38 vegetable crops are presently cultivated in the region as against the only 10 grown in sixties. The region is famous for producing giant size vegetables and some of them make national records (FRL, 1996). The crops like musk melon are far sweeter than those grown in Kashmir. The South and Akchamal (Kargil) areas are famous for growing musk melon and cucurbits. The other enlisted crops, despite mostly being introductions, have been cultivated on commercial scale. It is worthwhile to mention that trench or polyhouse cultivation practices are adopted for fast growth in the areas where summers are cooler, while there are areas (Batalik to Khalsi) with hot summers, having cropping season from third week of March to mid October, so are less harsh for cultivation. Saffron cultivation has recently been introduced in Kargil. The plant flowing in October and continues to do so till ending November, if climate remains mild. However, as is the case with other crops it requires assured irrigation.

Table 19.1: Crop diversity in cold arid eco-region of J&K

	Crop category		*Crops grown*
	1		*2*
1.	Cereals and millets	:	* Barley (*Hordeum Vulgare*); *wheat (*Triticum aestivum*) maize (*Zea mays*); *Buckwheat (*Fagopyrum esculenium*); Fox tailed millet (*Setaria italica*); Common millet (*Panicum melliaceum*)
2.	Oilseeds	:	*Mustard (*Brassica Juncea*); repeseed (*Brassica campestris*); *sunflower (*Helianthus annus*); linseed (*Linum usitatissimum*); safflower (*Carthamus tinctorius*)

	1		2
3.	Pulse and legumes	:	Gram (*Cicer arietinum*); *local peas (*Lathyrus* sp.); green gram (*Phaseolus aureus*); black gram (*Phaseolus mungo*); *lentil (*Lens esculenta*); soybean (*Glycine max*); *broadbean (*Vicia faba*); kidney bean (*Phaseolus vulgaris*); field pea (*Pisum sativum* var. *arvense*); *pea (*Pisum sativum*)
4.	Fodder	:	*Oats (*Avena sativa*); mustard (*B. juncea*); *alfalfa/lucerne (*Medicago falcata, M. sativa, M. varigata, M. media*); pea (*Pisum sativum*); *sorghum (*Sorghum vulgare*); maize (*Zea mays*); while clover (*Trifolium repens*); red clover (*Trifolium pratense*); sweet clover (*Melilotus officinalis*) gram (*Cicer microphyllum*); * Prangos (*Prangos pabularia*)
5.	Fruits	:	*Apricot (*prunus armeniaca*); *grape (*Vitis vinifera*); strawberry (*Fragaria grandiflora*); *walnut (*Juglans regia*); *apple (*Malus domestica, M. pumila*); almond (*P. dulcis*); peach (*Prunus persica*); pear (*Pyrus communis*); oleaster (*Elaeagnus hortiensis*); cherry (*Prunus cerasus: P. avium*); plum (*Prunus domestica*)
6.	Vegetables	:	*Onion (*Allium cepa*); leek (*Allium porrum*); garlic (*Allium sativum*); knol knol (*Brassica oleracea* var. *gongylodous*); Karan sag (*Brassica oleracea* var. acephala); * chinese cabbage (*Brasica chinensis*); cauliflower (*Brassica oleracea* var. *botrytis*); cabbage (*B. oleracea* var. *capitala*); brussels sprouts (B. oleracea var. gmmifera); broccoli (*B. oleracea* var. *italica*); beet root (*Beta vulgaris*); *turnip (*Brassica rapa*); *carrot (*Daucus carota*); *radish (*Raphanus sativus*); Jerusalem artichoke (*Helianthus tuberosus*); *potato (*Solanum tuberosum*); watermelon (*Citrullus lanatus*); *muskmelon (*Cucumis melo*); long melon (*Cucumis melo* var. *utilissimus*); pumpkin (*Cucurbita moschata*) summer squash (*Cucurbita pepa*); bottle gourd (*Lagenaria siceraria*); bitter gourd (*Momordica charantia*); edible amaranthus (*Amaranthus tricolor*); * vegetable mustard (*Brassica juncea*); *Indian spinach beet (*Beta vulgaris* var. *benghalensis*); coriander (*Corriandrum sativum*); mint (*Mentha piperita; M. longifolia*)* spinach (*Spinacea oleracea*); fenugreek (*Trigonella foenum–graecum*); celery (*apium graveolens*); lettuce (*Latuca sativa*); parsley (*Petroielinum crispum*); capsicum (*Capsicum annum*); tomato (*Lycopersicon esculentum*); Brinjal (*Solanum melangena*); chillie (*Capsicum frutescens*); okra (*Abelmoschus esculentus*).
7.	Forest	:	*Oleaster (*Elaeagans angustifolia*); *seabuckthorn (*Hippophae rhamnoides*); * poplars (*Populus ciliata, P. capsica, P. deltoides, P. euphratica, P. nigra P. sylvestris*); *willows (*Salix alba, S. angustifolia, S. babylonica, S. caesia, S. daphnoides, S. excelsa,*

1		2
		S. denticulata, S. hastata, S. sclerophylla, S. tetraspesperma); juniper (*Juniperus communis*); elm (*Ulmus wallchiana*); robinia (*Rubinia pseudoacacia*); *Myricaria (*Myricaria squamosa: M. germanica*).
8. Miscellaneous	:	Saffron (*Crocus sativus*), *mulberry* (*Morus alba*) hops (*Humulus lupulus*), betula (*Betula utilis, B. myricana*); black zeera (*Cuminum sativa*).

*Except these, rest are introduced cultivated crops.

The break-through in attaining maximum crop diversity has been due to the various projects undertaken by the State and Central Government which include, desert development, tribal development and command area development projects, through Departments of Agriculture and Horticulture, national agricultural Research programme through S.K. University of Agricultural Sciences and Technology and research and development programmes undertaken by Field Research Laboratory (Shafi, 1987; SKUAST, 1987; FRL, 1996). The protected production through trench cultivation (polyhouse cultivation) for raising early seedlings of vegetables helped in attaining transplantable size before the onset of cropping season. This technology has also made possible to raise crops under sub-zero ambient temperatures. The improved package of practice has helped in raising many crops in a systematic manner compared to conventional cultivation. In order to reduce soil moisture, soil mulching with organic polytheylene plastics has been found effective. The technology has increased the irrigation interval by 15 days against 5 in open. The most commercial crops like cabbage and cauliflower have shown 59 and 45% increase in yield when grown under protected environment (SKUAST, 1993). The creation and augmentation of irrigation network and adoption of water harvesting tanks and soil and water conservation measures have been instrumental in enhancement of area under cultivation.

CONCLUSION

About three fold increase in crop diversity during last three decades through untiring efforts by various agencies reveal the enormous potential the region has in respect of agriculture, horticulture and forest development. However, the problems in the region are diverse and complex, therefore integrated approach needs to be adopted to enhance crop diversity for sustainable production on arable as well as non-arable lands.

LITERATURE CITED

Bhatia, A.K., Singh, R.P. and Gaur, G.S. (1977). Important apricot varieties of Ladakh. *Progressive Horticulture*, 8 : 19-25.

Bhutt, G.M. (1966). *Soils in Jammu and Kashmir State-Their Origin, Fertility Status and General Management Recommendations*. Department of Agriculture, Jammu and Kashmir Govt., Srinagar. 96. pages.

Chaurasia, O.P. and Singh, B. (1996 a). *Cold Desert PlantsVol. 1 Leh Valley*, Field Research Laboratory, DRDO, Leh (J&K), 146 pages.

Chaurasia, O.P. and Singh, B. (1996 b). *Cold Desert Plants, Vol. II Nubra Valley*. Field Research Laboratory, DRDO, Leh (J&K) 56 pages.

Cunningham, A. (1854). *Ladakh-Physical, Statistical and Historical.* Sagar Publications, New Delhi, India, 483 pages.

Directorate of Economics and Statistics, J&K Govt., (1996). *Digest of Statistics, (1994-95).* Directorate of Economics and Statistics, Planning and Development Department, Jammu and Kashmir Govt., Srinagar (J&K), 367 pages.

FRL (1996). *Social Contributions to Ladakh.* Field Research Laboratory (FRL), DRDO, Leh, 12 Pages.

Shafi, M. (1987). Review of Desert development programme, Leh in relation to resource endowment and problems thereof. pages 60-63. In : *Workshop on Development of Cold Arid Areas.* July 28-30, 1987. Kargil, SK. University of Agricultural Sciences and Technology (SKUAST), Srinagar (J&K).

Sharma, H.C. (1987). Background paper on characteristics of cold arid region (Ladakh) constraints in agricultural production and research work proposed by SKUAST. pages 1-11. *In : Workshop on Development of Cold Arid Areas.* July 28-30, 1987. Kargil, SK University of Agricultural Sciences and Technology (SKUAST), Srinagar (J&K).

SKUAST (1987). Workshop on Development of Cold Arid Areas. July 28-30, 1987. Kargil, SKUAST, Srinagar (J&K), p. 133 pages.

SKUAST (1993a). *Status Paper for Special Evaluation Team (NARP) Visiting Jammu, May 10-13, 1993.* Regional Agricultural Research Station, (SKUAST), Leh, 28 pages.

SKUAST (1993b). *Third Meeting of Zonal Research and Extension Advisory Committee* (Cold Arid Zone). August 23-24, 1993. Leh, Regional Agricultural Research Station, SKUAST, Leh. 75 pages.

SECTION - VI

GENETIC RESOURCES AND BIODIVERSITY

Chapter - 20

Varietal Diversity in the Germplasm of Cherries Under Cultivation in Kashmir

Fayaz A. Lone and B. A. Wafal**

ABSTRACT

The cherries under cultivation in Kashmir (sweet cherries - *Prunus avium* L. and sour cherries - *P. cerasus* L.) exhibit remarkable diversity in their tree, foliar, floral and fruit characteristics, pollen production, harvesting time and shelf life. Analysis and assessment of this diversity which was much needed has been undertaken in order to evaluate the horde of the varieties that constitute this important germplasm in the valley. In light of this diversity the local nomenclature assigned arbitrarily to various varieties has been discussed. Makhmali Gilas, - the craze of the cherries and a prized possession of the valley although contributing its own share in cherry production of the state, calls for revival and large scale multiplication. The sizeable variation cherries exhibit here provide a strong edifice on which a breeding programme for cultivar improvement can be launched.

Key words: Sweet cherries (*Prunus avium* L.), sour cherries (*P. cerasus* L.), morphological diversity, harvesting time, shelf life.

INTRODUCTION

Belonging to genus *Prunus* L. (Rosaceae) cherries represent one of the most important fruit srops of the temperate world and are mainly produced by Germany, France, Britain, Czechoslovakia, Poland, Italy, Japan, Australia, New Zealand, Turkey and USA. Of a horde of cherry species growing all over the world, only three namely: sweet cherries (*P. avium* L.), sour cherries (*P. cerasus* L.) including their hybrids called Duke cherries and flowering cherries *(P. serrulata)* have been domesticated. Although their history goes back to Theopharastus (300 BC), it seems probable that these were domesticated centuries before in Asia Minor or Greece. The sweet cherries are believed to be indigenous to southern and central Europe and west Asia (Anon. 1969) from where they reached through early settlers to North America and spread throughout Roman Empire. A few got introduced into the

*Department of Botany, University of Kashmir, Srinagar-190006, J&K, India.

Himalayan regions where they settled down in Kashmir, Simla and Kullu valley. Kashmir happens to be the major cherry producing area in the region and the acerage under its cultivation here is roughly 1200 hectares with an annual production of approximately 600-1000 tonnes which accounts for 95% of the gross production in the entire country (Bhat, 1988). The annual return that it fetches to the state, is estimated to be around Rs. 4.5 crores on an average.

In Kashmir two species namely *P. avium* L. (sweet cherry) *and P. cerasus* L. (sour cherry) have got naturalized and according to an estimate made by the State Department of Horticulture (Anon. 1967-68; 1971-72) 11 varieties of sweet cherry are under cultivation in plains, foot hills and upland areas of the valley which include Bandipora, Lolab, Pattan, Tangmarg, Baramulla, Sopore, Narbal, Nishat, Shalimar Harwan. Zakura, Ganderbal, Kangan, Chararisherief, Pampore, Pulwama, Tral, Kellisr. Shopian, Bejbehara, Kulgam and Qazi Gund. Locally called Gilas, six varieties namely Double Gilas, Mishri Gilas, Gole Gilas, Awal Number Gilas, Makhmali Gilas and Tontal Gilas represent the most popular cultivars. Besides these the State Department of Horticulture is introducing newer varieties from time to time which at present are in the process of acclimatization and multiplication.

Sour cherry is believed to be a native of West-Asian and South East European regions (Anon. 1969) from where it reached the Western Himalayas where it has acclimatized in Kashmir, Himachal Pradesh, Kumaon and Garhwal at elevations upto 2300m. In Kashmir wherever sweet cherries are grown, sour cherries grow side by side. Locally known as "Aalich" it serves as poor man's table dish. On commercial scale, however, it is used for canning purposes.

The cherry germplasm inherited by Kashmir is replete with diversity which accumulates partly through hybridization and seed based propagation, bud mutations and selection and partly through continuous introduction of newer varieties from the western world. Genetic diversity encountered in these fruits promises a strong edifice on which a breeding programme on cultivar improvement can be launched. Various kinds of stresses such as selective and yield based propagation on the one hand and ruthless felling of the trees for baker's fuel on the other, have brought the fruit under stress and we are loosing trees which are well adapted to the local agro-climatic conditions and carry superior traits such as resistance to pests and pathogens and high yield. The natural consequence is that the cherry plantations of the valley are steadily shrinking and according to an official estimate the annual production of cherries showed an abrupt fall from 1242 tons in the year 1983-84 to mere 569 tons during 1985-96. It is need of the hour to launch a multi-dimensional programme not only to salvage this important natural resource but also to improve and upgrade it. This goal cannot be achieved unless the varieties we already have are censued, evaluated and standardized. Unfortunately, cherries have remained a neglected lot. In absence of knowledge about their diversity of habitat and fruit quality their genetic improvement can not be planned on sound lines.

Earlier, Kabu *et al.* (1976; 1979) studied the distribution and performance of cherries in Jammu and Kashmir state. The present communication analyses several varieties for their variation in the foliar, floral and fruit characters. This information will be of immense use for those interested in cherry cultivation, breeding and trade.

METHODS

Eight varieties of sweet cherry *(P. avium* L.) namely Awal Number Gilas, Double Gilas, Gole Gilas, Gogji Makhmali Gilas, Makhmali Gilas, Mishri Gilas, Tontal Gilas and

Acc CTH[1] and one'of sour cherry (*P. cerasus* L.) namely Aalich are grown in different orchards of the valley. One to five trees of each variety were selected at random from four orchards (Table 20.1) and studied for tree, foliar, floral and fruit morphology. The leaf size has been worked out from a random sample of 20 leaves per variety by marking the boundary of the leaf on the graph paper and counting the total number of squares involved.

Table 20.1: Cherry varieties presently scanned and their source.

Species	*Cultivar*	*Source*
Prunus avium (Sweet Cherry)	1. Awal Number Gilas	• Simon Orchards, Habak, Srinagar, Kashmir • Horticultural Nursery, New Thed, Harwan, Kashmir
	2. Double Gilas	"
	3. Gole Gilas	"
	4. Mishri Gilas	"
	5. Gogji Makhmali Gilas	Mohd. Ramzan's Orchard, New Thed, Harwan, Srinagar, Kmr.
	6. Makhmali Gilas	• Simon Orchard, Srinagar, Kashmir • M. Yusuf Horticultural Assistants' Orchard, New Thed, Harwan, Kashmir
	7. Tontal Gilas	Horticultural Nursery, New Thed, Harwan, Kashmir.
	8. Acc. CTH	M. Yusuf Horticultural Assistant's Orchard, New Thed, Harwan, Kashmir
P. Cerasus (Sour cherry)	Aalich	Horticultural Nursery, New Thed, Harwan, Kashmir

Celluloid replica method has been employed for working out the stomatal frequency. A thin film of quickfix was gently spread over some area of both the leaf surfaces. After about five minutes the replica was scraped and mounted in water on a glass slide and number'of stomata per field counted at 10 x. 40x magnification. The stomatal frequency was calculated as under:

$$\text{Stomatal frequency} = \frac{\text{Total No. of stomata counted}}{\text{No. of observations}}$$

1. Two trees bearing in exceptional fruit easily distinguishable from the varieties grown in M. Yousuf Horticultural Assistant's Orchard, New Thead, Harwan, Kashmir. Since these could not be assigned to any known variety, they were given an arbitrary Accession Number.

Guard cell length was determined at 10x . 40x magnification with the help of an ocular micrometer after staining the epidermal peeling in 1% safranin. Only guard cells of closed stomata have been considered.

The fruits were scanned for shape, size and colour of the skin, taste of the flesh and weight of the fruit, stone and kernel. Procedure developed by Beach (1905) and Taylor (1936) for nomenclature of fruit shape in apples has been adopted for cherries as well. Definitions of some of the uncommon terms pertaining to the fruit shape are as under:

oblate : diameter of the fruit more than its length;

ovoid : fruit intermediate in shape between spherical and conical;

lopsided : one side of the fruit larger and better developed than the other;

symmetrical : sides of the fruit equally developed.

RESULTS

1. Sweet cherries (*P. avium* L.)

1.A. Qualitative characters:

The crown is semi errect in all the varieties. Leaf may be oblong, obovate, elliptic or lanceolate; the base may be cuneate, obtuse or truncate and the apex acuminate or acute. The dentations vary from serrate to double serrate or dentate. (Table 20.2). In sweet cherries the petiole is green and pubes ecent and has a pair (rarely 1 or 3) of reddish glands at the petiole-lamina junction. The leaf is invariably dark green and glabrous on the adaxial surface and light green and pubescent on the abaxial surface.

1.B. Quantitative characters

1.B.1. Trunk height

The trunk height in well pruned varieties ranges from 1.0 to 1.5m and in unpruned or unattended trees, it may go even upto 3.5 m.

1.B.2. Leaf size. petiole length and stomatal size and frequency

In cherries, the leaves exhibit remarkable diversity in size from 8.1 x 4.8 to 18.7 x 8.3 cm and in leaf area from 33.79 to 109.13 cm^2 While the leaves of "Awal Number Gilas" are smallest (11.64 ± 0.63 cm long, 5.19 ± 0.4 cm broad) those of "Makhmali Gilas" are the largest (13.61 ± 0.72 cm long, 6.94 ± 0.41 cm broad) (Table 20.2). The leaf size varies a great deal even at the intravarietal level. On the basis of leaf size the present sample of varieties is divisible into two groups: "large" and "small" leaved. The former comprises varieties namely Gole Gilas, Mishri Gilas, Tontal Gilas, Double Gilas, Makhmali Gilas and Gogji Makhmali Gilas whose average leaf length ranges from 12.84 ± 0.64 to 13.61 ± 0.72 cm and leaf area from 52.82 ± 4.21 - 68.38 ± 7.02 cm^2. The latter include Awal Number Gilas and Acc.CTH, having an average leaf length of 11.64 ± 0.63 cm and 11.79 ± 0.65 cm and leaf area of 45.32 ± 3.97 cm^2 and 48.1 ± 4.25 cm^2 respectively.

The petiole length ranges from 2.8 to 6.5 cm. Gole Gilas, Makhmali Gilas, Gogji Makhmali Gilas and Awal Number Gilas, have long petioles (4.22 ± 0.34 to 4.76 ± 0.24 cm) and Mishri Gilas, Tontal Gilas, Double Gilas and Acc.CTH. have short petioles (3.5 ± 0.11. to 4.02 ± 0.21 cm).

Table 20.2: Follar features of sweet cherry varieties under cultivation in Kashmir Valley.

Leaf Characteristics	*Awal Number Gilas***	*Double Gilas*	*Gole Gilas*	*Gogji Makh-mali Gilas*	*Makhmali Gilas**	*Mishri Gilas*	*Tontal Gilas*	*Acc. CTH*
1	*2*	*3*	*4*	*5*	*6*	*7*	*8*	*9*
Qualitative								
Shape	obovate or lanceolate	Elliptic or broadly lanceolate	obovate or elliptic	broadly elliptic	broadly elliptic	oblong or lanceolate	broadly lanceolate or elliptic	obovate or elliptic
Leaf base	cuneate	Obtuse	cuneate or sometimes obtuse	truncate or obtuse	truncate or obtuse	obtuse	truncate or obtuse	obtuse
Leaf apex	acuminate	acuminate	acuminate	acuminate	acuminate or acute	acuminate	acuminate	acuminate
Nature of dentations	serrate	double serrate	double serrate	double serrate	double serrate	dentate	double serrate	double serrate
Quantitative								
Leaf length (cm)	11.64±0.63	12.84±0.64	13.52±1.08	13.57±0.64	13.61±0.72	13.48±0.74	13.01±0.66	11.79±0.65
Leaf breadth (cm)	5.19±0.04	6.06±0.13	6.2±0.33	6.94±0.40	6.94±0.41	3.5-6.9 5.4±00.36	4.0-8.0 6.01±0.358	4.9-6.8 5.67±0.19
Length ($\bar{x}$)/ Breadht ($\bar{x}$) ratio	1.92	2.11	2.18	1.99	1.92	2.49	2.16	2.05
Leaf area (cm2)	45.32±3.97	5.82±4.21	57.74±07.04	66.43±7.94	68.38±7.02	52.9±5.6	54.68±4.89	48.1±4.25
Petiole length (cm)	4.22±0.34	3.5±0.11	4.76±0.24	4.23±0.24	4.34±0.36	3.75±0.18	3.95±0.13	4.02±0.21

1	2	3	4	5	6	7	8	9
Stomatal frequency	26.56±0.28	35.37±0.47	30.44±0.28	32.89±0.63	32.33±0.49	52.81±1.44	38.19±0.79	31.43±1.10
Guard Cell length (μm)	27.72±0.29	26.5±0.25	26.7±0.4	25.79±0.21	25.86±0.23	22.25±0.31	24.43±0.01	21.86±0.39

Note : largest leaves (*) and smallest leaves (**) leaves in these varieties.

Figures in each column represent means ($\bar{x}$=S.E.) of a random sample of 20 leaves taken from different trees of a variety.

Cherries have hypostomatic leaves (stomata present only on the abaxial surface). At the intvarietal level they do not differ much in guard cell length which ranges from 21.86 ± 0.39 to 27.72 ± 0.29 μm (Table 20.2).

The stomatal frequency, however, varies considerably. While Mishri Gilas has 52.81 ±1.44 stomata per field, the figure works to only 26.56 ± 0.28 in Awal Number Gilas (Table 20.2).

1.C. *Inflorescence*

In sweet cherries the inflorescence is comprised of two types of buds, "simple" and "compound". The former open into a single flower and the latter into more than one (up to four) flowers. The buds are arranged in fascicled and simple racemes. However, in cvs. Double Gilas and Acc.CTH the inflorescence is only a fascicled raceme born on short peduncles.

1.D. *Flower*

1.D.1. *Qualitative characters*

The floral characters are relatively more conservative and exhibit marginal variations from variety to variety. In sweet cherries, the petals are invariably white and cream coloured: those of Double Gilas are, slightly dull. The stamens are arranged in 2 - 3 whorls. Anthers of the outermost whorl are in level with the stigma in Awal Number Gilas, Double Gilas, Mishri Gilas and Tontal Gilas, above the stigma in Makhmali Gilas and below it in Gole Gilas.

1.D.2. *Quantitative characters*

The pedical length averages from 2.26 ± 0.08 to 4.29 ± 0.12 cm. The longest pedical (3.5 - 5.0 cm) obtains in Mishri Gilas and shortest (1.8 - 2.6cm) in Double Gilas.

Flowers are 2.82 ± 0.1 to 3.91 ± 0.10 cm across when fully open. Largest flowers (3.91 ± 0.10 cm) are characteristic of Double Gilas and smallest (2.82 ± 0.10) those of Mishri Gilas (Table 20.3). Flowers are invariably pentamerous. The average petal length ranges from 1.44 ± 0.02 to 1.72 ± 0.03 cm and breadth from 1.15 ± 0.01 to 1.42 ± 0.02 cm. The data (Table 20.3) reveal that the cv Double Gilas has largest petals (1.72 ± 0.03 cm) while Tontal Gilas has shortest (1.44 ± 0.02 cm).

Since the number of stamens per flower ranges from 35 to 45, the length of their filaments varies a great deal within a flower. The longest stamens constitute the outermost whorl and the shortest ones the innermost whorl. At the intervarietal level, there is no significant variation in filament length. The length of the style ranges between 1.09 ± 0.02 to 1.42 ± 0.01 cm, the longest styles obtain in cv Mishri Gilas and shortest in Tontal Gilas (Table 20.3).

1.D.3. *Pollen*

In sweet cherry varieties the pollen yield does not show a remarkable variation at the intervarietal level and ranges between 1387 to 1765 per anther. Awal Number Gilas produces biggest pollen grains of an average diameter of 41.2 ± 0.76 μm and Gole Gilas the smallest, 26.33 ± 0.79 μm (Table 20.3).

Table 20.3: Intervarietal variability in quantitative floral and pollen characters of cherries grown in Kashmir Valley

Floral & Pollen Features	*Awal Number Gilas***	*Double Gilas*	*Gole Gilas*	*Gogji Makh-mali Gilas*	*Makhmali Gilas**	*Mishri Gilas*	*Tontal Gilas*
Floral & Pollen Features	2.99±0.09	3.91±0.10*	2.91±0.05	-	3.08±0.06	2.82±0.1	3.05±0.14
Petal length (cm)	1.65±0.03	1.72±0.03	1.52±0.02	-	1.59±0.03	1.61±0.01	144±0.02
Petal breadth (cm)	1.42±0.02	1.19±0.01	1.15±0.01	-	1.22±0.05	1.22±0.03	1.21±0.01
Filment length outer whorl (cm)	1.22±0.02	0.52±0.01	1.06±0.01	-	1.26±0.05	1.20±0.03	1.05±0.01
Filament length inner whorl (cm)	0.46±0.02	0.52±0.01	0.40±0.02	-	0.49±0.02	0.50±0.02	0.37±0.07
Style length (cm)	1.37±0.02	1.26±0.08	1.39±0.01	-	1.24±0.02	1.42±0.01***	1.09±0.02**
Pedicle length (cm)	3.91±0.07	2.26±0.08	3.95±0.10	-	3.07±0.11	4.29±0.12	3.91±0.07
No. of pollen grains per anther	1733±21.62	1765±24.81	1577±17.34	1387±26.38	1428±21.26	1444±20.62	1545±25.61
Maximum diameter of pollen (μm)	41.2±0.76	36.56±0.74	26.33±0.79	38.24±0.79	39.76±0.5	40.56±0.69	37.45±0.76
Pollen viability %	68.21	67.43	62.84	68.63	70.42	71.45	68.76

Note : *Largest flower (*), shortest styles (**) and largest styles (***) in these varieties.

Figures in each column represent means ($\bar{x}$ ± S.E.) of 30 fully anthesced flowers taken at random from different trees.

1.E. Fruit

Diagnostic features of the cherry fruit common to all the varieties.

Fruit 2.30 to 6.1 gm in weight, 1.50 - 2.0 cm long, 1.45 - 2.10 cm broad, 1.35 -1.85 cm thick; oblate, ovoid, ovate or spherical, symmetrical, sutured along the breadth, suture visible in the form of a line; pedicellate, pedicel long and thin, stem cavity obtuse, apex rounded or slightly depressed, skin thin and shiny, light to fast carmine, rarely blackish red, flesh dull white or light to fast carmine, clinging to stone, sub acidic and juicy, stone 0.21 to 0.51 gm in weight, 0.9 -1.25 cm long, 0.8 - 1.0 cm broad, 0.6 - 0.8 cm thick, surface smooth, stone oriented at right angles to the breadth of the fruit; kernel 0.01 - 0.18 gm in weight; 0.6 - 0.9 cm long, 0.3 - 0.6 cm broad, 0.1 - 0.4 cm thick, oily in taste, pungent and bitter at the end.

Fruit features characteristic of individual varieties.

The details of quantitative characters have been summarised variety-wise in Table 20.4 and those of specific qualitative characters briefly described below:

1.E.1. Awal Number Gilas

Fruit' oblate, broader than long, skin light to fast carmine, flesh dull white to light yellow, changing to light carmine 24 hours after harvest

1.E.2. Double Gilas

Fruit ovoid; skin light yellow, blushed with carmine on the exposed side, flesh light yellow,

1.E.3. Gole Gilas

Fruit ovoid, spherical, slightly broader than long; skin light to fast carmine, flesh light carmine, a shallow groove runs on the opposite side of the suture line, stone lies at the level of the stem cavity or 1 mm below.

1.E.4. Gogji Makhmali Gilas

Fruit spherical, flattish, skin intense carmine to blackish red, flesh light carmine, a shallow groove runs on the opposite side of suture line.

1.E.5. Makhmali Gilas

Fruit spherical, symmetrical, rarely lopsided, skin intense carmine to blackish red, flesh yellow to light carmine, a shallow groove runs on the opposite side of suture line.

1.E.6. Tontal Gilas

Fruit oblong, longer than broad and as broad as thick, skin intense carmine, flesh yellow, light or intense carmine.

1.E.7. Mishri Gilas

Fruit ovoid, tending to be oblate, skin light to intense carmine, slightly thicker than other varieties, flesh light yellow.

Table 20.4 : Diversity in fruit, stone and kernel characters of sweet cherry varieties under cultivation in Kashmir valley.

Fruit characters	*Awal Number Gilas*	*Double Gilas*	*Gole Gilas*	*Gogji Makhmali Gilas*	*Makhmali Gilas*	*Mishri Gilas*	*Tontal Gilas*	*Acc. CTH*
(a) Fruit								
Weight(gm)	3.07±0.05	5.41±0.13	3.04±0.08	3.57±0.08	4.43±0.08	4.02±0.06	3.16±0.09	4.68±0.10
Size								
legnth (cm)	1.6±0.01	2.10±0.01	1.62±0.01	1.62±0.04	1.87±0.01	1.79±0.01	1.78±0.01	1.78±0.01
max. breadth (cm)	1.78±0.01	2.01±0.02	1.79±0.02	1.94±0.01	2.0±0.01	1.85±0.01	1.66±0.01	1.95±0.01
thickness (cm)	1.56±0.01	1.86±0.01	1.49±0.1	1.62±0.01	1.77±0.01	1.73±0.01	1.6±0.01	1.69±0.01
(b) Stone								
Weight(gm)	0.27±0.00	0.32±0.007	0.36±0.009	0.31±0.00	0.33±0.00	0.31±0.00	0.3±0.00	0.32±0.05
Size								
length (cm)	1.01±0.00	1.13±0.008	1.10±0.00	1.04±0.00	1.10±0.00	1.17±0.00	1.13±0.00	1.08±0.00
max. breadth (cm)	0.83±0.00	0.83±0.015	0.87±0.01	0.86±0.00	0.90±0.00	0.83±0.01	0.87±0.00	0.77±0.00
thickness (cm)	0.67±0.00	0.69±0.00	0.7±0.00	0.69±0.00	0.67±0.00	0.62±0.00	0.67±0.00	0.62±0.00
(c) Kernel								
Weight(gm)	0.04±0.00	0.06±0.00	0.06±0.00	0.04±0.00	0.12±0.00	0.09±0.00	0.04±0.00	0.06±0.00
Size								
length (cm)	0.66±0.00	0.73±0.00	0.7±0.00	0.66±0.00	0.5±0.00	0.82±0.01	0.8±0.00	0.8±0.00
max. breadth (cm)	0.54±0.00	0.46±0.00	0.48±0.00	0.38±0.00	0.32±0.01	0.45±0.01	0.46±0.00	0.47±0.00
Thickness (cm)	0.35±0.00	0.36±0.00	0.28±0.01	0.21±0.01	0.16±0.01	0.37±0.02	0.32±0.01	0.37±0.00

Figures in each column represent means ($\bar{x}$ ± S.E.) of 20 healthy fruits taken at random from different trees of a variety.

1.E.8 Acc. CTH

Fruit oblate or ovoid; skin light to intense carmine, flesh light yellow.

2. SOUR CHERRIES (ALICH) : (*P. CERASUS* L.)

2.A. Qualitative

The crown wears a peculiar look because the branches do not elongate much, remain thin and grow more or less horizontally or droop downwards. The leaf is ovate or obovate, base broadly cuneate, apex acute or tending to become obtuse and margins double serrate. It is invariably dark green and glabrous on the adaxial surface and light green and pubescent on the abaxial surface, petiole green and pubescent, a pair of reddish glands is found at the petiole lamina-junction.

2.B. Quantitative characters

2.B.1 Trunk height

The trees of sour cherry are usually never pruned and remain un-attended; their trunk height ranges from 2.0 - 3.0 m.

2.B.2. Leaf size, petiole length and stomatal size and frequency

The leaf size ranges from 4.7 x 2.9 to 7.4 x 4.6 cm, the leaf area from 9.89 to 23.97 cm^2 ($\bar{x}$= 15.17 ± 1.38 cm^2)-and the petiole length between 1.1 and 1.8 cm ($\bar{x}$= 1.47 ± 0.38 cm^2). The leaves are hypostomous (stomata present only on the abaxial surface). The length of the guard cells varies from 27.4 to 37.5 µm ($\bar{x}$=32.33 ± 0.4 µm) and the stomatal frequency is 21 - 31 ($\bar{x}$= 25.83 ± 0.23) per field.

2.C. Inflorescence

The inflorescence is composed of simple and compound buds, the former sprout into a single flower and the latter into more than one (up to four) flowers. The buds are mostly arranged in simple racemes.

2.D. Flower

2.D.1. Qualitative characters

The petals and anthers are white and cream coloured respectively. The stamens are arranged in 2-3 whorls. Anthers of the outermost whori are in level with the stigma.

2.D.2. Quantitative characters

Flowers of the species compared to those of *P. avium* are distinctly smaller (almost half of the latter) measuring 2.2 - 2.9 cm across when fully open and are invariably pentamerous. The petal size ranges from 0.9 x 0.85 to 1.2 x 1.1 cm ($\bar{x}$= 1.11 ± 0.03 x 0.96 -± 0.02 cm). Each flower has 35-45 stamens whose outermost whorl has longest filaments (0.8 - 0.92 cm, $\bar{x}$ 0.84 ± 0.01 cm) and the innermost whorl the shortest (0.4 - 0.5 cm). The length of the stye is 0.8 -0.95 cm ($\bar{x}$= 0.85 ± 0.02 cm) and that of the pedicel is 2.7 - 3.9 cm ($\bar{x}$= 3.2 ± 0.11 cm). Pollen yield per anther is relatively low (1087) and the pollen grains measure 36-48 µm.

2.E. Fruit

Fruit distinctly small, 2.3 to 2.86 gm ($\bar{x}$= 2.6 ± 0.3 gm) in weight; 1.3 x 1.6 x 1.5 to 1.4 x 1.8 x 1.7 cm in size, tending to be oblate, broader than long, symmetrical, sutured along the breadth, suture visible in the form of a line, pedicellate, pedicel long and thin, stem cavity obtuse, apex rounded, skin thin and shiny, intense carmine to blackish red, (Fig. 20.6) , flesh clinging to stone, colour similar to skin, acidic and juicy, stone 0.37 to 0.57 gm ($\bar{x}$= 0.32 + 0.003 gm) in weight, 0.8 x 0.8 x 0.7 to 0.9 x 0.9 x 0.7 cm in size, surface smooth, stone oriented at -right angles to the breadth of the fruit;.kernel 0.09 to 0.12 gm ($\bar{x}$= 0.102 ± 0.001 gm) in weight., 0.6 x 0.5 x 0.4 to 0.7 x 0.5 x 0.5 cm in size, oily in taste, pungent and bitter at the end, usually all the kernels plump and healthy.

3. FLOWERING PHONOLOGY

Floral buds start appearing in the second week of August. Thickly covered with scale leaves they remain dormant till the last week of February when their activity revives. During this period which represents winter season in Kashmir valley these dormant buds are exposed to fairly low temperatures which seems to be a prerequisite for blossoming in the rosaceous fruit species. By the third week of March, individual floral buds peep out of the scales and within 8-19 days (i.e. 1st and 2nd week of April) they turn white due to the development of petals, most popularly called the "white bud" stage. The time of initiation and total span of a particular phenophase in the fruit fluctuate from year-to-year by 3-7 days as is evident from the data raised during two consecutive growing seasons, 1992 and 1993 (Table 20.5). The data raised during this period reveal that all the cultivars under reference enter the white bud stage from first to second week of April. Among sweet cherries, Awal Number Gilas is first to produce white buds (1st week of April) and Acc.CTH the last (2nd week of April). Aalich, also produces white buds in the second week of April. The overall period during which different varieties enter into white bud stage is 5 - 7 days. Majority of floral buds anthese in the morning. In a few flowers, however, opening may continue up to noon. The individual buds of a compound floral bud usually open 1-2 days after each other; occasionally they may open on the same day. On the basis of the number of flowers that open during a particular period in a tree or variety, the blooming period is divisible into 3 sub-stages: (i) initial bloom , when 10% of the floral buds of a tree are open, (ii) full bloom, when more than half of the floral buds of a tree are open and (iii) end of bloom, when 90% or more flowers of a tree have dropped their petals. Since flowering in one variety follows the other just in 1-3 days, these sub-stages overlap in different varieties. Awal Number Gilas blossoms first (1st to 2nd week of April), and there is a gap of only 4-6 days between the first and last blooming varieties. Table 20.5 shows that the duration of 'initial', 'mid' and 'end' of bloom varies from cultivar to cultivar. However, the varieties remain in full bloom for 4-6 days. Subsequently, the flowers drop their petals (end of bloom) within 3-4 days. The duration of total blossoming period in the cherries investigated now varies from 15-16 days. The stigmas start becoming receptive 1-2 days before anthesis and the receptivity lasts till 3 days after anthesis. The peak receptivity when maximum stigmas show germinating pollen is exhibited on the day of anthesis or one day after. The yield of functional gametes in different varieties ranges between 62.84% to 71.45%. The presently studied varieties can be grouped into "best pollinizers" (> 70% viable pollen viz; Makhmali Gilas and Mishri Gilas) and "medium pollinizers" (50 - 70% viable pollen viz; Gole Gilas, Gogi Makhmali Gilas, Awal Number Gilas, Tontal Gilas, Acc.CTH and Aalich).

Cherries are mostly entomophilous and honey bee (*Apis* sp.) is the chief pollinator. The pollinators are seen in the cherry plantations on sunny days between 100 to 1600 hrs

Table 20.5: The flowering behaviour of different varieties of cherry under cultivation in Kashmir Valley. (the data pertain to the year 1992 and 1993)

Species	*Variety*	*White Bud stage*		*No. of days*	*Intial bloom*		*No. of days*	*Full bloom*		*No. of days*	*End of bloom*		*No. of days*	*Totol blooming period*
		start	*end*		*start*	*end*		*start*	*end*		*start*	*end*		*(days)*
Prunus	Awal Number	5/4	10/4	6	11/4	12/4	2	16/4	19/4	4	24/4	26/4	3	16*
avium	Gilas	31/3	4/4	5	5/4	6/4	2	9/4	14/4	6	19/4	21/4	3	17**
	Makhmali Gilas	7/4	12/4	6	13/4	4/4	2	16/4	20/4	5	25/4	28/4	4	18
		2/4	7/4	6	8/4	9/4	2	12/4	15/4	4	19/4	21/4	3	14
	Tontal Gilas	7/4	12/4	6	13/4	14/4	2	19/4	19/4	4	22/4	25/4	4	17
		2/4	5/4	4	6/4	7/4	2	11/4	13/4	4	19/4	22/4	4	17
	Mishri Gilas	7/4	12/4	6	13/4	14/4	2	17/4	21/4	5	26/4	28/4	3	19
		¼	4/4	4	5/4	6/4	2	10/4	14/4	5	18/4	21/4	4	17
	Gogji Makhmali	7/4	12/4	6	13/4	15/4	2	17/4	20/4	4	24/4	27/4	4	18
	Gilas	2/4	7/4	6	8/4	9/4	2	13/4	15/4	3	20/4	23/4	4	16
	Gole Gilas	7/4	11/4	5	12/4	13/4	2	16/4	20/4	5	25/4	28/4	4	17
		3/4	7/4	5	8/4	10/4	3	17/4	17/4	4	21/4	24/4	4	17
	Double Glass	9/4	13/4	6	14/4	15/4	2	19/4	23/4	5	26/4	28/4	3	16
		5/4	9/4	5	10/4	11/4	2	15/4	18/4	4	22/4	25/4	4	16
	Acc. CTH	11/4	15/4	5	16/4	17/4	2	21/4	24/4	4	29/4	1/5	3	16
		7/4	11/4	5	12/4	13/4	2	16/4	20/4	5	26/4	27/4	3	16
P.	Aalich	10/4	15/4	6	16/4	17/4	2	21/4	24/4	4	20/4	1/5	4	16
cerasus		7/4	11/4	5	12/4	13/4	2	16/4	19/4	4	23/4	26/4	4	15

Figures in the table represent specific week/month of the years 1992 (upper line*) and 1993 (lower line**).

and are most active between 1200 to 1400 hrs. The pollinators get attracted towards the trees because of fascinating bloom and fragrance of flowers. The bee mostly visits the side branches and "walks" into the flower for 5-10 seconds thereby sucking nectar and collecting pollen. During this process the pollen get attached to its abdomen, limbs and proboscis and are deposited on the stigmas of the flower visited next.

4. HARVESTING TIME AND KEEPING QUALITY

Cherries begin to be harvested and appear in the market from last week of May. The data (Table 20.6) reveal that various varieties are harvested one after the other till mid July. On the basis of their harvesting time, the presently studied varieties can be grouped into "early", "mid" and "late" harvesting types.

Early harvesting varieties: Awal Number Gilas and Tontal Gilas are the first-to-harvest varieties and are picked from 1st week of May to 1st week of June.

Mid harvesting varieties: Five sweet cherry varieties namely Gole Gilas, Gogji Makhmali Gilas, Makhmali Gilas, Double Gilas and Mishri Gilas and Aalich - the sour cherry constitute this group. These are picked from 1st to 3rd week of June.

Late harvesting varieties: In the present sample, only one variety namely Acc.CTH is harvested late between 1st to 2nd Week of July.

Table 20.6 : Dates of picking and keeping quality of cherry varieties under cultivation in Kashmir valley.

	Date of picking		
Variety	*from*	*to*	*Shelf life (days)*
Early harvesting			
Awal Number Gilas	3w*-May	4w-May	5-6
Tontal Gilas	4w-May	1w-July	6-7
Mid harvesting			
Gole Gilas	1w-June	2w-June	5-7
Gole Gilas	2w-June	3w-June	9-10
Makhmali Gilas	2w-June	3w-June	8-10
Aalich	2w-June	3w-June	6-7
Double gilas	2w-June	3w-June	10-12
Mishri Gilas	3w-June	4w-June	13-15
Late harvesting			
Acc. CTH	1w-July	2w-July	11-13

* 1w, 2w, 3w, 4w = 1st, 2nd, 3rd, 4th week of a month.

The keeping quality of the fruits is determined under room conditions (temperature: max = 22-28°C. Min = 16-22°C). Under these conditions the shelf life of the cherries presently scanned ranges from 5-13 days (Table 20.5). While the fruits of varieties Mishri Gilas and Acc.CTH. remain healthy and edible for 11-13 days after harvest, that of Awal Number Gilas, Tontal Gilas and Gole Gilas survive only for half this period (5-6 days).

Discussion

Cherries belong to genus *Prunus* (sub family *Prunoideae)* of the family *Rosaceae.* The genus represents a renowned taxonomic complex and an assemblage of species exhibiting tremendous morphological diversity. Many workers viz.- Render (1954), Knight (1969) and Zylka (1970, 1971) attributed this diversity to frequent interspecific hybridization within the genus. Knight (1969), therefore, divided it into sub-genera and sections. Cherries were placed in subgenus *Cerasus* and section *Eucerasus.* Linneaus (1737, 1753) placed both sweet and sour cherries *in P. cerasus* and treated the former as a variety of this species, naming it var. *avium.* Subsequently, however, he (Linneaus, 1755) raised the status of this variety to that of a full fledged species *P. avium* L. and confined the name *P. cerasus* exclusively to sour cherries. *P. avium* (sweet cherry) is a highly polymorphic species and many workers (Rehder, 1954; Knight, 1969; Zylka, 1970, 1971) argue that this polymorphism owes its origin to rampant hybridizations within the species which at times renders it difficult to classify various varieties of this fruit.

All the cherries under cultivation in Kashmir except Aalich, the sour cherry (*P. cerasus)* are sweet and belong to *P. avium.* The former species differs from the later in its tree shape, floral and fruit features (Table 20.7).

Table 20.7: Morphology of *P. Avium* and *P. cerasus* compared.

Features		*P. avium L. (Sweet cherries)*	*P. Cerasus L. (Sour cherries)*
Tree			
	Trunk	Dwarf	Tall
	Crowb	Canopy of the crown large, branches upright	Canopy of the crown smaller, branches droop downwards
Leaf			
	Size	Large	Small
	Shape	Ovate	Oblong, obovate,
Flower		Large	Small
Fruit		Large	Small

Local nomenclature

The varieties presently studied are popularly known only by local Kashmiri names which find exclusive usage even in their trade both within the local as well as the outside markets all over the country. On this basis one would be tempted to speculate that these varieties might have originated n the valley's orchards. What casts doubt on this assumption is the fact that the State Department of Horticulture has put on record European name of each variety in some of its annual reports (Anon. 1967-68). These are shown in Table 20.8. Whether these exotic names are factual or not is difficult to say at the moment because the original description of the varieties known by these names is not becoming available. Should the exotic names be correct, it will speak of the cherries presently under cultivation here having got introduced into the valley rather than originated indigenously. If so, presumably the exotic names being difficult to pronounce by the local populace and fruit growers might have got replaced over the years by Kashmiri epithets. On the contrary, a few varieties exist that have no European name and are known by Kashmiri names only. These varieties may have originated in the valley's orchards from hybrid seed or as bud mutants which through selection over the years established as full fledged marketable cultivars. That it may be really so, is evidenced by the presence of certain elite trees in the orchards which bear high quality fruit and yet have no name. In course of present study such trees have been assigned specific Acc. Nos. Total neglect of these trees and lack of even the local name for them pointedly speaks of their recent origin in the orchards here. Such trees need to be standardized as distinct cultivars and assigned, proper names.

Table 20.8 indicates that majority of the local names assigned to these varieties reflect their harvesting time or shape, size and colour of the fruit. Some of the local names, viz. Mishri Gilas, connote no specific varietal attribute, hence seem to have been arbitrarily assigned. The taste of Mishri Gilas is not as sweet as the term Mnishri indicates. Seemingly, this name has been assigned to find good market for the variety under a superior trade name - Mishri.

Table 20.8 : Nomenclature of the cherry varieties under cultivation in Kashmir valley.

Kashmir name	*European name (Anon. 1967-68)*	*Attributes connoted by the Kashmir name*
Double Gilas	Biggarreau Napoleon	Fruit prominently large in size, sometimes almost twice the size of other cherries; hence the name Double Gilas.
Mishri Gilas*	Bigarreau Noir Grossa	Mishri in Oriental language refers to sweet. As the name indicates, the fruit of this variety ought to have been distinctly sweet in the present sample but in its taste this fruit resembles other sweet cherries.
Gole Gilas	Black Heart	Fruit spherical in shape. "Gole in vernacular refers to spherical shape.
Awal Number Gilas	Guigne Pourpera Precece	Harvested first of all. "Awal" in vernacular means first.
Makhmali Gilas	Guigne Noir Hative	The fruit resembles velvet in the softness of its flesh and intensity of carmine colour "Makhmal" in local dialect means velvet. The variety is a high quality cherry.
Tontal Gilas	Guine Noir Gross Lucenta	The fruit is longer than broad. In Kashmiri language, the term "Tont" refers to the bird's beak which is long and pointed.
Gogji Makhmali Gilas**		The name refers to the inferior quality of the fruit. The fruit compared to Makhmali Gilas is less sweet and flattish resembling a turnip which in local language is called "Gogji"
Acc. CTH.**		

* This name seems to be a misnomer.

** The varieties have no foreign names.

From the foregoing discussion it follows that while some Kashmiri names of various cherry varieties are meaningful and refer to certain specific qualitative character(s) of the fruit, others do not. Replacement of such names by more meaningful ones though highly warranted is extremely difficult at this stage because of their age old use.

Morphology and phenotypic variability

(a) Qualitative

(i) The crown

The crown is semi-errect in all the sweet cherry (*P. avium*) cultivars but resembles the crown of a palm tree in Aalich, (*P. cerasus*). Being species specific the shape of the crown among cherries seems to be genetically determined.

(ii) Foliar and floral features

The present sample of cherries revealed that the trees do not exhibit any marked variation in qualitative features such as shape and texture of the leaf, nature of leaf margin and floral architecture. Therefore, these features do not provide any clues of classifactory value. However, in terms of spatial relationships between anthers and stigmas this group of varieties can be distinguished into homo-and heterostylous forms. While in the homostylous cultivars anthers and stigma fall in level with each other, in the heterostylous ones they are spatially isolated, the stigma lying above or below the level of anthers respectively designated as "pin eyed" and "thrum eyed" forms. This positional isolation of the two sex organs in a flower is a highly stable feature at intervarietal level and has an important bearing on the nature of the breeding system of a cultivar.

(iii) Fruit

Among fruit trees, fruit is the only important part which is specific to a variety in terms of its qualitative features. In cherries also the sole parameter that has been utilized for classification and naming different varieties is the fruit quality which includes shape, colour and taste of the fruit. On this basis the present sample of cultivars, can be conveniently classified into some discrete groups as shown in the figure below:

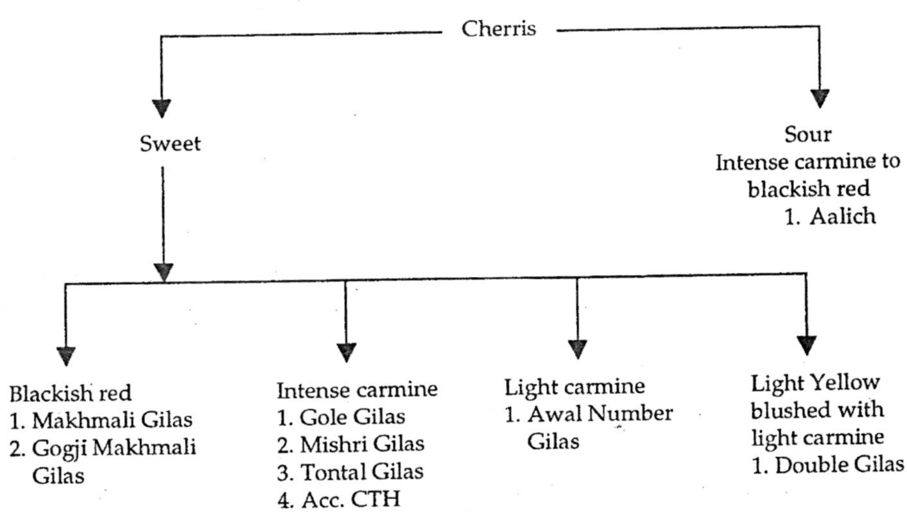

(b) Quantitative

Leaf dimensions

Although, the cherries presently investigated exhibit quantitative diversity in foliar dimensions, both at intra - as well as inter-varietal levels, but this diversity is only of marginal nature, hence has no utility in varietal delimitation and standardization. The sole exception is the foliage of Aalich which compared to the remaining varieties is distinctly smaller and exhibits apparent differences from them in length and breadth of the leaf and leaf area (Table 20.3). Foliar dimension in cherries, therefore, seems to be more important for classifactory purposes at inter-rather than intraspecific levels.

Stomatal frequency and guard cell length:

The stomatal frequency and guard cell length vary a great deal at inter varietal level and when compared from this stand point, the varieties exhibit marked differences (Table 20.2). Should the varieties be arranged in increasing order of stomatal frequency, they show a corresponding decrease in guard cell length. These two parameters, however, do not depict any clear relationship with the leaf area which fluctuates marginally or at times significantly between various varieties. This comparison depicts a general trend of inverse relationship between stomatal frequency and guard cell length among these varieties.

In order to find out the degree of association between leaf area, stomatal frequency and guard cell length, co-efficients of correlation were computed (Table 20.9) and the values obtained indicate that the leaf area and stomatal frequency, stomatal frequency and guard cell length and leaf area show low positive or low negative correlation in most of the varieties and thus vary independent of each other except in cv. Double Glass which exhibits a high positive correlation (0.698) between leaf area and stomatal frequency. This is the reason why with increase in leaf size there is a proportionate increase in the stomatal frequency in this variety. A high negative correlation (-0.934) exists between these two parameters in Makhmali Gilas and between stomatal frequency and guard cell length in Gole Gilas (-0.939). Thus, in the former variety larger leaves have lesser stomata per unit area and in the latter the stomata being relatively larger in size are lesser in frequency.

Table 20.9 : Co-efficients of correlation between some leaf variables of cherry varieties presently studied.

Variety	*Leaf area and Stomatal frequency*	*Stomatal frequency & guard cell length*	*Guard cell lenght and leaf area*
Sweet cherries			
Awal Number Gilas	-0.81	-0.271	-0.392
Double Gilas	0.698	0.496	0.486
Gole Gilas	0.584	-0639	0.587
Makhmali Gilas	-0.934	-0.058	-0.014
Mishri Gilas	-0.469	-0.59	0.445
Tontal Gilas	0.165	-0.532	-0.547
Sour cherry			
Aalich	-0.574	-0.29	0.103

Flower

Among sweet cherries Double Gilas has the largest flowers and Mishri Gilas the smallest. Size differences between flowers of these varieties are well pronounced and easily descemible (Table 20.3). The flowers of Aalich are, however, the smallest in the sample presently surveyed. The floral diameter of the remaining varieties exhibits marginal differences.

Fruit

Among sweet cherries the fruits of Double Glass are the largest and heaviest and those of Awal Number Gilas the smallest and lightest (Table 20.4). Awal Number Gilas differs markedly from all other varieties in respect of its stone and kernel characters also (Table 20.4).

The degree of association between weight of flesh, stone and kernel was computed and the values obtained (Table 20.10) reveal a high positive correlation between flesh weight and stone weight, stone weight and kernel weight and kernel weight and flesh weight in some varieties and a low positve correlation in others. In former varieties, variation in one parameter of the fruit is reflected commensurately in the other parameters and in the latter the quantitative features of the fruit vary almost independent of each other. Besides, a high negative correlation exists (-0.642) between flesh weight and stone weight, and flesh weight and kernel weight (-0.504) in Awal Number Gilas. Thus, increase in flesh weight in this variety is associated with decrease in stone weight. Therefore, fruits having larger quantities of flesh generally have lighter stones with shrivelled kernels. Reduced development of kernels in such fruits might have some genetic/embryological predesposition which may lead to the allocation of more resources towards the development of flesh.

Table 20.10. Coefficients of correlation of some fruit variables of cherries under cultivation in Kashmir Valley.

Variety	*Flesh weight and stone weight*	*Stone weight and kernel weight*	*Kernel weight and flesh weight*
Sweet cherries			
Awal Number Gilas	-0.642	0.82	-0.504
Double Gilas	0.893	0.057	0.224
Gole Gilas	0.256	0.725	0.364
Gogji Makhmali Gilas	0.723	0.828	0.644
Makhmali Gilas	0.365	0.412	0.191
Mishri Gilas	0.233	0.605	0.404
Tontal Gilas	0.841	0.717	0.346
Acc. CTH	0.357	0.456	0.087
Sour cherry			
Aalich	0.48	0.755	0.707

Flowering phonology

Singh *et al.* (1977), Dhillon *et al.* (1982), Fougat (1984), Singh (1985) and Soodan, *et al.* (1989) on the basis of blooming time classified apples, almonds and peaches into 'early', 'mid' and 'late' flowering types. This classification has been followed for the present sample of varieties too, which have been grouped into 'early', 'mid' and 'late' bloomers. Most of the varieties are 'mid' bloomers and their flowering periods overlap to a larger extent. Although blooming time is known to be under polygenic control (Kester, 1966), in the presently analyzed varieties, it is affected to a certain extent by the mean maximum temperature that prevails during the flowering period (data not shown), hence year to year fluctuations by 3-7 days in time of initiation and overall duration of a particular phenophase (Table 20.5). Bulatovic (1959), Lawrinowver (1968), Tabuenca (1977) and Fougat (1984) proposed that temperature prevailing a few days before anthesis holds key to the initiation of bloom in stone fruits which includes cherries too.

Among the three flowering groups, the 'late' bloomers are the most suited for local conditions in the valley because their late blooming habit helps them to evade the early spring weather inclemancies such as heavy snowfall and frequent rains and hailstorms which are so characteristic of this seaon in the valley. Such varieties escape premature floral drop that occurs in 'early' and 'mid' bloomers on this account. These varieties, however, have an edge over the late bloomers in that their fruit reaches the market at a time when no other fruit is available and thus sells readily. The presently studied varieties are mostly mid bloomers which are qualitatively good and give a better yield. The better yield is attributed to the fact that a wide range of pollinizer trees (trees providing large quantities of viable pollen) are available to them during mid blooming season. Cherries being largely outbreeding (Crane, 1923; Crane and Lawrence, 1929; Allard, 1960; Free and Spencer-Both, 1964; Kolesnikov, 1966 and Ortmayer, 1973) set more fruit through cross rather than self-pollination.

Cherries remain in bloom from 14 to 19 days and depending on the weather conditions such as temperature and period of sunshine, the longevity of bloom varies from year to year both within an individual variety and from variety to variety. Lawrinowver (1968) and Turovtsev (1976) argue that sunshine enhances the rate of transpiration, thus reducing the blooming period and that the calendar time of various phenological events varies according to the weather conditions. Delay in the process of flowering due to cold weather conditions in the rosaceous fruits has earlier been reported by Brown (1940) and Srivastava (1975) also. It is interesting to note that most of these varieties synchronize in their flowering behaviour which creates chances of gene flow between them. Most of these varieties, subject to their cross-compatibility status can be best and medium pollinizers, since they produce copious quantities of functional pollen. Under such circumstances in order to ensure higher yield polyculture of cross-compatible types is recommended for these varieites.

Future prospects

Owing to pest incidence and lack of adequate cultural practices the yield of the valley's cherry plantations is poor. Low yield coupled with poor shelf life and keeping quality of some cultivars has discouraged the valley's cherry growers with the result that the cherry plantations are not registering significant expandion. In order to revive interest in the cherries as a whole, it is imperative to exploit the diversity of the fruit to the fullest extent and select high quality cultivars which besides late blooming and frost evading are resistant to various pests and pathogens, efficient pollinizers, high yielders and whose fruit is attractive, relisheable, good in taste (such as Makhmali Gilas) and has a longer shelf life. The high yielding and resistance genes could be transferred from Aalich (*P. cerus*) through both conventional and non-conventional methods.

Improvement work on these lines is already underway in western countries. In Kashmir, however, cherry breeding has made no headway. The extent of diversity noted in the present studies indicates that there are enough opportunities for launching a well planned breeding programme to bring about cultivar improvement in this fruit.

LITERATURE CITED

Allard, R. W. (1960). Principles of Plant Breeding. Willey International Edition.

Anonymous (1967-68). Annual Progress Report of Fruit Research Station, Shalimar. Department of Horticulture, J&K Government.

Anonymous (1969). The Wealth of India, Raw Materials Vol. VIII. Information Directorate, CSIR, New Delhi.

Anonymous (1971-72). Annual Progress Report of Fruit Research Station, Shalimar. Department of Horticulture, J&K Government.

Beach, S. A. (1905). The Apples of New York, Vol. 2. Albary, J. B. Lyon Company Printers.

Bhat, A. A. (1988). Studies on cherry leaf spot caused by *Cercospora circumscissa* Sacc. M.Sc. Thesis, SK University of Agricultural Sciences and Technology, Shalimar, Srinagar, Kashmir.

Brown, A.G. (1940). The order and period of blossoming in apple varieties. Journal of Pomology and Horticultural Science 18: 68-74.

Bulatovic, S. (1959). A contribution to the study of flowering period and the causes of bud and flower drop in peaches in the Daubian region near Snederevo, Zborn. Rad. Polioresiv. Fak. Beograd, 7:1-17.

Crane, M. B. (1923). Report on tests of self-sterility and-cross-incompatibility in plums, cherries and apples at the John-lnnes Horticultural Institution. Journal of Pomology, 3: 67-84.

Crane, M.B. and W. J. C. Lawrence (1929). Genetical and cytological aspects of incompatibility and sterility in cultivated fruits. J. Pom. Hort. Sci., 7: 276-301.

Dhillon, A. S., A. S. Dhat and R. P. S. Gill. (1982). Pollination studies in almonds (*Prunus amygdalus* Batsch) growing under sub-tropical conditions. I. Floral Biology. Ind. J. of Hort., 39: 190-195.

Fougat, R.S. (1984). Assessment of the germplasm of apple grown in Kashmir Valley. Ph.D. thesis, University of Jammu.

Free, J. B. and Y. Spencer- Both (1964). The effect of distance from pollinizer varieties on the fruit set of apples, pear and sweet cherry trees. J. Hort. Sci., 39: 54-60.

Kabu, K. L. and Dalal, M. A. (1976). Introduction, assessment and performance of stone fruits in Jammu and Kashmir State. Res. Rept. Fruit Research Workshop. AICFIP. Ranikhet pp. 78-87.

Kabu, K.L.; Dalal, M.A. and Singh, T. (1979). A note on cherry distribution in Jammu and Kashmir. Hort. Res. Rep. 2: 48-50.

Kester, D. E. (1966). Inheritance of time of bloom in certain progenies of almond. Proc. Amer. Soc. Hort.Sci., 87: 214-221.

Knight, R. L. (1969) Abstract bibliography of fruit breeding and genetics (1965). *Prunus*. Tech. Comrnun. Commonw. Bur Hort. Plantn. Crops., 31: 649.

Kolesnikov, V. (1966). Fruit Biology. Mir Publishers, Moscow.

Lawrinowver, R. (1968). Investigations on the relationship between the flowering of fruit trees and weather. Ber. dtsc. Wettardvenst. 15:1-32.

Linnaeus, C. (1737). Genera Platarum. Lugd. Bot. Ed. 2, 1742; ed. 5, 1754; Ed. 6, 1764.

Linnaeus (1753). Species Plantarum. 2 Vols. Stockholm.

Linnaeus (1755). Flora of Suec. ed. 2: 165.

Ortmayer, C. (1973). Selecting Pollinators for H. P. apples. Indian Hort., 18: 7-9.

Rehder, A. (1954). Manual of Cultivated Trees and Shrubs. New York, 2nd edn.

Singh, G., R. S. Rana. D. K. Uppal and A. S. Rawat. (1977). Varietal wealth of almond in Kalpa valley of Himachal Pradesh. In NiJjar G. S. (Editor). Fruit Breeding in India. Oxford and IBH Publishing Co.

Singh, A. (1985). Assessment of the germplams of almond and peach under cultivation in Kashmir valley and its environs. Ph.D. thesis, University of Jammu.

Soodan, A.S.; A. K. Koul and B.A. Wafai (1989). A technology package for almond orchardists. Ind. Hort. 36: 4-9.

Srivastava, R. P. (1975). Apple varieties for Uttar Pradesh. Indian Hort., 19: 3-8.

Tabuenca, M. C. (1977). Winter chilling requirements in almonds. Ann. Aula. Die. 11: 325 329.

Taylor, H. V. (1936). The Apples of England, Crogby Lockwood and Son. Ltd., London.

Turovtisev, N. I. (1976). The use of phenological data for predicting the flowering and fruit bearing time of the sweet cherry. S-KH Biol., 11: 575-577.

Zylka, D. (1970). Die verwendung von wildarten der Gat tung *Prunus* in der Sortenzuchtung und als Unterlage. Wiesbaden.

Zylka, D. (1971). Die Verwendung Von Wilden Kirscharten in de Sortenzuchtung und als unterlage. Gartenbauwissenschaft, 36: 261-291; 417-444; 557-572.

Studies on Varietal Diversity in Indigenous and Exotic Germplasm of Almonds in Kashmir-I. Comparative Evaluation

M. Amin Dalal, M. Salman** and M. Ashraf Rather**

ABSTRACT

A cross-section of almond germplasm comprising 13 indigenous numbered selections and 10 exotic cultivars was evaluated for various traits of horticultural importance. Floral phenology examined on the basis of conventional bud development stages and bud stages indices revealed existence of late, mid and early blooming almond varieties with diverse characteristics of fruit setting, yield, cropping efficiency and nut quality. The units of degree day accumulation, using a base temperature of 4.5°C, need by these varieties to reach various stages of floral phenology were worked out. The implications of sizeable variability encountered in the germplasm in respect of blooming, production potential and nut quality characteristics have been highlighted.

Key Words : Almond (Prunus amygdalus Batsch), bloom span, bud stage development, bud stage indices, cropping efficiency, degree days accumulation, shelling percentage.

INTRODUCTION

The cultivated almond (*Prunus amygdalus Batsch*) belongs to the family Rosaceae, sub-family *Prunoideae*. The nut is typified by a drupe fruit structure. It is a native of central and south western Asia, from where cultivation spread along the Mediterranean Sea to countries like Italy, Spain and North Africa (Kester, 1978).

*Division of Pomology, S.K. University of Agricultural Sciences and Technology, Shalimar, Srinagar - 191 121 Kashmir

**Present Address : Department of Horticulture, Govt. of Jammu and Kashmir, Srinagar - 191 001

Major tonnage of almond produced in the world is mainly contributed by USA (California), Spain and Italy. In India almond is chiefly grown in Jammu and Kashmir State where 20,000 hectares under almond in Kashmir valley accounts for 11 per cent of the total area under fruits. There is much domestic demand for almond and the current level of almond production to the tune of 10,000 tonnes in the States does not meet the consumption requirement at national level. In-Shell ad Shelled almonds are imported mainly from USA to meet the short falls in production.

Most almond cultivars are self incompatible and this is of gametophytic type (Sociasi Company et al. 1976). Self incompatibility is an outbreeding mechanism which maintains high degree of heterozygosity in almond. The existing plantation of seedling almonds in Kashmir represent highly variable and heterogenous group as a result of high degree of heterozygosity. Members of the group are typified on the basis of Shell hardiness viz. hard-shelled (*duda*), medium shelled (*darmiani*) and paper shelled (*kagzi*). These almond types, representing assortment of traditional varieties, are regarded as cultivars despite the fact they lack uniformity. The lack of uniformity in almond produce, though a dominant feature of almond culture in Kashmir, poses many problems from commercial standpoint (Quershi and Dalal, 1985).

In contrast to traditional varieties, vegetatively propagated standard cultivars are adopted as modern varieties of almonds in the industrial production areas of California, France, Australia, Spain and Italy. Attempts were initiated to upgrade the genetic base of almond industry of J & K state from assortment of traditional varieties to one of modern varieties (Dalal, et al. 1979 , 1982a, 1982 b, 1988, 1993) which culminated into the release of four almond cultivars, viz. *Mukhdoom, Parbat, Waris* and *Shalimar* in 1995. This genetic upgrading has started setting in motion the process of rapidly replacing heterogeneous groups of non descript traditional varieties as a result of which genetic base of the almond industry, though upgraded, shall be considerably narrowed. The share of modern varieties in almond industry of Kashmir is bound to increase over a period of time as the development and diffusion of modern cultivars not only constitute the mainstay of increasing yields, production and quality but also generally improve yield stability (Byerlee, 1994). Nevertheless, traditional varieties remain equally important because of their characteristics as well as a source of genetic diversity. These varieties sometimes have superior performance in particular locations especially in marginal environments (Alexandratos, 1995). In order to circumvent loss of valuable genes as a result of erosion of genetic diversity *ex-situ* germplasm repository of almond has been established in SKUAST. This study features important horticultural characteristics in a cross section of this germplasm.

MATERIAL AND METHODS

The studies were carried out on indigenous germplasm of Kashmir valley and introduced material (exotic germplasm) of named almond varieties developed and adopted for cultivation elsewhere.

INDIGENOUS GERMPLASM

Extensive survey of mature standing trees of seedling origin in almond growing belts of Kashmir valley was undertaken in the sixties and early seventies. In course of initial survey 261 variants were earmarked. These variants were further screened on the basis of quality assessment (Dalal, *et al.* 1979) which led to identification of 45 selections. Amongst these numbered selections, 33 were found to have distinct characteristics of horticultural interest. Sweet almond rootstocks of seedling origin were raised and the bud wood of 33 selections (obtained from earmarked mother trees) was used for vegetative propagation by carrying out budding on almond rootstocks.

EXOTIC GERMPLASM

Rooted plants of 10 almond varieties were obtained through National Bureau of Plant Genetic Resources. The regions from where these varieties originated are as follows:–

	Variety	*Region of Origin*
1.	Afghanistan seedling	Afghanistan
2.	Chellestan	Australia
3.	I. X. L.	California
4.	Jordanolo	California
5.	Marcona	Barcelona
6.	Nikitskij	Russia
7.	Nonpareil	California
8.	Pethick's Wonder	California
9.	Pranyaj	Russia
10.	Primoskij	Russia

This material was propagated on sweet almond rootstock in a similar manner as that for indigenous germplasm.

GERMPLASM CONSOLIDATION

Vegetatively propagated rooted saplings of indigenous and exotic germplasm was planted in consolidated collection block at Shalimar. Planting was done with bud union above ground line so that scion rooting would not take place. Three single plant replications of each numbered almond selection (indigenous) or introduced variety (exotic) were set out in hexagonal system of lay out with spacing of 6 m from tree to tree.

Germplasm consolidation site was located at 38°5′ N latitude and 77°5′ E longitude with altitude of 1646 m. Soil type was silty clay loam, fairly deep and non-calcareous. Soil pH varied from 6.3 to 7.1 upto 30.5 cm depth. The north-west aspect, gentle slope, elevated site, good air and water drainage and readily available irrigation were the salient features of the site. During the entire course of germplasm maintenance vegetative growth under the trees was controlled to dripline by hoeing and rest of the cultural practices were similar for all trees.

DATA COLLECTION AND ANALYSIS

For the present study data were collected in respect of 13 numbered indigenous almond selections and 10 introduced almond varieties (Table 21.1 to 21.9). Statistical analyses were made by analysis of variance in a randomised block design (Snedecor and Cochran, 1980). Differences required for significance (C.D. values) are given to indicate the general magnitude of inter-varietal diversity in the data.

Growth measurements were obtained in the fall of 1992, when trees attained 20 years of age. Trunk circumference was recorded 30 cm above stock-scion union. Tree trunk cross-sectional area was calculated from trunk circumference. Cropping efficiency (kg cm^{-2}) was worked out by computing individual tree yield per unit trunk cross-sectional area (Westwood and Roberts, 1970). Data regarding phenology of bud development, cropping and nut characteristics were recorde during growing season of 1993. Nuts collected immediately after harvest were dehulled, dried and thereafter evaluated.

Two representative branches from replicated trees of each numbered almond selections (indigenous) and introduced varieties (exotic) were chosen and used for monitoring flower bud development. Fifty flower buds per branch were counted from the distal end of each branch. The buds were evaluated either daily or every alternate day depending on weather conditions. Bud development extending from bud swelling to post bloom with leafing out were studied under three distinct phenological stages, viz. pre-bloom stages (Table 21.1), blooming stages (Table 21.2) and post blooming stages (Table 21.3). The data so obtained were also utilized for recording the percentage of buds in each of 7 stages : (1) bud swell (2) green tip (3) initial pink bud (4) mid pink bud (5) advanced pink bud (6) full bloom (7) post bloom with petals shed. A bud stage index was calculated to provide one composite number for comparison among cultivars on a given day by modifying the procedure described by Lezzoni and Hamilton (1985) as follows:–

$$\text{Bud stage index} = \sum_{i=1}^{n} (i x p_i x 100)$$

Where

i = 1,2,3, ... 7 depicting stage 1, stage 2, ... stage 7 respectively.

P_i = proportion of bounds in the ith stage, and n = total number of stages.

A value of 100 indicates that all flowers were in stage 1 and a value of 700 indicates that all the flowers were in stage 7.

Maximum and minimum temperature were used to calculate accumulated degree days according to the Baskerville and Emin (1969) method using a base temperature of 4.5° C. The equation used to calculate degree day accumulation per day is as under :

$$\text{Accumulated degree days} = \sum_{i=1}^{n} \bar{X}\left[\frac{T_i \max - T_i \min}{2} - T_b\right]$$

where,

T_i (max) = maximum temperature on i^{th} day

Ti (min) = minimum temperature on i^{th} day

Tb = base temperature, and

n = total number of days.

Studies with regard to nut yield and cropping efficiency were also made for comparative performance. From a sample of 200 nuts a subsample of 20 nuts was drawn at random in each of the 3 replicates for recording physical characteristics of nut and kernel.

RESULTS

Floral phenology

The varieties of indigenous germplasm attained various stages of floral phenology relatively earlier than exotic cultivars (Table 21.1 to 21.3). The presently studied varieties can be divided into three groups on the basis of blooming behaviour. Almond genotypes that came into bloom at least 5 days before the beginning of HS-2 bloom were designated as early bloomers viz. HS-7, HS-26, HS-29 and K_1-77, whereas those varieties which bloomed at least 5 days after the beginning of HS-2 bloom were designated as late bloomers viz. Jordanolo, Marcona and Nikitskij. Almond genotypes with blooming period almost identical to HS-2 were designated as mid bloomers. Rest of the sixteen almond varieties comprising nine numbered selections and seven exotic cultivars (Table 21.2) behaved as mid bloomers.

When bud stage indices (Table 21.4) were examined on 26th February at 57 degree day accumulation, it was found that while flowers of HS-7 had reached to advanced pink bud stage, those of HS-1 HS-2 and HS-29 were in mid pink bud stage and those of HS-4, HS-8, HS-27 ad HS-39 in green tip stage. Exotic cultivars like Afghanistan seedling, Pethick's Wonder and Chellestan (Table 21.5) were in initial pink bud stage and in contrast to this I.X.L., Nonpareil, Pranyaj and Primoskij had attained green tip stage. Jordanolo, Marcona and Nikitiskij were yet to come in bud swell stage. By 18th March at 120 degree day accumulation, all almond selections had fully opened except HS-5, HS-6 and HS-9 which had just started their bloom, whereas, flowers of all exotic cultivars except Jordanolo, Marcona and Nikitiskij had fully opened. On 28th March at 127 degree day accumulation these varieties had progressed to full bloom stage and all other varieties were past the petal fall stage.

BLOOM SPAN IN RELATION TO METEROLOGICAL PROFILE

Various numbered almond selections had duration of blooming spread (Table 21.6) ranging from 12 to 15 days. In exotic cultivars bloom spread ranged from 16 to 19 days. Most of the exotic cultivars had bloom spread to the tune of 16 to 17 days with the exception of Jordanolo, Nikitiskij and Marcona. Bloom spread of Jordanolo and Nikitiskij was of 18 days duration whereas that of Marcona was of 19 days duration. Exposure of the bloom of indigenous selections to non-rainy days ranged from 5 to 9 days, whereas in exotic cultivars it ranged between 8 to 11 days. Relative percentage of exposure to non rainy days was, in general, higher in exotic cultivars as compared to indigenous selections. Wider bloom spread of exotic almond cultivars in comparison to numbered selections tended to expose the bloom to higher precipitation (rainfall) and relatively elevated mean maximum temperature.

PRODUCTION AND NUT CHARACTERISTICS

In exotic germplasm highest final fruit set (Table 21.7) was recorded in Nikitiskij followed by Marcona ad Jordanolo. Fruit set of I.X.L, Nonpareil, Pranyaj and Primoskij did not differ significantly with the exception of Pethick's Wonder which showed lowest fruit set. In numbered almond selections highest fruit set was recorded in HS-1 and HS-6 followed by HS-9 ad HS-7, whereas the fruit set in HS-39, HS-4, HS-5, HS-7, HS-8 and HS-27 was statistically of the same order. Among 23 almonds genotypes, Pethick's wonder and K_1-77

Table 21.1 : Pre-Bloom Stages of Various Almond Genotypes

Treatment	*Bud Swell*[1]	*Green tip*[1]	*Initial Pink Bud*[1]	*Mid pink bud*[1]	*Advanced pink bud*[1]
1	2	3	4	5	6
HS-1	10.00(9/2-11/2)	14.00(13/12-15/2)	18.00(17/2-19/2)	26.00(25/2-27/2)	34.00(5/3-7/3)
HS-2	11.00(10/2-12/2)	15.00(1/2-16/2)	19.00(18/2-20/2)	27.00(26/2-28/2)	35.006/3-8/3
HS-4	24.33(24/2-25/2)	26.00(25/2-27/2)	29.66(28/2-2/3)	32.00(3/3-5/3)	34.005/3-7/3)
HS-5	28.00(27/2-1/3)	30.33(2/3-3/3)	33.66(4/3-6/3)	36.33(8/3-9/3)	37.66(8/3-10/3)
HS-6	28.66(27/2-1/3)	30.00(1/3-3/3)	34.33(5/3-6/3)	36.00(7/3-9/3)	38.00(9/3-11/3)
HS-7	5.00(4/3-6/2)	9.00(8/2-10/2)	13.00(12/2-14/2)	18.00(17/2-19/2)	24.00(23/2-25/2)
HS-8	24.66(23/2-25/2)	2633(26/2-27/2)	29.00(28/2-2/3)	32.33(4/3-5/3)	34.00(5/3-7/3)
HS-9	28.66(27/2-1/3)	30.66(1/3-3/3)	34.00(3/3-5/3)	35.66(6/3-8/3)	39.33(11/3-12/3)
HS-26	20.00(19/2-21/2)	22.33(22/2-23/2)	25.66(24/2-26/2)	28.00(27/2-1/3)	30.66(1/3-3/3)
HS-27	23.00(22/2-24/2)	25.66(24/2-26/2)	28.33(28/2-1/3)	30.66(1/3-3/3)	32.33(4/3-5/3)
HS-29	19.33(19/2-20/2)	21.66(20/2-22/2)	24.00(23/2-25/2)	27.66(26/2-28/2)	30.00(1/3-3/3)
HS-39	24.33(24/2-25/2)	26.00(25/2-27/2)	28.66(27/2-1/3)	31.00(2/3-4/3)	33.66(4/3-6/3)
K_1-77	19.66(18/2-20/2)	21.00(20/2-22/3)	24.33(24/2-25/3)	28.33(28/2-1/3)	30.33(2/3-3/3)
Afghanistan Seeding	25.33(25/2-26/2)	28.33(28/2-1/3)	31.33(2/3-4/3)	33.33(4/3-6/3)	35.33(6/3-8/3)
Chellestan	27.00(26/2-28/2)	29.00(28/2-2/3)	33.00(4/3-6/3)	35.00(63-8/3)	37.66(8/3-10/3)
I.X.L	27.33(27/2-28/2)	30.00(1/3-3/3)	33.33(5/3-6/3)	36.66(7/3-9/3)	38.66(9/3-11/3)
Jordanolo	33.00(4/3-9/3)	37.00(8/3-10/3)	40.00(11/3-13/3)	42.33(13/3-15/3)	44.33(15/3-17/3)
Marcona	32.66(3/3-5/3)	36.33(7/3-9/3)	40.33(12/3-13/3)	42.00(13/3-15/3)	44.00(15/3-17/3)

1	2	3	4	5	6
Nikitiskij	33.33(5/3-6/3)	37.66(8/3-10/3)	40.66(11/3-13/3)	42.33(14/3-15/3)	43.66(14/3-16/3)
Nonpariel	28.00(27/2-1/3)	31.33(2/3-4/3)	33.66(4/3-6/3)	35.33(7/3-8/3)	37.66(8/3-10/3)
Pethick's wondar	25.66(24/2-26/2)	27.66(26/2-28/2)	30.66(1/3-3/3)	33.66(4/3-6/3)	35.00(6/3-8/3)
Pranyaj	28.66(27/2-1/3)	31.66(2/34/3)	34.00(5/3-7/3)	36.00(7/3-9/3)	38.00(9/3-11/3)
Primoskij	27.33(26/2-28/2)	30.66(1/3-3/3)	35.33(6/3-8/3)	37.33(8/3-10/3)	39.33(11/3-12/3)

C.D[1]					
P=0.05	1.49	1.49	3.17	3.29	2.32
P=0.01	2.07	2.19	4.64	4.84	3.41

1. Figures represent mean number of days from the reference date (February 1, 1993) to the date of bud stage indicated in parentheses.
2. Critical difference pertains to the number of days from the reference date.

Table 21.2. Blooming Stages of Various Almond Genotypes

Treatment	*Initial bloom*	*25% bloom*	*50% bloom*	*75% bloom*	*Full bloom*
1	2	3	4	5	6
HS-1	37.00(8/3-10/3)[1]	39.00(10/3-12/3)[1]	42.00(13/3-15/3)[1]	44.00(15/3-17/3)[1]	46.00(17/3-19/3)[1]
HS-2	39.00(10/3-12/3)	41.00(12/3-14/3)	43.00(14/3-16/3)	45.00(16/3-18/3)	47.00(18/3-20/3)
HS-4	36.33(8/3-9/3)	39.00(10/2-12/2)	41.33(13/3-14/3)	42.66(13/3-15/3)	44.33(16/3-17/3)
HS-6	39.66(10/3-12/3)	43.33(15/3-16/3)	45.66(16/3-18/3)	47.00(18/3-20/3)	49.66(20/3-22/3)
HS-6	40.33(12/3-13/3)	44.00(15/3-17/3)	46.33(18/3-19/3)	48.00(19/3-21/3)	50.00(21/3-23/3)
HS-7	28.00(27/2-1/3)	30.00(1/3-3/3)	32.00(3/3-5/3)	34.00(5/3-7/3)	36.00(7/3-9/3)
HS-8	37.00(8/3-10/3)	39.33(11/3-12/3)	41.00(12/3-14/3)	43.33(15/3-16/3)	45.66(16/3-18/3)
HS-9	41.00(12/3-14/3)	45.66(16/3-18/3)	46.00(17/3-19/3)	48.33(20/3-21/3)	50.33(22/3-23/3)
HS-26	32.33(4/3-5/3)	36.00(7/3-9/3)	38.66(9/3-11/3)	40.33(11/3-12/3)	42.00(13/3-15/3)
HS-27	35.66(6/3-8/3)	38.66(9/3-11/3)	40.66(11/3-13/3)	42.33(14/3-15/3)	44.33(16/3-17/3)
HS-29	32.00(313-5/3)	35.66(6/3-8/3)	37.33(9/3-10/3)	39.66(10/3-12/3)	41.33(13/3-14/3)
HS-39	36.33(8/3-9/3)	39.33(11/3-12/3)	41.00(12/3-14/3)	43.00(14/3-16/3)	44.00(15/3-17/3)
K_1-77	31.66(2/3-4/3)	36.33(8/3-9/3)	38.00(9/3-11/3)	40.00(11/3-12/3)	42.00(3/3-15/3)
Afghanistan Seedling	39.00(10/3-12/3)	41.00(12/3-14/3)	43..00(14/3-16/3)	45.33(16/3-18/3)	47.66(18/3-20/3)
Chellestan	40.00(11/3-13/3)	42.00(13/3-15/3)	44.00(5/3-17/3)	46.00(17/3-19/3)	48.00(19/3-21/3)
I.X.L	41.33(12/3-14/3)	43.00(14/3-16/3)	45.00(16/3-18/3)	47.00(18/3-20/3)	48.66(20/3-21/3)
Jordanolo	46.00(17/3-19/3)	48.33(19/3-21/3)	49.00(20/3-22/3)	50.00(21/3-23/3)	58.00(29/3-31/3)
Marcona	45.66(16/3-18/3)	46.66(17/3-19/3)	48.33(19/3-21/3)	49.00(20/3-22/3)	57.33(28/3-29/3)

1	2	3	4	5	6
Nikitiskij	44.33(15/3-17/3)	46.66(17/3-19/3)	47.66(18/3-20/3)	48.66(19/3-21/3)	57.66(23/3-30/3)
Nonpariel	40.66(11/3-13/3)	42.66(13/3-15/3)	44.66(15/3-17/3)	46.66(17/3-19/3)	18.33(19/3-21/3)
Pethick's wonder	39.33(10/3-12/3)	41.66(12/3-14/3)	42.33(13/3-15/3)	44.66(15/3-17/3)	46.33(17/3-19/3)
Pranyai	41.33(129/3-14/3)	43.33(14/3-16/3)	46.33(17/3-19/3)	47.33(18/3-20/3)	48.66(19/3-21/3)
Primoskij	41.00(12/3-14/3)	43.33(14/3-16/3)	45.33(16/3-18/3)	47.33(18/3-20/3)	48.33(19/3-20/3)
C.D.[2]					
P = 0.05	3.33	4.92	3.26	1.40	3.01
P = 0.01	4.90	7.25	4.80	2.07	4.44

1. Figures represent mean number of days from the reference date (February 1, 1993) to the date of bud stage indicated in parentheses.
2. Critical difference perteins to the number of days from the reference date.

Table 21.3 : Post Blooming Stages of Various Almond Genotypes

Treatment	*Initial petal fall*	*50% Petal fall*	*Complete patel fall*	*Leafing out*
1	2	3	*4*	5
HS-1	42.00(13/3-15/3)	49.00(20/3-22/3)	53.00(24/3-26/3)	42.00(13/3-15/3)
HS-2	43.00(14/3-16/3)	50.00(21/3-23/3)	53.00(24/3-26/3)	43.00(14/3-16/3)
HS-4	40.00(12/3-13/3)	49.33(21/3-22/3)	55.66(26/3-28/3)	39.00(10/3-12/3)
HS-5	44.66(15/3-17/3)	54.00(25/3-27/3)	61.00(1/4-3/4)	43.33(15/3-16/3)
HS-6	45.33(17/3-13/3)	55.33(27/3-28/3)	61.33(2/4-3/4)	46.00(17/3-19/3)
HS-7	34.00(5/3-7/3)	42.00(13/3-15/3)	48.00(19/3-21/3)	34.00(5/3-7/3)
HS-8	40.00(11/3-13/3)	50.66(21/3-23/3)	56.33(28/3-29/3)	40.33(12/3-13/3)
HS-9	45.00(16/3-18/3)	55.00(26/3-28/3)	61.00(1/4-3/4)	45.66(16/3-18/3)
HS-26	37.66(8/3-10/3)	47.00(18/3-20/3)	53.33(25/3-26/3)	36.00(7/3-9/3)
HS-27	39.66(10/3-12/3)	49.33(21/3-22/3)	55.00(26/3-28/3)	38.66(9/3-11/3)
HS-29	36.00(7/3-9/3)	46.00(17/3-19/3)	52.00(23/3-25/3)	37.33(9/3-10/3)
HS-39	40.00(11/3-13/3)	49.00(20/3-22/3)	56.33(28/3-29/3)	41.00(12/3-14/3)
K_1-77	37.33(9/3-10/3)	47.33(19/3-20/3)	53.33(25/3-26/3)	36.33(8/3-9/3)
Afghanistan Seedling	42.33(14/3-15/3)	59.00(30/3-1/4)	67.33(8/4-9/4)	41.33(13/3-14/3)
Chellestan	43.33(15/3-16/3)	60.33(1/4-2/4)	69.00(9/4-11/4)	44.00(15/3-17/3)
I.X.L.	44.00(15/3-17/3)	62.66(2/4-4/4)	68.33(9/4-10/4)	43.66(14/3-16/3)
Jordanolo	60.00(31/3-2/4)	68.33(9/4-10/4)	71.66(11/413/4)	59.33(31/3-11/4)
Marcona	59.33(30/3-31/3)	68.00(8/-10/4)	72.00(12/4-14/4)	60.00(31/3-2/4)

1	2	3	4	5
Nikitiskij	59.66(30/3-1/4)	67.66(7/4-9/4)	71.33(12/4-13/4)	59.66(30/3-1/4)
Nonpariel	43.66(14/3-16/3)	63.00(3/4-5/4)	70.33(11/4-12/4)	44.33(16/3-17/3)
Pathick's wonder	42.00(13/3-16/3)	59.33(30/3-31/3)	67.00(7/4/-9/4)	42.33(14/3-15/3)
Prannyaj	45.00(16/3-18/3)	63.33(3/4-4/4)	70.66(10/4-12/4)	45.33(17/3-18/3)
Primoskij	44.33(16/3-17/3)	62.66(2/4-4/4)	69.66(9/4-11/4)	44.66(15/3-17/3)
C.D.[2]				
P = 0.05	3.24	3.15	3.47	3.20
P = 0.01	4.77	4.64	5.10	4.71

1. Figures reprsent mean number of days from the reference date (February 1, 1993) to the date of bud stage indicated in parentheses.
2. Critical difference pertains to the number of days from the reference date.

Table 21.4 : Bud Stage Indices of Various Numbered Almond Selections

Date	Degree day accumulation from 1-2-93	HS-1	HS-2	HS-4	HS-5	HS-6	HS-7	HS-8	HS-9	HS-26	HS-27	HS-29	HS-39	K1-77	C.D.	
															P = 0.05	P = 0.01
26th Feb.	57	321.0	342.3	115.3	75.7	58.0	409.0	122.3	87.0	240.0	102.0	382.0	163.0	287.0	2.34	3.45
3rd Mar.	62	357.7	375.0	217.7	148.0	136.3	459.7	226.7	168.3	318.0	202.0	422.0	232.0	288.0	2.66	3.90
6th Mar.	76	394.7	408.0	285.7	228.3	215.3	510.7	290.7	234.7	415.7	273.0	506.3	292.7	427.0	2.58	3.80
8th Mar.	88	412.3	437.3	305.3	278.3	267.0	555.0	325.7	292.3	486.3	302.0	553.7	388.3	493.7	2.38	3.50
11th Mar	104	447.7	470.7	356.3	336.7	306.0	576.7	370.7	362.0	499.7	359.0	567.0	396.7	508.3	2.94	4.31
14th Mar.	106	482.3	504.0	456.7	448.7	428.0	603.0	478.0	467.3	526.0	480.0	583.7	488.3	572.3	2.13	3.14
18th Mar.	120	562.7	574.3	548.3	524.3	506.3	681.3	552.7	537.3	565.0	550.7	618.0	573.0	594.3	2.41	3.55
20th Mar.	125	656.0	673.3	580.3	563.0	528.0	693.3	587.7	573.7	612.0	588.7	670.0	594.3	635.3	2.45	3.60
24th Mar.	127	681.3	688.3	602.0	570.7	559.3	700.0	628.3	580.7	670.0	602.3	700.0	617.3	690.7	1.91	2.81
28th Mar.	129	696.7	699.7	681.3	602.3	601.3	700.0	687.7	638.7	700.0	670.3	700.0	695.3	700.0	2.96	4.34

Table 21.5 : Bud Stage Indices of Various Exotic Almond Cultivars

Date	*Degree day accumulation from 1-2-93*	*Afg. seedling*	*Pethick's wonder*	*Chellestan*	*I.XL*	*Non Pareil*	*Pranyaj*	*Primoskij*	*Jordanolo*	*Marcona*	*Nikitiskij*	*C.D.*	
												P = 0.05	*P = 0.01*
26 Feb.	57	283.0	229.7	216.7	185.3	175.0	186.0	190.0	83.3	54.7	64.3	2.91	4.29
3rd Marh.	62	297.0	256.7	241.3	238.7	227.7	248.7	258.7	16.3.7	138.3	151.7	1.68	2.47
6th Mar.	76	414.7	426.7	401.3	365.3	360.7	361.7	357.3	256.3	252.7	249.7	1.63	2.40
8th Mar.	88	476.0	446.0	419.0	415.3	414.0	417.7	404.7	308.3	302.7	306.7	1.72	2.53
11th Mar.	104	502.7	496.0	460.7	470.0	478.7	488.0	490.3	375.0	380.3	370.3	1.49	2.19
14th Mar.	106	518.7	505.7	491.7	491.0	499.7	495.7	500.0	460.7	450.7	440.3	1.27	1.66
18th Mar.	120	569.7	549.7	520.7	515.0	518.7	520.7	525.7	490.3	483.3	489.7	1.42	2.09
20th Mar.	125	696.3	666.7	618.3	606.7	622.3	613.7	610.7	557.3	458.7	556.0	1.37	2.02
24th Marh.	127	700.0	700.0	640.7	644.0	644.0	648.7	850.3	581.7	572.7	569.7	1.20	1.77
28th Mar.	129	700.0	700.00	700.0	700.0	699.7	690.3	688.7	635.7	625.3	618.3	1.12	1.52

Table 21.6 : Bloom Span of Various Almond Genotypes in Relation to the Meteorological Profile

Treatment	Stage and date of bloom		Number of days		Relative percentage of		Temperature °C			
							Maximum		Minimum	
	25% bloom	75% petalfall	Total	Rainy/ Non-rainy	non-rainy days	Rainfall (mm)	Range	Mean	Range	Mean
1	2	3	4	5	6	7	8	9	10	11
					NUMBERED SELECTIONS					
HS-1	11/3	23/3	3.7(13)	7.6	46.1	108.8	4.2 to14.6	10.5	0.0 to 5.3	2.3
HS-2	13/3	25/3	3.5(12)	7/5	41.6	148.9	2.0 to14.6	10.9	-0.2 to 5.3	2.3
HS-4	11/3	24/3	3.8(14)	8/6	42.9	92.1	2.0 to14.6	9.5	0.0 to 5.3	2.2
HS-5	15/3	29/3	3.9(15)	6/9	60.0	152.5	2.0 to 15.3	10.9	-0.2 to 5.3	1.4
HS-6	16/3	30/3	3.9(15)	5/10	66.7	148.0	2.0 to 15.3	11.0	-0.2 to 5.3	1.3
HS-7	2/3	16/3	3.8(14)	5/9	64.3	73.0	4.2 to 20.4	13.6	0.8 to 6.6	3.1
HS-8	11/3	25/3	3.9(15)	9/6	40.0	126.3	2.0 to 14.6	9.2	0.0 to 5.3	2.0
HS-9	17/3	30/3	3.8(14)	5/9	64.3	147.0	2.0 to 15.3	10.8	0.2 to 5.3	1.2
HS-26	8/3	22/3	3.9(15)	6/9	60.0	81.6	4.2 to 20.4	12.4	0.0 to 6.1	3.1
HS-27	10/3	24/3	3.9(15)	8/7	46.7	92.1	2.0 to 18.8	10.2	0.0 to 6.1	2.4
HS-29	7/3	21/3	3.9(15)	6/9	60.0	81.1	4.2 to 20.4	12.8	0.0 to 6.6	2.9
HS-39	11/3	25/3	3.9(15)	9/6	40.0	126.3	2.0 to 14.6	9.2	-0.2 to 5.3	2.0
K_1-77	8/3	22/3	3.9(15)	6/9	60.0	81.6	4.2 to 20.4	12.4	0.0 to 6.6	3.1

1	2	3	4	5	6	7	8	9	10	11
					EXOTIC CULTIVARS					
Afghanistan Seedling	13/3	28/3	4.1(16)	8.8	50.0	162.1	2.0 to 14.6	10.3	-0.2 to5.3	1.6
Chellestan	14/3	30/3	4.2(17)	7/10	58.8	157.5	2.0 to 15.3	11.0	-0.2 to5.3	1.5
I.X.L.	15/3	30/3	4.1(16)	6/10	62.5	152.5	2.0 to 15.3	11.1	-0.2 to 5.3	1.4
Jordanolo	20/3	6/4	4.3(18)	9/9	50.0	148.7	2.0 to 19.9	12.0	-0.2 to 5.3	1.6
Marcona	18/3	5/4	4.4(19)	9/10	52.6	150.7	2.0 to 19.9	12.1	-0.2 to 5.3	1.6
Nikitiskij	18/3	4/4	4.3(18)	8/10	55.6	150.7	2.0 to 19.9	11.7	-0.2 to 5.3	1.5
Non pareil	14/3	30/3	4.2(17)	6/11	64.7	157.9	2.0 to 15.3	11.2	-0.2 to 5.3	1.5
Pethick's	13/3	28/3	4.1(16)	8/8	50.0	162.1	2.0 to 14.6	10.3	-0.2 to 5.3	1.6
Pranyaj	15/3	31/3	4.2(17)	6/11	64.7	152.9	2.0 to 15.3	11.3	-0.2 to 5.3	1.7
Primoskij	15/3	31/3	4.2(17)	6/11	64.7	152.9	2.0 to 15.3	11.3	-0.2 to5.3	1.7

C.D.[2] $p = 0.05$ 0.33

$p = 0.01$ 0.48

1. Figures represent square root transformed values of total days of bloom span (25% to 75% petal fall) in parentheses.
2. Critical difference pertains to the transformed values of total days of bloom span.

Table 21.7 : Fruit set and productivity of various almond genotypes.

Varieties	*Fruit set*[1] *(%)*	*Nut yield per tree (kg)*	*Cropping efficiency (kg / cm2 x 1000)*
NUMBERED SELECTIONS			
HS1	27.97(22)	1.96	7.00
HS-2	25.10(18)	2.99	13.00
HS-4	23.58(16)	2.77	10.00
HS-5	25.10(18)	1.17	4.00
HS-6	27.97(22)	3.03	10.00
HS-7	26.56(20)	1.60	5.00
HS-8	25.10(18)	2.85	11.00
HS-9	26.56(20)	1.52	6.00
HS-26	20.27(12)	2.75	8.00
HS-27	25.10(18)	2.30	7.00
HS-29	21.97(14)	2.45	6.00
HS-39	23.58(16)	1.33	5.00
K_1-77	18.44(10)	1.40	6.00
EXOTIC CULTIVARS			
Afghanistan seedling	20.27(12)	1.93	9.00
Chellesta	21.97(14)	1.90	10.00
I.X.L.	23.58(16)	1.37	7.00
Jordanolo	26.56(20)	4.09	13.00
Marcona	27.97(22)	1.08	7.00
Nikitikij	29.23(24)	3.75	17.00
Nonnpareil	23.58(16)	3.15	11.00
Pethick's Wonder	16.4(38)	2.50	12.00
Pranyaj	25.10(18)	3.32	11.00
Primoskij	25.10(18)	2.87	8.00
C.D. (P = 0.05)	3.39	2.48	2.69
(P = 0.01)	4.98	3.59	3.96

1. Figures represent are sine transformed values of fruit set in parenthesis and C.D. pertains to the transformed values.

had lowest fruit set as compared to Nikitiskij, HS-6 ad HS-1 which ranked high in percentage fruit set.

The nut yield per tree (Table 21.7) among numbered almond selections ranged from 1.17 kg to 3.03 kg. The highest yield of 3.03 kg and 2.99 kg of dried nuts per tree was recorded in HS-6 and HS-2 respectively, whereas HS-39 and HS-5 registered lowest yield of 1.33 and 1.17 kg of dried nuts respectively. The highest nut yield 4.09 kg of dried nuts was recorded in Jordanolo followed by Nikitiskij, Pranyaj, Nonpareil, Primoskij, Pethick's Wonder, Afghanistan Seedling and I.X.L.

Highest cropping efficiency of 17 was found in Nikitiskij, whereas the lowest cropping efficiency of 4 was found in HS-5 (Table 21.7). Among numbered almond cropping efficiency of HS-7, HS-8, HS-29, HS-39 and K_1-77 were intermediate in magnitude and statistically of the same order but superior to HS-5. Cropping efficiencies of Afghanistan Seedling and Pethick's Wonder did not differ significantly but were superior to I.X.L. and Marcona which showed lowest cropping efficiency among exotic cultivars. The cropping efficiency of exotic was, in general, higher in magnitude as compared to numbered almond selections.

Table 21.8 : Production parameters of various almond blooming groups averaged over genotypes

	*Bloom groups**		
Production parameter	*Early (14)*	*Mid(16)*	*Late(3)*
Final fruitset (%age)	14.00	16.87	22.00
Nut yield per tree (kgs)	2.05	2.31	2.98
Cropping efficiency	6.25	8.81	11.66

*Figures in parenthesis indicate number of genotypes.

PHYSICAL CHARACTERS OF NUT AND KERNEL

Pronounced inter-varietal differences were observed in respect of weight and size of nuts and kernel (Table 21.9). Exotic germplasm contained higher proportion of bolder nuts than indigenous selections, similarly shelling percentage was wider in range and higher in magnitude in the former than in the latter. The shelling percentage of various numbered almond selections ranged from 24.05 to 56.49 per cent.

DISCUSSION

Almond was probably domesticated during third millennium BC (Spigel Roy, 1986). Traditional almond culture primarily utilized open pollinating seedlings (Grasselly 1972, Rikhter 1972). Thus, from the seedling almond population which represented earliest stage of evolution in cultivar development, varieties began to be selected and became more or less established in a given area. Vegetatively propagated cultivars that are grown extensively in specific production areas of industrially advanced countries originated from open pollinated group as chance seedlings or from well planned breeding programmes. Such cultivars represent most advanced stage of evolution in almond cultivation. In contrast to traditional varieties, these highly evolved varieties can be termed as 'modern varieties' (Alexandratos, 1995). Trends in eliminating unsatisfactory cultivars and replacing them with modern varieties were followed in traditional areas of almond growing around

Mediterranean Sea and the Middle East, although a few regions still plant seed propagated almonds e.g. Middle East (Grasselly, 1972), Canary Islands (Socias i Company, 1990) Morocco (Janick, 1989), Afghanistan (Kester and Asay, 1975) and Kashmir (Qureshi annd Dalal, 1985). These regions are replete with diversity in almond germplasm which need to be exploited in the first place by local selection and upgraded by introduction of exotic germplasm. Present study attempts to achieve this objective by comparative testing of important characteristics among various varieties in almond germplasm. Significant ($P > 0.05$) inter-varietal differences (Table 21.1-21.9) as observed in the present studies, are indicative of the remarkable diversity present in the germplasm.

Flowering phenology is one of the most important features in almond cultivar selection because it indicates chilling requirement during dormancy (Rattigan and Hill, 1986), modifies vulnerability to frost (Mustafa and Kester, 1976) and determines yield stability (Dalal et al., 1993). Studies on floral phenology (Table 21.1-21.3) indicate that during 1993, 50 per cent bloom in early flowering variety like HS-7 started from 3rd to 5th March, whereas in late blooming cultivar like Jordanolo it was 20th to 22nd March. The actual time of bloom, however, depends upon the interaction of specific genotypes and temperature pattern of both the previous dormant season and that immediately preceding the bloom (Tabuenca, 1977; Rattigan and Hill, 1987). The sequence of bloom remains relatively constant from year to year but actual dates may vary by several weeks in different season (Kester and Asay 1975). Consequently classification of varieties as early, mid or late bloomes in present studies have been made in relation to the bloom date of an indigenous almond selection HS-2 which has been recently released and christened as *Mukhdoom*.

Flowering phenology of various varieties (Table 21.1-21.3) is based on predominant floral bud development and does not take into account various overlapping sub-stages on any given date. This limitation was overcome by examining floral phenology on the basis of bud stages indices adopted by lezzoni and Hamilton (1985) in cherry. The procedure was modified by using bud stage classification appropriate for almond. The attainment of bud stage indices (Table 21.4, 21.5) in various almond varieties was also examined in relation to prevailing temperatures worked out on the basis of degree day accumulation (Baskerville and Emin 1969). Some discrepancy was noticed in the phenology based on conventional predominent bud development stages (Table 21.1-21.3) and that based on bud stages index (Table 21.4, 21.5) . For instance bud stage index of about 550 indicates attainment of stage corresponding to 50% bloom which was attained from 3rd to 5th March in HS-7 (Table 21.2), whereas the equivalent bud stage index (555) was attained by HS-7 on 8th March (Table 21.4). The same stage in Jordanolo was attained between 20th to 22nd March on conventional system (Table 21.2) and on the basis of bud stage index (552.7) it was attained on 18th March (Table 21.4). This discrepancy is expected because unlike conventional system, the bud stage index integrates various sub-stages into a one composite number. Considering degree day accumulation (Table 21.4, 21.5), it was noticed that late blooming cultivars needed relatively more units of degree days to reach a given phenological stage. Late blooming attribute of Jordanolo, apart from high chilling requirements, was due to the fact that it needed 32 units of degree days more than that of HS-7 to attain 50% bloom. Late blooming cultivars were also found to have relatively wider duration of bloom span (Table 21.6) than early blooming cultivars which imparts the former greater chances of pollination during warm and non-rainy days.

Table 21.9 : Almond Genotypes Physical Characters of Nut and Kernel of Various Almond Genotypes

Treatment	*Average weight of nut (g)*	*Nut size (cms)*		*Average weight of kernel*	*Kernel size*		*Shelling %age*
		Length	*Breadth*		*Length*	*Breadth*	
1	2	3	4	5	6	7	8
NUMBERED SELECTIONS							
HS-1	1.31	2.88	1.53	0.74	2.12	0.96	56.49
HS-2	2.05	3.11	1.91	0.79	2.13	1.06	38.54
HS-4	1.63	3.01	1.62	0.46	1.55	0.92	29.69
HS-5	1.55	3.01	1.62	0.46	1.55	0.92	29.69
HS-6	1.38	2.97	1.73	0.59	1.92	0.86	42.75
HS-7	1.08	2.51	1.26	0.52	1.81	0.88	48.15
HS-8	1.43	2.83	1.31	0.62	1.85	0.98	43.36
HS-9	2.30	3.35	1.93	0.58	2.07	1.03	25.22
HS-26	1.78	2.99	1.85	0.60	2.00	1.03	33.71
HS-27	1.38	3.07	1.62	0.41	2.38	1.53	29.71
HS-29	1.50	2.59	1.53	0.63	1.60	1.00	42.00
HS-39	1.15	2.54	1.61	0.39	1.78	0.85	33.91
K_1-77	1.58	2.54	1.81	0.38	1.53	0.83	24.05
EXOTIC CULTIVARS							
Afghanistan seedling	1.43	4.57	1.83	0.91	2.27	1.21	63.64
Chellestan	1.10	3.67	1.93	0.85	2.16	1.02	77.27

1	2	3	4	5	6	7	8
I.X.L.	1.26	3.29	1.96	0.52	2.18	1.10	41.27
Jordanolo	3.40	5.26	1.92	2.00	2.72	1.94	58.82
Marcona	3.10	2.98	2.17	0.38	1.71	0.95	12.26
Nikitiskij	1.76	3.68	2.18	0.82	2.34	1.25	46.59
Nonpareil	1.93	3.41	1.93	1.20	2.53	1.60	62.18
Pethick's wonder	1.61	3.31	1.82	0.58	2.20	1.05	36.02
Pranyaj	2.12	3.12	2.16	0.84	2.23	1.19	39.62
Primoskij	1.77	3.64	2.03	0.87	2.56	1.93	49.15
C.D.							
P = 0.05	0.28	0.24	0.22	0.17	0.20	0.16	4.20
P = 0.01	0.40	0.34	0.32	0.25	0.29	0.23	6.00

Fruit set, nut yield per tree and cropping efficiency (Table 21.7) were various parameters utilized to evaluate production potential of varieties in almond germplasm. The ratio of individual tree yield to that of tree trunk cross-sectional area was calculated to obtain a measure of cropping efficiency in terms of yield per unit square centimeter of cross-sectional area. Fruit yield per trunk cross-sectional area is a most useful estimate of cropping efficiency (Preston 1967), (Westwood and Robert, 1970). It indicates the extent to which potential bearing surface is usually being used to produce fruit and this parameter integrates yield with tree size. Considering the production parameters on the basis of blooming behaviour of almond genotypes, it was found that late and mid bloomers had relatively better production potential as compared to early bloomers. When fruit set, nut yield and cropping efficiency of different blooming groups in the germplasm was averaged over various genotypes, the striking difference between these blooming groups was apparent (Table 21.8). High nut yield and high cropping efficiency in late and mid blooming group is expected as the fruit set in these almond groups was relatively higher then the early group.

Nut weight and size have great economic importance in almonds. Most of these characters are quantitatively inherited and their expression is highly influenced by environment. Among 23 almond genotypes (Table 21.9) nut weight varied from 1.08 g to 3.40 g. Almond varieties with high sheiling percentage assume importance as these types can yield higher kernel tonnage per unit area. Selection of varieties with higher shelling percentage generally leads to selection of types with thinner shells. Consequently, better shell seal characters need to be emphasized in almond types having higher shelling percentage. The shelling percentage of various varieties in our study ranged from 24.05 to 77.27 per cent. Considering the average shelling percentage of 25 to 30 per cent in almonds produced in Kashmir (Dalal et al. 1979), selection of proper varieties can lead to remarkable genetic gains in kernel recovery of the shelled nuts.

CONCLUSION AND FUTURE OUTLOOK

Consolidation, evaluation and enhancement are the essential features of dynamic germplasm systems which has the potential of saving the traditional varieties from extinction and making available wide range of variability within almond species, much of which has not yet been utilized or systematically exploited. Germplasm pools could be considered as the feed-stock for varietal development either by way of selection from open pollinated seedlings or planned breeding programmes. Concerted and continuous efforts are needed to rectify the deficiencies in almond germplasm and integrate some exotic germplasm into adapted germplasm so that hidden variability is increased to meet the challenges of cultivar improvement. Breeding material is available in the germplasm to bring about remarkable improvement in production potential and cropping efficiency. There are considerable prospects for producing later blooming and hardier type of almonds. The sizeable variability in nut and kernel characteristics holds the promise of bringing about profound improvement in qualitative aspects of almond culture.

LITERATURE CITED

Alexandratos, N (ed.) (1995). World Agriculture Towards 2010, an FAO Study. Jhon Wiley, England.

Baskerville, G.L. and emin, P (1969). Rapid estimation of heat accumulation from maximum and minimum temperature. *Ecology* 50 : 514-17.

Byerlee, D. (1994). Modern Varieties, Productivity and Sustainability. Recent Experiences and Emerging Challenges, CIMMYT, Mexico.

Dalal, M.A., Rashid, A and Raina, B.L. (1979). Studies on qualitative assessmennt of various almond selections. Fruit Symposium on Possible Improvement in Temperature Fruit Culture in J & K state. Department of Horticulture, J & K Govt. September 12-13 Lalmandi, Srinagar.

Dalal, M.A., Raina, B.L. and Bhat, A.R. (1982a). Growth and production performannce on almond selections budded on various seedling rootstocks. *Horticulture Research Reporter* 2 : 1-7.

Dalal, M.A., Raina, B.L. and Bhat, A.R. (1982b). Nut and kernel characteristics of almond selections budded on various seedling rootstocks. Tec. Doc. 17. Fruit reseearch workshop in guava, grape and temperate fruits. Nagpur. Feb. 9-14, pp. 221-225.

Dalal, M.A., Qureshi, A.S., Want, M.S. and Paul, T.M. (1988). Pollination studies in almond. In : Chada, T.R., Bhutani, V.P and Kaul, J.l. (eds.). Advances in Research on Temperate Fruits. Dr. Y.S. Parmar University of Horticulture & Forestry, pp. 135-138.

Dalal, M.A., Mir, N.A., Bhat, A.R. and Wani, M.S. (1993). Studies on stability and inter-relationship of blominng and fruit set in various almond selections. Golden Jubilee Symposium on Horticultural Research-Changing Scenario. Horticulture Society of India, Banglore. May 24-28, pp. 49 (Abstr).

Grasselly, C. (1972). The almond : morphological and physiological characters of varieties, measurements and their transmission to the first generation hybrids (in French). Thesis, Univ. of Bordeaux.

Iezzoni, A.F. and Hamilton, R.L. (1985). Differences in spring floral bud development among sour cherry cultivars. *HortScience* 20 : 915-16.

Janick, J. (1989). Horticulture in Morocco : North Africa's California, *HortScience* 24 : 18-22

Kester, D (1978). Variety improvement and genetic problems. In : W. Micke and D. Kester (eds.). Almond Orchard Management. Div. of Agri. Sci. Univ. of Calif, pp. 10-20.

Kester, D.E. and Asay, R. (1975). Almonds. In : Janick, J and Moore, J.N (eds). Advances in Fruit Breeding Pudue Univ. Press, Lafayette, pp. 387-414.

Mustafa, B. and Kester, D.E. (1976). Comparative hardiness of flower buds and blossoms of some almond genotype in relation to time of bloom and leafing. *Journal of the American Society for Horticultural Science* 101 : 344-47.

Preston, A.P. (1967). Apple rootstock studies. 15 years results with some mixed crosses. *Journal of Horticultural Science* 42 : 41-50.

Qureshi, A.S. and Dalal, M.A. (1985). Status of nut crop in Jammu and Kashmir state. *Progressive Horticulture* 17 : 197-205.

Rattigan, K. and Hill, S.J. (1986). Relationship between temperature and flowering in almond. *Australian Journal of Experimental Agriculture* 26 : 399-404.

Rattigan, K. and Hill, S.J. (1987). Relationship between temperature and flowering in almond : Effect of location. *Australian Journal of Experimental Agriculture* 26 : 399-404.

Rikhter, A.A. (1972). Biological basis for the creation of almond cultivars and commercial orchards (in Russian) Ed. An USSR, Glavny. *Bot. Sad.* Moscow.

Snedecor, G.W. and Cochran, W.G. (1980). Statistical Methods. The Iowa State Univ. Press, Ames.

Socias i Company, R. (1990). Breeding self-compatible almonds. *Plant Breeding Review* 8 : 313-338.

Socias i Company, R., Kester, D.E. and Bradley, M.V. (1976). Effects of temperature and genotype on pollen tube growth in some self incompatible and self compatible almond cultivar. *Journal of the American Society for Horticultural Science* 101 : 490-493.

Speigel-Roy, P. (1986). Domestication of fruit trees. In : C. Barigozzi (ed.) The Origin and Domestication of Cultivated Plants. Elsevier, Amesterdam, pp. 201-211.

Tabuenca, M.C. (1977). Winter chilling requirements in Almonds. Am. Aula. Dic. 11 : 325-329.

Westwood, M.N. And Roberts, A.N. (1970). The relationship trunk cross-sectional area and weight of the apple tree. *Journal of the American Society for Horticultural Science* 95 : 23-30.

Chapter - 22

Studies on Varietal Diversity in Indigenous and Exotic Germplasm of Almonds in Kashmir-II. Variability Analysis

M. Amin Dalal, M. Salman and M.A. Rather*

ABSTRACT

Variability analysis was carried out in a cross-section of almond germplasm, comprising 13 indigenous selections and 10 exotic cultivars. Phenotypic variance of production parameters, time of blooming and nut and kernel characteristics was partitioned into genotypic and environmental components of variance. Low heritability estimates in broad sense were obtained in production related characters of tree yield (0.56) and cropping efficiency (0.38). High heritability estimates of blooming time (0.88) and shelling percentage (0.81) were associated with low genetic gain. Linear measurements comprising length and breadth of nut and kernel had heritability estimates ranging from 0.93 to 0.96 with moderate expectations of genetic gain. Heritability estimates of nut weight and kernel weight were recorded as 0.83 and 0.85 respectively.

Key Words : *Variance components, heritability, genetic advance, genetic gain.*

INTRODUCTION

The cultivated almond (*Prunus amygdalus* Batsch) is believed to have originated from three wild species–*Prunus bucharica, P. fenziliance* and *P. ulmifolia* (Kester and Asay, 1977). This tree nut species represents a wide range of morphological and geographical forms and by virtue of the polymorphism, one encounters remarkable diversity in an assemblage

Division of Pomology S.K. University of Agricultural Sciences and Technology, Shalimar, Srinagar - 191 121 Kashmir

*Present Address : Department of Horticulture, Govt. of Jammu and Kashmir, Srinagar - 190 001

of almond varieties. This diversity is basic to the continued improvement of almond, whether through natural selection or directed plant breeding. The heritable and the environmental components of variability contribute to observable diversity. In almond diversity in time of bloom, productivity and nut and kernel characteristics have great economic importance. Most of the traits are quantitatively inherited (Kester & Asay, 1975; Kester *et al.*, 1977) and their expression is influenced by the environment. Thus, it becomes difficult to judge whether the observed variability is heritable or environmental in nature. It is therefore, necessary to partition the observed variation into heritable and non-heritable components. This communication assesses the extent of genotypic and environmental variability in a cross-section of almond germplasm established in SKUAST.

MATERIAL AND METHODS

The studies were carried out on almond germplasm comprising 13 indigenous numbered selections and 10 exotic varieties. Each variety and numbered selections were represented by three single tree replications. Details of collection acquisition and consolidation of germplasm have been described previously (Dalal *et al.*, MS).

Partitioning of various components of variance was done on the basis of expected mean sum of squares in analysis of variance as follows :

Source of variation	*df*	*M.S.S.*	*Expected M.S.S.*
Replications	r - 1	M_r	$\sigma^2_e = v\,\sigma^2_r$
Varieties	v - 1	M_v	$\sigma^2_e + r\,\sigma^2_g$
Error	(r-1) (v-1)	Me	σ^2_e

where,

df = Degree of freedom

M.S.S.= Mean sum of squares

M_r = Mean sum of squares due to replication

M_v = Mean sum of squares due to varieties

M_e = Mean sum of squares due to error

r = Number of replications

v = Number of varieties (genotypes)

σ^2_g = Genotypic variance

σ^2_e = Environmental variance

Phenotypic variance (σ^2_p), genotypic variance (σ^2_g) and error variance (σ^2_e) were worked out by the following relationships

$$\sigma^2_p = \sigma^2_g + \sigma^2_e$$

$$\sigma^2_g = [M_v - M_e] / r$$

$$\sigma^2_e = M_e$$

Coefficients of variations were worked out by the method given by Burton and De Vane (1953); and heritability and genetic advance was estimated in accordance with the Jhonson *et al* (1955). Confidence limits of mean values were determined as per the method described by Panse & Sukhatme (1985).

RESULTS AND DISCUSSION

The magnitude of variability in various characters is shown in Table 22.1. The varietal differences in respect of all characters were highly significant (P <0.01). The wider range of mean values of various attributes in different varieties was exhibited by shelling percentage, followed by time of full bloom, cropping efficiency and yield per tree. The precision within which general mean values lie between minimum and maximum values of the range can be gauged by examining lower and upper confidence limits. The narrower the confidence limits the better the precision. Of all characters yield per tree exhibited the least precision, as the values of confidence limits of mean tended to overstep the extreme values of range. Narrow to moderately narrow confidence limits of general mean values were observed in nut and kernel characteristics.

Table 22.1. Range, mean and confidence limits for various characters in almond.

Character	*Oberved range*		*General mean + S.E.*	*Expected confidence limits of mean (P=0.05)*		*C.D. (P=0.01)*
	Min.	*Max*		*Lower*	*Upper*	
Days to full bloom[1]	36.00	57.33	47.66 ± 1.06	45.54	49.78	4.44
Yield per tree (kg)	1.17	4.09	2.35 ± 0.87	0.61	4.09	3.59
Cropping efficiency[2]	5.00	17.00	8.82 ± 0.95	6.92	10.72	3.96
Nut weight (g)	1.08	3.40	1.72 ± 0.09	1.54	1.90	0.40
Kernel weight (g)	0.38	2.00	0.71 ± 0.06	0.59	0.83	0.25
Shelling percentage	28.22	77.27	41.85 ± 1.48	38.89	44.81	6.00
Nut length (cm)	2.51	5.26	3.23 ± 0.08	3.07	3.39	0.34
Nut breadth (cm)	1.26	2.18	1.79 ± 0.07	1.91	2.19	0.29
Kernel length (cm)	1.53	2.72	2.05 ± 0.07	1.91	2.19	0.29
Kernel breadth (cm)	0.83	1.94	1.13 ± 0.06	1.01	1.25	0.23

1. Days from reference date (February 1, 1993)
2. Yield per tree trunk cross-sectional area (kg cm^{-2} x 1000)

Range is a crude measure of variability as it does not take into account the values falling between lower and upper limits. Variance estimates were therefore, worked out to quantify variability and are presented in Table 22.2. Phenotypic and genotypic components of variance in respect of shelling percentage and time of bloom were relatively high. These variance estimates were intermediate in magnitude in case of yield and cropping efficiency whereas rest of the characters exhibited low magnitude of variance. Though the magnitude of phenotypic and genotypic variances were almost analogous to the magnitude of width observed in minimum and maximum values of range, yet these measures of variability amongst different characters are not comparable as these estimates are dependent on scale of measurement used. For such a comparison co-efficient of variation, being independent of scale of measurement, is appropriate.

Examining of components of co-efficient of variation made the picture of comparative variabilities, among different characters in almond varieties, quite clear. Interestingly, shelling percentage and time of bloom exhibited appreciably lower magnitude of phenotypic and genotypic co-efficient of variation as compared to other characters. Yield per tree was characterised by highest values of phenotypic and genotypic co-efficient of variation and in rest of the characters these co-efficients of variation were low to moderate in magnitude. The estimates of environmental co-efficient of variation were also found to be high in tree yield and cropping efficiency. This is in conformity with findings of Shah (1997) who noted relatively high magnitude of environmental co-efficient of variation in these characters in pear. Higher environmental co-efficient of variation in tree yield and its related parameter of cropping efficiency can be attributed to polygenic inheritance of yield (Alston and Watking, 1973) as polygenic inheritance renders the yield highly susceptible to environmental variation.

Table 22.2 : Estimates of components of variance and co-efficient of variation for various characters in almond.

Character	*Components of variance*			*Components of co-efficient of variation*		
	Phenotypic (σ^2_p)	*Genotypic* (σ^2_g)	*Error* (σ^2_e)	*Phenotypic* (P.C.V.)	*Genotypic* (G.C.V.)	*Environmental* (E.C.V.)
Days to full bloom	27.95	24.60	3.352	11.18	10.55	3.84
Yield per tree	5.17	2.90	2.275	96.76	72.46	64.18
Cropping efficiency	4.33	1.65	2.677	23.59	14.56	18.55
Nut weight	0.18	0.15	0.029	24.66	22.51	9.90
Kernel weight	0.07	0.06	0.011	37.26	34.50	14.77
Shelling percentage	34.39	27.97	6.526	14.01	12.64	6.10
Nut length	0.35	0.33	0.21	18.31	17.78	4.49
Nut breadth	0.37	0.35	0.18	33.98	33.05	7.49
Kernel length	0.22	0.20	0.015	22.88	21.81	5.97
Kernel breadth	0.23	0.22	0.009	42.44	41.51	8.39

Phenotype results from pattern of development as conditioned by environment and success in selection for cultivar improvement depends upon the correspondence between genotype and phenotype. Heritability (Table 22.3), which measures the relationship between genotype and phenotype is an important consideration for success in a breeding programme. Low heritability estimates were obtained for production parameters of yield and cropping efficiency. This is expected as both these parameters are easily influenced by environmental variation. Nut and kernel characters, shelling percentage and blooming time were characterised by high heritability estimates. High heritability estimate of bloom time agrees with a study carried out by Kester *et al.* (1973). Inheritance of bloom in almonds follows essentially the same pattern as that for cherry, peach and walnut (Hansche) *et al.*, 1966, 1972a, 1972b). Heritability estimate along with estimate of expected genetic gain is more useful than the estimate of heritability value alone in predicting the resultant effect for selecting the best individual of desired character (Johnson *et al*, 1955). High heritability values of nut and kernel characters were associated with high genetic gain which is probably due to additive gene effects (Panse, 1957) and thus selection for these characters would be more effective. Estimates of high heritability in nut and kernel characters, as obtained in the present studies, are in accordance with findings of Singh and Uppal (1977) and Singh and Chaudhary (1993). High heritability estimate of shelling percentage was, however, associated with relatively low genetic gain which may be attributed to non-additive (dominance and epistasis) gene effects (Panse, 1957).

Table 22.3 : Heritability and genetic advance estimates of various characters in almond.

Character	*Heritability (Broad sense)*	*Genetic Advance*	*Genetic gain (Gennetic advance as per cent of mean)*
Days to full bloom	0.88	9.51	19.95
Yield per tree	0.56	2.62	111.48
Cropping efficiency	0.38	1.63	18.48
Nut weight	0.83	0.73	42.44
Kernel weight	0.85	0.46	64.79
Shelling percentage	0.81	9.79	22.39
Nut length	0.94	1.15	35.60
Nut breadth	0.95	1.19	66.48
Kernel length	0.93	0.90	43.90
Kernel breadth	0.96	0.95	84.07

It seems evident from the variance and heritability estimates that within the assemblage of varieties in a cross-section of this germplasm selection for late blooming almond genotypes with high shelling percentage and appropriate traits of nut and kernel should produce sizeable genetic gain. This genetic gain can be achieved either by direct adoption of varieties from the germplasm or selection of offsprings in a planned breeding programme.

LITERATURE CITED

Alston F.H. and Watkin, R. (1973). Apple breeding at East Malling. Proceedings of Eucarpia Top Fruit Breeding Symposium Sept. 11-14, pp. 14-29.

Burton, G.W. and De Vane, E.M. (1953). Estimating heritability in tall fescue (*Festuca arundinacea*) from replicated clonal material. *Agronomy Journal* 45 : 478-83.

Dalal, M.A., Salman, M and Rather, M.A. (MS). Studies on varietal diversity in indigenous and exotic germplasm of almonds in Kashmir-I. Comparative evaluation. In : Khan, M.A. and Farqooq, S. (Editors), Environment, Biodiversity and Conservation, Department of Botany, University of Kashmir (in press).

Hansche, P.E., Beres, V. and Brooks, R.M. (1966). Heritabiliy and genetic correlation in the sweet cherry. *Proceedings of American Society for Horticultural Science* 88 : 173-183.

Hansche, P.E. Beres, V. and forde, H.I. (1972a). Estimates of qualitative genetic properties of walnut and their implication in cultivar improvement. *Journal of the American Society for Horticultural Science* 97 : 279-284.

Hansche, P.E., Hesse, C.O. and Beres, V. (1972b). Estimates of genetic and environmental effects on several traits in peach. *Journal of the American Society for Horticultural Science* 97 : 76-79.

Johnson, H.W., Robinson, J.F. and Comstock, R.E. (1955). Estimates of genetic and environmental variability in soyabeans. *Agronomy Journal* 47 : 314-318.

Kester, D.E. and Asay, R. (1975). Almonds pages 387-414. In : Janick, J and Moore, J.N. (Editors)Advances in Fruit Breeding, Purdue Univ. Press, Lafayette.

Kester, D.E. and Asay, R.N. (1977). Germplasm sources in almonnds. *California Agriculture* 31 : 20-21.

Kester, D.E., Hansche, P.E., Beres, V. and Asay, R.N. (1977). Variance components and heritability of nut and kernel traits in almond. *Journal of the American Society for Horticultural Science* 102 : 264-266.

Kester, D.E., Raddi, P. and Asay, R. (1973). Correlation among chilling requirements for germination, blooming and leafing in almond (*Prunus amygdalus Batsch*). *Genetics* (suppl.) 74 : 5135 (Abstr).

Panse, V.G. (1957). Genetics of quantitative characters in relation to plant breeding. *Indian Journal of Genetics* 17 : 318-328.

Panse, V.G. and Sukhatme, P.V. (1985). Statistical Methods for Agricultural Workers. Publ. & Inform. Div. I.C.A.R. New Delhi.

Shah, M.N. (1997). Comparative performance and variability studies of various pear cultivars. M.Sc. thesis, Div. of Pomology, Faculty of P.G. Studies SKUAST,

Singh, G. and Uppal, D.K. (1977). Variability in nut characters in almonds (*Prunus dulcis* Miller) pages 126-131. In : Nijjar, G.S. (Editor) Fruit Breeding in India. Oxford & IBH Publishing Co. New Delhi.

Singh N.B. and Chaudhary, V.K. (1993). Variability, correlation and path analysis between kernel yield and other nut characters in wild apricot. *Advance in Horticulture and Forestry* 3 : 61-67.

Chapter - 23

Conservation of Biodiversity : Indigenous Rice Needs special Attention

*B. Neog**

ABSTRACT

Twenty seven varieties of indigenous deep water rice (*Oryza sativa* Linn) cultivated in the Dibrugarh district, Assam, were studied during their growing seasons and a detailed scrutiny of the quantitative and qualitative traits and agronomic requirements was made. These varieties have been evaluated on the basis of land under cultivation and quality quantity indices, the ethnological significance and economic viability. Also, several conservation measures have been suggested.

Key Words : Biodiversity, conservation, germplasm, indigenous rice.

INTRODUCTION

The rice *(Oryza sativa* Linn.) germplasm of North-Eastern India is natural heritage. Indigenous rice varieties of (local land races) are very rich in genetic diversity and valuable gene systems. This genetic variability is, however, seriously threatened by persistant propaganda for growing high yielding varieties (Das *et al.* 1981). This has prompted the present author to make a scientific stocktaking of these varieties and to evaluate their ethnological importance and economic viability. Since most of these varieties are getting sacrified at the alter of yield based cultivation of exotic varieties having uniform genetic base, their conservation measures also have been suggested.

MATERIALS AND METHODS

The varieties selected in the present study alongwith their local names are enlisted in Table 23.1. The selected plants represent a random sample taken from a wide range of habitats and distributed in different localities of Dibrugarh district. Information about local names of various varieties and their ethnological uses was obtained from the farmers after a thorough survey. Data on quantitative characters were recorded from fully grown, mature

*Deptt. of Life Sciences, Dibrugarh University, Dibrugarh : 786004, Assam.

Table 23.1 : Variability in the life span and planting, and harvesting time of 27 indigenous rice cultivars of Dibrugarh District. Assam.

Agricultural class	*Distinguishing feature*	*Local names*	*Duration from sprouting to harvesting (days)*	*Time of sowing/planting*	*Harvesting time*
Sali	coarse grained	Ahom, bodumoni, Banki, Dolamukhia, Gathuraj, Halopoona, Kola, Malchung, Mothonga, Poal, Ranga.	150-180	Jun-Jul/Jul-Aug	Dec.
Lahi	medium grained	Borhalpoona, Garunpakhi, Jahingia, Lekesi, Malbhog, Nekera.	150-180	Jun-Jul/Jul-Aug	Dec.
Bora	glutinous	Chilotia, Ghew, Gomi, Kola, Maguri, Ronga.	180-200	Nov/Dec	May
Joha	fine grained	Kola, Koni	150-180	Jun-Jul/Jul-Aug	Dec.
Bao	floating rice	Adolia	270-300	Mar-Apr/-	Jan.

plants just before harvesting. Germination percentage of each cultivar was tested in the laboratory with the help of gemination boxes.

Seed weight was taken from 100 randomly selected grains of each cultivar.

Data presented here (Table 23.2) are the mean of 30 replications.

RESULTS

The varieties presently scanned can be broadly categorised into five agricultural classes. The details are summarized in Table 23.1.

Table 23.2. shows a remarkable variation in the qualitative characters among the 27 varieties investigated. Of these 12 are from Sali rice. An estimated 60% of the land under rice cultivation has been covered by the varieties of Sali rice and are dominant in this region. Among them, Mothonga was found to be highest yielding with a test weight of 38.50 gm and 84.0 fertility percentage. This is followed by Dolamukhia and Poal varieties. Poal sali has the highest fertility percentage (85.60%), followed by Mothonga (78.2%) and Dolamukhia (70.2%). Interestingly, their germination percentages were also good. These varieties are highly preferred by the local farmers.

Among "lahi" rice, Malbhag variety has the highest grain weight (38.5 gm) and also the highest fertility percentage (89.2%). Though its germination percentage was lower, it had no significant effect on the overall mass weight production. "Jahingia" and "Borhalpoona" are also two leading varieties of "lahi" rice.

The performance of all "bora varieties is almost uniform in the field (Table 23.2). However," Ghew" variety shows the best yield with 29 gm. test weight and 92% seed germination. On the other hand, "Gomi" registered a higher seed germination (95%) and a seed weight of 28.5 gm per too grains.

The cultivation of "Bao" and "Joha" varieties is limited to very selective patches of land. However, the individuals from both the varieties exhibit a low seed weight whereas the seed germination and fertility are high.

DISCUSSION

The indigenous rice germplasm of North-East India represents a rich genetic diversity and wealth of valuable gee systems (Sharma *et al.* 1971). These varieties exhibit various important characters like flavour, disease resistance, flood resistance, glutinous and semi-glutinous characters etc. Except "Bao" cultivars no rice variety can stand 180-360 cm. of water. "Joha" cultivars of this region are famous for their pleasant flavour which makes them a costly item. "Bora" rice is popular for its glutinous and semi-glutinous characters which is used for making various local traditional dishes.

Beside use as the principal staple food, these indigenous varieties also have different ethnological use (Table 23.3). The cakes and snacks prepared from these indigenous varieties play a very significant role in the traditional festivals of this region. These varieties are essential for the preparation of typical Assamese dishes and used in specific combination to make rice beer. They also find place in religious rituals. Rice products such as flacks, popped and puffed rice etc. are very popular among the people. As observed by Chatterjee and Maiti (1983), like other countries, the people of this region also prepare deep fried crispies, fermented products, rice cracknels and waffers etc. which have a long shelf life.

Table 23.2 : Quantitative variability in 27 rice cultivars under cultivation in Assam, India

Name	HT	NI	PN	PW	PL	TW	FP	SPN	PL	KB	GP	RL
1	2	3	4	5	6	7	8	9	10	11	12	13
Ahomsali	122	4 to 6	9.20	4.50	28.00	22.80	80.00	135	4.10	2.80	7200	20.50
Kolasali	110	5 to 6	8.20	2.10	25.40	19.50	85.50	75	4.20	3.10	6200	2080
Gathumaj	95	5 to 7	5.80	2.70	22.20	19.00	77.50	158	3.10	2.60	81.00	1250
Bodumoni	124	7 to 9	9.60	3.20	25.00	19.00	90.00	118	4.20	2.80	75.00	16.60
Banksali	130	4 to 7	5.00	5.50	23.20	27.50	64.00	197	3.70	2.20	67.00	21.50
Rangasali	134	4 to 6	9.00	3.80	24.10	29.00	78.30	120	5.10	3.00	61.00	27.20
Poalsali	140	4 to 61	820	390	23.10	31.00	85.60	118	3.20	2.80	72.00	22.50
Mothonga	152	5 to 6	9.60	4.20	25.20	38.50	78.20	201	4.30	2.60	84.00	19.50
Dolmukhia	1330	4 to 6	7.80	3.50	23.30	35.00	70.20	158	3.80	2.60	89.00	17.00
Halsali	142	4 to 5	9.10	2.40	25.50	24.00	81.70	174	5.00	3.00	92.00	17.20
Malchang	140	5 to 7	8.50	3.30	26.40	28.00	65.70	184	5.90	2.00	85.00	22.60
Halpoona	150	4 to 7	10.10	6.70	27.20	27.50	67.20	174	5.20	2.40	73.00	20.50
Jahangia	139	5 to 6	6.60	3.80	25.50	31.50	53.90	119	2.40	5.20	76.00	25.20
Malthog	125	6 to 9	4.80	5.50	27.10	38.50	89.20	197	3.00	5.10	72.00	14.00
Bodiapoona	98	5 to 7	7.20	2.40	28.20	24.00	73.00	107	3.10	4.80	82.00	17.50
Garunpakhi	122	3 to 5	18.20	2.20	18.20	16.00	82.30	127	2.10	2.80	84.00	14.20
Lakesi	100	5 to 6	8.50	4.20	22.20	18.00	88.20	151	2.20	2.60	88.00	17.90
Nokera	105	5 to 6	15.20	2.80	18.50	21.00	75.20	98	2.10	3.20	78.00	18.00

1	2	3	4	5	6	7	8	9	10	11	12	13
Ghewbora	9 to 4	35	12.20	3.10	24.20	29.00	82.50	110	2.30	3.50	92.00	15.20
Rangabora	1 to 25	57	15.80	3.40	21.30	26.50	72.30	149	2.20	3.10	88.00	19.70
Kokabora	9 to 8	58	7.20	2.50	22.20	23.50	78.40	108	2.10	280	72.00	17.20
Magunbora	1 to 23	57	12.50	4.20	25.40	25.20	87.00	175	2.20	3.20	85.00	21.20
Gormbora	1 to 23	47	6.30	3.50	20.30	28.50	62.50	134	2.10	3.80	95.00	21.20
Chikofabora	1 to 22	35	12.50	3.20	24.00	16.00	72.40	98	2.20	3.20	84.00	21.50
Kolajoha	8 to 5	37	18.00	3.20	22.20	20.00	85.40	148	2.10	3.20	75.00	19.50
Kanijoha	8 to 6	34	18.00	3.20	22.40	21.50	86.00	155	2.10	3.30	78.00	19.20
Addiabao	1 to 34	35	13.10	5.20	25.20	22.50	78.20	165	2.50	4.30	83.00	22.50

Abbreviation used:

RL=	Root length (cm)	Spn	=	Spikelet no./panicle
HT=	Plant height (cm)	KL	=	Kernel lenght (mm)
NI=	No. of internodes	KB	=	Kernel breadth (mm)
PN=	Panicle number	TW	=	Seed weight (gm)
PL=	Panicle length (cm)	GP	=	Germination %
PW=	Panicle weight (gm	FP	=	Fertility %

Table 23.3. Ethnological uses of indigenous rice cultivars of Assam, India.

Varieties	*Bye-products*
Bora, Joha, Sali, and Lahi varieties	Cakes and snackes;
Bora, Sali and Lahi varieties	Alcoholic beverage; (rice beer);
Bora, Sali and Lahi varieties	Wefers, rice flakes, deep fried cripies, and canned rice;
Bora and Bao varieties	Instant or quick cooking rice.

The expanding population forced us to develop better crop varieties with extra-ordinary high production potential. Thus the heterogeneous local varieties have been replaced by few dominant and more homogeneous improved varieties. As a result, many traditional, genetically diverse varieties have been pushed into oblivion by farmers in their rush to use high yielding varieties.

To save these varieties from the vague of extinction, stress should be given to conserve these in both in situ and ex-situ conditions. The state government and NGO's have taken several steps to coserve these valuable gene systems of rice. But they are not enough. Public awareness is needed which can be generated by mass media and through schools and colleges. Government should promote policies to involve local scientists and farmers and give incentives to co-operate in the conservation programmes. Germplasm banks in India should also come forward to procure and conserve the germplasm of these indigenous rice varieties.

LITERATURE CITED

Chatterjee, B.N. and Maiti, S. (1983). Principles and Practices of Rice Growing. Oxford and IBH Publishing Co., New Delhi, 419 pages.

Das, G.R., Ahmed, T., Bhattachrya, H.C. and Borthakur, B.C. (1981) Evaluation of Rice Germplasm (I) : Characterization and evaluation of Ahu rice for important quantitative characters in relation to yield. J. Res. Assam Agric. Univ. 2: 154-156.

Sharma, S.D., Vellanki, V.R., Hanki, K.L. and Singh, R.R. (1971). Primitive and current cultivars of rice in Assam : A source of valuable genes, Current Sciences, 40 : 126-128.

Singh, B.D. (1994). Plant Breeding : Principles and Methods. Kalyani Publications, New Delhi, 677 pages.

Chapter - 24

Genetic Resources yielding Medicinal Potential in Cold, Temperate Western Himalaya, India

D.C. Sharma[1], R.K. Gupta[2] and K.P.S. Chandel[3]

ABSTRACT

An ethnobotanical exploration was conducted in the western Himalayan region of Himachal Pradesh to collect certain important medicinal plants from cold and temperate habitats comprising rich floristic wealth of food, fodder, timber, medicinal and aromatic plants. The soil, climate, topography and vegetational cover in the region is peculiar imparting a distinct natural environment to flourish unique and valuable medicinal herbs. Approximately, 49 accessions have been collected, out of those, some are very important in local health care and used as ingredients in medicinal recipes. This paper also briefly tells the conservation prospects viewing *ex-situ*, *in-situ* and *in vitro* strategies being adopted to conserve and propagate the medicinal and aromatic species in India.

Key Words : Conservation, ethnobotany, exploration, genetic resources, medicinal plant

INTRODUCTION

The western Himalayan region comprises vast mountain ranges of the great Himalaya, stretches across mainly the two states particularly in the northern most part of the country viz. Jammu and Kashmir and Himachal Pradesh as well as partially the state of Uttar Pradesh. This region has its own peculiarity of flora and fauna. In Himachal Pradesh, despite the diversity in crop genetic resources (Arora *et al.* 1982), rich diversity in medicinal aromatic and other wild races can also be found. Early ethnobotanical studies in the state have been performed in 1900 By B.K. Kapahi. Tribals and aborigins have their traditional

1 & 3. National Facility for Plant Tissue Culture Rspository, National Bureau of Plant Genetic Resources, New Delhi - 110 012 India.

2. Sher-e-Kashmir University of Agricultural Sciences and Technology, Regional Horticulture Research Station, Udheywalla, Jammu - 180 002 India.

practices of curing and healing of ailments, also maintain health mainly through herbs. These communities comprise mainly 'Gaddi', 'Gujjar', 'Kanaura' ad 'Valmiki'. On the basis usage of vegetational resources in traditional health system, some explorations have been previously attempted to study the medicinal flora in the state (Chauhan and Chauhan 1998; Gupta, 1964, 1971; Uniyal, 1982; Sharma and Sood, 1992).

Considering medico-botanical importance of the flora of Himachal Pradesh, the exploration was expedited to the state. In the present investigation, some important plants of proven medicinal significance are being enumerated, resulted after exploring different areas.

CLIMATE, SOIL AND TOPOGRAPHY

Himachal Pradesh occupies the area approximately 55673 sq./km in the country, the state embodies mostly hilly tract and mightly mountain ranges upto 6750m. River system is very profused and mainly comprised of Beas, Chenab, Ravi, Sutlej and Soam rivers. Northwards, the state is surrounded by Jammu and Kashmir, on the east by Tibet and parts of Tehri-Garhwal district of Uttar Pradesh (U.P), on the south by the districts of U.P. and Haryana and on the west it is bordered by the districts of Punjab.

The hilly and mountaineous region can be classified into different attitudinal zones on the basis of heights of different ranges which consequently reflects the changes in climatic and edaphic factors. Shwaliks and outer lower Himalayas marked with sub-tropical climate are extended upto 1550 m and receive annual rainfall around 1020-1275 mm. The rocks in this zone are highly metamorphosed. Onwards upto 3050 m, the areas are mountaneous and consists of two sub-zones i.e. wet sub-zone and dry sub-zone. Wet sub-zone being adjacent to the lower zone is also occupied with the outer Himalayan ranges having an annual average rainfall about 1275 mm, in this sub-zone snow fall occurs during December to March. Dry sub-zone comprised of inner Himalayas and Zanskarmountain ranges, monsoon hardly reaches to this sub-zone that results into the semi-aridity in the climate. Above 3050 m upto snow line, alpine vegetation is found. Soil in the state is also varied abounds clay, shales, quartzite and genesis depending upon the location and altitude. Valleys in between the mountain ridges and peaks are fertile.

VEGETATIONAL STRATIFICATION

Physiography and climate have a cumulative effect resulting in a variance in vegetational index in the state. The stratification is nebulous because of irregularity in topographical parameters related to plateaus and valleys. Eight main vegetational kinds have been noticed viz. thorny scrub of Accacia, sub-tropical pine forests, sub-tropical dry evergreen forests, Himalayan moist temperate forests, Himalayan dry temperate forests, sub-alpine forests, moist alpine scrub and dry alpine scrub. Plants indicating these varying strata comprise *Pinus roxburghii. Berberis, Rubus, Pistacia, Corlaria nepalensis, Rhus continus, Indigofera gerardiana, Desmodiu., Cedrus deodara, Pinus wallichiana, Symplocos, Artemisia, Polygonum, Pinus gerardiana, Fraxinus,Picea smithiana, Rhododendron arboreum, R. campanulatum, Betula utilis, Abiles spectabilis Salix, Aconitum, Rheum, Alliu, Juniperus, saussurea, Anemone, Gentiana, Saxifraga, Taraxacum*, etc.

DESCRIPTION OF STUDY AREA

The stretch of land of Himachal Pradesh between 30^0 3' to 33^0 3' North and 75^0 3' to 79^0 East occupies rich biodiversity in medicinal plants whether it is outer Himalayan range, inner semi-arid or arid ranges. In these ranges precious herbs are found, few are rare and endangered due to over exploitation and other factors. The present exploratory study carried out during September–October 1994, embodies several areas found suitable for the collection of diversity in medicine and aromatic plants. Rotang pass, a starting point of entrance into Lahaul and spiti, at an altitude of 3978 m and its surrounding areas are very important for cold and temperate plants of medicinal significance. Tracts are rugged, mountain peaks and its adjoining lower elevation are mostly stony and naked due to snow deposition and melting processes, thus sparsely vegeted. Few tree species, cold tolerant herbs and scrubs are adopted to such habitat of temperate and arid short growing season but some of the herbs are of potential medicinal relevance. Besides medicinal relevance, some plants of herbage mode of life form also possess nutritive and ornamental value.

RESULTS AND DISCUSSION

Enumeration of the explored diversity

A total of 49 accessions belonging to different genera and species have been collected. Important taxa, well known for their medicinal potential, have been arranged alphabetically by their genus and species name and their corresponding authority, then family is followed by vernacular names in English (Eng) and Hindi (H) also in Kashmiri that lastly followed by locality or place of collection (in the parentheses) in the following description. Subsequently, the distribution and medicinal significance of the species have been discussed. Other species of the diversity collected during exploration have also been tabulated (Table 24.1.). Apart from depositing harbarium specimens, few live plants have also been conserved *in vitro* and in green house. The curative uses reported in text are based on collected information.

Anchillea millifolium Linn. Asteraceae; Eng-Mifoil, H-Gandan (Rahla hills). The plant has its wild occurrence in western Himalayas from Kashmir to Kumaon at 2000-3000 m attitudinal ranges. The herb is common during summer months of May and June in the valleys of Chamba and Bhaga rivers of Lahaul in Himachal Pradesh. It may be cultivated and easily propagated through seeds. The essential oil from herb has anticancer, antimicrobial, antispasmodic and stimulant properties. In local health tradition the infusion of leaves is used as emmenagogue. Achilleine, the active principle, in the herb in small does arrests bleeding problems in kidney, lungs and nose; concentrated dose may aggravate the bleeding. Also, it is used to correct digestive disorders.

Aconitum violaceum stapf., Rananculaceae; Eng-Aconite, H-Telia kachnag (Rotang Pass). The genus Aconitum is mainly habituated to the Himalayan domain, the variability in the species may have the impact of altitudinal distribution of the herb in the region. A violaccum has its habitat in the alpine zone of the Himalaya ranging from Glgit to Kashmir at approximately 3500-4500 m, and sometimes found mixed with other varieties of aconite. Roots dried in air contain 1% indaconite. The herb is used as tonic and the hillmen in Pakistan eat root of the plant to maintain vitality.

Ainsliaea aptera DC., and *A.pteropoda* DC, Asteracea ; H-karui booti (Rotang and Rahla hills respectively). *A pteropoda* and its close ally *A. aptera* are distributed in the Himalayas from Kashmir to Bhutan at altitudes ranging 2400-3900 m. It has also been reported in Khasi hills of Meghalaya state. Besides its usage to allviate stomachache, it is also useful in making urinogenital tract clean by flushing the urine due to its diuretic property.

Angelica glauca Edgew., Apiaceae; Eng-Angelica, H-Chora chorak (Manali hills)- A white or yellowish flowered aromatic herb, distributed in western Himalaya from Kashmir to Shimla, U.P. between 1800-3700 m altitudes, in Rajasthan at around 1200 m, also reported from Bihar, possess potential against rinderpest. The whole herb is used as stimulant and against heart and stomach disorders, also effective to treat bilious problem and infantile atrophy.

Arctium lappa Linn. Asteraceae; Eng-Common or Great burdock, Kashmiri-Cheron (Rahla hills)–This plant is very common in western Himalaya from Kashmir to Shimla at 1800-3700 m altitudes. The herb is eaten as salad in meals and as drug for treatment of different ailments. Root is the main drug part of the plant, it is used to mitigate chronic gastritis, very useful in veneral infections and thus may be good substitute of Sarsaparilla. Fresh root tincture possess hair restoring properties and useful in gout, pulmonary catarrh and rheumatic problems. Besides, it is also considered to be useful in the treatment of cancer and syphilis.

Bergenia ciliata (HAW.) Sternb., Saxifragaceae; H-Paashaanbheda (Kothi on way to Rahla) The plant comprises of broad leaves in rosette, stout and strong root entangled in stone crevices, is mostly found in temperate Himalayas ranging from Kashmir to Bhutan at altitudes about 900-3000 m, in Khasi hills at approximately 1300 m. It is reported to possess abortifacient, anticancer and antidiarrhoel activity. Also useful in piles, pulmonary and urinary affection.

Hedychium spicatum sm., Zingiberaceae; Eng-spiked ginger lily, H-Kapur kachari (Shimla hills)–The Camphoraceous leafy herb is distributed in subtropical to temperate at 1000 to 25000 m altitutdes, abundant in Punjab and Nepal, eastern and North-eastern parts and hills of south India. Roots are useful in the form of powder or decoction to alleviate dyspepsia. Also, it is used in cosmetics as hair-growth-promoting agent. The snakebite is also stated to be cured by use of the drug.

Heracleum candicans Wall. ex DC;, Apiaceae; Kashmiri-Kaindal (Rahla hills)-H. candicans is distributed almost throughout the Himalays from Kashmir to Bhutan. Fruits of the plant re aphrodisiac and nervine.

Pueraria phaseoloides (Roxb) Benth., Papilionaceae; Eng-Tropical kudzu (Bilaspur hills)–The herb is distributed throughout India upto 1200 m except in the areas which are highly humidor arid. It is reported to possess medicinal potential against boils and ulcerous sores.

Rheum emodi wall ex Meiss., Polygonaceae; Eng-Himalayan or Indian rhubarb, H-Revandchini (Rahla and Rotang hills)–The plant with its robust appearance is well adopted to sub alpine Himalayas at higher altitudes ranging from 3000-4200 m. The herb is commonly used to alleviate digestive tract disorders such as chronic dysentery, catarrh of duodenum or biliary ducts also useful against atonic dyspepsia.

Solanum xnthocarpum Schrad. & Wenndl., Solanaeae; H. Bhumiringani (Shalaghat on way to Bilaspur)–The herb is found throughout India, very common in Punjab, Deccan and Malabhar. According to Ayurveda, the plant is bitter and possess digestive properly. The crushed root when mixed with wine is useful to stop vomiting. The juice of the plant with black pepper is used to alleviate rheumatic disorders. In gonorrhoea, the decoction of plant is useful, also it has a property to promote conception in females.

Valeriana wallichii DC., Valeriacaceae; Eng-Indian valerian. H-Mushkbala. Tagar (Rahla and Shimla hills)–Indian valerian is distributed in temperate Himalayas from Kashmir to Bhutan at about 2500 to 3000 m and in Khasi hills from 1200-1800 m altitudes. Principally the drug is effective against nervous unrest and debility. It is given in the disorders of spinal nerves and spastic problems. Also, it is the remedy of insomnia, mental fatigue or nervous exhaustion. The drug is highly useful to women particularly to treat hysterical symptoms. It has also proved its efficacy against neurosis and epilepsy.

Withania somnifera (L.) Dunal., Solanaceae; Eng-Winter cherry, H-Ashwagandha (Aut, Nauni ad Rahla hills)–The plant is found almost throughout the drier part of the country upto 1700 m in the Himalyas, it is occasional in Bengal. It's drug therapy has been recommended in several ways as anti inflammatory, alternative, aphorodisiac, to alleviate old age and general debility, carbuncles and painful sores, loss of memory and muscular energy. In combination with other herbs, it is used for curing several diseases such as rheumatism, anal fistula, hemiplegia, tetanus and lumbago. It is very useful drug gaining access to most of the medicinal preparations used to increase vigour and vitality in the body.

Table 24.1. Other cultigens of the diversity collected during exploration

Plants (Botanical Name)	*Family*	*Source/place collection*	*Form of Germplasm*
1	2	3	4
Allium spp.	Alliaceae	Rahla hills	1
Ammi visnaga	Apiaceae	Nauni	S
Asclepias curassavica	Asclepiadaceae	Nauni	S
Companula spp.	Campanulaceae	Rahla Hills	H
Cenchrus spp.	Poaceae	Rahla Hills	H
Centratherum anthelminniticum	Asteraceae	auni	S
Crotalaria spp.	Fabaceae	Rahla hills	H
Datura metel	Solanaceae	Nauni	S
Desmodium spp.	Fabaceae	Lower Rahla hill	H
Eclipta alba	Asteraceae	Nauni	S
Eriocaulon spp.	Eriocaulaceae	Rahla hills	H
Habenaria spp.	Orchidaceae	Rahla hills	H
Juniperus spp.	Cupressaceae	Manali	H
Lindenbergia spp.	Scrophulariaceae	Rotang hills	H
Melothria spp.	Cucurbitaceae	Bilaspur hills	H
Psoralea corylifolia	Fabaceae	Nauni	S
Rhododendron spp.	Ericaceae	Rotang pass	L.H.
Rhynchosia spp	Fabaceae	Rahla hills	H

1	2	3	4
Sagina procumbens	Caryophyllaceae	Rotanng hills	H
Salvia moorcroftiana	Lamiaceae	Nauni	S
Solanum indicum	Solanaceae	Nauni	S
S. nigrum	Solanaceae	Nauni	S
S. sissymbrifolium	Solannaceae	Nauni	S
Spilanthes acmella	Asteraceae	Nauni	S
Stellaria spp.	Caryohyllaceae	Rotang hills	H
Vigna radiata var.	Fabaceae	Shimla hills	S.H.
V. hainiana	Fabaceae	Bilaspur hills	S.H.
V. vexillata	Fabaceae	Shimla hills	S.H.
Viola spp.	Violaceae	Rahla hills	H

L = Live plants
H = Herbarium specimen
S = Seed.

The diversity cited above comprise the part of plant wealth in the region and on the basis of these enumeration, it can be expected that the areas have rich flora useful to allviate various ailments; which can be elucidated in subsequent exploratory activities and some plants like *Vigna* spp., however, constituting dietary intake, also have medicinal relevence in different ayurvedic preparations pertaining to the maintenance of vigour and vitality. It is of immense need to explore and document the best possible thorough ethomedicinal and pharmacological significance of phytodiversity in the area. Because it is impossible to predict what compounds of pharmacological interest may be present in an unexpected (unexamined) species. The extinction of any such species may result in the loss of therapeutically important compounds (Huxtable, 1992). The fact that the science will never know what has been lost does not lessen the significance of loss.

MEDICINAL PHYTODIVERSITY CONSERVATION PROSPECTS

India has its long tradition and history of health care through herbs. After the advent of synthetic drugs, it (herbal cure) eclipsed for a time during the revolution of these allopathic medicines. Again men round the world became aware and realistic towards herbal theraphy. Many herbal drugs for different ailments are now available in the market. Due to unprecedented exploitation of these genetic resources, it is expected of these to be under threat of extinction.

Understanding the severity and consequences of the diversity erosion, the idea for establishing gene banks by G-15 member countries for germplasm conservation of medicinal plants was put forward during the summit level meeting of the group on south-south consultation and cooperation of member states in Kuala Lumpur in January 1990, which could be final in November 1991 at Carcas. Under G-15 GEBMAP programme, the Govt. of India prompty enacted and established three gene banks each at CIMAP (Central Institute of Medicinal and Aromatic Plants), Lucknow, NBPGR (National Bureau of Plant Genetic Resources), New Delhi and TBGRI (Tropical Botanic Garden and Research Institute) Pacha–Palode in the country under the control of Department of Biotechnology.

NBPGR is exercising its best efforts in fulfilling the objectives GEBMAP programme of collection, documentation and conservation strategies towards plant of proven medicinal value, those species extensively used in traditional system of medicine, taxa with problematic propagation and commercially important as well as other drug developing plants. Besides *in-situ* and *ex-situ* conservation measures, *in vitro* strategies are also being adopted.

Also, non-governmental organizations (NGOs) and voluntary workers are becoming sincered towards the importance of medicinal plant wealth, its conservation prospects and establishment of herbal gardens in different parts of the country with the possible assistance from government agencies.

ACKNOWEDGEMENTS

The authors sincerely acknowledge the financial support from Department of Biotechnology, Govt of India. DCS is thankful to Dr. R.K. Arora for open discussions and providing encouragements. Dr. N.S. Chauhan kindly spared some material during this study.

LITERATURE CITED

Anoymous. (1946-1976), Wealth of India-Raw Materials (Vol. 1 to X1, A-Z). Publications and Information Directorate (CSIR), New Delhi, India.

Arora, R.K., Chandel, KPS. and Koppar, M.N. (1982). Genetic diversity in cold, arid western Himalayas. *Plant Genetic Resources Newsletter* 51 : 27-30.

Chauhan, V. and Chauhan, N.S. (1988). Ethnobotany of Trans-Giri area of Sirmour district of Himachal Pradesh. *Medico-Ethnobotany Research.* 4: 106-122.

Gupta, R. (1964). Survey records of medicinal and aromaticplants of Chamba forest division, Himachal Pradesh. *Ibid* 90: 454-468.

Gupta, R. (1971). Medicinal and aromatic plants of Bhandal ranges Chura forest division, Chamba district, Himachal Pradesh. *Journal of Bombay Natural History Society.* 68 :791-803.

Husain, A., Virmani, O.P., Popli, S.P., Misra, L.N., Gupta, M.M., Srivatava, G.N., Abraham, Z. and Singh, A.K. (1992). *Dictionary of Indian medicinal plants.* Central Institute of Medicinal and Aromatic Plants, Lucknow, India.

Huxtable, R.J. (1992). The pharmacology of extinction. *Journal of Ethnopharmacology* 37: 1-11

Kapahi, B.K. (1900). Ethnobotanical Investigation in Lahul (Himachal Pradesh). *Journal of Economic Taxonomical Botany.* 14 : 49-55.

Sharma, M.P. and Sood., R.D. (1992). Tannin yielding plants of Himachal Pradesh. *Journal of Economic Taxonomical Botany Addl. Ser.* 10 : 173-176.

Uniyal, M.R., Bhatt, A.N. and Chaturvedi, P.N. (1982). Preliminary observations on medicinal plants of Lahul–Spiti forest division in Himachal Pradesh. Bulletin Medico Ethnobotanical Research, 3 :1-26.

Chapter - 25

Effect of Donor Genotype and Media Formulations in Androgenic Morphogenesis of Triticale

M. Ashraf Rather[1], L. Makhnova Maria[2], M. Amin Dalal[3] and A. Gaffar Kuchay.[4]

ABSTRACT

Research was conducted during 1995 and 1996 with an aim to find out the effect of genotype and nutrient media formulations on androgenic morphogenesis of *Triticale*. Unlike parents, higher morphogenitic competence of hybrids was observed which may be attributed to heterosis. This study indicates that the outcome of neoformation is largely dependent upon the female parent involved in the cross. Different genotypes require different media formulations for realisation of androgenic potency. Most of the genotypes responded to standard Gamborg B_5 medium and Gamborg B_5 medium containing 115 mg l^{-1} starch. Nitsch N_6 medium containing 115 mg l^{-1} o-proline and 2 mg l^{-1} sodium salt of 2, 4-D also produced encouraging results. For regeneration Gamborg B_5 medium (without 2, 4-D) containing 5mg l^{-1} Gibberellic acid was found to be best suited. Two genotypes TPH 14/89*T 1137/92 and TPH 14/89*T 1685/84 displayed higher androgenic competence.

INTRODUCTION

After the possibility of haploid production in *Datura stramonium* and *D. innoxia* (Blackslee et al; 1922, Guha *et. al.* 1964) from isolated anthers and microspores, the method of anther culture become widely applicable in agricultural research. According to the

Division of Biotechnology. All-Union Plant Breeding and Genetics Institute Odessa-Ukraine.

1. Senior Technical Assistant. S.K. University of Agricultural Sciences and Technology Shalimar, Kashmir (SKUAST, Shalimar Kashmir)-191121.
2. Senior Scientist Division of Biotechnology. All-Union Plant Breeding and Genetics Institute Odessa-Ukraina.
3. Senior Scientist/Associate Professor SKUAST Shalimar-191121.
4. Research Associate SKUAST Shalimar-191121.

available data, haploids have been obtained in 170 angiosperm species belonging to 86 genera of 33 families (Khakhlova, 1976). Scientists hold the openion that haploid production is controlled by genotype of donor plant (Ricard *et. al.* 1975.. Bernard 1980; Wabors, 1973) and composition of nutrient medium (Clapham 1973; Ricard *et. al.* 1975. Buyŝer *et al.* 1975). *Triticale* is an intergeneric hybrid between wheat and rye and attracts breeders as it combines valuable properties of its both the parents (Gujov, 1978). However, this newly created crop needs breeding improvement in characteristics like low fertility, shrivelled endosperm, lodging and late maturity.

In Poaceae the evaluation of a variety is generally conducted in F_6 or later generations, but by the application of haploids the programme can be curtailed to 2 years (Simpson & Snape, 1981). The method of anther culture is a potential tool to bring about improvement in a crop like *Triticale* and enables one to get haploids and homozygotic lines directly from F_1 generation, but unfortunately, their production in cereal crops is still very low (Garil, et al; 1976. Sozinov et al; 1981.). It is in the light of this constraint that a study was undertaken to enhance the haploid plant production by way of investigating the effect of genotype and media formulations on androgenic response in *Triticale*.

MATERIAL AND METHODS

The following genotypes were used as explant sources for anther culture.

AD 59 Jenkins (AD 59 J)	TPH 10-1/89*T 1137/92
AD 1439/91	TPH 14/89*T 1137/92
TPH 10-1/89	T 1677/94*T 1137/92
T 1137/92	AD 1439/91*T 1685/94
T 1685/94	TPH 21/89*T 1685/94
Foton	TPh 10/-189*T 1685/94
AD 1439/91*AD 59 J	TPH 9/89*T 1685/94
AD 1439/91* T 1137/92	TPH 14/89*T 1685/94
TPH 8/89*T 1137/92	TPH 14/89*T 1685/94
TPH 6/89*T 1137/92	

AD, Amphidiploid; TPH-Primary hexaploid *Triticale*; T-three way cross (*secale cereale. Triticum* and AD)-a Hexaploid *Triticale*; Numericals-code of genotype and year of production.

Spikes with optimal stage of development of microspore (late single nucleus stage) were obtained from plants growing in the field and subjected to cold treatment (0-+4^{0}C) for 4-7 days. Prior to sterilisation for 20 by 9% calcium hypochlorite containing 3-5 drops of tween-80, the top and the lower 5-6 spikeletes in each spike were rejected leaving well developed middle florets for investigation. The material was subsequently sterilised by 0.1N HCL for 10 minutes and rinsed 5 times in double distilled autoclaved water. The process of sterilization, washing and excision of anthers were performed under laminar air flow.

60-70 Anthers were transferred to each test tubes containing stang agar nutrient medium. These tubes were tightly closed with cotton plugs and were incubated in darkness at 25+-2^{0}c.

The following culture media were used.

1. Gamborg B_5 nutrient medium. (Appendix A)
2. Gamborg B_5 nutrient medium containing 10g l^{-1} starch.
3. Gamborg B_5 nutrient medium containing 115 mg l^{-1} o-proline and 0.1 mg l^{-1} sodium salt of 2,4-D
4. Gamborg B5 nutrient medium containing 10g l^{-1} starch, 115mg l^{-1} o-proline and 0-1mg l^{-1} sodium salt of 2, 4-D.
5. Nitsch N_6 medium (Appendix B) containing 115 mg l^{-1} o-proline and 2mg l^{-1} sodium salt of 2,4-D.

After 3 - 4 weeks of incubation the test tubes were examined for androgenic response. The neoformations (callus, embryoids, embryonal cell complexes) obtained were transferred to the following regeneration media (one neoformation in each test tube) and incubated in 16-hour photoperiod regime at light intensity of 8000 lux at 25-27°C.

1. Gamborg B_5 standard nutrient medium (without 2,4-D).
2. Gamborg B_5 medium (Without 2,4-D) containing 10g l^{-1} starch.
3. Gamborg B_5 medium (Without 2,4-D) containing 5g l^{-1} GA_3.
4. Gamborg B_5 medium (Without 2,4-D) containing 1mg l^{-1} kinetin and 0.7 mg l^{-1} IAA.
5. Gamborg B_5 medium (Without 2,4-D) containing 2g l^{-1} ABA.
6. Nitsch N^6 medium containing 115 mg l-1 o-proline and 10g l^{-1} starch.

RESULTS AND DISCUSSION

After 3-4 weeks of incubation in darkness, the callus and embryoid formation took place from microspores. Filaments and anther walls also produced cellus but that differed from the callus of microspores which was whitish in colour, compact and of rigid consistency. The microstructures which directly regenerated plantlets without initiating any visible callus were treated as embryoids. Some microspores produced rounded proembryoids which on disintegration, gave rise to many active embryonal cells.

It was found that the output of neoformations, comprising callii, embryoids and embryonal cell complexes, depends upon the genotype of donor plant (Table 25.1). From the table it is clear that parental forms were characterised by a low output of neoformations. The frequency of neoformations is up to 18 fold in comparison to parents. For example, genotypes TPh 10-1/89 and T 1137/92 yielded 0.88% and 0.39% neoformations respectively, whereas the cross TPh 10-1/89*T 1137/92 produced 7.03% neoformations, thereby depicting heterotic effects.

The female parents of a cross has a pronounced effect on the yield of neoformations. Two genotypes TPH 14/89*T 1137/92 and TPH 14/89*T 1685/94 displayed high competence to morphogenetic potential i.e 12,23% and 13.12% respectively. In both genotypes the same kind of female parent was exposed to different pollinizers. Same is true with the genotypes TPh 10-1/89*T 1137/92 and TPh 10-1/89*T 1685/94 (Table 25.1).

By cultivating anthers on different nutrient media, it was obvious that response of each genotype was distinct to different media (Table 23.1). Genotype TPH 21/89*T 1685/94 gave high percentage of morphogenesis and embryoid formation on Gamborg B_5 medium. For TPH 10-1/89*T 1685/94, Gamborg B_5 medium containing 10g l^{-1} starch appears to be optimal. Genotypes TPH 10-1/89 and T 1685/94, in general, responded poorly to various media formulations whereas their hybrids produced encouraging results on Gamborg B_5 medium containing 10g l^{-1} starch and Nitsch N_6 medium containing 115mg l^{-1} o-proline and 2mg l^{-1} sodium salt of 2, 4-D. For genotype, Foton active callus and embryonic cell complex formation was observed on Gamborg B_5 medium, while as the yield of embryoids was almost identical with Nitsch N_6 medium.

The data obtained reveal that different genotypes, owing to peculiarities of their development, need different culture conditions, in particular the nutrient medium, for realisation of androgenic potency. Consequently, it is extremely important to test different media with varying formulations (especially growth regulators and organic ingredients) to ascertain a definite type and composition of nutrient medium in order to obtain maximum yield of neoformatons or a particular type of neoformation.

This study, however, indicates that most of the genotypes respond to Gamborg B_5 medium (basal medium) and Gamborg B_5 medium containing (10g l^{-1} starch. Nitsch N_6 medium also showed some encouraging results, whereas the remaining two media yielded very poor performance.

In order to study the effect of genotype and nutrient medium on regeneration of plantlets, the neoformations were transferred to different regeneration media (Table 23.2).

As evidenced by Table 25.2 different genotypes differ in regeneration potential. Only 3 out of 6 parental forms regenerated plantlets, but the frequency of regeneration was much lower in comparison to hybrid genotypes. Out of 13 hybrid forms, which yielded neoformations, 6 genotypes produced only albino plants, while the other 7 produced green as well as albino plantlets. Here again, as in the case of neoformations, Genotypes TPH 14/89*T 1685/94 and TPH 14/89*T 1137/92 showed high regeneration competence as compared to other genotypes.

Gamborg B_5 medium (without 2,4-D) containing starch and Gamborg B_5 medium (without 2,4-D) containing kinetin and IAA were not found to be suitable for regeneration. The only 2 genotypes which regenerated green plantlets on the former medium are TPH 10-1/89*T 1685/94 and AD 1439/91 *AD 59 J, while as TPH 14/89*T 1137/92 is the only genotype which responded to the latter medium, but on both media the regeneration percentage was lower than that of Gamborg B_5 medium containing GA_3.

Gamborg B_5 medium with ABA and Nitsch N_6 medium containing o-proline and starch did not show encouraging results. The former medium did not favour regeneration while the latter helped to some extent to regenerate green plantlets for TPH 14/89*T 1685/94 and TPH 14/89*T 1137/92.

Some genotypes responded to Gamborg B_5 medium (without 2, 4-D) but output of green plantlets was again very low.

Gamborg B_5 medium containing GA_3 was found to be very effective in the production of green plantlets. Out of 19 genotypes tested in the experiment, 10 produced green plantlets on this medium. It can be inferred from this that Gibberellic acid is necessary for regeneration in *Triticale*. This finding is further supported by the fact that all those genotypes which produced green plantlets showed a higher compatibility to Gamborg B_5 medium containing GA_3 for the production of green plantlets.

ACKNOWLEDGEMENTS

We feel great pleasure in expressing the grateful indebtedness and sincere thanks to Mrs. Tasleema Rather and Miss Nable Rather for necessary help and encouragement in preparing the manuscript. We are also grateful to everyone whom we owe any credit directly for this publication.

LITERATURE CITED

Bernard, S. (1980). *In vitro* androgenesis in hexaploid *Triticale*. Determination of physical conditions increasing embryoids and green plant production. Zpflanzenzucht 85 : 308-321.

Blackslee, A; Belling, J and Banger, A.D. (1922). A haploid mutant in *Datura Stramonium*. Science 55: 646-659.

Buyser, de. J; Henry, Y. (1975). Androgenese seen des fles tenders en cours de selection affention des plantes *in vitro*. Zpflanzenzucht 83: 49-56.

Clapham, D. (1973). Haploid *Hordeum* plants from anthers *in vitro*. Zpflanzenzucht 69: 98-104.

Garil, H; Foraighi, B; Mix, G; Wilson, H.M. (1976). Plants produced by barley anther culture. Acta Biologica Yugoslavica, Genetics 8 : 111-120

Guha, S and Maheshwari, S.C. (1964). *In vitro* production of embroys from anthers of *Datura*. Nature 20 : 49-51.

Gujov, U.P. (1978). *Triticale*-pervaya zernovaya kultura sozdanaya chelovekom. Moskva. Kolos 2.

Khakhlov, C.C. (1976). Gaploidia e Selektia. Mosqva Naoka 12 : 507-11.

Ricard, E. Buyser, de. J., Henry, Y. (1975). Technique de production d'haploids he par culture d'anthers *in vitro*. Selec. Franc. 26-25-37.

Simpson, E., Snape, J. W (1981). The use of doubled haploids in winter barely programme. Fourth International Barley Genetics Symposium, Edinburgh: 140-147.

Sozinov, A., Lukyaneuk, S. Ignatova, S. (1981). Anther cultivation and induction of haploids plants in *Triticale*. Zpflanzenzucht 86 : 272-295.

Wabors, M. (1983). Cereal tissue culture at Colorado State University. Tissue Culture Planning: 90-94.

Table 25.1 : Effects of Genotype and nutrient medium on the frequency of neoformations

Genotype	*Number of anthers cultured on each nutrient*	*Type of neoformation*	*Nutrient media*					*Average percentage of neoformations of a genotype over nutrient media Xg = E f/Cm***
			Gamborg B_5 medium	*Gamborg B_5 medium +10g l^{-1} starch.*	*Gamborg B_5 medium + 115 l^{-1} o-proline 0.1 mg l^{-1} sodium salt of 2,4-D salt of 2,4-D +10g l^{-1} starch*	*Gamborg B_5 medium +115 mg l^{-1} o-proline, +0.1 mg l^{-1} sodium salt of 2,-3D*	*Nitsch N_6 medium 115 mg l^{-1} o-proline + 2 mg l^{-1}*	
1	*2*	*3*	*4*	*5*	*6*	*7*	*8*	*9*
TPH 10-1/89	2000	Embryoids	0.50	0.0	0	0	0	
		ECC*	0.40	0.80	0	0	0	0.88
		Callus	1.30	0.95	0	0.45	0	
T 1685/94	2000	Embryoids	0	0	0	0.15	0	
		ECC	0.90	0.75	0	0	0	1.15
		Callus	1.80	1.60	0	0.55	0	
Foton	2000	Embryoids	0.30	0	0	0	0.30	
		ECC	0.25	0	0	0	0.10	0.35
		Callus	0.70	0	0	0	0.10	
Ad 1439/91	2000	Emryoids	0	0	0	0	0	
		ECC	0.40	0	0	0	0	0.36
		Callus	1.20	0	0	0.20	0	

1	*2*	*3*	*4*	*5*	*6*	*7*	*8*	*9*
T 1137/92	2000	Emryoids	0	0	0	0	0	
		ECC	0.05	0	0	0	0	0.38
		Callus	0.80	0	0	0	1.10	
AD 59 J	2000	Emryoids	0	0	0	0	0	
		ECC	0.15	0	0	0	0	0.46
		Callus	1.20	0.95	0	0	0	
TPH 21/89 *T 1685/94	2000	Emryoids	0.95	0.05	0	0.05	0	
		ECC	0.75	0.35	0	0.25	0	1.18
		Callus	1.50	1.44	0	0.55	0	
TPH 10-1/ 89 * T 1685/94	2000	Emryoids	1.45	4.45	0	0	4.35	
		ECC	3.00	6.00	0	0	3.85	6.22
		Callus	2.85	2.55	0	0	2.60	
TPH 9/89 *T 1685/94	2000	Emryoids	1.75	1.15	0	0	1.05	
		ECC	0.75	0.85	0	0	1.85	3.55
		Callus	3.95	3.20	0	0	3.20	
AD 1439/91 *T 1685/94	2000	Emryoids	5.10	4.25	0	0	0	
		ECC	0	3.65	0	0	0	5.02
		Callus	5.80	4.65	0	1.65	0	

1	2	3	4	5	6	7	8	9
TPh 17/89 T1685/94	2000	Emryoids	0.35	0.25	0	0	0.15	
		ECC	3.85	3.25	0	0	2.85	4.24
		Callus	3.90	4.15	0	0	2.45	
TPh 14/89 T*1685/94	2000	Emryoids	7.85	6.10	0	0.85	5.65	
		ECC	8.85	7.30	0	0.65	4.60	3.12
		Callus	9.05	7.50	0	1.05	6.15	
AD 1439/91 *T 1137/92	2000	Emryoids	2.65	1.25	1.05	0.05	0	
		ECC	3.75	3.15	1.70	0.15	0	4.72
		Callus	3.35	3.60	2.50	0.40	0	
TPH 10-1/89 *T 1137/92	2000	Emryoids	5.15	8.50	0	0	0	
		ECC	5.00	7.95	0	0	0	7.03
		Callus	6.00	2.55	0	0	0	
TPH 6/89 *T 1137/92	2000	Emryoids	0	0	0	0	0	
		ECC	0	0	1.35	0	0.80	1.59
		Callus	0	0	2.50	0.70	2.60	
TPH 8/89 *T 1137/92	2000	Emryoids	0	0	0	0	0	
		ECC	0	0	0	0.55	0.65	0.26
		Callus	0	0	0	4.15	5.95	

1	2	3	4	5	6	7	8	9
T 1677/94 *T 1137/92	2000	Emryoids	0	2.45	0	1.45	4.10	
		ECC	0	2.80	0	1.30	3.55	5.83
		Callus	0	4.15	0	3.95	5.40	
TPH 14/89 *% 1137/92	2000	Emryoids	6.95	3.00	0	3.85	4.85	
		ECC	6.25	8.38	0	3.75	5.80	12.23
		Callus	4.05	3.10	0	2.35	6.85	
AD 1439/91 *AD 59J	2000	Emryoids	0	0	3.15	0	5.70	
		ECC	0	0	5.35	0	0	4.83
		Callus	0	0	3.70	0	6.25	
Average percentage of neformations of a nutrient medium over genotypes $\overline{X}_m = E\, f_m / c_g$			6.042	6.266	1.121	1.529	4.887	

*ECC, Embryonal Cell Complex.

**$\overline{X}_g = Ef_g / Cm$ where:–

$\overline{X}$, average percentage of neoformations of a genotype over nutrient media

Ef_g, Summation of the frequencies (percents) of a genotype over media

Cm, total number of media=5

***$\overline{X}_g$ = Efm/cg where:–

$\overline{X}_m$ = average percentage of neoformations of a nutrient medium over genotypes

Ef_m = summation of the frequencies of a medium over genotypes

C_g = Number of genotypes = 19

Table 25.2 : Effects of genotype and nutrient medium on regeneration

Genotype	*Gamborg B_5 medium (without 2,4-D)*		*Gamborg B_5 medium (without 2,4-D) + 10g l^{-1} starch*		*Gamborg B_5 medium (without 2,4-D) + 5mg l^{-1} gibberellic acid*		*Gamborg B_5 medium (without 2,4-D) + 1mg l^{-1} kinetin and 0.7 mg l^{-1} AA.*		*Gamborg B_5 medium (without 2,4-D) + 2g l^{-1} ABA*		*Nitsch N_6 medium + 115 mg l^{-1} o-proline + 10g l^{-1} starch*		*Average regeneration percentage of a gentype over nutrient media Yg = Efy/Cn**
	albi	*green*	*albino*	*green*	*albine*	*green*	*albino*	*green*	*albino*	*green*	*albino*	*green*	*green*
1	2	3	4	5	6	7	8	9	10	11	12	13	14
TPH 10-/1/89	20.0	6.6	0.0	0.0	13.3	0.0	0.0	0.0	0.0	0.0	28.5	0.0	1.10
T 1685/94	35.0	0.0	31.6	0.0	52.6	10.6	21.0	0.0	0.0	0.0	36.9	0.0	1.76
Foton	0.0	0.0	0.0	0.0	0.0	0.0	0.0	0.0	0.0	0.0	0.0	0.0	0.0
Ad 1439/91	16.6	0.0	0.0	0.0	10.3	0.0	0.0	0.0	0.0	0.0	33.3	0.0	0.0
T 1137/92	0.0	0.0	0.0	0.0	24.6	16.6	0.0	0.0	0.0	0.0	14.3	0.0	2.76
AD 59 J	0.0	0.0	10.5	0.0	20.3	12.5	0.0	0.0	0.0	0.0	0.0	0.0	2.08
TPH 21/89*T 1685/94	45.0	0.0	40.0	0.0	50.7	0.0	40.1	0.0	0.0	0.0	47.4	0.0	0.0
TPH 10-1/89*T 1685/94	70.2	10.0	77.0	3.8	49.0	25.0	72.1	0.0	0.0	0.0	38.3	0.0	4.80
TPH 9/89*T 1685/94	70.0	0.0	42.4	0.0	64.4	0.0	45.7	0.0	0.0	0.0	49.2	0.0	0.0
AD 1439/91*T 1685/94	0.0	0.0	0.0	0.0	69.0	24.8	77.4	0.0	0.0	0.0	0.0	0.0	4.13
TPH 17/89* T 1685/94	70.9	0.0	0.0	0.0	0.0	0.0	0.0	0.0	0.0	0.0	42.9	0.0	0.0
TPH 14/89*T 1685/94	67.3	22.7	0.0	0.0	50.5	46.8	0.0	0.0	0.0	0.0	26.6	10.1	13.27
AD 1439/91*T 1137/92	50.6	0.0	0.0	0.0	35.4	25.3	0.0	0.0	0.0	0.0	0.0	0.0	4.22
TPH 10-1/89*T 1137/92	70.3	0.0	72.0	0.0	80.5	0.0	42.7	0.0	0.0	0.0	34.0	0.0	0.0
TPH 6/89*T 1137/92	37.0	0.0	0.0	0.0	0.0	0.0	40.3	0.0	0.0	0.0	0.0	0.0	0.0
TPH 8/89*T 1137/92	44.7	0.0	0.0	0.0	73.7	0.0	0.0	0.0	0.0	0.0	37.9	0.0	0.0
TPH 1677/94*T 1137/92	59.2	9.2	74.2	0.0	76.3	17.8	43.3	0.0	0.0	0.0	72.2	0.0	4.50
TPH 14/89*T 1137/92	35.3	11.8	0.0	0.0	53.9	43.3	44.1	11.8	0.0	0.0	25.6	9.8	12.78

1	2	3	4	5	6	7	8	9	10	11	12	13	14
AD 1439/91*AD 59 J	56.8	7.2	46.9	6.9	67.9	13.6	87.5	0.0	0.0	0.0	72.5	0.0	4.62
Average regeneration percentage of a nutrient medium over genotype over genotypes $\overline{Y}_n = E\ fn/Cg$ **	-	3.55	-	0.59	-	12.44	-	0.62	-	0.0	-	1.47	

* $\overline{Y}g$ -= $E\ f_n/c_n$ where:–

$\overline{Y}_g$, average regeneration percentage (only green plants) of a genotypes of over nutrient media.

Ef_y, summation of the frequencies of a genotype over nutrient media

C_n, total number of media = 6

** $\overline{Y}_n = Ef_n/c_g$ where

$\overline{Y}_n$, average regeneration percentage of a nutrient medium over genotypes.

Ef_n, summation of the frequencies of a nutrient medium over genotype.

C_g, total number of genotypes = 19

Appendix A Composition of Gamborg B5 medium (mg I_{-1})

NaH_2 Po_4 H_2O	150.0
KNo_3	2500.0
$(NH_4)_2$ So_4	134.0
$MgSo_4$ $7H_2O$	250.0
$CaCL_2$ $2H_2O$	150.0
Na_2EDTA	37.23
$FeSo_4$ $7H_2O$	28.0
KJ	0.75
$MnSo_4$ H_2O	10.0
$Cocl_2$ $6H_2O$	0.025
CuSo4 $5H_2O$	0.025
ZnSo4 $7H_2O$	2.0
H_3 Bo_3	3.0
Na_2MoO_4 $2H_2O$	0.25
Inositol	100.0
Thiamine HCL	10.0
Peredoxine HCL	1.0
Nicotinic acid	1.0
2,4-D	1.0
Kinetin	1.0
Sucrose	30000.0
Agar	7000.0
(pH = 5.5)	

Appendix B Composition of Nitsch N6 medium (mg L^{-1}).

K NO_3	950.0
NH_4 No_3	720.0
Mg So_4 $7H_2O$	185.0
Ca Cl_2 $2H_2O$	166.0
K H_2 Po_4	68.0
$MnSo_4$ $4H_2O$	25.0
H_3Bo_3	10.0
Zn So_4 $7H_2O$	10.0
Na_2 $Mo0_4$ $2H_2O$	0.25
Fe So_4 $7H_2O$	27.8
Na_2 EDTA	37.23
Cu So_4 $5H_2O$	0.025
Myco-Inositol	100.00
Nictonic acid	5.0
Peredoxine HCL	0.5
Thiamine HCL	0.5
Folic acid	0.5
Biotin	0.05
Glycine	2.0
I AA	0.1
Sucrose	2000.0
Agar	7000.0

(Ph = 5.5)

Chapter - 26

In Vitro Green Pod Culture of *Dendrobium transparens*

Syed Arshad, Azra. N.* Kamili and A.M. Shah**

ABSTRACT

Immature seeds from young green unbursted pods of the orchid *Dendrobium transparens* have revealed different germination responses under the influence of different growth additives. Coconut milk when added to the MS basal medium produced complete seedlings with profuse rooting. Multiple protocorms were obtained when basal medium was supplemented with BA and 2,4-D+BA. Higher concentration of 2, 4=D inhibited the germination. MS basal medium alone led to the formation of rooted as well as rootless seedlings. The shoots developed profuse rooting in different rooting media especially with NAA at a low concentration. Profuse green callus was obtained when the seeds were cultured on basal medium augmented with coconut milk and a cytokinin.

Key Words *: multiple protocorms, Seed callus, Seedlings,*

Abbreviations

MS-Murashige and Skoog, 1962; BA–6 benzylaminopurine; IAA–inndole–3- acetic acid; RH–Relative humidity; CM-Coconut milk, NAA-Naphthalene acetic acid; 2, 4-D-2,4 Diclorophenoxy acetic acid.

INTRODUCTION

Orchids, mostly used as ornamental plants, are now becoming vulnerable to extinction mainly because of habitat destruction and over-collection. Efficient propagation techniques are nowadays developed to check the extinction of this group of plants and in this direction the technique of *in vitro* propagation has become a potential tool to produce these in large numbers. The orchids, although produce their seeds in large numbers (1300 to 400,000 per capsule), but only few (approx. 5%) germinate under natural conditions due to their fragile structure, absence of endosperm and specific mycorrhizal association (Rao,

*Centre of Research for Development. University of Kashmir, Srinagar 190006, J&K India.

1977). *In vitro* seed germination of a number of archids has been possible and this is an efficient method to produce orchids in large numbers.

The present paper report the *in vitro* culture of immature seeds obtained from green pods of orchid *Dendrobium transparens* and 100% germination frequency was accomplished in completely defined media.

MATERIAL AND METHODS

Unripe green pods/capsules of *Dendrobium transparens*, approximately 2-3cm long were collected from mature flowering plants (procured from local nurseries in New Delhi). The pods were first washed under running tap water for twenty minutes and then rinsed twice in double distilled water, followed by surface sterilization with chlorine water for ten minutes and four washings in sterile distilled water. These pods were then dipped in 95% ethanol and flamed. The pods were then opened under sterile conditions and seeds were carefully dusted on the MS basal medium gelled with 0.8% agar and supplemented with different growth harmones namely CM, AA and 2,4-D used at different concentrations separately as well as in combinations. The pH of the medium was adjusted at 5.8 with 0.1N NaOH or 0.1N HC1. 30ml and 75ml of nutrient medium were dispensed in each 100ml-culture tube and 250ml-Erlenmeyer flask respectively. The medium was autoclaved at 1.05Kgcm-2 for 15-20 minutes at 121^0 C. The cultures were incubated under continuous cool white flourescent light (3000 lux) at $25^0 \pm 5^0$ C with 60-65% RH.

Subculturing of the rootless shoots and friable green callus obtained from initiating cultures was done in MS basal medium (1962, revised) supplemented with different auxins-2, 4-D, NAA, IAA at different concentrations.

RESULTS

The seeds showed varying responses on different combinations and concentrations of growth adjvants (Table 26.1). Seeds cultured on MS basal medium alone produced green rooted and rootless seedlings. The medium supplemented with CM-15% (v/v) produced fullfledged seedlings (Plate I). However when the medium was supplemented with CM at 15% (v/v) + BA (1mg/1) immense green callus was formed besides multiple protocorms which later on differentiated into shoots (Plate II). Multiple protocorms were also obtained in MS basal medium supplemented with BA (1mg/1) (Plate III). Low protocorm formation was recorded when 2,4-D (0.5 mg/1) in combination with BA (0.5mg/1) were used in the MS basal medium. Seed germination was completely inhibited on MS=2, 4-D (1 mg/1).

The rootless shoots produced from seed culture were subcultured in rooting media (Table 26.2) High growth of long green and thick roots was observed on MS + AA (0.5 mg/1) (Plates IV, V). Others growth additives like NAA (1mg/1), IAA (0.5mg/1) ad IAA (1mg/1) when used independently showed moderate growth of root formation. No rooting was observed when 2,4-D (0.5mg/1) was added to the basal medium.

The friable green callus obtained from the seeds was subcultured on MS basal medium supplemented with different auxin and cytokinin concentrations. No differentiation of organs was observed here, but instead it continued to proliferate as callus only.

The *in vitro* born seedlings were allowed to grow upto 7cm high in culture vials after which they were transplanted into thumb pots containing a mixture of sterile ground charcoal and brick chips and were kept in growth chamber maintained at 80-90% R.H. and

IN VITRO SEED PROPAGATION DENDROBIUM (DENDROBIUM TRANSPARENT ORCHID AS INFLUENCED BY DIFFERENT MEDIA

Orchids are an excellent item for garden and for indoor decoration. They exhibit a wide range of diversity in form, size, colour and texture of flowers beyond the imagination of human mind (Mukherjee, 1983). Among the flowering plants more than 25000 species and 700-800 genera are under orchidaceae family (Begum, 2000). Several orchids have been used as food in different parts ot the world. The stem of *Dendrobium* species are used in making baskets in the Philippines, Indonesia and New Guinea and pseudobulbs of *D. tokai* are used as oral contraceptives (Bose and Bhattacharjee, 1999). Commercial cultivation of orchids both for plant sale as well as cut-flower production has been developed into sizeable industries in many countries and the sale of flower runs in millions of dollars. The orchid flower trade bring about two hundred million dollars a year (Singh, 1998). In Bangladesh, the orchid trade is still in its nascent stage. But there is a great scope of orchid trade in Bangladesh considering their availability and climatic conditions.

Orchids grow in nature through seeds but in absence of appropriate hosts they don't germinate in adequate numbers, so it remains as a rare species. These obstacles may be over come by adopting tissue culture technique. For appropriate germination of orchid seeds, *in vitro* propagation is imperative for multiplication rather than *in vivo* (Arditti, 1979). A good number of orchids with beautiful flowers are available in Bangladesh. Like other countries, appropriate propagation techniques for large-scale production of native orchids may be a profitable source of earning foreign exchange in Bangladesh. But, there is no well-recognized report on standardization of *in vitro* suitable techniques for seed propagation of native orchids. Besides the media reported earlier were suitable for exotic varieties of orchids, which may not be suitable for our native orchids. *Dendrobium transparens* is one of the important fascinating native orchid. Therefore, the present investigation was undertaken to develop a low cost commercially applicable medium for our native orchid, *Dendrobium transparens.*

MATERIAL AND METHODS

Studies on the seed propagation of native orchid species, *Dendrobium transparens* were carried out in the Tissue Culture Laboratory of Bangabandhu Sheikh Mujibur Rahman Agricultural University (BSMRAU), Salna, Gazipur during February, 1998. Seeds of the mentioned orchid species were collected form Khrishibid Orchid and Cactus Nursery, Khamarbari, Dhaka through personnel contact. Four different media were used as treatment and these were:

M_1 = Hyponex medium (3 g L^{-1} Hyponex + 8 g L^{-1} peptone + 2 g L^{-1} charcoal + 8 g L^{-1} Agar + 30 g L^{-1} Sucrose)

M_2 = MS medium, Murashige and Skoog (1962)

M_3 = OKF_1 medium (3 g L^{-1} OKF_1 + 8 g L^{-1} + Peptone + 2 g L^{-1} Charcoal + 8 g L^{-1} Agar + 30 g L^{-1} Sucrose)

M_4 = Knudson C medium (Knudson, 1964), used as check

Completely randomized design (CRD) was followed with ten replication. Pods were collected and surface sterilization was done by 1% $HgCl_2$, 70% ethanol and dissected inside the clean bench and collected seeds were sown on to the medium. The explanted seeds were incubated at 22°C under 16 h photoperiod illuminated with fluorescent light of 3000 lux.

The development of plantlets (Plate 1a), NAA (1mg L^{-1}) also applied in each medium except check. Data on percentage of seed germination, required days to seed germination, protocorm and plantlet formation; length of plantlet, number of roots and leaves at different dates and also plant survivability percentage were recorded, analyzed statistically and means were separated using DMRT.

RESULTS AND DISCUSSION

In vitro seed germination: There was highly significant variation in percent seed germination among different media (Table 1). Germination of seeds in different media were shown in Plate 1b. The highest seed germination (78%) was observed in MS medium, which was followed by Hyponex medium (73%) while OKF_1 gave the poor performance (58%). This result is in agreement with that of Ismat (1982) who conducted an experiment with *Dendrobium pieradii* on MS and Hyponex media and observed percentage of seed germination was 79 and 70 respectively. More or less similar results (percentage of seed germination was 81 and 74 in MS and Hyponex media respectively) were found by Hoque (1993) in case of *Dendrobium formusum*.

The present work is different from Ismat (1982) and Hoque (1993) by orchid species only. The contents of each media were same in all cases. But NAA (1 mg L^{-1}) was applied in each media after plantlet development in the present study.

Required days to germination: Highly significant variation among different media had been observed in required days to seed germination (Table 1). Maximum days (49 days) to seed germination was recorded by OKF_1 medium while the least period (50 days) was required by MS medium. Ismat (1982) and Hoque (1993) found that required days for seed germination of *Dendrobium* sp. In MS medium is 51 and 55 days respectively. The present findings have similarity with their results.

Days required to protocorm formation and plantlet development: The required days to protocorm formation and plantlet development were significantly influenced by different media (Table 1). In case of protocorm formation, maximum duration (48 days) was required by OKF_1 medium which was significantly similar to Knudson C medium (43 days) while MS medium took the least period (36 days). Similar trends were also observed in plantlets development (Table 1). Hoque (1993) stated that protocorms differentiated into plantlets within 72-78 days in different media which is closely to the present findings. Developed protocorms in different media were shown in Plate 1c.

Plantlet height: Highly significant variation among different media at different dates

Table 1: Effects of different media on percentage of seed germination and required days to seed germination, protocorm formation and plantlet development of Dendrobium transparens

		Days required to		
Treatments	% Seed germination	Seed germination	Protocorm formation from seed germination	Plantlet development from protocorm formation
M_1 (Hyponex)	73a	55ab	39bc	83bc
M_2 (MS)	78a	50b	36c	79c
M_3 (OKF_1)	58b	59a	48a	92a
M_4 (Knudson C)	66a	56ab	43ab	86ab
CV(%)	4.46	4.36	6.14	2.70

Different letters indicate significant ($p \le 0.01$) variation

Table 2: Effect of different media on plantlet height (cm) and number of leaf per plantlet at different weeks after plantlet development (WAPD)

	Plantlet height (cm) at WAPD				Number of leaves per plantlet at WAPD			
Treatments	3	6	9	12	3	6	9	12
M_1 (Hyponex)	1.57b	2.36b	2.97b	3.59b	3.00ab	4.20a	4.90ab	5.50b
M_2 (MS)	2.22a	2.92a	3.40a	4.06a	3.50a	4.60a	5.40a	6.40a
M_3 (OKF_1)	1.26c	1.68d	2.06d	2.52d	1.70c	3.10b	4.00c	4.50c
M_4 (Knudson C)	1.34c	1.97c	2.44c	2.96cb	2.40bc	3.40b	4.70b	5.20bc
CV(%)	10.07	10.10	10.14	8.60	20.80	14.60	10.70	11.10

Table 3: Effects of different media on number of roots per plantlet at different weeks after plantlet development

Treatments	At 3 WAPD	At 6 WAPD	At 9 WAPD	At 12 WAPD
M_1 (Hyponex)	2.10b	3.30b	4.40b	5.40b
M_2 (MS)	3.20a	4.40a	6.10a	7.80a
M_3 (OKF_1)	1.10c	1.90c	2.30c	3.00c
M_4 (Knudson C)	1.60b	2.40c	4.20b	5.30b
CV(%)	20.00	15.50	12.90	10.80

Different letters indicate significant ($P \leq 0.01$) variation

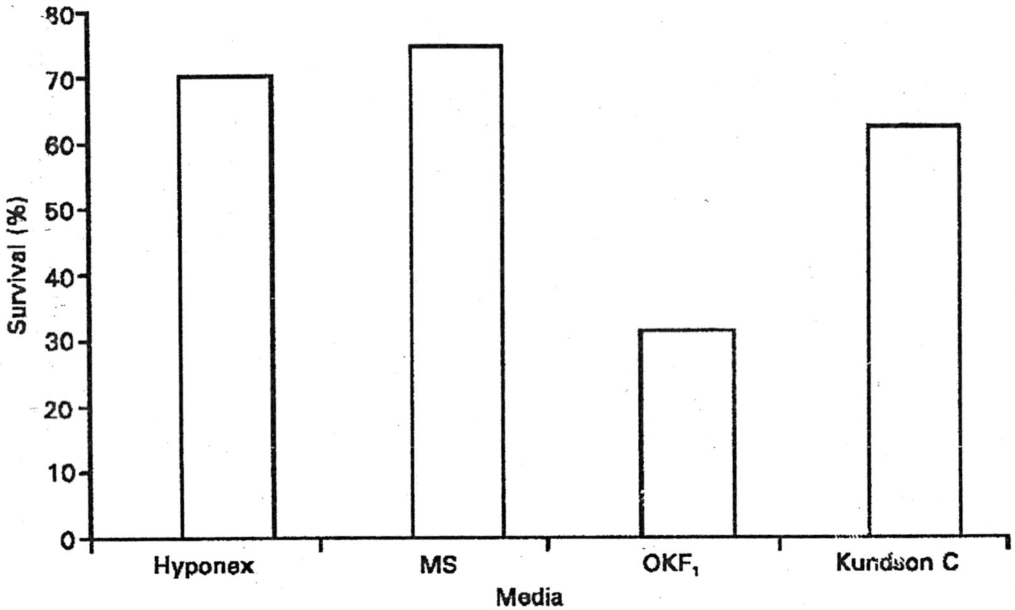

Fig 1: Effect of different media on survivility (%) at 12 weeks after plantlet development

had been observe in plant height (Table 2). The tallest plantlets (2.22, 2.92, 3.40 and 4.06 cm at 3, 6, 9 and WAPD respectively) were observed in MS medium, followed by Hyponex medium while OKF_1 medium produced the dwarf plantlets (1.26m 1.68, 2.06 and 2.52 cm at 3, 6, 9 and 12 WAPD respectively).

Number of leaves: The number of leaves per plantlet was significantly influenced by different media (Table 2). The highest number of leaf was observed in MS medium which was statistically similar to Hyponex medium at different dates except at 12 WAPD and the least number was found in OKF_1 medium at all counting dates. This results is in agreement with that of Davidson (1994) who stated that the number of leaves on the plantlet of *Dendrobium* species showed better development in MS medium than Knudson C medium.

Number of roots: At all dates, number of roots per plantlet were significantly higher in MS medium which was followed by Hyponex medium while OKF_1 produced the lowest number of leaves (Table 3) whish is consistent with that of Davidson (1994). Root length was also measured at 12 WAPD. The MS medium gave the longest root (1.40 cm) which was statistically similar to Hyponex medium (1.20 cm) while OKF_1 gave the shortest root (0.50 cm).

Plant survivility of percentage: Plant survivility percentage also showed significant differences among treatments at 12 WAPD (Fig. 1). The maximum survival plat (76%) was observed in MS medium which was statistically similar to that of Hyponex medium (71%) whereas OKF_1 medium gave the lowest survival percentage (30%).

From the above discussion, it may be suggested that the MS medium was better for in *vitro* seed propagation of *Dendrobium transparents.*

REFERENCES

Arditti, 1979. Aspects of the physiology of rochids. Adv. Bot. Res., 7: 421-655.

Burgeff, H. (1936). Die anmenkeimung her orchideen *Fischer jena*. 63

Curtis, J. T. (1947). Undifferentiated growth of orchid embroys on media containing barbiturates. *Science*. 105 : 128.

DePauw, M.A; Remphrey, W.R and Palmer, C.E. (1995). The cytokinin preference for *in vitro* germination and protocorm growth of *Cypripedium candidam*. *Annals of Botany*, 75 : 267-275.

Goh, C.J. (1970). Tissue culture of *Vanda* Mis Joaquim. *Journal of Singapore National Academy of Sciences*. 2 : 31-33.

Ito, I. (1955). Germination of seeds from immature pod and subsequent growth of seedlings in *Dendrobium nobile* lindl. *Science Reports of Saikyo University of Agriculture*. 7 : 35-42.

Kerbauy, G.B. and Handro, W. (1981). Culture of orchid embroys in liquid medium. *Orchid Review* 89 (1056): 316-318.

Mathews, V.H. and Rao, P.S. (1980). Invitro multiplication of Vanda hybrids through tissue culture techniques. *Plant Science Letters*. 17 (3) : 383-389.

Mercier, H. ad Kerbauy, G.B. (1991). Effects of nitrogen source on growth rates and levels of endogenous cytokinins and chlorophyll in protocorms of epidendrum fullgens. *Journal of Plant Physiology*. 138 : 195-199.

Mitra, G.C. (1971) Studies on seeds, shoot tips and stem discs of an orchid grown in aseptic culture. *Indian Journal of Experimental Biology*, 9 : 79-85.

Mitra, G.C. Prasad, R.N. and Roychowdhury, A. (1976). Inorganic salts and differentiation of protocorms in seed callus of an orchid and correlated changes in its free aminoacid content. *Indian Journal of Experimental Biology*. 14 (3) : 350-351.

Murashige, T and Skoog, F. (1962). A revised medium for rapid growth and bioassays with tobacco tissue cultures. *Physiological Plantarum*. 15 :473-497.

Rao, A.N. (1977). Tissue culture in the orchid industry. In Applied and Fundamental Aspects of Plant Cell, Tissue & Organ Culture. Ed. Reinert, J. and Bajaj, Y.P.S. Springer, Verlag, Berlin, Hiedelberg, New York. pp 44-65.

Rao, A.N. and Avadhani, P.N. (1963). Effect of chlorox on the germination of *Vanda* seeds. *Current Science* 32 : 467-468.

SECTION - VII

IMPACT OF ANTHROPOGENIC PERTURBATIONS ON BIODIVERSITY

Photosynthetic Performance of Some Dicotyledonous Tropical Plants under Degraded Environment

Muhammad Iqbal, S. Tanvir Ali* and Mahmooduzzafar**

ABSTRACT

Photosynthesis is affected by various environmental factors including the pollutants. This study investigates the effect of coal-smoke pollution on pigment concentration as well as the rate of photosynthesis in 31 dicotyledonous species growing around a thermal power plant in Delhi. Degree of sensitivity of different species to the air pollutants varies considerably: most of them are adversely affected, some have shown improvement, while others remain by and large unaffected or only slightly disturbed. Tolerant species have been indicated on the basis of their carbon fixation capacity under the pollution stress. The mechanism of injury and photosynthetic inhibition has been discussed and tentative conclusions are drawn regarding the relative performance and survival chances of the species studied in condition of their being consistently exposed to the degraded environment.

Keywords: *air pollution, chlorophyll, photosynthesis, plant growth*

INTRODUCTION

Biodiversity is the outcome of the evolutionary plasticity of living organisms, and has multiplied geometrically during about 2.5 billion years course of the organic evolution on the earth. The number of species existing is said to have been fluctuating widely during this long time period due to a variety of reasons ranging from the gradual climatic changes to the huge catastrophic disasters (Fosberg 1986).

During the first millenia of its history, the species *Homo sapiens* (man) behaved as a normal member of the faunas of ecosystems in which it occurred but later it started

* Department of Botany, Faculty of Science, Hamdard University, Hamdard Nagar New Delhi-110062

increasing its geographic range, improving its life style and of late even trying to modify the environment so as to make it suit its requirements better. After having struggled for over 100, 000 years, man has established a world-wide dominance and, in the wake of progressing more and more, his activities now pose a serious danger to the marvellous biological diversity that has evolved over two and a half billion years (Fosberg 1986).

Biodiversity is identified as (i) ecosystem diversity which is related to the variety of habitats, (ii) genetic diversity which relates to the variation in genes within a particular species, and (iii) species diversity, which indicates the varieties of the living beings on the earth. Ours is a tropical country with markedly heterogeneous environmental set ups, ranging from tropical rainforests of Andaman and Arunachal Pradesh to the hot deserts of Rajasthan and the cold deserts of Ladakh. Consequently, it has a rich biological heritage (about 46 thousand plant species and 50,000 animal species) which places it among the twelve megadiversity nations of the world (Gadgil 1992). The approximate number of plant species (including monera and fungi) may be expressed as 15000 angiosperms, 65 gymnosperms, 1000 pteridophytes, 2600 bryophytes, 1600 lichens, 23000 fungi and 2500 algae. This constitutes about 11% of the known plant species of the world (Nayar 1977; Rao 1994).

The exact rate of disappearance of the genetically-distinct populations and species of other organisms is not easy to determine but it is certainly far too high and is continuously accelerating. This is due mostly to the destruction and conversion of the natural habitats, which has its main origin in the fast expanding processes of urbanization, industrialization and mechanization. The rate of tropical deforestation in 1989 was almost double the one in 1979, with roughly 1.8% of the remaining forests disappearing every year (Ehrlich 1991). Human beings have greatly increased the capacity to alter the rate and extent of habitat change. It has been estimated that 140 species of plants and animals are lost every day in the world (Agarwal 1996).

There is a synchronised relationship between environment and the organisms it encompasses. The organisms are either filled in or filtered out by the environmental parameters. In the modern technological society, however, this natural dynamic system has been disturbed due to overconsumption of resources and the environmental degradation. As a result, many species are becoming endangered or getting extinct.

Plants growing in the open, near urban establishment or around industries, are exposed to a variety of air pollutants usually emanating from fossil fuel combustion. When various pollutants interact, they may have synergistic, additive or antagonistic influence on plant performance, depending on various plant as well as environmental factors. Exposure of eastern white pine to a mixture of SO_2 and O_3 produced injury symptoms at a concentration much lower than the threshold for SO_2 or O_3 alone (Dochinger *et al.*1970). The same combination of gases, however, caused less injury than each of the gases did to the tree species studied by Kress (1980). SO_2 and HF together caused greater reduction in CO_2 uptake by *Picea abies* than they did individually (Keller 1980). Many subsequent studies have brought out the damaging potential of the various pollutants in combination or alone (for review see Yunus and Iqbal 1996).

The most frightening data regarding the pollution effect on plants pertain to the reduction in net primary productivity (NPP), i.e., the energy fixed by photosynthesis in addition to the requirement of the plant itself for maintaining its metabolic activities. NPP may be considered roughly as the total food supply of all animals and decomposers and has immense significance in view of the fast multiplying human population. Decline in photosynthesis rate due to toxic pollutants also hampers the growth of the plant. Given this, the present study investigates the variation in rates of CO_2 uptake and photosynthesis in a variety of woody plants growing in normal condition and under air pollution stress. This provides an idea of the degrees of resistance/sensitivity of the different plant species to the environmental insult which they consistently have to experience.

STUDY SITE AND METHODOLOGY

Delhi is situated in north of India, 160 km south of Himalayas. It has a tropical semi arid climate with extremely hot summers and moderately cold winters. July to September is the typical monsoon season. Badarpur Thermal Power Station (BTPS), the main source of power supply to the city of Delhi, is located at the Delhi-Haryana Border along the Badarpur-Faridabad road in South Delhi at 77°22' E longitude and 28°25' N latitude, 220 m above sea level. The soil is loam and clayey loam type with a pH 8-9. The relief of the land is uniform. The trees and shrubs are not shielded by buildings from the effect of smoke (gases+particulates) released by the stacks of the power station. The power plant has 5 units; three units of 100 MW capacity each and two of 210 MW capacity each (Table 27.1). The station requires 12,000 metric tonnes of coal everyday and has the capacity to generate 720 MW electricity. The chief chemical constituents of the bituminous coal used are 2.90% moisture, 40% ash, 31.86% volatile matter, including 0.48% sulphur, 5.61% hydrogen 5.23% nitrogen and 20.3% oxygen, and 42.47% fixed carbon on an average. The power plant is a point source of air pollution, its emissions (flyash and gases) are primarily dispersed in the prevailing wind direction affecting the soil and vegetation. The power stations in the city emit 1450 tonnes of flyash and 600 tones of SO_2 each day and contribute up to 40% of the total air pollution of Delhi.

Table 27.1. Technical details on the power plant units of Badarpur thermal power station, New Delhi

Specification	*Stack I*	*Stack II*	*Electrostatic Precipitator*
Type	3 × 100 MW	2 × 210 MW	
Height	150 m	150 m	
Exit gas velocity	10-12 m/sec	15-18 m/sec	
Exit flue gas temperature	150-160°C	120°C	
Flue gas density	0-8 kg/m^3	0-8 kg/m^3	
Collection efficiency			99.61%
Temperature of flue gas at inlet			150°C
Dust concentration of flue gas at outlet			150 mg/nm^3

Hamdard University is located on Badarpur-Mehrauli road about 10 km west from the power station. The soil at both the sites is saline and alkaline and has a similar composition.

Table 27.2 : Plant species examined for their growth performance at Hamdard University campus (reference site) and around the Badarpur thermal power station (polluted site) on the basis of photosynthesis rate.

	Species	*Family*	*Habit*
1.	*Abutilon indicum* (Linn.) Sweet	Malvaceae	Herb
2.	*Achyranthes aspera* Linn.	Amaranthaceae	Herb
3.	*Ailanthus excelsa* Roxb.	Simaroubaceae	Tree
4.	*Amaranthus viridis* Linn.	Amaranthaceae	Herb
5.	*Awdirachta indica* A. Juss	Meliaceae	Tree
6.	*Bougainvillea spectabilis* Choisy	Nyctaginaceae	Shrub
7.	*Calotropis procera* (Ait.) R. Br.	Asclepiadaceae	Herb
8.	*Cannabis sativa* Linn.	Cannabinaceae	Herb
9.	*Cassia occidentalis* Linn.	Caesalpiniaceae	Shrub
10.	*Cichorium intybus* Linn.	Asteraceae	Herb
11.	*Cordia dichotoma Forst.*	Boraginaceae	Tree
12.	*Croton bonplandianum* Baill.	Euphorbiaceae	Herb
13.	*Dalbergia sissoo* Roxb.	Papilionaceae	Tree
14.	*Datura innoxia* Mill.	Solanaceae	Herb
15.	*Eclipta alba* (Linn.) Hassk.	Asteraceae	Herb
16.	*Eucalyptus citriodora* Hook.	Myrtaceae	Tree
17.	*Euphorbia hirta* Linn.	Euphorbiaceae	Herb
18.	*Hibiscus rosa-sinensis* Linn.	Malvaceae	Shrub
19.	*Mangifera indica* Linn.	Anacardiaceae	Tree
20.	*Morus alba* Linn.	Moraceae	Tree
21.	*Peristrophe bicalyculata* (Retz.) Nees.	Acanthaceae	Herb
22.	*Pongamia pinnata* (Linn.) Pierre	Papilionaceae	Tree
23.	*Prosopis spicigera* Linn.	Mimosaceae	Tree
24.	*Sida acuta* Burm. f.	Malvaceae	Herb
25.	*Sida rhombifolia* Linn.	Malvaceae	Herb
26.	*Solatium nigrum* Linn.	Solanaceae	Herb
27.	*Syzygium cumini* (Linn.) Skeels	Myrtaceae	Tree
28.	*Tephrosia purpurea* (Linn.) Pers.	Papilionaceae	Herb
29.	*Withania somnifera* (Linn.) Dunal	Solanaceae	Herb
30.	*Xanthium strumarium* Linn.	Asteraceae	Herb
31.	*Zizyphus mauritiana* Lamk.	Rhamnaceae	Tree

Pigment concentration and photosynthesis rate were measured in the leaves of 31 dicotyledonous plants (Table 27.2) growing around the BTPS (the polluted site) and also in the Jamia Hamdard Campus (the reference site) which is relatively free from the vehicular pollution.

Chlorophylls *a* and *b* as well as carotenoids in the leaves of normal and polluted samples were analysed by digesting fresh leaf material (0.1 gm) in 7 ml of dimethyl sulfoxide (DMSO) at 65°C. One ml aliquot added with 3 ml DMSO was taken in a tube and vortexed. Optical density was recorded at 480, 510, 645 and 663 nm wave length with the help of Beckman DU 640 B spectrophotometer following Hiscox and Israelstam (1979). The chlorophyll and carotenoid contents were estimated by the following formulae of Duxbury and Yentsch (1956) and MacLachlan and Zaiik (1963), respectively.

$$\text{Chlorophyll } a \text{ (mg/g fresh wt)} = \frac{12.30\text{x}OD_{663} - 0.86\text{x}OD_{645}}{\text{dx}1000\text{xw}}\text{xV}$$

$$\text{Chlorophyll } b \text{ (mg/g fresh wt)} = \frac{19.30\text{x}OD_{645} - 3.60\text{x}OD_{645}}{\text{dx}1000\text{xw}}\text{xV}$$

$$\text{Total chlorophyll (mg/g fresh wt)} = \frac{20.20\text{x}OD_{645} - 8.02\text{x}OD_{663}}{\text{dx}1000\text{xw}}\text{xV}$$

$$\text{Carotenoid (mg/g fresh wt)} = \frac{7.6\text{x}OD_{480} - 1.49\text{x}OD_{510}}{\text{dx}1000\text{xw}}\text{xV}$$

Where V = volume of chlorophyll solution, d = length (cm) of light path, and w = fresh weight (gm) of leaves.

Infra Red Gas Analyser (LI 6200 portable photosynthesis system) was used for measuring photosynthesis rate, stomatal conductance and intercellular CO_2 concentration in leaves. Gas-exchange measurements included the transient (< 1 min) exchange rates of H_2O vapour and CO_2 in a closed system. Each mature and fully expanded leaf of the selected species was enclosed in a 250 cm^3 leaf chamber for 30 to 60 seconds during which depletion of CO_2 and increase in H_2O vapour over time were monitored in conjunction with leaf temperature, air temperature and flow rate at the control as well as the polluted site. All measurements were made between 09.00 hr to 12.00 hr under field conditions, with air temperature ranging from 24 to 32°C, photosynthetic active radiation (PAR) 900 to 1300 μ mol s^{-1} m^{-2}, and relative humidity 30-40 per cent. Mean of each set of observations and the per cent variation between control and polluted populations are presented in Tables 27.3-27.8.

OBSERVATIONS AND DISCUSSION

Since the anthropogenic environmental stresses have increased drastically during the current century, concern regarding the environmental health, conservation of natural resources and sustainability of living beings on the earth has also begun being felt more seriously for the last few decades. The emerging paradigms of change, complexity, diversity and sustainability reflect the enlarging scale of our perception and understanding of the world, as well as an awareness of the great power that human kind possess to influence its environment and modify the course of its evolution (Younes 1996). The new key issues of biological diversity, global climate change and sustainable development constitute the triad upon which enthusiastic inter governmental debates and deliberations have been held at Rio-de-Janeiro, Brazil (1992), followed by the sittings at Nassan, The Bahamas (1994), Jakarta, Indonesia (1995) and Chonju, South Korea (1996). Of late, issues relating to the ecosystem function of biodiversity were focused (di Castri and Younes 1990; Solbrig 1991). It is admitted that the events of speciation and extinction have a strong ecological context, and predominantly occur as an outcome of a changing environment, especially when this involves loss of habitats. Studies on the structural/functional relationships among systems have revealed that plant size and relative growth rates are the characteristics that quantitatively affect ecosystem processes by controlling rates of acquisition (Younes 1996) and that ecosystem services include, among several other things, maintenance of the gaseous quality of the atmosphere. A disbalance in the air composition affects plant growth rate, as is evident from the present study also. Not only the net primary productivity is affected but even the total biomass of the individual plants is often reduced and the plant remains from achieving the normal size characteristic of the species.

The environmental stress may also cause mutations at the genie level and the mutation may become a permanent and inheritable character if the given individual and its offsprings continue to be exposed to the stressed condition for long. If a given species has been made deprived of its original habitat and environmental support, it is bound to fall in line of the endangered species and initiate the journey to extinction in case the poor and suffocative living conditions persist for good. The present study, like many others, also demonstrates, however, that there are certain species which know how to live in distress and adjust maximally with the adversities.

Of the species studied, a few herbs and shrubs and several trees showed tolerance of varying degree to the pollutants in the atmosphere and managed to thrive or even do better under the degraded environment. However, most of them were adversely affected by the pollution, showing differential extent of sensitivity. Variation occurred in terms of chlorophyll content and the ultimate photosynthetic activity. Among the herbs/shrubs, the amount of chlorophyll *a* was significantly reduced in 13 species (with maximum reduction in *Peristrophe bicalyculata, Cannabis sativa, Croton bonplandianum,* and *Solanum nigrum),* increased in *Achyranthes aspera, Bougainvillea spectabilis* and *Withania somnifera* and varied only marginally in four of them (Table 27.3). Chlorophyll *b* content was reduced significantly in the polluted atmosphere in 15 species (with a maximum in *Cichorium intybus, Tephrosia purpurea* and *Croton bonplandianum),* increased in *Achyranthes aspera* and *Bougainvillea spectabilis* and remained almost constant in *Eclipta alba, Withania somnifera* and *Xanthium strumarium* (Table 27.4). Photosynthesis rate declined under pollution stress in 13 species (with a maximum in *Amaranthus viridis, Cichorium intybus* and *Hibiscus rosa-sinensis),* improved in *Bougainvillea spectabilis* and *Withania somnifera,* and was little disturbed in five species (Table 27.5).

Among the eleven tree species under study, *Ailanthus excelsa, Cordia mixa, Dalbergia sissoo, Mangifera indica* and *Moms alba* exhibited reduction in pigment concentration (Tables 27.6 & 27.7) and photosynthesis rate (Table 27.8) under air pollution stress. *Morus alba* experienced the greatest decline in photosynthesis rate. On the other hand, *Azadiracta indica, Eucalyptus citriodora, Pongamia pinnata, Prosopis spicigera, Syzygium cumini* and *Zizyphus mauritiana* were capable of withstanding the hostile environment and grow more vigorously in the presence of coal-smoke pollutants which have SO_2 as a major constituent (Table 27.8).

Table 27.3 : Concentration of chlorophyll *a* (mg/g fr. wt.) in certain dicotyledonous herbs and shrubs growing in the reference and the polluted environmental conditions. The values indicate the mean of 10 readings obtained with the help of spectrophotometer.

Species	*Reference site Mean ± SD*	*Polluted site Mean ± SD*	*Per cent variation*
Damaged			
Abutilon indicum	0.89±0.05	0.66±0.02	24.84*
Amaranthus viridis	0.67±0.04	0.51±0.03	23.88*
Cannabis sativa	0.49±0.001	0.07±0.005	85.71**
Cassia occidentalis	0.47±0.02	0.23±.01	51.06**
Croton bonplandianum	0.88±0.08	0.30±0.02	65.90*
Datura innoxia	0.38±0.01	0.31±0.04	18.42**
Euphorbia hirta	0.79±0.02	0.36±0.06	54.43**
Hibiscus rosa-sinensis	0.82±0.02	0.43±0.08	47.56**
Peristrophe bicalyculata	1.17±0.14	0.61±0.09	91.04**
Sida acuta	0.90±0.05	0.53±0.06	41.11**
Sida rhombifolia	0.67±0.11	0.32±0.02	52.23**
Solanum nigrum	1.39±0.12	0.48±0.07	65.46**
Tephrosia purpurea	1.24±0.12	0.61±0.08	50.80**
Least affected			
Calotropis procera	0.41±0.06	0.34±0.02	17.07NS
Cichorium intybus	0.53±0.11	0.47±0.06	11.32NS
Eclipta alba	0.62±0.01	0.49±0.11	20.96NS
Xanthium strumarium	06.39±1.25	08.05±2.11	25.97NS
Improved			
Achyranthes aspera	06.69±2.01	11.38±4.14	41.21*
Bougainvillea spectabilis	03.70±.64	05.40±0.77	45.94**
Withania somnifera	14.09±2.52	24.40±2.46	73.17**

* = Significant at 5% ** = Significant at 1% NS = Non-sigificant

Table 27.4 : Concentration of Chlorophyll b (mg/g fr. wt.) in certain dicotyledonous herbs and shrubs growing in the reference and the polluted environmental conditions. The values indicate the mean of 10 readings obtained with the help of spectrophotometer.

Species	*Reference site Mean ± SD*	*Polluted Site Mean ± SD*	*Per cent variation*
1	*2*	*3*	*4*
Damaged			
Abutilon indicum	0.55±0.08	0.40±0.10	27.27*
Amaranthus viridis	0.47±0.05	0.34±0.06	27.65*

1	2	3	4
Calotropis procera	0.64±0.19	0.28±0.03	56.25**
Cannabis sativa	0.60±0.1	0.50±0.05	16.60**
Cannabis occidentalis	0.24±0.02	0.15±0.01	37.50**
Cichorium intybus	0.44±0.01	0.05±0.01	88.63**
Croto bonplandianum	0.57±0.01	0.21±0.01	63.15**
Datura innoxia	0.31±0.03	0.27±0.01	12.90**
Euphorbia hirta	0.52±0.05	0.26±0.04	50.00**
Hibiscus rosa-sinensis	0.56±0.03	0.32±0.02	42.85**
Peristrophe bicalyculata	1.01±0.15	0.59±0.11	41.58**
Sida Acura	0.57±0.05	0.32±0.06	43.85**
Sida rhombifolia	0.30±0.01	0.16±0.02	46.66**
Solanum nigrum	0.53±0.02	0.29±0.04	45.28**
Tephrosia purpurea	2.66±0.10	0.49±0.05	81.57**
Least Affected			
Eclipta alba	0.38±0.03	0.35±0.05	07.10^{NS}
Withania somnifera	0.36±0.08	0.48±0.08	33.33^{NS}
Xanthium stramarium	0.46±0.01	0.51±0.02	10.86^{NS}
Improved			
Achyranthes aspera	0.30±0.01	0.70±0.01	133.33**
Bougainvillea spectabilis	1.56±0.02	3.23±0.05	107.05**

* = Significant at 5% ** = Significant at 1% NS = Non-significant

Table 27.5. Photosynthesis rate (μ mol m^{-2} s^{-1}) in certain dicotyledonous herbs and shrubs growing in the reference as well as the polluted environmental conditions. Values represent the mean of 10 readings obtained on sunny days in the field with the help of infra-red gas analyzer.

Species	*Reference site Mean ± SD*	*Polluted site Mean ± SD*	*Per cent variation*
1	2	3	4
Damaged			
Abutilon indicum	08.61±0.20	06.22±0.41	27.74
Amaranthus viridis	22.70±7.63	03.36±1.13	85.21*
Cannabis sativa	08.67±4.10	05.40±1.13	37.71**
Cassia occidentalis	10.17±1.76	07.21±3.33	29.07**
Cichorium intybus	08.36±0.67	03.22±1.10	61.48**
Croton bonplandianum	12.81±0.43	10.14±0.83	20.84**
Datura innoxia	15.32±2.41	07.48±3.54	51.17*

1	2	3	4
Eclipta alba	12.93±0.75	07.19±2.94	44.39**
Euphorbia hirta	16.03±0.83	14.51±1.03	09.48*
Hibiscus rosa-sinensis	07.29±2.11	02.42±0.15	66.80**
Peristrophe bicalyculata	10.72±2.94	06.15±1.89	42.63*
Sida rhombifolia	20.80±1.30	14.30±1.50	31.25*
Solanum nigrum	16.10±1.57	11.19±2.30	30.50**
Least affected			
Achyranthes aspera	06.69±2.01	11.38±4.14	70.10^{NS}
Calotropis procera	15.90±0.79	12.66±0.88	20.37^{NS}
Sida acuta	07.20±3.04	06.88±1.38	04.44^{NS}
Tephrosia purpurea	15.37±3.97	10.64±3.19	30.77^{NS}
Xanthium strumarim	06.39±1.25	08.05±2.11	25.97^{NS}
Improved			
Bougainvillea spectabilis	03.70±0.64	05.40±0.77	45.94**
Withania somnifera	14.09±2.52	24.40±2.46	73.15**

* = Significant at 5% ** = Significant at 1% NS = Non-significant

Table 27.6 : Concentration of chlorophyll *a* (mg/g fr. wt.) in certain dictyledonous trees growing in the reference and the polluted environmental conditions. The values indicate the mean of 10 readings obtained with the help of spectrophotometer.

Species	*Reference site Mean ± SD*	*Polluted site Mean ± SD*	*Per cent variation*
Damaged			
Ailanthus excela	6.27±1.73	4.89±0.53	22.01**
Cordia dichotoma	0.6±0.07	0.42±0.04	30.00**
Dalbergia sissoo	4.60±0.42	2.62±0.44	43.03**
Mangifera indica	4.14±0.16	2.58±0.18	37.68**
Morus alba	7.50±1.32	1.63±0.35	78.26**
Improved			
Azadirachta indica	3.74±.04	7.86±0.44	110.16**
Eucalyptus citriodora	3.82±0.85	9.43±0.73	146.85**
Pongamia pinnata	5.06±0.29	7.34±0.59	45.06**
Prosopis spicigera	3.41±0.50	9.38±0.47	175.07**
Syzygium cuminii	2.56±0.75	3.29±1.20	28.51**
Zizyphus mauritiana	2.26±1.10	7.03±3.12	211.06**

* = Significant at 5%

** = Significant at 1%

Table 27.7 : Concentration of chlorophyll *b* (mg/g fr. wt.) in certain dicotyledonous trees growing in the reference and the polluted environmental conditions. The values indicate the mean of 10 readings obtained with the help of spectrophotometer.

Species	*Reference site Mean ± SD*	*Polluted site Mean ± SD*	*Per cent variation*
Damaged			
Ailanthus excelsa	0.44±0.04	0.33±0.02	25.00*
Cordia dichotoma	0.46±0.14	0.37±0.05	19.56*
Dalbergia sissoo	0.66±0.08	0.25±0.02	62.12**
Mangifera indica	0.53±0.03	0.29±0.01	45.28**
Morus alba	0.72±0.05	0.43±0.03	40.27**
Improved			
Azadirachta indica	1.54±0.02	2.89±0.04	87.66**
Eucalyptus citriodora	0.72±0.03	2.01±0.3	179.16**
Pongamia pinnata	0.19±0.02	0.27±0.01	42.10**
Prosopis spicigera	1.56±0.02	2.21±0.03	41.66**
Syzygium cumini	0.13±0.01	0.21±0.01	61.53**
Zizyphus mauritiana	0.33±0.06	0.82±0.06	148.48**

* = Significant at 5% ** = Significant at 1%

Table 27.8 : Photosynthesis rate (μ mol m^{-2} s^{-1}) in certain dicotyledonous trees growing in the reference and the polluted environmental conditions. The values represent the mean of 10 readings obtained on sunny days in the field with the help of infra-red gas analyzer.

Species	*Reference site Mean ± SD*	*Polluted site Mean ± SD*	*Per cent variation*
Damaged			
Ailanthus excelsa	6.27±1.73	4.89±0.53	22.01**
Cordia dichootoma	24.5±4.34	11.63±2.37	52.53**
Dalbergia sissoo	4.60±0.42	2.62±0.44	43.03**
Mangifera indica	4.14±0.16	2.58±0.18	37.68**
Morus alba	7.50±1.32	1.63±0.35	78.26**
Improved			
Azadirachta indica	3.74±0.04	7.86±0.44	110.16**
Eucalyptus citridora	3.82±.85	9.43±0.73	146.85**
Pongamia pinnata	5.06±0.29	7.34±0.59	45.06**
Prosopis spicigrera	3.41±0.50	9.38±0.47	175.07**
Syzygium cuminii	2.56±0.75	3.29±1.20	28.51**
Zizyphus mauritiana	2.26±1.10	7.03±3.12	211.06**

* Significant at 5% ** Significant at 1%

The magnitude of photosynthetic activity is based on the abundance of chloroplasts and the amount of green area of leaves. The amount of total chlorophyll in *Peristrophe bicalyculata* was relatively low under the pollution stress, the maximum loss being at the pre-flowering stage. Chlorophylls *a* and *b* of different plant species may react differently to the coal-smoke pollutants (Nighat 1998). Chlorophyll content may be low in leaves of plants treated with heavy metals (Veer and Lata, 1989; Gangopadhyay and Santra 1996; Saxena *et al.* 1991; Ajay and Subrahmanyam 1996), gases (Sasek and Richardson 1989; Khan *et al.* 1990; Shan *et al.* 1996) and N,P and K (Vij *et al.* 1981; Dubey *et al.* 1982; Khan and Khan 1994). It is known that N deficiency reduces chlorophyll content (Lawlor *et al.* 1989). Chloroplast damage due to coal-smoke pollutants might cause reduction in chlorophyll content in the polluted leaves (Pandey *et al.* 1991). Cement dust also has similar effect on agroecosystem (Singh and Rao 1978; Pandey and Sinha 1990). Presence ofPb is suggested to inhibit chlorophyll biosynthesis (van Assche and Clijsters 1990; Sinha *et al.* 1993).

Chlorophyll *a* was more affected than chlorophyll *b* in *Mangifera indica* and *Psidium guajava* (Kumarand Singh 1988), *Shorea robusta* (Williams and Banerjee 1995), Norway spruce (Mandre and Tuulmets 1997) and *Triumfetta rhomboidea* (Nighat, 1998). On the contrary, chlorophyll *b* was more severely affected in *Abutilon indicum, Achyranthus aspersa. Cassia occidentalis* and *Xanthium strumarium* (Amani, 1982), *Lycqpersicon esculentum* and *Chlorella pyrenoidosa* (Ajay and Subramanyam 1996), *Peristrophe andRuellia* (Nighat 1998), *Oryza sativa* (Nandi *et al.* 1986) and *Sorghum bicolor* (Joshi *et al.* 1993). The high SO_2 sensitivity of chlorophyll *b* was suspected to be due to induced chlorophyllase activity or inhibition of chlorophyll *b* synthesis (Joshi *et al.* 1993). However, in some plant species chlorophylls *a* and *b* are equally susceptible (see Singh *et al.* 1990a).

Production of superoxide radicals due to reaction of sulphite with chlorophyll under illumination might reduce chlorophyll content (Shimazaki and Sugahara 1980; Williams and Banerjee 1995). According to Singh *et al.* (1990b), the total chlorophyll and carotenoid content continued to decrease with increasing sodium meta bisulphate ($Na_2S_2O_3$) concentration in tomato leaves. SO_2 is believed to degrade chlorophyll *a* to phaeophytin by replacing $Mg2^{2+}$ ions from chlorophyll molecules (Domes 1932) but chlorophyll *b* degradation leads to the formation of chlorophyllide *b* as SO_2 removes the phytol group of the chlorophyll *b* molecules (Rao and LeBlanc 1966). These pigments, accumulated at the expense of chlorophyll, cause chlorosis and retard photosynthesis.

Continuous application of particulate matters clogs the stomata and interferes with the gaseous exchange.This may raise leaf temperature which in turn may hamper synthesis of chlorophyll (Singh and Rao 1981). At pH above 3.5, SO_2 may affect thylakoid membrane of the chloroplast by oxidation of carotenoid through generation of O_2 (superoxide redicals) from HSO_3 (Peiser and Yang 1978). This reduces quantity of chlorophyll and hence the photosynthetic ability of the plant.

Photosynthesis rate is affected severely in the polluted air (Schmidt *et al.* 1990; Takahama *et al.* 1992; Chappelka and Freer-Smith 1995; Sane *et al.* 1996); it may decline as much as up to 65% in *Peristrophe* and 75% in *Ruellia* (Nighat 1998). The net photosynthetic rate decreased in 03-exposed *Pinus armandi* seedlings (Shan *et al.* 1996) and *Lupinus albus* fed with high Cd concentration (Costa and Spitz 1997). SO_2 doses below those causing visible symptoms may reduce photosynthesis (Eckert and Houston 1980). Sometimes, photosynthesis rate temporarily increases after SO_2 fumigation; it recovers when fumigation is discontinued but then drops further when fumigation is resumed (Winner and Mooney

1980). Post-fumigation detoxification, as evident by the recovery of photosynthesis, emanates from chloroplast reactions (Veljovic Jovanobic *et al.* 1993). High concentration of SO_2 suppressed photosynthesis in *Sorghum bicolor* (Joshi *et al.* 1993) and *Gylcine max* (LeSuer-Brymer and Ormrod 1984), etc. Net photosynthesis in branches of mature red spruce trees (*Picea rubeus* Surg.) fumigated with SO_2 was depressed nearly uniformly regardless of branch position (Meng *et al.* 1995).The decline in photosynthesis is due to damage to the intersystem electron transport between photosystems, which retards the rate of electron transport through the whole chain (Ishibashi *et al.* 1997). SO_2 may change the levels of metabolic intermediates (Malhotra and Hocking 1976; Soldatini and Ziegler 1979) and inhibit photosynthetic electron transport (Shimazaki and Sugahara 1979). Stomatal closure is not a major factor contributing to the SO_2-induced reduction of photosynthesis; it is caused primarily due to breakdown of chloroplast activity rather than through changes in mitochondrial or peroxisomal activity or stomatal closure (Furukawa *et al.* 1980). Inhibition of photosynthesis in isolated spinach chloroplast was due to competition between bicarbonate and sulphite for the active site on ribulose-1, 5-diphosphate carboxylase, the central enzyme for photosynthetic CO_2 fixation (Ziegler 1972; 1973).

The absorbed SO_2 is converted into sulphite which is known to inhibit irreversibly both cyclic and non-cyclic photophosphorylations by affecting the coupling factor of chloroplast thylakoids (Ryrie and Jagendrof 1971) and Calvin cycle enzymes in leaves (Ziegler 1972; Ziegler and Libera 1975; Paul and Bassham 1978; Tanaka *et al.* 1984), with or without disturbance in the photosynthetic electron transport (Silvius *et al* 1975; Plesnicar and Kalezic 1980; Daniell and Sarojini 1981). Shimazaki and Sugahara (1980) suggested that SO_2 inhibits electron transfer at the site close to the reaction center of photosystem II and also the O_2 evolution in presence of silicomolybdic acid (Shimazaki and Sugahara 1980). At high concentrations (1 or 2 ppm) of SO_2, the PS II is inhibited irreversibly and the possible site of inhibition is close to the reaction centre (Shimazaki and Sugahara 1980; Schmidt *et al.* 1988). It increases total glutathione level and reduces ATP formation in plants (Madamanchi and Alscher 1991). Adams and Rinne (1989) eliminated PS II as the possible primary site of damage by SO_2 and suggested that appreciable acidification of stroma and accumulation of H_2O_2 could lead to the inactivation of several key enzymes necessary for CO_2 fixation (see Alscher *et al.* 1987; Pfanz *et al.* 1992). Although such an inhibition is likely to affect the photosynthetic CO_2 fixation first, it could also lead to reduction in quantum yield of O_2 evolution and the energy storage (Veeranjaneyulu *et al.* 1991).

The decrease in phospho-enolpyruvate (PEP) activity and concentration as a result of hydrolysis and mobilization from leaves must inhibit the rate of photosynthesis (Joshi *et al.* 1993). Moreover, reduction in leaf area may be an additional factor responsible for the decline (Joshi *et al.* 1993). This idea gets support from the work of Makino *et al.* (1983) who reported that the loss of chlorophyll is not always correlated to that of photosynthesis. This holds for certain species of the present study also. Decreasing photosynthesis with O_3 exposure may be caused by a decrease in stomatal conductance (Reich 1987; Hassan *et al.* 1994), oxidant damage to the biochemical processes of light harvesting and carbon fixation (Floyd *et al.* 1989; Sasek and Richardson 1989; Sanders *et al.* 1992; Pell *et al.* 1994), or damage to membranes and other cellular components (Heath 1987; Matyssek *el al.* 1991).

In the leaves of *Pelargonium zonale* L. and *Spinacia oleracea* L., fumigated with high concentrations of SO_2 for very short duration, CO_2 assimilation in the light was inhibited, while reduction of the quinone acceptor (QA) of photosystem II and oxidation of the electron donor pigment P-700 of photosystem I increased. Accumulation of dihydroxy acetone phosphate, fructose-1, 6-phosphate and ribulose -1, 5-phosphate and a decrease of 3-phosphoglycerate and hexosephosphate indicated that SO_2 inhibited enzymes of the Calvin cycle. Detoxification of SO_2 and reactivation of the photosynthetic apparatus were slow or absent in the dark. Light initially increased the extent of inhibition of assimilation, but finally it reversed the processes. SO_2 was inhibitory due to chemical reactivity of its hydration products rather than to cellular acidification by the acid produced (Veljovic Jovanovic *et al.* 1993).

Thus upon entering the leaf cell, and the chloroplast in particular, sulphur dioxide as sulphite can be oxidised to sulphate. The oxidation pathway is an aerobic process which involves light and the photosynthetic electron transport chain (Asada 1980). A chain reaction involving the production of oxyradicals terminates through the action of superoxide dismutase (SOD). The product from SOD reaction, HO_2, can itself inhibit photosynthesis at low levels through oxidation of thiol-mediated stomatal enzymes. Therefore, since carbon assimilation is inhibited by H_2O_2, the photosynthetic capabilities of the plant will be impaired if H_2O_2 is not efficiently removed (Okpodu *et al.* 1996).

SO_2 inhibits the rate of photosynthesis more in the SO_2 tolerant than in susceptible plants (Lorenc-Plucinska, 1982). In *Phaseolus vulgaris,* older healthy leaves and the severely injured leaves showed significantly lower photosynthetic rate than the younger and less injured leaves (Salam and Saja 1995). Net photosynthesis of young foliage of spruce was not affected by SO_2 significantly but decreased with the increasing needle age and/or SO_2 deposition rate (Meng *et al.* 1995). The O_3 induced decrease in photosynthesis rate was significantly less at high CO_2 concentration (Kellomaki and Wang 1997). Gaseous air pollutants influence stomatal conductance in trees (Freer-Smith and Taylor 1992; Chappelka and Freer-Smith 1995; Salam and Saja 1995; Chandrashekar 1997). A significant decrease in stomatal conductance was observed in ozone-fumigated trees (Eamus and Murray 1991). Exposure to O_3 + CO_2 also had suppressive effects (Pye 1988; Field *et al.* 1995; Kull *et al.* 1996; Kellomaki and Wang 1997). Combination of O_3 and SO_2 had a greater influence than the single pollutants, causing decrease in leaf conductance (Steubing *et al.* 1989). According to Nakano *et al.* (1997), photosynthesis was suppressed because of decrease in nitrogen content in the leaf of rice.

An increase in the atmospheric CO_2 will shift the activity of ribulose-1,5 bisphosphate carboxylase and oxygenase (Rubisco) in favour of carboxylation (Bowes 1991; Stitt 1991) and depress stomatal conductance (Morrison 1987; Ceulemans and Mousseau 1994). The low stomatal conductance decreases the intercellular concentration of CO_2 in green parts of leaves in the polluted samples resulting in a reduced net photosynthesis (Bazzaz 1990). On the contrary, lowering in the stomatal conductance in many species is caused by decreased photosynthetic rate and increased intercellular CO_2 concentration (Farage *et al.* 1991).

Sometimes pollutants raise the stomatal conductance (Steubing *et al.* 1989, Nyman *et al.* 1990; Nighat 1998). *In vitro* culture of *Luplnus albus* showed increased stomatal conductance due to Cd up to 0.1 μM concentration (Costa and Spitz 1997). The intercellular carbon dioxide was significantly increased in *Peristrophe bicalyculata, Ruellia tuberosa* and *Triumfetta rhomboidea* (Nighat 1998). Pandey and Pandey (1996) observed greater carbon allocation in *Carissa carandas* which was due to photosynthetic acclimation of plants to the existing pollution load. There was no interactive effects of O_3 and simulated acid rain on carbon allocation of armand pine seedlings (Shan *et al.* 1995).

The foregoing discussion suggests that the process of carbon assimilation, the basic phenomenon for maintenance of growth in plants, is affected, mostly adversely, by the various air pollutants. It also addresses the mechanism of injury and photosynthetic inhibition. The continued exposure of our flora to the atmospheric pollution thus puts a question mark about the maintenance and survival of those species which are sensitive and have shown high degree of susceptibility to the environmental degradation. However, the species that put up resistance to the hostile condition and those capable of utilizing the specific pollutants for their own benefit and sustenance carry special significance in the global scenario of the present industry-oriented world. Such species may be screened and used for combating the air pollution menace. The present study identifies some species that seem to improve their photosynthetic performance and may thus assimilate more carbon in the degraded environment.

ACKNOWLEDGEMENT

We thank, Ms Anjum Ara, our office assistant, for help with computer typing and appreciate Ms Farah Nighat and Ms Tarannum Khudsar for arranging the bibliography of this manuscript.

LITERATURE CITED

Adams, L. A. and Rinne, R. W. (1969). Influence of age and sulphur metabolism on ATP sulphurylase activity in soybean and a survey of selected species. Plant Physiology, 44: 1241-1246.

Agarwal, S. K. (1996). Biodiversity, pages 1-5. In: Agarwal, S. K., Tiwari, S. and Dubey, P. S. (Editors) Biodiversity and Environment. APH Publishing Corporation, New Delhi.

Ajay and Subrahmanyam, D.(1996). Heavy metal induced changes in chlorophyll pattern in the tomato and *Chlorella pyrenoidosa*. Annals of Plant Physiology, 10: 17-22.

Alscher, R. Franz, M. and Jeske, C.W. (1987). Sulphur dioxide and chloroplast metabolism. pp. 1-28. In: Saunder, A. Kosak-Channing, L.and Conn E. E. (Editors). Phytochemical Effects of Environmental Compounds. Plenum Publishing Corporation, New York.

Amani, A. Z. (1982). Studies on the population dynamics of some common weeds under the stress of environmental pollution. Ph. D. Thesis, AMU Aligarh, India.

Asada, K. (1980). Formation and scavenging of superoxide in chloroplasts with relation to injury by sulphur dioxide, Research Report, National Institute of Environmental Studies, Japan, 11: 165-179.

Bazzaz, F. A. (1990). The response of natural ecosystems to the rising global CO_2 levels. Annual Review of Ecology and Systematics, 21: 167-196.

Bowes, G. (1991). Growth at elevated CO_2: Photosynthetic responses mediated through Rubisco. Plant, Cell and Environment, 14: 795-806.

Ceulemans.R. and Mousseau, M. (1994). Effects of elevated atmospheric CO_2 on woody plants. New Phytologist, 127: 425-446.

Chandrashekar, T. R. (1997). Stomatal responses of *Hevea* to atmospheric and soil moisture stress under dry sub humid climatic conditions. Journal of Plantation Crops, 25: 146-151.

Chappelka, A. H. and Freer-Smith, P.H. (1995). Predisposition of trees by air pollutants to low temperatures and moisture stress. Environmental Pollution, 87: 105-117.

Costa, G. and Spitz, E. (1997). Influence of Cd on soluble carbohydrates free amino acids, protein content in vitro cultured *Lupinus albus*. Plant Science, 128: 131-140.

di Castri, F. and Younes, T. (1990). Ecosystem Function of Biological Diversity. Biology International (Special Issue, No. 22).

Daniell, H. and Sarojini, G. (1981). On the possible site of sulphite action on the photosynthetic electron transport chain and the light modulation of enzyme activity. Photobiochemistry and Photobiophysics, 2: 61-68.

Dochinger, L. S., Bender, F. W., Fox, F.O. and Heck, W. W. (1970). Chlorotic dwarf of eastern white pine caused by ozone and sulphur dioxide interaction. Nature (London), 24: 225-226.

Domes, W. (1932). Uber die Brauchberkeit der spectroscopic Pheophytinprobe is der Rauchshaden Diognostic, Zeischrift Pflanzenkr. Cepllenk. 421: 257-273.

Dubey, P. S. Trivedi, L. and Shringi, S. K. (1982). Pollution studies of Betu! Forest area due to Satpura Thermal Pawer Station Air Discharge. Final Report, Department of Environment, Project No. 19/27/78. Env. 53-58.

Duxbury, A. C. and Yentsch, C. S. (1956). Plankton pigment nomography. Journal of Air Pollution Control Association, 16: 145-150.

Eamus, D. and Murray, M. (1991). Photosynthetic and stomatal conductance responses of Norway spruce and beech to ozone, acid mist and frost - a conceptional model. Environmental Pollution, 722: 23-44.

Eckert, R.T. and Honston, D.B. (1980). Photosynthesis and needle elongation response *of Finns strobus* clones to low level sulphur dioxide exposures. Canadian Journal of Forest Research, 10: 357-361.

Ehrlich, P. R. (1991). Biodiversity and humanity. Science and public policy. Environmental Awareness, 14: 27-35.

Field, C.B., Jackson, R. B. and Mooney, H. A. (1995). Stomatal responses to increased CO_2, implications from the plant to the global scale. Plant, Cell and Environment, 18: 1214-1225.

Floyd, R. A., West, M. S., Hogsett, W. E and Tingey, D. T. (1989). Increased 8-hydroxyguanine content of chloroplast DNA from ozone treated plants. Plant Physiology, 91: 644-647.

Forage, P. K., Long S. P., Lechner, E. G. and Baker, N. R.(1991). The sequence of change within the photosynthetic apparatus of wheat following short term exposure to ozone. Plant Physiology, 95: 529-535.

Fosberg, F. R. (1986). Biodiversity. Environmental Awareness, 9: 125-129.

Freer-Smith, P. H. and Taylor, G. (1992). Comparative evaluations of the effects of gaseous pollutants, acidic deposition and mineral deficiencies on gas exhange of trees. Agriculture, Ecosystems and Environment, 42: 321-332.

Furukawa, A., Natori, T. and Totsuka, T. (1980). The effect of SO_2 on net photosynthesis in sunflower leaf. pages 1-8 In: Sasa, M. (Editor) Studies on the Effects of Air Pollutants on Plants and Mechanisms of Phytotoxicity. Research Report National Environmental Studies. No. 11.

Gadgil, M. (1992). Biodiversity: Time for bold steps, pages 21-23. In: Ravi, N. (Editor) Survey of the Environment 1991. The Hindu, Madras.

Gangopadhyay and Santra S. C. (1996). Effect of heavy metal on chlorophyll content and nitrogenase activity of *Azolla pinnuta* R. Br. Pollution Research, 15: 95-97.

Hassan, I. A., Ashmore, M. R. and Bell, J. N. B. (1994). Effects of O_3 on the stomatal behaviour of Egyptian varieties of radish *(Ruphanus sativus* L. cv Baladey) and turnip *(Brassica napa* L. cv Sultani). New Phytologist, 128: 243-249.

Heath, R. L. (1987). The biochemistry of ozone attack on the plasma membrane of plant cells. Recent Advances in Phytochemistry, 21: 29-54.

Hiscox, J. D. and Israelstam, G. F. (1979). A method for the extraction of chlorophyll from leaf tissue without maceration. Canadian Journal of Botany, 57: 1332-1334.

Ishibashi, M., Sonoike, K. and Watanabe,A. (1997). Photoinhibition of photosynththesis during the rain treatment: Intersystem electron. Plant Cell Physiology, 38: 96-146.

Joshi, U. N., Arora, S. K. and Luthra, Y. P. (1993). SO_2 induced changes in CO_2 fixation and photosynthetic pigments in *Sorghum bicolor* L. leaves. Annals of Biology, 9: 102-108.

Keller, T. (1980). The simultaneous effect of soil-born NaF and air pollutant SO_2 and CO_2-uptake and pollutant accumulation. Oecologia, 44: 283-285.

Kellomaki, S. and Wang, K. Y. (1997). Effects of elevated O_3 and CO_2 on chlorophyll fluorescence and gas exchange in Scots pine during the third growing season. Environmental Pollution, 97: 17-27.

Khan, M. R. and Khan, M. W. (1994). Single and interactive effects of O_3 and SO_2 on tomato. Environmental and Experimental Botany, 34: 461-0469.

Khan, A. M., Pandey, V., Shukia J., Singh, N., Yunus, M., Singh, S. N. and Ahmad, K. J. (1990). Effects of thermal power plant emission on *Catharanthus roseous* L. Bulletin of Environmental Contamination and Toxicology, 44: 865-870.

Kress, L. W. (1980). Effects of O_3 and O_3+NO_2 on growth of tree seedlings. Proc. Symposium on Effects of Air Pollutants on Mediterranean and Temperate Forest Ecosystems, page 239. Pacific South-West Forest and Range Experiment Station, Berkeley, California.

Kull, O., Sober, A., Coleman, M. D., Dickson, R. E., Isebrands, J. G., Gagnon, Z. and Karnosky, D. F. (1996). Photosynthetic responses of aspen clones to simultaneous exposures of ozone and CO_2. Canadian Journal of Forest Research, 26: 639-648.

Kumar, N. and Singh, V. (1988). Sensitivity of *Mungifera indica* and *Psidium guajana* plants to SO_2 pollution. National Academy of Sciences, Letters, 11: 167-172.

Lawlor, D. W., Konttusi, M. and Young, A.T. (1989). Photosynthesis by flag leaves of wheat is relation to protein, ribulose bisphosphate carboxylase activity and nitrogen supply. Journal of Experimental Botany, 40: 43-52.

Le-Suer-Brymer, N. M. and Ormrod, D. P. (1984). Carbon dioxide exhange rate of fruiting soyabean plants exposed to ozone and sulphur dioxide single or in combination. Canadian Journal of Plant Science, 64: 69-75.

Lorenc-Plucinska, G. (1982). Effects of sulphur dioxide on CO_2 exchange in SO_2 - tolerant and SO_2-susceptible Scots pine seedlings. Photosynthetica, 16: 140-144.

MacLachlan, S. and Zaiik, S. (1963). Plastid structure, chlorophyll concentration and free amino acid composition of a chlorophyll mutant of barley. Canadian Journal of Botany, 41: 1053-1062.

Madamanchi, R. N. and Alscher, G. R. (1991). Metabolic bases for difference in sensitivity of two pea cultivars to sulfur dioxide. Plant Physiology, 97: 88-93.

Makino, A., Mae, T. and Obita, K. (1983). Photosynthesis and ribulose 1,5-diphosphate carboxylase in rice leaves. Plant Physiology, 73: 1002-1003.

Malhotra, S.S. and Hocking, D.(1976). Biochemical and cytological effects of sulphur dioxide on plant metabolism. New Phytologist, 76: 229-237.

Mandre, M. and Tuulmets, L. (1997). Pigment changes in Norway spruce induced by dust pollution. Water, Air and Soil Pollution, 94: 247-258.

Matyssek, R., Gunthardt-Goerg, M. S., Keller, T. and Scheidegger (1991). Inpairment of gas exchange and structure is birch leaves *(Berula pendula)* caused by low ozone concentrations. Trees, 5: 5-13.

Meng, F.R., Bourque, C.P.A., Belczecoski, R.F., Whitney, N.J. andArp, P.A. (1995). Foliage responses of spruce trees to long-term low-grade sulphur dioxide deposition. Environmental Pollution, 90: 143-152.

Morrison, J. I. L. (1987). Intercellular CO_2 concentration and stomatal response to CO_2. In: Zeiger, E., Farquhar, G. D. and Cowen, I. R. (editors). Stomatal function: Stanford University Press, Stanford, C. R.

Nakano, H., Makino, A. and Mae, T. (1997). The effect of elevated partial pressure of CO_2 on the relationship between photosynthetic capacity and N content is rice leaves. Plant Physiology, 115: 191-198.

Nandi, P. K., Agrawal, and Rao, D. N. (1986). Effects of fumigating rice plants with sulphur dioxide on photosynthetic pigment and nonstructural' carbohydrates. Agriculture, Ecosystems and Environment, 18: 53-62.

Nayar, M. P. (1977). Changing patterns of Indian flora. Bulletin of Botanical Survey of India, 19: 145-154.

Nighat, F. (1998). Growth performance of certain annuals in relation to coal smoke pollution Ph. D. Thesis, Jamia Hamdard, Hamdard Nagar, New Delhi.

Nyman, L. P., Benzing, D. H., Temple, P. J. and Arditti, J. (1990). Effects of ozone and sulphur dioxide on two epiphytic orchids. Environmental and Experimental Botany, 30: 207-213.

Okpodu, C.M., Alscher, R.G., Grabau, E.A. and Cramer, C.L. (1996). Physiological, biochemical and molecular effects of sulphur dioxide. Journal of Plant Physiology, 148: 309-316.

Pandey, D. D. and Sinha, A. K. (1990). Effect of cement kiln dust on chlorophyll in maize leaf. Goebios, 17: 40-41.

Pandey, **D. D.,** Sinha, C. S. and Tiwari, M. G. (1991). Impact of coal dust pollution on biomass chlorophyll and grain characterstics of rice. Journal of Biology, 3: 51-55.

Pandey, J. and Pandey, U. (1996). Adaptational strategy of a tropical shrub *Carrisa carandas* L. to urban air pollution. Environmental Monitoring Assessment, 43: 255-265.

Paul, J.S. and Bassham, J.A. (1978). Effects of sulfite on metabolism in isolated mesophyll cells from *Papaver somniferum.* Plant Physiology, 62: 210-214.

Peiser, G. D. and Yang, S. F. (1978): Chlorophyll destruction in the presence of bisulphite and linoloic acid hydroperdxide. Phytocheinistry, 17: 79-84.

Pell, E. J., Landry, L. G., Eckardt, N. A. and Glick, R. E. (1994). Effects of gaseous pollutants on ribulose bisphosphate carboxylase/oxygenase and associated net photosynthesis. New Phytologist, 120: 397-406.

Pfanz, H., Lourth, G., Oppmann, B. and Schultz, G. (1992). Sulphite oxidation in, and sulphate uptake from the cell wall of leaves. In muro studies. Phyton, 32: 95-98.

Plesnicar, M. and Kalezic, R. (1980). Sulphite inhibition of oxygen evolution associated with photosynthetic carbon assimilation. Period. Biol. 82: 297-301.

Pye, J. M. (1988). Impact of ozone on the growth and yield of trees: A Review. Journal of Environmental Quality, 17: 347-360.

Rao, R. R. (1994). Biodiversity in India: Floristic Aspects. Bishen Singh Mahendra Pal Singh, Dehradun.

Reich, P. B. (1987). Quantifying plant response to ozone a unifying theory. Tree Physiology, 3: 63-91.

Ryrie, I.J. and Jagendrof, A.T. (1971). Inhibition of photophosphorylation in chloroplasts by inorganic sulphate. Journal of Biological Chemistry, 246: 582-588.

Salam, M.A. and Saja, G. (1995). Bush bean (*Phaseolus vulgaris* L.) leaf injury, photosynthesis and stomatal functions under elevated ozone levels. Water, Air and Soil Pollution, 85: 1533-1538.

Sanders, G.E., Coils, J.J. and dark, A.G. (1992). Physiological changes in *Phaseolus vulgaris* in response to long term ozone exposure. Annals of Botany, 69: 123-133.

Sane, V.P., Yunus, M. and Tripathi, R.D. (1996). Impact of ozone on carbon metabolism in plants, pages 295-318. In: Yunus, M. and Iqbal, M. (Editors). Plant Response to Air Pollution. John Wiley and Sons, Chichester.

Sasek, T.W. and Richardson, C.J. (1989). Effects of chronic doses of ozone on loblolly pine: Photosynthetic characteristic in the third growing season. Forest Science, 35: 745-755.

Saxena, K., Rao, M.V. and Dubey, P. S. (1991). Lead accumulation and its effect on Sorghum, maize and wheat grown adjacent to highway. Indian Journal of Environmental Health, 33: 74-79.

Schmidt, W., Neubauer, C., Kolbowski, J., Schreider, and Urbach, W. (1990). Comparison of effects of air pollutants (SO_2O_3 and NO_2) on intact leaves by measurments of chlorophyll flurescence and P700 absorbance changes. Photosynthesis Research, 25: 214-248.

Shan, Y., Feng, Z., Izuta, T., Aoki, M. and Totsuka, T. (1996). The individual and combined effects of ozone and simulated acid rain on growth, gas exhange rate and water use efficiency of *Pinus armandi* Franch. Environmental Pollution, 91: 355-361.

Shimazaki, K. and Sugahara, K. (1979). Specific inhibition of photosystem II activity in chloroplasts by fumingation of spinach leaves with SO_2. Plant Cell Physiology, 20: 26-35.

Shimazaki, K. and Sugahara, K. (1980). Inhibition site in electron transport system in chloroplast by fumigation of lettuce leaves with SO_2. Research Report, National Institute of Environmental Studies, No. 11: 79-89.

Silvius, J.E., .Ingle, M. and Bear, C.H. (1975). SO_2 inhibition of photosynthesis in isolated spinach chloroplasts. Plant Physiology, 56: 434-437.

Singh, N., Singh, S. N., Srivastava, K., Yunus, M., Ahmad, K. J., Sharma, S. C. and Sharga, A. N. (1990a). Relative sensitivity and tolerance of some gladiolus cultivars to sulphur dioxide. Annals of Botany, 65: 41-44.

Singh, S. N. and Rao, D. N. (1978). Effect of cement dust pollution on soil properties and on wheat plants. Indian Journal of Environmental Health, 20: 258-267.

Singh, S. N. and Rao, D. N. (1981). Certain response of wheat plants to cement dust pollution. Environmental Pollution, 24: 75-81.

Singh, S. N., Yunus, M. and Singh, N. (1990b). Effect of sodium metabisulphite on chlorophyll, protein and nitrate reductase activity of tomato leaves. Science of Total Environment, 81: 269-274.

Sinha, S. K., Srivastava, H. S. and Tripathi, R. D. (1993). Influence of some growth regulators and cations on the inhibition of chlorophyll biosynthesis by lead is maize. Bulletin Environmental Contamination and Toxicology, 51: 241-246.

Solbrig, O.T. (1991). From Genes to Ecosystems; A Research Agenda for Biodiversity. IUBS Monograph Series No. 17, Paris.

Soldatini, G.F. and Ziegler, I. (1979). Induction of glycotate oxidase by SO_2 in *Nicotiana tabacum.* Phytochemistry, 18: 21-22.

Steubling, L., Fangmeier, A., Both, R. and Frankenfeld, M. (1989). Effects of SO_2, NO_2 and O_3 on population development and morphological and physiological paramaters of native herb layer species is beech forest. Environmental Pollution, 58: 281-302.

Stitt, M. (1991). Rising CO_2 levels and their potential significance for carbon flow is photosynthetic cells. Plant Cell and Environment, 14: 741-762.

Takahama, U., Veljovic-lovanovic, S. and Heber, U. (1992). Effects of air pollutant SO_2 on leaves. Plant Physiology, 100: 261-266.

Tanaka, K., Mitsubashi, H.; Kondo, N. and Sugahara, K. (1984). Further evidence for inactivation of fructose-1, 6-bisphosphatase at the beginning of SO_2 fumigation. Increase in fructose-1,6-bisphosphate and decrease in fructose-6-phosphate in SO_2 fumigated spinach leaves. Research Report, National Institute of Environmental Studies, Japan, 65: 213-216.

Van Assche, F. and Clijsters, H. (1990). Effects of metals on enzyme activity in plants. Plant, Cell and Environment, 13: 195-206.

Veer, B. and Lata, K. (1989). Responses of green gram to Zn treatment. Acta Botanica Indica, 17: 192-195.

Veeranjaneyulu, K., N' Soukpoe-Kossi, C. N. and LeBlanc, R. M. (1991). SO_2 effect of photosynthetic activities of intact sugar maple leaves as detected by photoacoustic spectroscopy. Plant Physiology, 97: 50-54.

Veljovic-Jovanovic, S., Bilger, W. and Heber, U. (1993). Inhibition of photosynthesis, acidification and stimulation of zeaxanthin formation in'leaves by sulphur dioxide of these effects. Planta, 191: 365- 376.

Vij, B. M., Trivedi, L. Shevade, A. and Dubey, P. S. (1981). Chlorophyll damage in tree species due to air pollution. Biological Bulletin, India, 3: 193-194.

Williams, A. J. and Banerjee, S. K. (1995). Effect of power plant emissions on the metabolic activities of *Mangifera indica* and *Shore a robust a.* Environment and Ecology, 13: 914-919.

Winner, W.E. and Mooney, H.A. (1980). Ecology of SO_2 resistance II. Photosynthetic changes of shrubs in relation to SO_2 absorption and stomatal behaviour. Oecologia, 44: 296-302.

Younes, T. (1996). Research priorities for biodiversity conservation and management, pages 11-24. In Lee, B. H., Kirn, T. H. and Sun, B. Y. (Editors) Biodiversity Research and its Perspectives in the East Asia. Korean Institute for Biodiversity Reasearch, Chonju, South Korea.

Yunus, M. and Iqbal, M. (1996). Plant Response to Air Pollution. John Wiley and Sons Chichester, U.K.

Ziegler, I. (1972). Effect of SO_2 on the activity of ribulose 1,5-diphosphate carboxylase in isolated spinach chloroplasts. Planta, 103: 155-63.

Ziegler, I. (1973). Effect of sulphite on phosphoenol pyruvate carboxylase and malate formation in extracts of *Zea mays.* Phytochemistry, 12: 1027-1030.

Ziegler, I. and Libera, W. (1975). The enhancement of CO_2 fixation in isolated chloroplasts by low sulphite concentrations and by ascorbate. Zeischrift Naturforsch, 30 C: 643-647.

Chapter - 28

Changing Environmental Patterns and Health

*Rais Akhtar**

ABSTRACT

Environmental conditions; physical and socio-economic are undoubtedly causal factors affecting both health and disease, and have therefore, been studied in search of understanding by a variety of disciplines from time immemorial. Hippocrates' *Air, Water* and *Places* written in the fifth century B.C. shows much concern about the environmental bases of human health.

Attempt has been made in this chapter to study the web of environmental correlations of health and disease with focus on changing environmental conditions with specific examples.

Key Words : environmental change, health hazard, population impact, socio-economic factor

INTRODUCTION

A recurring theme throughout history and continuing today is the notion of dynamic equilibrium between man and his total environment (Hunter and Akhtar, 1991). Thus the environment holistically ranges from the elemental, physical environment of rocks, soils, trace-elements, climate and weather through human constructed city and sub-urban landscapes and economic environments, to less tangible but equally potent social, political and cultural environments (WHO, 1972).

As the environmentally oriented approaches were in keeping with development of epidemiology and public health, so were studies geared toward understanding disease problems in more industrialized parts of the world. Social and cultural differences were taken as determinants of the spatially variability of disease and ill health in industrialized nations (Banks, 1959). Geography, as a humanistic science examines relationships between peoples and their environments with focus upon these interactions and relationships that

*Department of Geography & Regional Development University of Kashmir, Srinagar-190006

bear specifically upon human health (Hunter, 1988). The fact that endemic diseases maintain themselves in dynamic equilibrium over-time is consonant with the idea of disease as an agent-host-environment interaction and with the wider concept that all organisms, man and pathogens included, strike to maintain an equilibrium or homeostatic relationship with their environment. Thus it has been rightly pointed out that disease is the outcome of environmental and behavioural interactions.

CHANGING ENVIRONMENT

Concept of multifactorial associations and ecosystem replaced single factor hypothesis for disease etiology which by definition draw in biological and cultural environments and behaviours as atleast the context for disease processes (Armstrong, 1985). It follows that any processes of change, whether of the environment per se or of man's adaptational behaviour, will disrupt the balance and among other consequences, new diseases or health patterns will emerge. Such adaptational reactions follow when carriers regionally introduced "New" diseases or susceptible hosts. Major environmental changes are affected; and human life styles are significantly altered (Hunter, 1988). The introduction and spread of Acquired Immuno Deficiency Syndrom (AIDS) both in developed and developing countries is due to changing lifestyles particularly sexual habits. The substantial growth in the incidence of cancer in developing countries including India is yet another example of changing life styles especially the diet pattern. The outbreak of plague in India in 1994 and the spread of Abiola virus disease from Zaire (now Democratic Republic of Congo) to west African countries in 1995 are other examples of changing pattern of man's adaptational behaviour. With regard to environmental changes, and the rise of new or increased disease hazards, some of the clearest contemporary examples, include the incidence of malaria and schistosomiasis due to irrigation schemes in Africa and other Third World countries and the health hazards as a result of environmental pollution, and the process of urbanization and industrialization.

Within the broader dimension of environmental changes greater emphasis is being placed on climatic changes occurring in different parts of the world-increased incidence of skin cancer in north-western Europe due to ozone depletion, as a result of global warming. Besides the sudden occurrence of rainfall in dry and semi-dry areas led to the outbreak of malaria in desert part of Rajasthan. Similarly climatic warming also resulted in the high incidence of malaria in Rwanda.

POPULATION CHANGES

The marked population changes which often appear to accompany 'modernization' are now widely recognized. Epidemiological transition implies a range of changes; in attitude, education, diet, aspirations, urbanization, public health, and health care and its technology (Phillips, 1990). Thus there is a sequence linked with socio-economic development in which its burden of disease shifts from the infectious to the broadly degeneration ones (Learmonth, 1988). Momiyama Sakamoto argued that a good index of such changes may be found in decreasing seasonal variations in mortality as diet, clothing and domestic heating insulate people from adverse effects of the seasons (Sakamoto, 1977).

Three main successive phases, as propounded by Omran in his theory of Epidemiological Transition can be identified; the age of pestilence and famine, the age of receding pandemics and the age of degenerative or man-made diseases (Phillips, 1990). According to Learmonth, the epidemiological transition idea runs parallel to the demographic transition; that in demographic trends overtime, low-technology societies tend to have high birth rates and high death rates (usually with fluctuations but maintaining some sort of ecological equilibrium in the long term); with economic development there tends to be a phase of continuing high birth rates but such lower death rates and so of rapid population growth related to curative and especially preventive medicine, improved hygiene, nutrition and housing and that this is followed by a phase of renewed population stability or slow increase, linked with low birth rates and low death rates (and probably with increased proportions of old people) (Learmonth, 1988). It is argued that most western countries have undergone changes and are now experiencing the third and last phase of epidemiological transition with the dominance of degenerative diseases (cancer, heart and other cardiovascular diseases and AIDS) and man-made diseases related to technological hazard (chemical and nuclear leakages). However, most Third World countries experiencing both second and third phases with both infectious diseases-outbreaks of malaria and cholera, as well as increased incidence of degenerative diseases are widely reported.

STUDY OF SELF STUDY ENVIRONMENT

In the context of general environmental conditions and disease association, Armstrong provided a new concept and the methodology in the study of geographical perspective of cancer (Armstrong, 1976). According to him the environment and cancer studies must be based at the primary epidemiological analysis, namely to the individual patient and control subject. Case and control studies of the environments specific to individuals (self specific environments) can be conducted that explore differences indicative of associations between disease and environment. Armstrong used this methodology in a case among 53 Chinese patients with nasopharyngeal Carcinoma and 53 non disease subjects in Selangor, Malaysia. Daily activity patterns are used to describe the use of time and space in terms of selected activities, selected settings, census districts, travel and air pollution zones. The study noted that differences between patients and controls exposure time to zones of air pollution were not significant during week days; there was a slight tendency toward higher exposure for patients than non-patients control during week ends. The methodology described in this study establishes a time and space framework for self specific environments for subsequent geographical analysis.

Thus there are opportunities for medical geographers to play a more active role in policy formulation and in the maintenance (or creation) of a congenial environment.

ENVIRONMENTAL HEALTH AS A FIELD OF STUDY

Environmental health is concerned with assessing and managing the impacts of human population on their environment and the impacts of environment on populations (Moeler, 1992). The basis for assessment is epidemiological, that is, identification of the agents of ill-health, the populations affected; and the environments of exposure (Fig. 28.1), while the current scope of environmental health gives emphasis, to the physico-chemical and biological spheres of the environment, the cultural sphere is seen to be increasingly important (Armstrong, 1993).

According to Armstrong, the impacts of human populations on the environment, and vice-versa, have accelerated in the last 100 years because of increased numbers and densities of people, increased per-capita consumption of environmental resources, and increased per capita production of energy, goods and services (Fig. 28.2.) These activities have led degradation of ecosystems, depletion of resources, and deterioration of environmental quality through pollution. One severe example of a pollution impact is the depletion of ozone in the upper atmosphere with consequent increase in ultra-violet radiation on the earth surface and a rising incidence of skin cancers and other ill-affects in human population.

Environmental health seeks to define and manage those qualities of the environment that support optimal human health, which essentially means conservation of the environment to which human beings become adapted through evolution. These optimal environmental qualities for human health and continued evaluations are changing under the impact of current forms of socio-economic development, because high rates of population growth multiply this impact (Armstrong, 1993). According to Armstrong, the environmental diseases can only be avoided by changing present human behaviour primarily in the economy and in rate of reproduction. In this sense environmental health is a determinants of the nature and rate of socio-economic development (Armstrong, 1985).

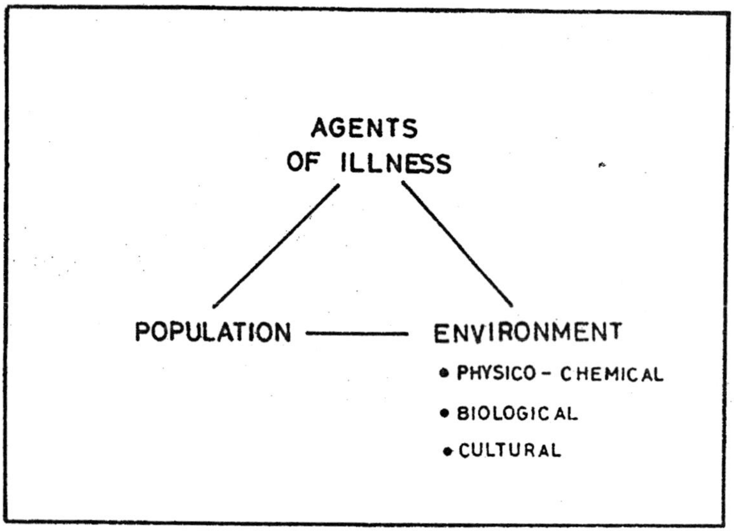

Fig. 28.1: The Epidemiological Triange
Source : Armstrong, 1993

Similar arguments concerning environmental disaster were put forward in an editorial entitled" The Demographic Trap and Sustainable Health" published in the 1991 issue (Vol. 62, January/February) of Candian Journal of Public Health. It says that the earth cannot any longer afford its human occupants. The carrying capacity of the earth is already exceeded in many places where famine is an actuality or a potential threat in the very near future. World grain reserves and other food stocks have been falling for the last four years, the current supply would last less than 60 days. There is simply not enough to feed the expected numbers by the end of this decade, and no way at all of reaching the necessary amount. Global warming seems certain to reduce harvests in many places, notably the grain belt of north America. If we do not control our own numbers very soon, they will be controlled for us by mass starvation. Because hungry people do not die quietly, starvation will be accompanied by desperate struggles-regional and global wars-for diminishing

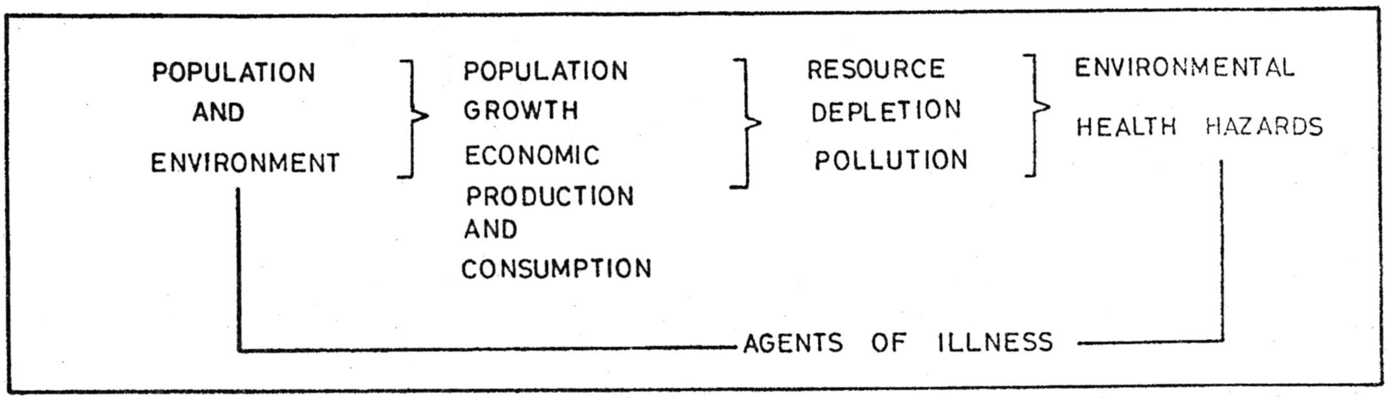

Fig. 28.2 : Cycle of Major Environmental Health Concerns.
Source : Armstrong, 1993.

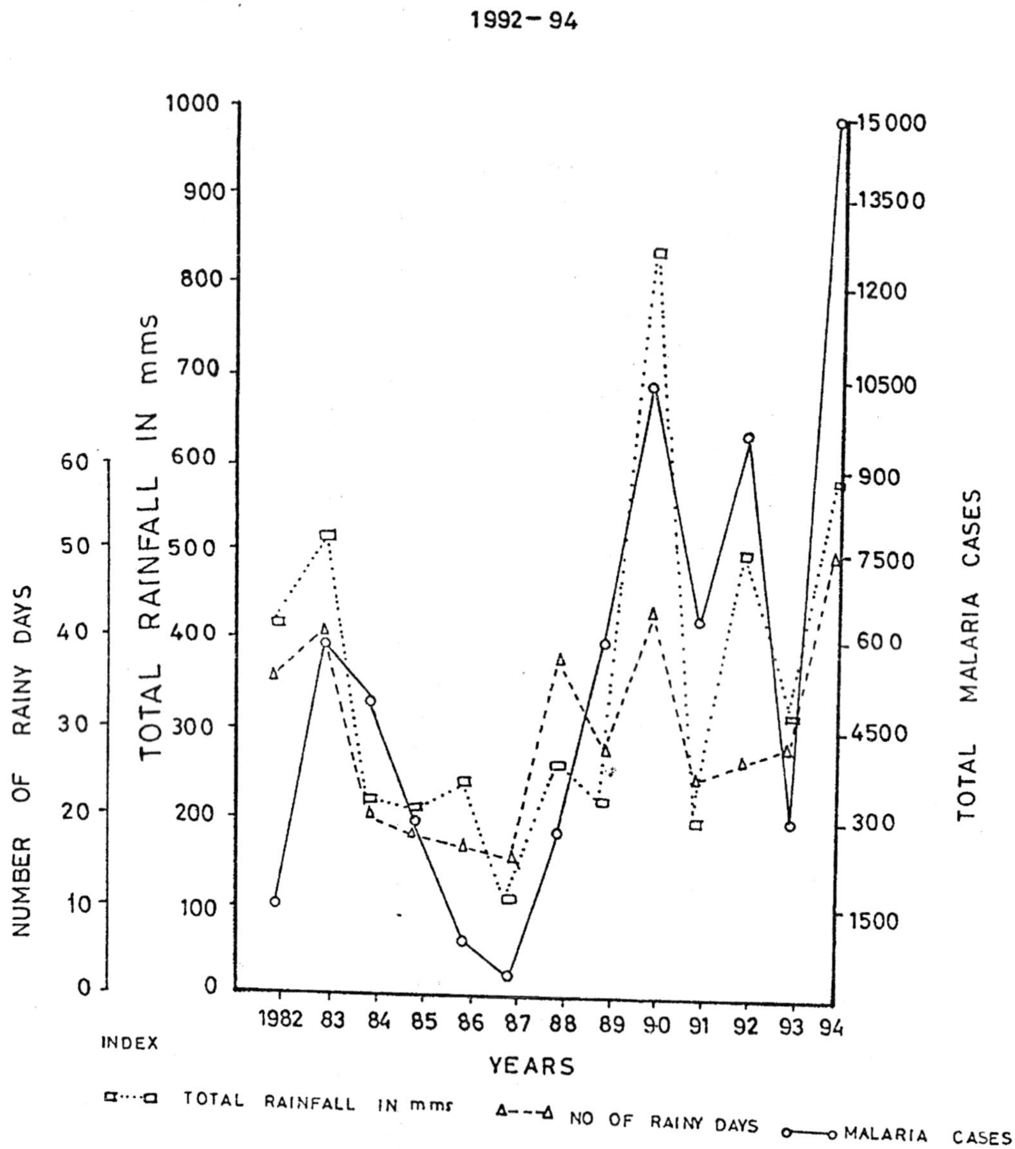

Fig. 28.3 : Malaria, rainfall and rainy days in the thar desert.

supplies; and by epidemics among population weakended by malnutrition. In these ways the population explosion might crash into reverse long before numbers reach the 14 billion predicted by the middle of the next century. Such a population crash would be an appaulling catastrophe, sweeping away with much of what we value about our civilization. The only way to prevent it, is for us all to work together to control the fundamental cause of over-population (Doll 1992).

RAINFALL DISTRIBUTION IN THAR DESERT (1982-94)

Based on thirteen years data, attempt has been made to study the association between rainfall conditions and the incidence of malaria. Figure 28.3 shows that both the total annual rainfall and the number of rainy days are positively correlated with the total positive cases of malaria in the epidemic years of 1983, 1990, and 1994. Thus the hypothesis of meteorological correlation is proved positively.

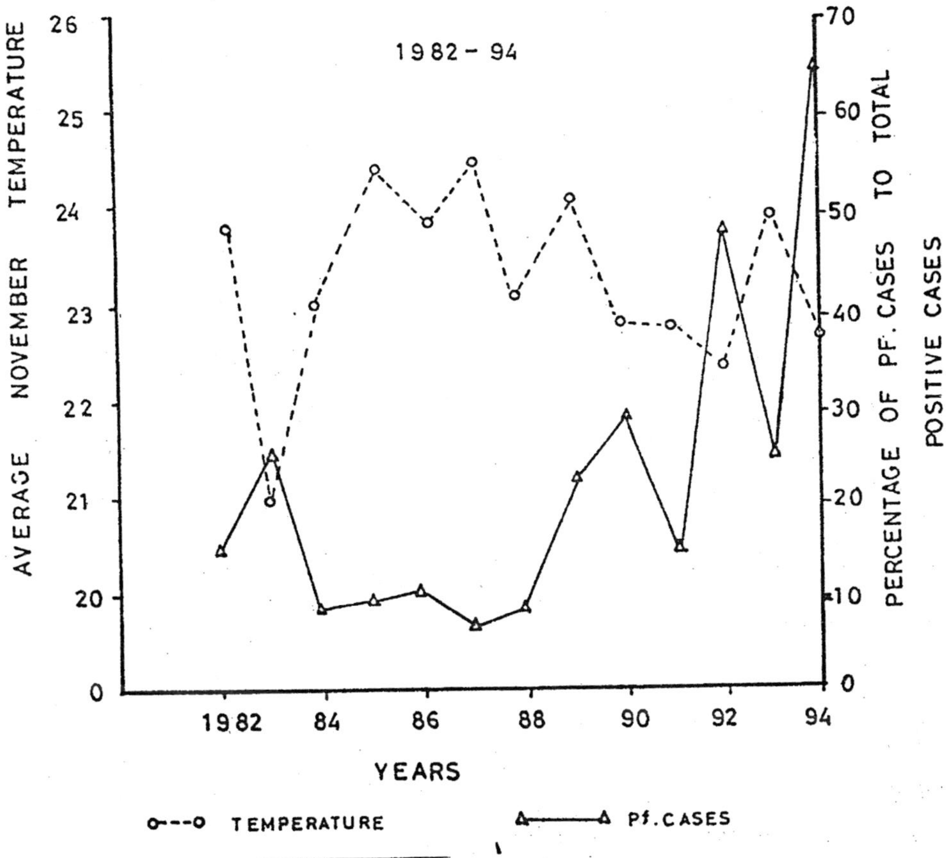

Fig. 28.4: P. Falciparum cases and average November temperature in the that desert

Attempt has also been made to correlate the average November temperature (average mean minimum and maximum) with the incidence of p. falciparum cases during 1982-94, the correlation has been negative, as there is hardly any positive correlation in any year except during 1988 and 1989. Regarding the epidemic correlation with the rainfall in 1994, it should be noted that the total monsoon rainfall (June to September) in the year 1994, was one of the four years which experience highest rains since 1871 (other years being 1917, 1944, 1990). Besides, Table 28.1 reveals that since 1982, the highest number of rainy days occurred (i.e. 51) only in 1994 (Akhtar & McMichael, 1996 and Akhtar, 1995). Patrick Hehir in his book : *Malaria in India*, published in 1927, identified the desert region with moderate epidemic zone (Hehir, 1927). He further said that malaria is markedly in character (usually autumnal) and moderately prevalent. It may be mentioned that no study was conducted in the desert area in the past except the one by Green who studied the distribution of malaria parasites in Nasirabad (near Ajmer), a border zone with desert area (Knowles & White, 1930). Green stated that in 1909, 90 per cent of malaria occurred due to P. viviax, with the remaining 10 per cent due to P. falciparum.

Table 28.1 : Average November temperature, total annual rainfall number of rainy days, total malaria cases, and percentage of *P. falciparum* cases to total cases (Jodhpur)

1982 1983 1984 1985 1986 1987 1988 1989 1990 1991 1992

	1982	1983	1984	1985	1986	1987	1988	1989
Average Nov. Temperature (⁰C)	24.0	21.0	23.0	24.5	24.0	24.6	23.2	24.2
Total annual rainfall (mm).	417	513	231	214	249	111	270	230
Number of rainy days	36	41	21	20	17	16	38	28
Total malaria cases	1509	5934	4803	2918	1034	399	2877	6011
Percentage of *P. falciparum* cases to total cases	15	25	9	10	11	7	9.5	23

	1990	1991	1992	1993	1994
Average Nov. Temp. (⁰C)	23.0	23.0	22.4	24.1	22.7
Total annual rainfall (mm).	844	204	526	326	595
Number of rainy days	44	25	27	29	51
Total malaria cases	10462	6374	9685	3096	14919
Percentage of *P. falciparum* cases to total positive cases	28.7	16.3	48.3	26.0	64.2

Source of data : Meteorological data obtained from Arid Research Centre, Jodhpur.

Malaria data obtained from National Malaria eradication Programme, New Delhi (Malaria data pertains to Jodhpur District).

CONCLUSION

The above discussion suggests that health encompasses multidisciplinary dimension. The theme environmental impact on health has become important in the context of health problems arising out of environmental pollution. Attempts should also be made to undertake both macro and macro level regional studies. However, emphasis should also be placed on the relationship between environmental change and changing health and disease pattern. It is important to predict the environmental change based on the knowledge of history of the natural and human-induced factors that have caused environmental change. The rapid urbanization, population growth and unplanned industrialization are important socio-economic factors which cause environmental change. Besides the natural phenomenon i.e. El-Nino (warm sea temperature) in the southern and eastern pacific ocean causes temperature and rainfall changes globally with heavy impact on Southern and Eastern Asia and Africa. It is expected that these aspects would be among most pressing research problems in the 21st century.

LITERATURE CITED

Akhtar, R. and McMichael, A.J. (1996). Rainfall and malaria outbreaks in Western Rajasthan, *The Lancet*, 348, November 23, 1457-1458.

Armstrong, R.W. (1976). The geography of specific environment of patients and non-patients in cancer studies, with Malaysian example, *Economic Geography*, 52, 161-170.

Armstrong, .W. (1985). *Medical geography*, In Ruffini, J.L. (editor) *Advances in Social Sciences*, Vol. I, Gordon & Breach New York, 173.

Armstrong, R.W. (1993). Environmental health as a determinant of socio-economic development: The need to modify contemporary Malaysian strategies, *Akademika*, 42&43, 125-142.

Banks, A.L. (1959). The study of geography of disease, *Geographical Journal*, 125, 119-210.

Doll, R. (1992). Health and environment in the 1990s, *American Journal of Public Health*, 82, 933-941.

Hehir, P. (1927). Malaria in *India, Humphrey* Milford, Oxford University Press, London.

Hunter, J.M. (1988). Geography and public health, *In Earth' 88: Changing Geographic Perspectives*, National Geographic Society, Washington D.C., 119.

Hunter, J.M. and Akhtar, R. (1991). The challenge of medical goegraphy, In Akhtar, R. (Editor). Environment and Health *Themes in Medical Geography*, Ashish Publishing House, New Delhi, 6.

Knowles, R. and White, R.S. (1930). Studies in the Parasitology of Malaria, Indian Medical Research Memoirs, Supplement Series, *Indian Journal of Medical Research*, 49.

Learmonth, A. (1988). *Disease Ecology*, Basil Blackwell, Oxford, 25-26 Moeller, D.W. (1992) *Environmental Health*, Harvard University Press, Cambridge, Massachussets,

Phillips, D.R. (1990). *Health and Health Care in the Third World*, Longman, London, 41.

Sakomoto, S.M. (1977). Seasonality in Human Mortality: A Mediogeographical Study, Tokyo University Press, Tokyo

WHO (1972). Health Hazards of Human Environment, WHO, Geneva.

Chapter - 29

Impact of Man on Fish Diversity in Kashmir

*A.R. Yousuf**

ABSTRACT

The fish fauna of Kashmir Himalayas is very peculiar in having the dominance of schizothoracine rather than cyprinine group as is generally found in other parts of the subcontinent. Because of geographical barriers the schizothoracine group has evolved a number of endemic species in the valley. During the past several decades the endemic fish fauna of the valley, including several schizothoracines, cobitids and sisorids, has been under tremendous stress because of human activities like introduction of exotic species, pollution and encroachment of aquatic habitats and construction of hydro-electric power stations. Population density of several endemic species has declined appreciably in the valley and from certain aquatic systems some of them have almost completely disappeared. In order to save the schizothoracine fish from further decline in the aquatic habitats of the valley some immediate conservation measures like check on pollution and encroachment of the aquatic systems and construction of hatcheries need to be taken. Ban on introducing any more exotic species like grass and silver carps without studying their impact on the local fish fauna is also needed.

***Key Words** : Conservation, human interference, ichthyofauna, Kashmir, population decline*

INTRODUCTION

The valley of Kashmir (33°- 35′N and 73°, 48′ - 75°, 30′ E) is a great synclinal trough in the northwestern complex of the Himalayan ranges and is surrounded by an almost unbroken ring of mountains which represent its bordering anticlines giving it the character of an enclosed vale (Kanth & Amin, 1996). There is left only one gap, the gorge near Baramulla on the north-western end through which the river Jhelum, being the line drainage system of the region, breaks and pursues a tortuous course to the plains of the Punjab (Yousuf, 1996). The Jhelum river system comprises mountain tarns, torrential streams, flatland lakes and springs and forms an important highland fishery resource of the region.

* P.G. Department of Zoology The University of Kashmir Srinagar - 190 006

Prominent streams of the valley include Bringhi, Lidder, Vishav, Rambiarrah, Romshu, Dudhganga, Sindh, Pohru, Mandri, Boniyar, Dagwan, Erin, Madhumati, Ningal and Kishanganga. All of these join the river Jhelum directly or indirectly. While in the upper reaches these streams are poor in nutrients and show circum-neutral pH (7.20 - 8.52), containing high oxygen content (9.4 - 12.5 mg/I), concentration of different ions increases towards their lower reaches and in the river Jhelum appreciable quantities of various plant nutrients are observed throughout the year.

The concentration of different nutrients increases in the iotic habitats in direct relation with the increase in the human settlements. The plant community in these habitats includes mainly the sessile algae. Bacillariophyceae is the most dominant component contributing as high as 90% of the algal community (Kumar and Bhagat, 1978). Prominent diatoms include *Navicula, Gomphonema, Cymbella, Amphora, A. Canthidium* and *Cyclotella,* Chlorophyceae, Desmidaceae and Cyanophyceae play very little role in the algal diversity of the streams. *Oscillatoria* and *Merismopedium* are some of the genera found in the polluted areas of these streams. The zooplankton is very low and is represented by rotifers (*Keratella, Polyarthra, Philodina,* etc.) and a few crustaceans.

The benthic life has been investigated in some iotic habitats (Kumar & Bhagat, 1978; Lingdell & Engblom, 1991; Yousuf, 1996) and the species diversification has been found to decrease with increase in the pollution level. Main contributors of the bottom community are insects, crustaceans and oligochaetes. *Baetis, Ephemerella, Heptagenia, Epeorus, Caenis, Amphinemura, Perla, Nemurella, Hydropsyche, Dytiscus, Antherix, Chironomus,* etc. are some of the prominent benthic organisms. Due to pollution stress the benthic community in the Jhelum and its tributaries shows appreciable diversity. *Baetis* and *Hydropsycha* and Chironomids are well represented in unpolluted habitats and oligochaetes, crustaceans (like *Gammarus fossarum, G. lacustris*) and bivalves are dominant in polluted water (Lingdell & Engblom, 1991; Yousuf, 1996).

The lentic habitats are very diversified, some being in the ultra-oligotrophic category (like Kousarnag, Alapathar, Shishnag, and many other mountain lakes and tarns of the valley). Many lakes like Manasbal, Nilnag, Nageen etc. fall in the mesotrophic category, while others like Dal, Anchar, Khushalsar etc. have already entered the eutrophic and hypertrohic conditions. The variations in the water characteristics of different lentic habitats are closely related with the human impact. The mountain lakes and ponds, which are situated in areas least visited by human populations, receive very little pollutants except, of course, the ones which may find their way with rain or snow. Their catchment area is also little disturbed and as such the water quality is good. All these habitats are characterized by circum-neutral pH with a tilt towards acidic side (6.5 - 7.5). Nutrient load is very low in such waters. The valley lakes, ponds and wetlands show varying concentration of nutrients in close association with the nature of human society. The rural water bodies like Manasbal, Malpursar, Nilnag etc., show much lower nutrient concentration than the suburban waters and peak pollution load is found in urban lakes and ponds like Dal, Khushalsar etc. (Table 29.1). In oligotrophic waters the simplicity of habitat has resulted in low diversity of plants and hence animals. The algal community of the lentic habitats also shows variations in close association with the water quality. The algal community of such waters is represented mainly by *Navicula, Fragillaria, Staurastrum, Cosmarium, Pandorina,* etc. Macrophytes are totally absent. The meso - and eutrophic waters of the valley have fair to dense growth of macrophytes which has resulted in the complexity of the habitats. Here the algae is represented by *Pediastrum, Scenedesmus, Cosmarium, Spirogyra, Staurastrum, Epithemia,*

Navicula, Gomphonema, Microcystis etc. The species diversity gets again decreased in hypertrophic environments where dense populations of only a few species are observed. In this type of habitat *Chlamydomonas, Synedra, Pinnularia, Euglena, Phacus* and *Spirulina* are dominant plankters.

Table 29.1: Physico–chemical characters of some typical water bodies of Kashmir Himalaya (mean values).

	Lakes					
Parameter	*Gangabal*	*Nundkol*	*Manasbal*	*Malpursar*	*Anchar*	*Khushalsar*
Water temp	10.9	11.7	15.4	14.1	13.5	13.8
pH	6.85	6.98	7.94	7.8	8.0	7.5
Transparency	167	-	201	46	52	42
Dis. Oxygen (mg. l^{-1})	12.4	12.0	7.5	10.8	8.6	5.3
Bicarbonates (mg. l^{-1})	18	19	127	138	137	179
T.P. (ug. l^{-1})	8.5	10	23.5	23.3	32.0	135
T.I.N. (ug. l^{-1})	154	175	339.5	533.5	652.3	2870
Trophic status	Oligo-trophic	Oligo-trophic	Meso-trophic	Meso-trophic	Eutro-phic	Hyper-trophic

T.P = Total phosphorus, TIN = Total inorganic nitrogen

Consequent to limited algal diversity in oligo - and hypertrophic waters the zooplankton is also rather limited. In the former the zooplankton is represented mainly by *Arctodiaptomus, Canthocamptus, Cyclops* (Copepoda), *Alona, Daphnia, Kellicotia* (Cladocera), *Lecane* and *Monostyla* (Rotifera), while in the latter *Hexarthra, Filinia, Brachionus* (Rotifera), *Moina* (Cladocera) and *Cyclops* (Copepoda) are the dominant genera. The meso–and eutrophic lakes and wetlands of the valley have a diverse zooplanktonic associations. Main contributors include *Keratlla, Polyarthra, Lepadella, Anuraeopsis, Brachionus, Mytilina, Lecane, Monostyla* (Rotifera), *Ceriodaphnia, Daphnia, Diaphanosoma, Bosmina, Chydorus, Eurycerus* (Cladocera), *Cyclopds, Mesocyclopds, Eucyclops, Acanthodiaptomus, Canthocamptus* (copepoda). The bottom fauna of the oligotrphic mountain lakes is yet to be investigated. The benthos of valley lakes and wetlands includes oligochaetes (*Limnodrilus, Tubifex, Aelosoma*), Crustaceans (*Gammarus*), insects (*Chironomus*) and molluscs (*Lymnaea*).

DIVERSITY IN ICHTHYOFAUNA

The ichthyofauna of the aquatic habitats of Kashmir valley is quite distinct from that of the plains of India. So far 37 species of fishes have been reported from different aquatic systems of the valley (Table 29.2). As the taxonomy of many a species is still under scrutiny, the total number may increase or decrease with further investigations in the field. However, whatever the number, one thing is clear that a number of species are endemic to the region. It is also evident from the list of species that the sub family schizothoracinae of family Cyprinidae dominates other groups–cyprininae and garrinae of family Cyprinidae, botinae and nemachilinae of family Cobitidae, and families Poecilidae, Sisoridae and Salmonidae. As has been shown in the Table 29.2, some of the reported species have never been subsequently recorded and their presence in the valley is rather doubtful, especially that of *Ompok bimaculatus* (Bloch), which is distributed in the plains of India in habitats not characteristic of Kashmir. The list also includes names of five species (two cyprinine, one poecilid and two salmonids), which have been introduced from outside. The introduction of exotic species started at the turn of the present century when two salmonids, *Salmo trutta fario* and *S. gairdneri*, were introduced in the streams of the valley. Introduction of the salmonids was to attract European tourists to the valley who loved to go for angling expeditions in the valley. Both the species got well established in the cold water streams of the valley by 1912. In 1955-56 three varieties of *Cyprinus carpio*, viz., scale carp, miror carp and leather carp, were introduced into the valley. This species was introduced for augmenting the fish production from the water bodies of the valley. Because of its eurytopic nature it served this purpose very well and two varieties, viz., mirror carp (*C. c. specularis*) and scale carp (*C. c. communis*) got established within a short period of time. By 1963 the fish contributed about 75% of the commercial catches from the flatland lakes and the slow flowing areas of the river Jhelum (Das & Subla, 1963). The purpose of introduction of *Gambusia affinis* (in 1920's) and *Carassius carassius* (probably in 1950's alongwith *C. carpio*) is not known. However, both these fishes established their populations in the flatland lakes, ponds and wetlands and even in the slow flowing areas of the river Jhelum. Presently *C. carassius* is contributing significantly to the fish catch from lakes of the flatland area. In early 1970's two more salmonids, Salmo salar and Salvelinus fontinalis were introduced on experimental basis by CIFRI (Central Inland Fishery Research Institute, Barrackpore, West Bengal) (Jhingran, 1975). However, both these species failed to established their stocks in the valley.

Table 29.2 : List of fish species reported from Kashmir (From Yousuf, 1996)

	Zoological name	*Kashmiri name*
	1	*2*
ORDER CYPRINIFORMES		
Family Cyprinidae		
Sub-Family Schizothoracine		
1.	*Schizothorax curvifrons* Heckel	Sater gaad
2.	*S. esocinus* Heckel	Chhuroo
3.	*S. hugeli* Heckel	
4.	*S. longipinnis* Heckel	
5.	*S. micropogon* Heckel	Ramgaad
6.	*S. nasus* Heckel	
7.	*S. niger* Heckel	Ale gaad
8.	*S. planifrons* Heckel	Chosh
9.	**S. progastus* Heckel	
10.	**S. punctatus* Heckel	
11.	*Oreinus Plagiostomus* Mclelland	Khont

	1	2
12.	**O. sinutatus* (Heckel)	Khont
13.	**Ptychobarbus conirostris* Steind.	
14.	*Diptychus maculatus* Steind.	
15.	**Schizopygopsis stoliczkae* Steind.	
Sub-family Cyprininae		
16.	#*Cyprinus carpio* Linn.	Panjaeb gaad
17.	*Labeo dero* (Heckel)	Roput
18.	*L. dyocheilus* (McClelland)	Heol, roput
19.	#*Carassius carassius* (Linn.)	Gang
20.	*Puntius conchonius*	Safed bacha
Sub-family Garrinae		
21.	**Gara gotyla* (Gray)	
22.	*Crosocheilus diplochilus* Heckel Tetthur	
ORDER SILURIFORMES		
Family Cobitidae		
Sub-family Botinae		
23.	Botia birdi Chaudhury	Rama gurun
Sub-family Nemachilinae		
24.	*Nemachilus gracilis* Day	Ara gurun
25.	*N. vitatus* Heckel	Ara garun
26.	*N. kashmirensis* Hora	Ara gurun
27.	*N. marmoratus* Heckel	Ara gurun
28.	*N. rupicola* McClelland	Ara gurun
29.	*N. yasinensis* Alcock	Ara gurun
30.	*N. stoliczki* Steind.	Ara gurun
Family Sisoridae		
31.	*Glyptothorax kashmirensis* Hora	Anuir
32.	*G. reticulatum* McClelland	Anuir
33.	*Exostoma stoliczki* Day	
Family Siluridae		
34.	**Ompok bimaculatus* + (Bloch)	
ORDER CYPRINODONITIFORMES		
Family Poecilidae		
35.	*Gambusia affinis* (Baird & Girard)	Mahe gaad

1		2
ORDER SALMONIFORMES		
Family Salmonidae		
36.	#*Salmo trutta* Linn.	Troutth
37.	#*S. gairdnri* Richardson	Troutth

* = Not recorded subsequently # = Exotic species

\+ = Probably reported due to labelling mistake

HUMAN IMPACT ON ICHTHYOFAUNA

The population of native fish in the valley, especially the schizothoracine species, has considerably decreased over the years. Several species have become rare, having been driven out of some of their original habitats. For example, *S. micropogon* is no more found in Anchar and Manasbai lakes and its population in the Dal lake is negligible. The population of *S. esocinus* and *Oreinus plagiostomus* has also been affected in the lacustrine habitats. *Botia birdi*, was a common feature of the flatland lakes about a decade back and was easily found in the commercial catches, is now very rare. Similarly the schizothoracine fish have altogether been exterminated from the Khushhal sar lake. While looking for the possible reasons for this trend one is struck with several factors which seem to have worked synergistically to influence the changes in the fish species composition in the aquatic habitats of Kashmir. The important factors which seem to be responsible for the decline in the native fish population include: (a) Introduction of exotic fish species; (b) Eutrophication and Pollution of water bodies; (c) Government Policy towards Fishery Management and (d) Hydro - power barrages.

(a) Introduction of exotic species and Competition for food

Introduction of exotic species from other parts of the world not only led to change in the fish species diversity of the region, but also resulted in a change in the pattern of feeding competition between different species. Yousuf (1993, 1995 and 1996), Yousuf and Firdous (1994) and Firdous (1995) have studied the feeding habits of several species. A perusal of the data reveals that the similarity in the food items of six studied fish varies from 31.3% between *Schizothorax esocinus* and *Cyprinus carpio specularis* to 51.7% between *S. niger* and *S. curvifrons* (Table 29.3). The data also make it clear that the competition between the fish is especially severe for some important components of the diet. All the six fishes compete very severely for insect and crustacean food. Copepods including *Cyclopds, Mesocyclopds, Macrocyclops, Eucyclops*, and *Canthocamptus* formed 10.14% - 26.45% of the food of these species. Similarly insect larvae, especially Chironomus, formed a significant portion of the diet of most of these fish. Oligochaetes formed about 39% of the live food of *C.c. communis*. The fish competed for this item with *C. c. specuilaris, S. niger* and *Carassius carassius. S. esocinus, S. curvifrons* and *S. niger* were also found to be competing for macrophytes with *Cyprius carpio* varieties (Table 29.4).

S. esocinus was noticed to have appreciable amounts of its food (40.8%) composed of fish tissue. Fish tissue was also found in the gut contents of *S. niger*, although in lesser quantities (21.4%). *Salmo trutta* and *S. gairdnerii* are true carnivores and consume significant quantities of fish tissue. This leads one to conclude that both these species not only compete with *S. esocinus* and similar other native species for food items of animal origin, but may also be devouring fry and fingerling stages of the local fish, especially the schizothoracines, as their breeding grounds lie in the areas inhabited by the trout. The competition for food among the exotic and native fish and the predation of trouts on the schizothoracine juveniles has significantly affected the population of the latter group.

Table 29.3: Food similarity index of six cyprinid fishes from Anchar lake, Kashmir (Adapted from Firdous, 1995)

	Schizothorax curvifrons	*Schizothorax niger*	*Cyprinus carpio communis*	*Cyprinus carpio specularis*	*Carassius carassius*
Schizothorax niger	51.7				
Cyprinus carpio communis	37.2	39.1			
Cyprinus carpio spècularis	37.4	38.1	46.3		
Carassius carassius	39.3	43.9	40.2	46.8	
Schizothorax esocinus	41.6	34.6	36.6	31.3	37.0

Table 29.4: Food items of six commercially important fishes in Anchar lake, Kashmir (Adapted from Firdous, 1995).

Food item	*S. curvfrons*	*S. esocinus*	*S. niger*	*C. c. specularis*	*C. c. communis*	*C. carassius*
Insects	33.0	7.0	24.0	23.0	11.5	34.0
Crustacea	24.5	0.5	8.0	32.0	27.0	41.5
Fish tissue	2.0	70.0	21.5	4.0	2.5	2.5
Mollusc tissue	1.5	-	4.5	2.0	3.0	1.0
Annelids	3.5	-	11.5	21.0	39.0	10.0
Planarians	0.5	-	0.5	-	-	4.0
Algae	2.0	-	1.0	-	-	-
Macrophytic tissue	31.0	10.5	19.5	18.0	13.0	1.5
Unidentified	2.0	12.0	9.5	-	3.0	5.5

The breeding behaviour of the local fish have added to their difficulties. The fecundity data of some of the schizothoracine fishes of the valley together with that of *C. carpio* are given in Table 29.5. From the data it is apparent that the exotic carp is three to five times more fecund than the endemic carps. The low fecundity of the schizothoracine fish result in the production of low number of young ones as compared to the common carp. The problem is further compounded by the spawning behaviour in these fish. Except for *S. niger*, all other schizothoracine species undergo spawning migration from lakes and lower reaches of rivers to the high altitude torrential streams. Their ideal breeding grounds lie at places where the bottom is gravelly and the water and the water is cool and well oxygenated.

There has been increased human activity in the vicinity of these habitats during the past several decades resulting in siltation of these areas and hence in reduction in the breeding grounds of the fish. The spawning migration takes place during spring/early summer at a time when the melting of snow increases the water flow in the streams significantly. The fish are therefore able to go upstream easily and lay their eggs high up in the upper reaches. However, after thawing of ice is over the water level recedes and the fish larvae get stranded in side pools leading to high mortality and hence low recruitment of the young ones in the water systems.

Table 29.5: Comparative fecundity of some fishes of Kashmir

Species	*Absolute Fecundity*	*Relative Fecundity*	*Authority*
Schizothorax esocinus	3910—53106	39	Raina, 1976
S. curvifrons	4843—31221	45	Sunder & Subla, 1984
S. longipinnis	5748—34709	45	Sunder, 1986 (c.f. Yousuf & Pandit, 1992)
S. niger	1296—23776	53.2	Yousuf & Pandit, 1992
Oreinus plagiostomus	2598—27846	29.6	Qadri et al, 1981
Cyprinus carpio	17450—312,460	166.6	Subla & Sunder, 1981

(b) Eutrophication and environmental Pollution of aquatic habitats

During the last several decades there has been significant increase in the trophic level of the valley lakes and the rivers are also carrying heavy load of sewage (Table 29.1). The native fish, especially the schizothoracines, being stenotopic with respect to the environmental conditions are finding it very difficult to cope up with this problem and are suffering losses in the population density. The exotic carps, on the other hand are able to adjust to wide fluctuations in their environment and have therefore easily adapted to the changing water quality. Whether a lake or a river, everywhere the environmental degradation has reached significant levels. Those schizothoracine fry, that are fortunate enough to reach the downstream from the hatching grounds, are faced with the problem of environmental pollution and become easy prey to the pollution. The exotic carps, on the other hand, have great adaptability to the ecological stress and because of high fecundity have higher recruitment rate. Their breeding grounds are located close to their feeding areas and as such they have not to face any problems associated with migratory habit.

S. niger, the only truly lacustrine schizothoracine species, has the least fecundity among the group. It breeds in such perepheral areas of the lakes as have gravelly or sandy bottom and some spring water oozing out. In the past such areas were very abundant in the flatland lakes of Kashmir like Dal, Anchar, Manasbal and Wular. However, because of deforestation and other detrimental human activities in the catchment areas of the lakes the scenario is totally changed. Most of these areas have got silted up and springs have dried. This has directly resulted in a decline in the spawning grounds of the fish. The increased trophic level of the lake water has further aggravated the situation for the fish and its population has as a consequence greatly decreased.

(c) Government Policy towards Management of Fishery Resources

The Government of Jammu and Kashmir is spending large sums of money on projects concerned with the management of Fishery Resources of the State. However, the focal point has been the trout and common carp farms only. A number of trout and carp farms have been established in different parts of the State and a good progress has been made in rearing these fish under human control. But the schizothoracine group has been totally neglected. While every year millions of trout fry are raised in farms and released in the streams and other trout waters of the valley, not a single schizothoracine fish has so far been selected for artificial rearing.

(d) Hydroelectric Power Projects and Barrages

The construction of Power projects is, without any doubt, an indication of the development of a country or a State. However, while the construction of the barrages and dams is undertaken, attention needs to be given to the ecological impact of such man-made structures. In the past several barrages have come up on the river Jhelum–Lower Jhelum and Uri Barrages in Kashmir and Mangla Dam in Pakistan. All these have greatly influenced the ecology of the river in their vicinity. The construction of Mangla dam has resulted in the total loss of mahseer (*Tor putitora*) from Kashmir, which upto 1950's used to reach as upstream in the river as Wular lake and contributed a significant proportion of the commercial catches. Although the two barrages constructed in Kashmir–Lower Jhelum and Uri Barrage–have been provided with ladders, these structures have also affected the schizothoracine fish to a great extent. As already noted most of the schizothoracine species are migratory in habit and move regularly upstream during spring and summer for spawning in the tributaries of the river Jhelum. Many fish are unable to negotiate the ladder and hence do not reach their spawning grounds upstream of the barrages. This has considerably affected the rate of recruitment of the young fish in the waterbodies of the valley.

CONSERVATION MEASURES

In order to save the schizothoracine fish from further decline in the aquatic habitats of the valley some immediate conservation measure need to be taken.

(a) *Pollution Control* : With the increasing population, the quantum of untreated sewage and other pollutants entering the water courses, has also increased significantly. This has affected the water quality of the aquatic systems and many of them have entered the meso–and eutrophic state. Immediate steps need to be taken to stop further pollution of these habitats in order to save the existing schizothoracine population there. If this is not done in the right earnest then the day will not be far off when the group will be totally driven out of the lakes of the valley as is already the case with Khushhal sar lake and Brari Nambal basin of Dal lake.

(b) *Check on Encroachment* : During the last ninety or so years large areas of perepheral areas of flatland lakes have been reclaimed and these water bodies have lost a considerable part to the terrestrial habitat. With this, many spawning grounds for fish like *S. niger* and *Labeo dero* and *L. diocheilus* have been lost. If this trend continues then the existing spawning grounds of these fish will also get affected. Therefore the encroachment of the shallow perepheral zones of the lakes, particularly Dal, Anchar and Wular needs to be checked.

(c) *Construction of Hatcheries* : If we want to conserve the schizothoracine fish in their existing habitats, construction of hatcheries is an important requisite. The Department of Fisheries of the State has the know-how for rearing the trout in captivity. This can be utilised with necessary modifications for the rearing of schizothoracine species. For the moment we need to concentrate only on the rearing upto fry stage. The fry can then be released in the natural habitat so as to augment the wild population. When success has been attained in this, it may not be difficult to make the schizothoracine culture economically viable keeping in view the progress made in the biotechnological field throughout the world. If these steps are taken it is possible not only to conserve the schizothoracine population of our aquatic habitats but also to increase the production of these fish.

ACKNOWLEDGEMENTS

The author would like to thank C. S. I. R., I. C. A. R. and N. H. P. C. for sponsoring research projects during the tenure of which data presented in the article were collected. He is grateful to his students, viz., Drs. Masood H Balkhi, Muni Parveen, Anil Pandit, Gazala Firdous and M. Farlooq Mir for collecting the field data. Thanks are also due to Swedish International Development Agency for providing an opportunity to make a Fishery survey of river Jhelum and some of its tributaries.

LITERATURE CITED

Das, S.M. and Subla, B.A. (1963). The ichthyofauna of Kashmir: Part I. History, topography, origin, ecology and general distribution. Ichthyologica, 2 : 87-106.

Firdous, G. (1995). *Fish and fisheries of Anchar lake, Kashmir*. Ph. D. Dissertation, University of Kashmir.

Jhingran, V. G. (1975). *Fish and Fisheries of India*. Hindustan Publ. New Delhi.

Kanth, T.A. and Amin, S. M. (1996). Drainage system in Kashmir Vallye pages. 185-197 : In : Khan, A.H. And Pandit, A.K. (editors) *Ecology, Environment and Energy*. University of Kashmir.

Kumar, K. and Bhagat, M.J. (1978) Observation on the ecology of two trout streams in Kashmir and its possible effects on the brown trout (*Salmo trutta fario* Linnaeus) catches. *Journal of the Inland Fishery Society of India*, 10: 1-8.

Lingdell, Par–Erik and Engblom, Eva (1991). Bottom fauna collected in Kashmir 1990. SIDA project "Aquatic Studies on river Jhelum "Technical Report to National Hydroelectric Power Corporation of India. 6 pp.

Qadri, M.Y., Mir, S. and Yousuf, A.R. (1983), Breeding biology of *Schizothorax richardsoni* Gray & Hard. Journal of Indian Institute of. Science 64: 73-81.

Raina, H. S. (1976) Seasonal histological changes in the ovary of *Schizothorax esocinus* Heckel. Matsya 2: 66-71.

Subla, B.A. and Sunder, S. (1981). Biology of mirror carp (*Cyprinus carpio*) in a streth of river Jhelum with reference to certain hydrobiological parameters. Scientific Report S.P. College, Srinagar.

Sunder, S. and Subla, B. A. (1984). Fish and Fisheries of river Jhelum. Kashmir. Zoologica Orientalis. 1 : 34-39.

Yousuf, A.R. (1992), Biotic communities and their role in the trophic condition in Kashmir Himalayan lakes. Technical Report on CSIR Research Project, University of Kashmir.

Yousuf, A.R. (1995). Nutritional regime and its impact on the fishes of Anchar lake, Kashmir: Ecology and feeding biology of cyprinid fishes. Technical Report on ICAR Research Project, University of Kashmir.

Yousuf, A.R. (1996). Fishery resources of Kashmir, pp. 87-128: in Khan, A.H. and Pandit, A.K. (Editors) *Ecology, Environment and*). University of Kashmir.

Yousuf, A.R. and Pandit, Anil, K. (1992). Breeding biology of *Schizothorax niger* Heckel: fecundity pages. 55-62: In : Yousuf, A.R., Raina, M.K. and qadri, M.Y. (Editors) *Current trends in fish and fishery biology and aquatic ecology*. University of Kashmir.

Chapter - 30

Anthropogenic Causes of species Extinction

*Zafar A Reshi**

ABSTRACT

The paper reviews the causes of present spasm of species extinction. Impacts of anthropogenic factors like unsustainable exploitation, habitat destruction conversion, introduction of alien species and other related factors on biodiversity are highlighted.

***Key Words :** alien species, anthropogenic factors, biodiversity, habitat loss, over exploitation, species extinction.*

INTRODUCTION

Extinction refers to an irrevocable loss of some taxon over its entire distribution range. During the course of life on Earth, there appear to have been two types of extinction. One is the continuous, a low-level extinction of species, sometimes called *background extinction*. The second is the mass extinction, when numerous species and taxa disappear. From the geological record it is well established that five or six great waves of mass extinction have embarrassed the planet Earth. The causes of such extinction, particularly mass extinctions, are not well understood. Natural environmental and biological factors seem to be involved.

At present we are again confronting an episode of species extinction greater than anything the world has experienced for the past 65 million years. According to some estimates we are losing species at between 1,000 and 10,000 times the natural rate (Raven, 1994) Judged from the fossil record, the background rate of extinction can be calculated to about four (4) species a year. At a moderate estimate, we are now likely to lose around 50,000 species a year over the next decade (Raven, 1994). However, according to a recent study (Pimm et al. cf. Stevens, 1995) the rate of species extinction is 100 to 1,000 times as high as the prehuman background extinction rate. This is about one-tenth the rate of the most pessimistic global estimates. The present spasm of extinction is solely the handiwork of man. The most important anthropogenic factors that are behind the present great wave

* Department of Botany University of Kashmir, Srinagar - 190 006, India

of extinction are excessive exploitation, habitat destruction and conversion and introduction of predators, competitors, or diseases. Of late environmental factors like global warming, acid deposition, stratospheric ozone depletion and heavy metal pollution are believed to be contributing to species extinction. Over the past four centuries, there has been on global scale, a total of more than 700 known extinction, including 100 species of mammals and 160 birds, all because of anthropogenic influences (Fitter, 1968; Wood, 1972; Soule, 1983; Reid et al., 1992).

The present article is an attempt to consolidate information on species extinction caused by some of the anthropogenic factors.

OBSERVATIONS AND GENERAL COMMENTS

Species Richness and loss

Globally about 1.7 million species have been identified and given bionomial name. Of these identified species, 6 percent live in boreal or polar latitudes, 59 percent in the temperate zones, and remaining 35 percent in the tropics (WRI, 1986). Knowledge of species richness is woefully incomplete, and according to some estimates the global species richness could range between 30-50 million species (WCMC, 1992), with 90 percent of species inhabiting tropical areas (Wilson, 1988; Erwin, 1991).

Every species has a finite life span and many of the species that once inhabited Earth are now extinct. The extinction can be a natural process caused by random catastrophic events, biological interactions and physical stresses. In the recent past many a species have become extinction not because of above cited natural factors but because of unsustainable human activities and rate of present extinction far exceeds any reasonable estimate of background extinction rates. Though precise or even approximate dimensions of the loss of biodiversity are elusive but nonetheless conservative estimates of species loss (Table 30.1) indicate that 384 plant species, 23 fish species, 2 amphibian species, 21 reptile species, 113 bird species and 83 mammal species have gone extinct while 1869 plant species, 320 fish, 48 amphibian, 1355 reptile, 924 bird and 414 mammal species are threatened with extinction. These facts and figures do point to fact that we are in the midst of one of the great extinction spasms of geological history. The anthropogenic factors leading to such unprecedented species loss include:

(a) Unsustainable exploitation

(b) Introduction of fallen species

(c) Habitat destruction and conversion.

These factors have not only condemned many species to extinction but have also reduced populations of many species thus exposing them to risk of extinction (Table 30.1).

EXTINCTION DUE TO UNSUSTAINABLE EXPLOITATION

The most prominent and potent cause of extinction and endangerment of a relatively large number of species is exploitation or harvesting at a grossly unsustainable scale. Though species were exploited by man for various purposes during earlier times but his power to damage was limited. With the advent of industrial civilization, man's greed and capability to exploit natural resources including biodiversity increased manifold. This inevitably leads to collapse of natural resources, extinction and endangerment of a number of species. Some of the species lost as a result of over-exploitation and over-harvesting are given in Table 30.2. Data summarized in the table point towards great waves of extinction of large animal species of different places with the end of the Pleistocene Epoch. Many of these mass extinction events coincide with the colonisation of landmasses that were uninhabited by humans previously. For example, in North America the wave of extinction about 11 thousand years ago, is almost coincident with the colonisation of the continent by humans. Likewise, in New Zealand and Australia, mass extinctions coincide with the arrival of aboriginal hunter. The immediate fallout of these human colonisations of landmasses has been overhunting and over exploitation causing or contributing to the extinction of species of horse, antelope, mastedon, lion, pelican, eagle, flightless goose etc. Extinction of dodo, great auk, passenger pigeon, among other species, are examples of more recent extinctions caused due to unsustainable exploitation.

Table 30.1 : Current status of plant and animal species

Taxonomic Group	*Number of species*	*No. of Extinct species*	*Percentage*	*No. of threatened species*	*Percentage*
Plants	250,000	384	0.15	1869	1.4
Fish	19,056	23	0.12	320	1.6
Amphibians	4,184	2	0.05	48	1.1
Reptiles	6,300	21	0.33	1355	21.5
Birds	9,198	113	1.23	924	10.0
Mammals	4,170	83	1.99	414	10.0

(*Source* : McNeely et al.. 1990)

Commercial exploitation of a number of species has brought them to the brink of extinction. Notable example is that of Whales. It has been estimated that during the commercial hunts of the last 400 years more than 2.5 million Whales of all species were killed (NMFS, 1991). As a result of this over-exploitation, populations of blue whale, gray whale, bow head whale, humpback whale, sel whale have been considerably reduced.

EXTINCTION OF COMPETITORS OF HUMANS

During the course of evolution, man has come to consider many species to be his competitors in some economically important resource. The competitors include herbivores, agricultural pests and predators. For example, the carolina parakeet was considered the agricultural pest causing damage to fruit orchards and gain fields. For this reason the bird species was persecuted by humans and became extinct in its entire original range of distribution largely through pest control efforts (Table 30.3).

Tasmanian wolf, is another species which incurred wrath of humans, because it predated on sheep and chicken that were raised on Tasmanian ranches. This species was hence relentlessly persecuted by humans and became extinct over its entire range by 1933.

Table 30.2 Some examples of species extinction by over-exploitation and over-harvesting

Extinct Taxa	*Time/Period of Extinction*	*Native place*	*Reference(s)*
Horse (*Equus*, 10 species) Saiga antelope (*Saiga, tatarica*) Mastodon (*Mammut, americanum*) American Lion (*Panthera leo atrox*)	11 thousand years ago	North America	Martin, 1967, 1984; Diamond, 1982.
Moa (*Dinormis maximus*) Flightless goose (*Cnemionris calcitrans*) Pelican (*Pelecanus novaezealandiae*) Eagle (*Harpagornis moorel*)	Less than one thousand years ago	New Zealand	McCullock, 1982; Diamond, 1984; Holdaway, 1989; McGlone, 1989
Pink headed duck (*Rhodonessa caryophyllacea*) Jerdon's Courser (*Cursorius bitorquatus*) Himalayan mountain quail (*Ophrysia superciliosa*) Forest spotted owlet (*Athene blewitti*)	300 years ago	Asia	Fisher et al., 1969 King, 1981.
Dodo (*Raphus cuculiatus*)	1681	Mauritius	McClung, 1969; Campbell and Lack, 1985
Great Auk (*Pinguinus impennis*)	early 1800	Funk island	McClung, 1969; Nettleship and Evans, 1985
Passenger Pigeon (*Ectopistes migratorius*)	300 years ago	USA and Canada	Schorger, 1955; Blockstein and Tordoff, 1985
Steller's sea cow (*Hydrodamalis stelleri*)	1768	Bering Sea	Fitter, 1968.
Carribean monk Seal (*Monachus tropicalis*)	18th century	Carribean sea and Gulf of Mexico	Thomback and Jenkins, 1982.

Table 30.3. Examples of extinctions of competitors of humans.

Extinct Taxa	*Time/Period of Extinction*	*Nattive place*	*Reference(s)*
Coroline paraket (*Conuropisis carolineusis*)	1914	Southeastern United States	Ziswiler, 1967
Tasmanian Wolf (*Thylacinus cynocephalus*)	1933	Tsmania, Australia	Freedman, 1995

Many of the large predators were also persecuted because they were considered to be important competitors and some were even predators of humans themselves. Examples of such animals that are now extirpated or endangered over much of their distribution range include the wolf (*Canis lupus*), brown bear (*Ursus arctos*), cougar (*Felis concolor*) and tiger (*Panthera tigris*).

EXTINCTION DUE TO INTRODUCTION OF ALIEN SPECIES

The history of introduction of alien species is as old as man himself. The species are introduced either deliberately or accidentally. Some examples of such introductions are *Oryctolagus cunninculus* in Australia, *Cervus elaphus* in *New Zealand, Herpestes auropunctatus* in Hawaiian archipelago, *Eichomia crassipes, Lanatana camara, Parthenium hyserophorus* and *Salvinia molesta* in tropical and sub-tropical regions, *Anthemis cotula, Robinia pseudoaccacia* and *Xanthium spinosum* in Kashmir.

The introduced alien species may escape and then spread with debilitating consequences for structural and functional attributes of local ecosystems (MacDonald. et al. 1989). One of the deleterious effects is acceleration in the extinction of species at local or global level. The recent, tragic events of mass extinction of native fish community in Lake Victoria, Africa by the introduction of Nile perch; extinction of a mistletoe in New Zealand by an introduced generalist herbivore and extinction of American chestnut in North America by chestnut blight fugus are a few outstanding examples of the many extinctions caused by introduced species (Table 30.4).

Table 30.4 : Examples of Extinctions caused by introduced species.

Extinct Taxa	*Introduced Species*	*Time/Period of Extinction*	*Native place*	*Reference (s)*
Native cichlids Native minnow (*Restrinneobola argentea)*	Nile perch (*Lates nilotius*)	Late 1980's	Lake Victoria, Africa	Baskin, 1992 Kaufman, 1992
Endemic mistaletoe (*Trillipidea adamsil*)	Generalist herbivore (*Trichosurus vulpecula*)	Recent	New Zealand	Norton, 1991
American Chestnust (*Castanea denfata*)	Chestnut blight fungus (*Cryphoneicria parasitica*)	1950's	North America	Von Broem-bhsen, 1989

The potential role introduced species might play in future extinctions is indicated by the data on the verteberate species currently considered to be threatened with extinction by introduced species (Table 30.5). This dimension is further substantiated by one well-documented case where invasions by introduced plants pose a major threat to native plant species diversity. In the fynbos and karoo biomes of South Africa there are 1808 threatened plant species and just over 50 percent are threatened by the spread of introduced trees and shrubs, mainly species of *Acacia, Hakea* and *Pinus*. (Hall and Ashton, 1983; Hall and Vldhuis, 1985).

Table 30.5 : The percentage of terrestrial vertebrate species threatened by introduction of alien species.

Taxonomic group	*Percentage of threatened species affected by introduction*	
	Mainland areas	*Insular areas*
Mammals	19.4 (283)	11.5 (61)
Birds	5.2 (250)	38.2 (144)
Reptiles	15.5 (84)	32.9 (76)
Amphibians	3.3 (30)	30.8 (13)

Total number of threatened species is given in parentheses.

Source: MacDonald et al., 1989.

EXTINCTION DUE TO HABITAT DESTRUCTION

Since the advent of agriculture some 12,000 years ago, natural habitats have been cleared and converted into farmland, rangeland, and human and industrial settlements. This trend accelerated significantly in the 20th century as human population tripled. Increase in cropland area between 1700 to 1980 is given in the Table 30.6. The data are suggestive of large scale conversion of forests and wetlands to croplands. These conversions lead to habitat destruction and fragmentation with consequent threat to biodiversity. In this context tropical deforestation assumes significance as these forests are believed to inhabit about 90 per cent species. Annual tropical deforestation has been recently estimated to be about 0.9 percent (Table 30.7) with highest rate of deforestation (1.2 percent) in Asia followed by Latin America (0.9 percent) and Africa (0.8 per cent).

Table 30.6 : Land use pattern from 1700-1980.

	Area (10^4 km²)		
Vegetation Types	*1700*	*1980*	*Percentage change*
Forests and Woodlands	6215	5053	-18.7
Grasslands and Pastures	6860	6788	-1.0
Croplands	265	1501	+ 466.4

Source : Richards,1990

Table 30.7 : Estimates of Tropical Forest area and rate of deforestation from 1981-1990.

Region	*Forest Area*		*Area deforested annually 1981-91*	*Annual rage of change 1981-90 (percent)*
	1980	*1990*		
Latin America	923,000	839,900	8,300	-0.9
Asia	310,800	274,900	3,600	-1.2
Africa	650,300	600,100	5,000	-0.8

Source : Food and Agriculture Organisation of the United Nations, 1991.

According to Reid (1992) these rates of tropical deforestation translate into extinction of 4 to 8 percent of rainforest species by 2015 and 17 to 35 percent by 2040. Assuming a conservative global total of 10 million species, 20 to 75 species per day would be condemned to extinction by 2040.

Besides endangerment, the conversion and destruction of natural habitats has already caused extinction of some species (Table 30.8). Notable examples include ivory billed woodpecker, dusky seaside sparrow and black-footed ferret.

EXTINCTION DUE TO ENVIRONMENTAL POLLUTION

Environmental pollution, besides other deleterious effects, is being shown to have adverse effect on biodiversity. For example the increase in pesticide use (ten-fold increase in global sales from 1975 to 1990) has been implicated in 10 percent reduction in population of barn owls (McNeely, 1992), premature births in Californian sea lions (DeLong et al., 1973), and death of about 30,000 birds in Spain (Mackenzie, 1986).

Acid deposition is another aspect of concern with regard to its impact on species loss (Schindler, 1988). It has been found involved in reduced breeding success in terrestrial birds in the Bunderkamp forest (Drent and Woldendrop, 1989); the destruction of seven major Salmon fisheries in Sweden (Walgage, 1986); the extinction of 30 percent fish, 60 percent amphibians, 30 percent of reptiles, 30 percent of birds and 35 percent of mammals in Czechoslovakia's Great Bohemian Wood (Csepel, 1984).

Climatic changes brought about by increase in levels of greenhouse gases are being examined for their impact on biodiversity (Crowley and North, 1988). Stratospheric ozone depletion and heavy metal pollution are being actively considered for their role in loss of species diversity.

CONCLUSION

The anthropogentic causes of species extinction described above do not operate in isolation, but rather tend to act with and exacerbate one another resulting in fragmentation of habitats and populations. The small fragmented populations are then exposed to deleterious effects of inbreeding and damographic instability. Consequently the populations decline further and in course of time all the small populations disappear and with them the species. Many a species have followed this path to extinction, and many more are on this path. This syndrome has appropriately been called "extinction Vortex" (Freedman, 1995).

Table 30.8 : Some examples of species extinction by habitat destruction.

Extinct Taxa	*Habitat*	*Time/period of Extinction*	*Native place*	*Reference (s)*
Ivory billed Woodpecker (*Campephilus principalis principalis*)	Angiosperm and cypress Swamps	1960's	North America	McClung,1969; King, 1981.
Dusky Seaside sparrow (*Ammodramus maritimus*)	Salt marshes	1980's	Eastern Coast of Florida	King, 1981
Black-footed ferret (*mustela nigripes*)	Short and mixed grass prairies	Not Known	North America	Thomback and Jenkins, 1982.

ACKNOWLEDGEMENT

Thanks are due to all the agencies and workers who made their works available to me.

LITERATURE CITED

Baskin, Y, (1992). Africa's troubled waters. Bioscience, 42 : 476-481;

Blockstein, D.E., and Tordoff, H.B. (1985) Gone forever. A contemporary look at the extinction of the passenger pigeon. Am. Birds 39 : 845-851;

Campbell, B., and Lack. A (1985). A Dictionary of Birds Omithologists Union, London.

Crowley, T.J. and North G.R. (1988) Abrupt Climate Change and extinction events in early history. Science 240: 996-1002.

Csepel, A. (1984), Czechs and the ecological balance, New Scientists 27: 10-13.

DeLong R. Gilmartin, W.G. and Simpson, J.G. (1973). Premature births in California sea lion: association with high organochlorine pollutant residue levels. Science 181: 1168-70.

Diamond, J.M. (1982), Man the exterminator. Nature (London) 298: 787-780.

Diamond, J.M. (1984). Historic Extinctions: a Rosetta Stone for understanding prehistoric extinctions pages. 824-862. In : Martin,P.S. and Klein, R.G. (Editors). Quaternary Extinctions, a Prehistoric Revolution. University of Arizona, Press, Tucson.

Drent, P.J. and Woldendrop, J.W. (1989) Acid rain and eggshells. Nature 239: 431.

Erwin, T.K. (1991) How many species are there? revisited Conservation Biology 5: 1-4.

Fisher, J., Simon, N and Vincent, J. (1969) Wildlife in danger. Viking Press, New York.

Fitter, R. (1968). Vanishing Wild Animals of the World. Kaye and Ward, London.

Food & Agricultural Organisation (FAO) of the United Nations (1991). Second Interim Report on the State of Tropical Forests. Paper presented at the 10th World Forestry Congress, Paris.

Freedman, B (1995) Environmental Ecology. Academic Press, New York.

Hall, A.V. and Ashton, E.R. (1983). Threatened plants of the Cape Peninsula. Threatened Plants Research Group, University of Cape Town, Cape Town 22 pp.

Hall, A.V., and Veldhuis, H.A. (1985). SouthAfrican Red Data Book: Plants-Fynbos and Karoo Biomes. South National Sci. Prog. Rep., 117: 1-160.

Holdaway, R.N. (1989). New Zealand's pre-human avifauna and its Vulnerability. New Zealand Journal of Ecology 12: 11-25.

Kaufman, L. (1992). Catastrophic change in specie-rich freshwater ecosystems. Bioscience 42: 846-858.

King, W.B. (1981). Endangered Birds of the World: The ICBP Bird Red Data Book. Smithsonian Institute Press, Washigton, D.C.

MacDonald, I.A.W., Loofe, L.I. Usher, M.D., and Hamann, O. (1989), Wildlife Conservation and the invasion nature reserves by introduced species: a global perspective pages. 215-255. In : Drake, J.A., Moony, H.A., dicastri, F, Groves, R.H., Kruger, F.H., Rajmanek, M., and Williams, M. (Editors) Biological Invasion: A Global Perspective. John Wiley and Sons, New York.

MacKenzie, D. (1986). Crayfish pesticide decimates Spanish birds, new Scientist 16 : 24.

Martin, P.S. (1967). Prehistoric overkill 75-120. In : Martin, P.S., Wright, H.E. Pleistocene Extinctions; The search for a cause. Yale Univ. Press, New Haven, CT.

Martin, P.S. (1984). Catastrophic extinctions and late Pleistocene blitzkrieg: Two radiocarbon tests pages 1530-189. In : Nitecki, M.H. (Editor). Extinctions. University of Chicago Press, Chicago.

McClung, R.M. (1969). Lost Wild America. The story of our extinct and vanishing wildlife. Marrow, New York.

McCullock, B. (1982). No Moa. Canterbury Museum, Christchurch, N. Zealand.

McGlone, M.S. (1989). The Polynesian settlement of New Zealand in relation to environmental and biotic changes. New Zealand. Journal of Ecology 12: 115-164.

McNeely, J.A. (1992). The sinking ark; Pollutioo and the worldwide loss of Biodiversity. Biodiversity and Conservation 1 : 2-18.

McNeely, J.A., Miller, K.R., Reid, W.V., Mittermeier, R and Werner, T. (1990) Connserving the world's biological diversity. Washington D.C, IUCN, World Resources Institute, World Bank, WWF-US and Conservation International.

National Marine Fisheries Service (NMFS) 1991. Endangered Whales: Status update NMFS, National Oceanic and Atmospheric Administration, U.S. Department of Commerce, Silver Spring.

Nettleship, D.N. and Evans, P.G.H. (1985). Distribution and status of the Atlantic Alcidae pp. 54-154 In : Nettleship, D.N., and Birkhead, T.R. (ed.). The Atlantic Alcidae. Academic Press, New York.

Norton, D.A. (1991). *Trilepidea adamsii* : An obituary for a species. Conservation Biology 5 : 52-57.

Raven, P. (1994). Why it matters. Our Planet, 6 : 5-8.

Reid, W.V. (1992). How many species will there be pp. 55-73. In : Whitmore, T.C. and Sayer, J.A. (Eds. Tropical Deforestation and Species Extinction. Chapman Hall, London, U.K.

Reid, W., Barker, C. and Miller, K (1992). Global Biodiversity Strategy. World Resources Institute, Washington, D.C.

Richards, J.F. (1990). Land Transformation pp. 163-178. In : Turner, B.L. (Ed.) The Earth as transformed by Human Action.

Schindler, D.W. (1988). Effects of acid rain on freshwater ecosystems. Science, 293: 149-157.

Schorger, A.W. (1955). The Passenger Pigeon. University of Wisconsin Press, Madison.

Soule, M.E. (1983). What do we really know about extinction pages 111-124 In : Schonewald-Cox, C.M;' Chambers, S.M., MacBryde, B; Thomas, W.I. (Editors). Gennetics and Conversation. Benjamin / Cumanings, Meulo Park, CA.

Stevens, W.K. (1995). How to measure species as they disappear. Times of India, August, 1.

Thomback, J. and Jenkins, M (1982). The IUCN Mammal Red Data Book. Part 1. IUCN. Gelannd, Switzerland.

Von, Broembsen, S.L. (1989). Invasions of Natural Ecosystems by Plant Pathogens 77-83. In : Drake, J.A., Mooney, H.A., diCastri, F., Groves, R.H., Kruger, F.J., Rejmanek, M., and Williamson, M. (Editors) Biological Invasions: A Global Perspective. John Wiley and Sons, New York.

Walgate, R. (1986). UK Denies responsibility for Scandinavian acid rain. Nature 323: 191.

Wilson, E.O. (Ed.) 1988). Biodiversity. National Academy Press, Washinngton.

Wood, G.L. (1972). Animal Facts and Figures, Guinness Superlativs Limited Enfield, U.K.

World Conservation Monitoring Centre (WCMC) 1992). Global Biodiversity; Status of the Earth's Living Resources. Chapman and Hall, Lonndon, U.K.

World Resources Institute (1986) World Resources. International Institute for Environment and development. New York.

Ziswiller, V. (1967). Extinct and Vanishing animals. Springer-Verlag, New York.

SECTION - VIII

BIODIVERSITY IN RELATION TO ECOSYSTEM DEVELOPMENT AND MODELLING

Chapter - 31

Biodiversity of Edaphic Ecosystem and its Conservation

*A.Q. Rather and Gh. Hassan Dar**

ABSTRACT

Soil ecosystem is comprised of great many microenvironments which are interrelated and always in a state of equilibrium under undisturbed conditions. Approximately 95% of terrestrial net primary production reach the soil in the form of dead organic matter and is subject to decomposition. Soil microflora and fauna function in consortium during organic matter turnover thereby play a key role in maintaining environmentally sound ecosystem. Microflora and fauna constitute approximately 2-25 t live weight ha^{-1} furrow slice. More than 70% soil flora and fauna is confined to the upper 6 cm soil layer rich in partially decomposed litter and organic matter. The diverse organisms through their activities also improve soil physico-chemical properties. The greater diversity of biotic components in soil ecosystem ensures better exploitation of nutritional resources, regulates geochemical cycling and allows continuous energy flow through edaphic ecosystem as well as soil humus formation. However, ruthless use of agrochemicals in crop production and disposal of pollutants in soil have not only impaired the soil quality but also decreased the diversity of its resident organisms and their population dynamics. Soils devoid of meso-and macro-fauna are vulnerable to soil degradation. Biotic interactions arising from input of potential organic wastes and biofertilizers into soil have significant impact on system function and in the conservation of edaphic biodiversity.

Key Words : biodiversity, biofertilizers, conservation, edaphic ecosystem, energy flow, nutrient cycling, organic farming, pollution, soil fertility.

INTRODUCTION

Soil is a dynamic entity, a biotechnically weathered and synthetic product of nature, which serves as a natural habitat for plants, paedo-and geo-bionts. The wide array of diverse

Directorate of Research, * Division of Environmental Sciences, S.K. University of Agricultural Sciences and Technology, Shalimar, Srinagar, Kashmir - 191121

organisms inhabiting soil ecosystem, of which more than 70% are confined to upper 6 cm layer, maintain the soil in living and enriched condition through various activities. The most important being organic matter decomposition, recycling of nutrients, humus formation and soil development, besides increasing aeration and water percolation in the soil. These processes, catalysed by micro-and macro-organisms, strongly influence the development of plant communities hence are crucial for the final appearance of vegetation.

The anthropogenic activities especially industrialization and indiscriminate use of agro-chemicals in modern agriculture have resulted in the dispersal of millions of tonnes of synthetic compounds throughout the biosphere, which adversely affect the biodiversity in edaphic ecosystem. Conservative estimates indicate that at least 10% of living species have become extinct due to the increased human activities. The conservation of soil ecosystem and biodiversity needs special attention in view of the growing concern to preserve the high soil fertility and crop productivity levels. The present paper reviews the existing information on soil biodiversity and its role in maintaining soil health and vigour vis-a-vis the measures to conserve soil biodiversity especially through organic farming.

BIODIVERSITY IN SOIL BIOTA AND ITS FUNCTION

The soil represents a high complex ecosystem composed of a great many micro-environments, interrelated with each other directly or indirectly. Though environmental factors determine the abundance, nature and composition of micro-flora and fauna present at any given time in a particular soil, quantitatively the soil on an average in one square meter surface area (upto 15 cm soil depth) contains fungi 112-1120 g, bacteria and actinomycetes 45-450 g each, algae 6-56 g, earthworms 10-120 g, protozoa 2-18 g, nematodes 1-11 g and other macro fauna 2- 18 g; whereas amongst macro-fauna, herbivores and detritivores constitute 11-17 and 4-172 g m^{-2} of soil (Brady, 1990). The live weight of microflora and fauna ranges from 2 to 25 t ha^{-1} of furrow slice.

In soil ecosystem the most active part is upper 30 cm layer wherein maximum metabolic activities occur. Most of these activities are due to the microflora, mainly fungi and bacteria. The distribution of microbial biomass viz. fungi, bacteria, algae, protozoa etc. in top 15 cm soil layer is governed by the metabolic potential and versatility of these organisms. However, the species diversity in soil is influenced by soil reaction, moisture, temperature, organic matter content, aeration, cation exchange capacity etc. as well as the microclimate and cultural practices adopted. The world wide figure of major soil dwelling species are fungi 70000, bacteria 4000, earthworms 60000, crustaceans 321, pseudoscorpions 3000, mites 27000, millipedes 7000, centipedes 3000, springtails 2000, earwigs 1200, termites 2000, cockroachs 3500, ants 10000 and beetles 67000, in addition to the thousands of algal and protozoan species (Hazra and Sanyal, 1997). There are also numerous species of spiders and macro-and mega-fauna like diptera larvae, moles, squirrels, hares, snakes, rodents and other soil dwelling animals active in soil ecosystem. This vast population includes both geobionts as well as paedobionts. Paedobionts are complex assemblages of species supported primarily by detritus derived from natural turnover of organic material present in the edaphic ecosystem. Their diverse feeding habits help in microbial dissemination, mixing of abiotic components of soil, transmission of cestodes/ pathogens and keeping nematodes and weed population in check, thereby enhance heterocity and productivity.

The fungi (mold, mushrooms, mycorrhizae) are most prominent soil dwellers mostly organotrophic in nature. The fungi are either saprophytic or parasitic, mostly involved in nutrient transformation. For instance, basidiomycetous fungi are extremely vigrous decomposers of woody materials. Slime molds engulf bacteria and flagellate fungi. As per the estimates one gram of soil visually contains 10-100 m of mold filament (O'Dell *et al.*, 1992). Mycorrhizae in soil ecosystem increase biodiversity of microflora and fauna through their mutualistic interactions. Plant species diversity in natural forests, grassland and agriculture ecosystem is usually correlated with the presence of different types of mycorrhizae (Zargar *et al.*, 1998). Moreover, their role in soil conservation through extensive mycelial network (12-19 m g^{-1} soil or 17-45 m cm^{-3} soil) and soil development through macroaggregate formation can not be under estimated.

The bacteria in soil may either be autochthonous or allochthonous (zymogenous) in nature. Ecological significance of bacteria lies in fixation of C and N, nutrient release through immobilization and mineralization of organic matter and their synergestic and antagonistic associations. Besides, in soil ecosystem actinomycetes (thread bacteria) produce numerous antibiotics which help in keeping the undesired populations under check. Cyanobacteria occupy unique inter-linking position in the continuum of life, especially through their phototrophic work and dinitrogen fixing potential.

Algae in general occupy ecologically diverse niche and may in some cases grow endosymbiotically in plants and animals. Their CO_2 fixing attribute increases the organic matter content in soil, besides serving as food for many micro-and macro-fauna. A symbiotic association between algae and fungi, forming are undifferentiated body called 'lichen', helps each partner to derive their energy needs. At least 30 genera of algae are known to form symbiosis with fungi resulting in more than 18,000 species of lichens, distributed in about 500 genera (Kumar and Rai, 1986).

Soil fauna perform basic ecological function by feeding. They search out and ingest carbon containing tissues to sustain metabolic activities and in the process return nutrients via defecation and excretory process (Berthet, 1963). Soil protozoa and other micro-fauna exhibit some feeding selectivity and increase diversity of bacteria by limiting densities of some species. It is believed that protozoan grazing keeps the microbial population in soil physiologically young and allows more specific forms of microbes to coexist, hence more efficient in soil decomposition. Their active motility in soil water helps the bacteria with dissolved oxygen and nutrients. About 5000 bacterial cells are endeavoured by nematodes per minute and approximately 8300 bacteria consumed for each new amoeba produced (Coleman *et al.*, 1984). Mites increase the inhibitions of mycostasis and bacteriostasis and accelerate the microbial growth by grazing on senescent colonies (Butcher *et al.*, 1971). Moreover, total consumption of microflora by bacteriovores and fungivores is as high as 50% of estimated production of microbial biomass. The soil fauna do have a major impact on litter decomposition rates in temperate sites whereas in tropical sites, though litter-fall and decomposition is greater, animal population tend to be low (Lal, 1987). Hot and dry climates generally suppress the soil faunal diversity and activity.

In general mites and collembola constitute about 77-88% of total micro-fauna population in different habitats. Mites and other arthropods provide a suitable physical substrate for microbial growth during initial fragmentation of litter. Earthworms and other micro-arthropods produce an intimate mixing of organic and mineral matter during engulfment of decomposing mass alongwith mineral soil particles with net result of humification of organic matter through their mixing activities, excretions and castings (Dash and Senapati, 1986).

NUTRIENT CYCLING AND ENERGY FLOW

The phytosynthesis of green plants on land results in fixation of 129 x 10^9 t atmospheric CO_2 annually. This amount is balanced through annual production of 37 x 10^9 t CO_2 due to the respiration of plants and animals and 92 x 10^9 t CO_2 by microbial mineralization (Stolp, 1988). Thus soil biota contribute largely (about 70%) in cycling of CO_2 on earth. The fate of all natural organic molecules (detritus), irrespective of their complexity, is final degradation by micro-organisms. The organisms involved in mineralization represent a pool of microbial biomass in agricultural soil contain upto 5% of the total soil carbon and upto 15% of total soil nitrogen (Paul and Clark, 1989).

Global net primary production is 8.5 x 10^{10} t yr^{-1}. Approximately 5% of terrestrial net primary production is consumed by herbivores and rest organic matter (95%) becomes litter (detritus). In edaphic system litter serves as source of energy for decomposing microbes. Even most resistant materials like lignin succumb to enzymatic actions and physico-chemical exploration, and net effect being release of energy as heat, CO_2 and water. Soil microflora and fauna work in consortium during decomposition processes. While some mites and other arthropods skeletonize decaying leaves or burrow into conifer needles in forest litter, others prefer to decompose outer sheaths of roots. The densities of various micro-arthropods are usually low in hot deserts (2100-3200 m^{-2}) and increase with increasing rainfall, decreasing aridity and increasing amount of plant litter (Lal, 1987). The microflora act upon the comminuted litter and release nutrients which are added to soil and either taken up by plants or leached by rainfall. Some mites pass the litter through their midgut and excrete them as faecal pelletes thereby increase litter surface area and make them more susceptible to microbial action. In xylophagous orbatids mostly spore bearing Gram positive bacteria and *Aspergillus* fungi, present in their gut, cause digestion of litter through enzymatic action.

Earthworms constitute more than 80% of the invertebrate biomass in most arable soils in both temperate and tropical regions and process about 20% of the net primary production. Earthworm activities improve soil fertility. Earthworms (*Lumbricus* sp.) annually deposit castings to the tune of more than 10 t ha^{-1} in surface soil under temperate conditions. These casts, rich in N,P,K, Ca and Mg, support growth of various microbial species. The bacterial communities, about 400 g^{-1} cast, constitute about 60% of microbial activity. Soils on ant and termite hills are rich in organic matter with high cation exchange capacities so stimulate luxuriant flora around their inhabitations. The amount of soil turnover by ants and termites at peaks range from 12 to 317 kg ha^{-1} (Arun *et al.*, 1988). Soil fungivorous collembola are reported to catalyze nitrogen mineralization activity of microbes in deciduous and coniferous forests. The interaction between saprophytic and mycorrhizal fungi plays a key role in the effectivity of collemboles. Millipedes feed litter alongwith microbial communities and thus microorganisms also get digested in the process. The excretory wastes or faecal materials of small animals like rodents, moles, rabits etc. influence the growth of certain bacteria in soil. The material gets mixed during soil excavation for burrow or nest making for egg laying by soil fauna thereby help in maintaining biodiversity. The dead bodies of animals and their carcasses pass through the detritus food chain and are acted upon by microflora and coprophagous and carnivorous anthropods releasing the energy and nutrients, bound in their carcasses and tissues (Cornaby, 1977).

Apart from carbon and water, nitrogen is the nutrient required in large quantities by plants and animals. In cycling of nitrogen non-biological transformations have little importance. The biological nitrogen fixation (BNF) is of fundamental importance in recycling of nitrogen. Only procaryotes are capable of fixing atmospheric nitrogen. *Azotobacter, Azospirillum, Clostridium* and *cyanobacteria* (all free living) and Rhizobium and *Frankia* (symbiotic) are involved in BNF to the tune of about $10^8 \times 10^6$ t N yr^{-1} The BNF accounts 65-67% of total global N fixed through various means. Ammonification has been estimated to amount 3×10^{10} t yr^{-1}. Bacterial nitrification by chemolithotrophic nitrifiers requires aerobic conditions. Interestingly most of the identification, responsible for nitrogen loss in soil, is also biologically-mediated.

The autotrophic bacteria bring about nutrient transformations in soil. For instance, some organisms cause NH_4, NO_2^-, NO_3^-, PO_4^- release and some bring about S, Fe and Mg oxidations. The plant rhizosphere is considered most active zone of nutrient transformation in soil and usually have 10 to 100 times greater biota population in comparison to non-rhizosphere area. The population of different organisms in rhizosphere is influenced by the chemical activities of the plant. The major effect observed is a quantitative one. Both arable and non-arable soil ecosystems are strongly influenced by the biotic associations within them. Although the function of soil fauna may be similar in the two types of systems, the relative proportion of specific group of organisms associated with a particular process e.g. N mineralization may vary.

Not more than 1% of the radiation energy, entering the biosphere, is conserved in the gross primary production. Some of this energy is reserved in organic matter that is produced by photosynthetic organisms and dessipated by microflora and other primary producers themselves as a results of their respiration. The rest of the chemically bound energy is taken over by micro–and macro–phytophages and microbes involved in the decomposition processes. The energy flow passes through food chain and at each step some energy is lost or dessipated (Figure 31.1). Each intersecting food chain consists of a number of species, the population of which reach equilibrium in terms of rate of reproduction and destruction. This essential equilibrium shifts abruptly if any one if its complex determinant changes, e.g. food supply, agro-climatic conditions, long term geochemical evolution or competition with other predators and parasites. The energy continuously flows through an ecosystem. The chemical energy is lost as heat from the ecosystem whereas bioelements do not get lost but are recycled. The role of micro-organisms in energy flow is mainly related to decomposition of detritus and as ultimate decomposers. Temporal and spatial variations in soil organisms, particularly microbes, effect the energy flow and so is in case of succession of different populations. In moving from the base of a trophic pyramid to the higher levels there is a progressive increase in calorific values (Wallwork, 1971). For instance,

$$\underset{(2665\,\text{cal}\,g^{-1})}{\text{Fungi}} \longrightarrow \underset{(4140\,\text{cal}\,g^{-1})}{\text{Cryptostigmatids}} \longrightarrow \underset{(5242\,\text{cal}\,g^{-1})}{\text{Mesostigmatid mites}}$$

The weakly sclerotized young animals are more vulnerable to predators and may transfer more energy and material than their adults to higher trophic levels in the edaphic community.

INDIRECT ROLE

(a) Transport of detritus

Fauna in soil ecosystem accelerate the decomposition process by promoting growth and distribution of microbes, especially fungi, and by transporting the decomposed products

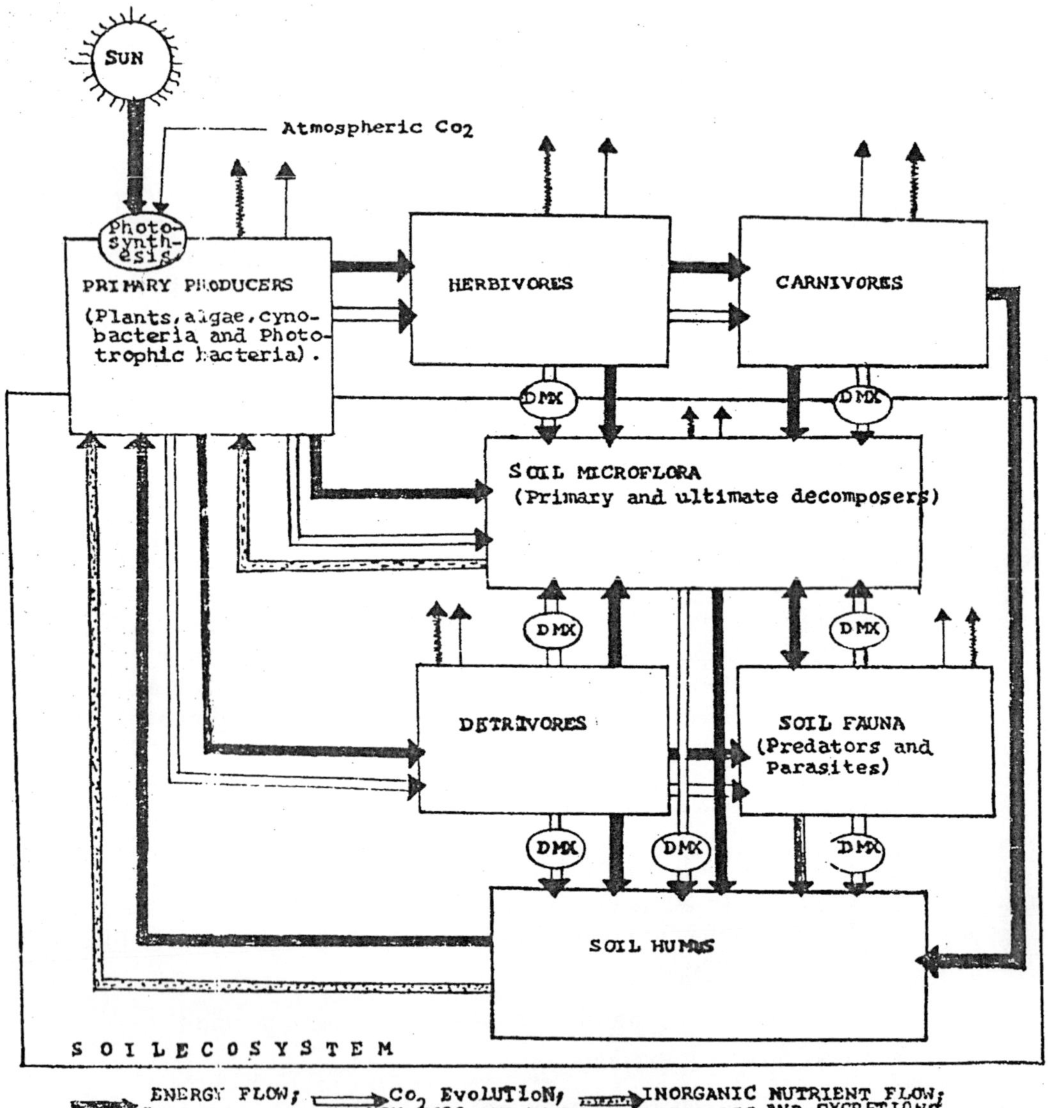

Fig. 31.1: Food chain and energy flow through edaphic ecosystem (simplified representation).

to lower layers especially deep root zones in soil profile. It is reported that a considerable volume of cryptostigmatid excrement accumulate in the root zone at a depth of 7 to 12 cms. During transportation of decomposed products they help in deep mixing of organic material and dissemination of fungal and bacterial cells/spores, which readily adhere to their body surface or while being in their gut come out through faecal pellets. The fungal and bacterial spores on reaching the litter organic residues initiate and enhance the process of degradation by creating a softening effect. This is a prerequisite for the feeding activity of the secondary decomposers like soil arthropods including mites.

(b) Soil aeration, drainage and water percolation

A system of capillary pores and passages of varied dimensions is created by the movement of soil fauna which air and water circulates. This increases soil aeration and water movement. Soil fauna like arthropods, mites, ants, nematodes and earthworms etc. possess suitable body size which help in the aeration of organic matter thus increase favourable sites for fungl and bacterial attack. The carabids, scarabaeid and scolytid beetles, feeding on inner bark or phloem of trees, serve as hosts for a great variety of mites especially tarsonemus and lasioseius species which are mostly seen to remain as symbionts. These mites act as sanitizers in the galleries by feeding on fungal excrements (Rishi and Rather, 1983).

SOIL POLLUTION

Demographic pressure has put soil under enormous stress by the sprawl in agriculturization which poses threat to its productivity. Pesticides, fertilizers, herbicides, sewage sludge, industrial wastes, detergents, garbage, synthetic polymers etc. find their way in soil ecosystem directly or indirectly. In India use of pesticides and fertilizers in crop production is showing an increase with 9.0×10^4 t pesticides and 13.5×10^6 t of fertilizer consumption during 1994-95. Only 0.1% of aerially sprayed pesticides actually reach the targeted organisms and rest reach the soil environment. Similarly only 40 and 15% of nitrogen and phosphorus applied is utilized by intended crop. Therefore, the accumulations enhance soil pollution. Agro-chemicals indiscriminately kill all insects and micro-organisms both beneficial as well as pests. For instance, the use of benzimidazole fungicides significantly reduce mycorrhizal, fungal and bacterial populations and species diversity. Similarly, in chlorinated hydrocarbon treated soils the population of lycosid spiders, ants, coleoptera, millipedes and mites get appreciably reduced. About 99% of microarthropod population is usually killed by fumigant DD and Vapam and it may take as long as two years for the population to recover fully (Brady, 1990). Fortunately, the recovery time for the microflora is generally much less. The effects on non-target organisms have adverse impact on soil health in terms of reduction in decomposition activity, predation, hyperparasitism etc., alteration in microbial succession and reduction in the diversity of soil biota (Domsch, 1984). The recurring pesticide-use over last 20 years resulted in 5, 10 and even 12% reduction in the number of different insect species per annum (Webb, 1996). Biocides disturb rhizosphere population and composition as these besides influencing resident organisms change the plant physiology and type and amount of root exudation. The decomposed products of pesticides, accumulated by plants, exert a negative influence on the activity of soil bionts. For instance, in methyl parathion treated soils earthworms disperse to deeper layers even upto 20 cms.

The disposal of the sewage sludge, usually on land, results in environmental destabilization as these are often rich in heavy metal and other contaminants/recalcitrants.

So prolonged sludge use enhance the concentration of contaminants in soil, decrease microbial biomass C and impede various biological processes (Dar, 1995). Similarly, heavy metal pollution induces shift towards Gram-negative bacteria compared to Gram-positive organisms. Frequently sewage sludge-amended soils have abundance of *Pseudomonas* sp. The polyethylene waste has direct bearing on physical conditions of soil and influences aeration. Moreover, these are resistant to degradation so persist in soil for quite longer periods.

The other concern is that crop-land is fast losing fertile top soil. In India 1.39 x 10^8 ha of land is under crops and about 4.7 x 10^9 t of soil is lost annually through erosions (Brown and Wolf, 1984). There is reduction in organic matter content in soil, its flora and fauna, with tilth and aeration with adverse impact on soil structure and ecosystem. The erosion losses force increased fertilizer use for greater productivity thus increasing pollution hazards. The soils devoid of meso–and macro-fauna are vulnerable to rapid soil degradation because they do not receive the beneficial effects of these animals (Lal, 1987).

CONSERVATION OF BIODIVERSITY IN SOIL ECOSYSTEM

Jammu and Kashmir state is showing an increasing trend of fertilizer use with NPK consumption of 4.2, 7.3 and 1.2 thousand tonnes in 1994-95, an 7.5% increase over previous year. Thus the State is likely to face the serious environmental consequences where water bodies like Dal, Wular lakes etc. are already depicting hazardous symptoms. While agricultural practices have different effects on different soil organisms, a few generalization can be made. Pesticides, soil erosion, monoculture, land clearing, excessive tillage and industrial sewages generally reduce species diversity as well as population densities of soil organisms. Other practices such as adding lime, balanced fertilizers, organic matter enrichment, proper soil and water conservation measures, organic farming etc. promote sustained crop production and protection and result in improvement of sil flora and fauna both in terms of species diversity and overall population dynamics (Brady, 1990). Due to the environmental and economic constraints more emphasis is laid on sustainable agriculture practices like organic farming and use of biofertilizers which can reduce costs vis-a-vis protect the environmental qualities by enhancing beneficial biological interactions and natural processes in soil ecosystem.

Return of all unconsumed crop residues and natural manures to soil to support crop and animal production and to maintain proper soil health constitutes organic recycling or organic farming. The beneficial effects of organic farming (use of green manures, organic manures, composts, crop residues, biogas slurry, sewage sludge, bio-fertilizers etc.) predominantly lies in furnishing humus forming materials to bring about improvement in soil structure, water holding capacity, microbial populations and their activities, base exchange capacity, aeration and resistance to soil erosion thereby create ideal conditions for sustained biodiversity. However, the activity of soil microflora and fauna is strongly modified by soil temperature, pH, water potential etc. The enhanced mineralization of organic residues is simply related to biomass of animals, irrespective of animals, irrespective of taxonomic groups, whereas the activity of animals is clearly complementary to that of microbes. For example, the conditioning of litter by fungi facilitates the attack by fauna probably another manifestation of a succession. The faunal consumption liberates N at a C:N ratio that lead to its immobilization by microbes. Therefore, organic matter and its diversity and supply are obviously important factors in determining faunal activity and species diversity.

(a) Organic wastes

(i) *Crop residues* : In India crop residues about 2-3 t ha^{-1} are produced so a total of 1.8 x 10^8 t yr^{-1} are available for recycling. Population densities of a number of microbial genera especially cellulolytic organisms, *Azotobacter, Rhizobium*, mycorrhizae and microbial metabolitic processes all respond to the increase in organic carbon input. The rate of various conversions in the soil ecosystem depends upon the interactions of soil microbes, nematodes and microarthropod communities. About 25-40% of plant material remains in soil as underground materials (roots) at harvest time. These are subject to degradation so play vital role in the maintenance of native soil fertilty.

(ii) *Organic manures* : It is estimated that about 1.1 x 10^9 t animal dung is available annually from 237 billion cattle population in India (ABE, 1985) and 2.27 x 10^8 t organic manures are used for crop production producing 3-4 million tonnes of N,P and K. These, besides providing macro–and micro-nutrients, promote root growth and improve soil physical, chemical and biological properties. The soil enrichment with organic manures under nutrient-limiting conditions is ecologically advantageous for diverse resident flora and helps in continuous production of all sorts of enzymes to utilize the substrate with the help of diverse microbial and faunal growth. Mishra and Kapoor (1986) obtained good control of root rot and wilt caused by *Fusarium solani* by binding fertilizer with millet straw and organic manures. The mixing of worm casts increase the nitrogen utilization efficiency of nitrogenous fertilizer from 22 to 28% and that of P fertilizers from 33 to 50%. The earthworms and some species of mites and beetles flourish well in soils enriched with organic manures.

(iii) *Composts* : Composts are decomposed and stabilized organic wastes obtained through enhanced activities of aerobic/anaerobic micro-organisms. Cellulolytic and ligninolytic fungi and bacteria viz. *Trichurus, Aspergillus, Trichoderma, Chaetomium, Cytophaga, Cellulomonas* etc. are inoculated to organic residues and heaped in layers in pits of 2.0 x 1.0 x 2.7 m. The organic residues have a C:N ratio of 30:1 whereas microbes and soil have a ratio of 5:1 and 10:1, respectively. The fermentation, induced by microbes, decreases C:N ratio to 15-20:1 with organic matter content of 7-10% and moisture 50-70% in compost. When applied as mulch, the compost inhibits weed emergence and disease development and yet often stimulates crop establishment (Lee, 1993). Composts provide suitable substrate for the growth of aerobic and anaerobic bacteria and cellulolytic and ligninolytic fungi. Organic manures, compost and farm yard manure together contribute 36 kg 'NPK' t^{-1} applied.

(iv) *Biogas slurry* : Biogas slurry contains enriched nutrients as most of easily degradable carbon is lost during biomethanation. During anaerobic fermentation C:N ratio of crop and animal wastes is reduced to 9 - 12:1 (Dar and Tandon, 1987). One ton of agricultural waste produces 175 m^3 biogas and 2 t moist bio-fertilizers, suitable for application in soil for enriching its fertility. The use of biogas slurry reduced the exploitation of forest trees for firewood and reduces exclusive dependence on chemical fertilizers.

(v) *Vermicompost* : Composts prepared from crop residues animal wastes, garbages etc., in conjunction with soil, with the help of worms and bacterial/ fungal action are known as vermicomposts. These composts are nutrient-rich and improve soil humus formation and fertility. During composting earthworm casts are reported to increase soybean yield by 30-50% and also enhance efficient phosphorus intake. Biologically earthworms enhance the activity of micro-organisms and micro-fauna through prevention of bacteriostasis and mycostasis. The inoculation of organic wastes, usually rich in decomposer populations, results in stimulated growth of earthworms through various biotic interactions. Earthworms are known to decrease the plant parasitic and non-parasitic nematode population and enhance N_2 fixing bacteria in soil. Mixing of castings alongwith phosphorus fertilizers is a convenient mean of enhancing phosphorus utilization in dry lands.

(vi) *Green manures* : It is the practice of growing nitrogen-rich plants like legumes and ploughing them into the soil usually after 45 days at (flowering stage), just to add some organic matter. Green manuring provides nutrients (about 44 kg N t^{-1} + other nutrients) to the succeeding crop, improves soil structure and minimizes erosion losses. Green manuring increases yield by 8-12%. The cowpea intercroped with millets add 30 kg N ha^{-1} alongwith other nutrients while rice straw with water hyacinth provides 20 kg N t^{-1} besides other nutrients to the soil. Application of green manures reduces the hazards of chemical fertilizers and conserve soil biodiversity.

(b) Bio-fertilizers

These are microbial inoculants, artificially multiplied cultures of beneficial organisms, introduced into soil or plant to enhance soil fertility and crop productivity. These organisms in addition to supplementing nutrient needs, synthesize organic substances, growth promoting chemicals and antibiotics, suppress diseases and bind soil aggregates through their mucilages and other excretions. Presently about 2900 t *Rhizobium, Azotobacter, Azospirillum* and *cyanobacterial* inoculants are produced annually and the Rhizobium requirement has been estimated to 15,000 t to cover 30-40 million ha area under legumes.

(i) *Rhizobia* : These symbiotic bacteria are present in soil and form nodules on legume roots which are protective structures for nitrogen fixation. Besides nitrogen supplementation to a tune of 50-100 kg ha^{-1}, they secrete slime (extra-cellular polysacharides etc.) so help in soil binding.

(ii) *Cyanobacteria* : These phototrophic nitrogen fixing procaryotes are ubiquitous in distribution and may be single-called or filamentous. Mostly used in wet-lands (rice), the cyanobacteria cause NH_4 accumulation in soil through their N_2 fixation process upto 20 kg ha^{-1}. Besides, these organisms excrete vitamin B_{12}, auxins, ascorbic acids and other plant growth promoting substances. Their mucilages help in soil aggregation. In J&K a yield increase of 10-15% has been reported in rice through cyanobacterial inoculation (Dar and Ganai, 1990). This biofertilizer has a potential to provide about 7.0 x 10_5 t biological nitrogen to rice crop area per season. *Azolla* and *Frankia*, the other symbiotic nitrogen fixers, contribute 3-4 kg N t^{-1} applied and 10-20 kg N ha^{-1} to soil, respectively.

(iii) *Free living N_2 - fixing bacteria* : *Azotobacter, Azospirillum* and *Beijrinkia* are free living non-symbiotic nitrogen fixers in soil. Besides nitrogen fixation (20-30 kg N ha-1), these improve P and K supply and secrete auxins, antibiotics and antifungal substance. These are widely used in cereals, legumes, vegetables, fruit crop nurseries etc. and cause 10-30% increase in crop yield.

(iv) *Phosphate solubilizing bacteria* : Hardly 0.1% of total P present in soil is available to crops. The bound or fixed phosphorus is solubilized or mobilized by various bacteria and fungi and made available to crop. Phosphobacterin (bacterial preparation of

Bacillus megaterium var. Phosphaticum) is commercially used in various crops and is reported to enhance crop yield from 5 to 30%.

(v) *Mycorrhizae* : These fungi form symbiotic associations with the plant roots and increase absorption of nutrients through increase in surface area of roots, phosphorus-solubilizing enzyme production, auxins and plant growth promoting metabolite secretions and help the plant to overcome stress conditions. VAM hyphae respond to microsite accumulations of organic matter by proliferating these sites. The activities and by-products of the soil faunal community influence both hyphal growth and the creation of organic microsites (Jeffrey, 1987). Evidences suggest that soil micro-fauna like collembola graze and disrupt the VAM colonization and soil hyphal network by feeding on them. If disruption or grazing is significant, the associate hyphal loss have detrimental effect on physical entaglement mechanisms. Alternately, grazing of external hyphae by micro-fauna stimulate hyphal proliferation by creating nutrient-rich microsites of fecal pellets (Tate, 1987). The role of mycorrhizae in weathering of parent materials and providing more sites for diversity of organisms in soils cannot be ignored. The analysis of microbial composition of soil aggregates created by VAM fungi have revealed the presence of a range of fungi, bacteria, actinomycetes and algae including cyanobacteria (Jeffrey, 1987). Mycorrhizae cause a saving of 30-50% fertilizer, compensate 20-30 P ha^{-1} and enhance yield potential by 30-60%.

ORGANIC FARMING AND CONSERVATION OF BIODIVERSITY

Organic farming is the practice of restoring organic matter content in the soil to its original level or even enhance it for sustainable crop production. The judicious cycling of organic wastes is a sound low-cost practice having beneficial impact on physico-chemical and biological properties of soil. Since nutrient bases in natural communities are largely derived from host plants, therefore specialized abilities of their associated beneficial microflora and fauna to release, mobilize and transport key nutrients help in the conservation of biodiversity within edaphic ecosystem. It has been observed that about 20-50% of nutrients supplied by the animal manures and organic residues are recovered by the first crop following the application (Tisdale and Nelson, 1978). The rest of the nutrients, held in humus-like compounds, are released slowly at the rate of 2-4% yr^{-1} (Brady, 1990). Soil aeration and water percolation increase under organic farming system.

Since organic matter turnover in soil involves diverse organisms, therefore fluctuations in the composition, abundance and functioning of beneficial organisms usually depict the changes in various biochemical transformations and that of soil health. At community level these organisms being greater diversity and stability in primary producers and ultimately influence other biota as well. For instance, the mycelial network of mycorrhizae interlink plant and soil biota at both intra-and inter-community (specie) levels (Read, 1990). Moreover, physical entanglement, either through mucilages/excretions or hyphal network, enhances macroaggregate formation thereby reduces erosion losses and improves residence of nutrients within the system. The organic recycling has greater significance in minimizing soil ecosystem destabilization induced by various pollutants. The role of organic farming in general and biofertilizers in particular in understanding ecosystem functioning and maintenance of sound environment through conservation of biodiversity and stability justify intense research in this direction.

CONCLUSION

The changes in soil physico-chemical properties influence the community structure via mutualistic and competitive processes. Biotic interactions arising from input of potential organic wastes and biofertilizers into soil have major impact on system functioning particularly nutrient cycling and energy flow. The functional diversity among different soil communities demands understanding the soil ecosystem holostically rather than organisms in isolation. Therefore, effective management of soil ecosystem is to be adopted to have long term gains. The excessive use of fertilizers and pesticides for greater productivity need to be lessened through adoption of organic farming and integrated nutrient/pest management for the conservation of soil resources and greater diversity.

ACKNOWLEDGEMENT

Authors are grateful to Dr. G. M. Wani, Director Extension Education, SKUAST, Shalimar for his valuable comments and encouragement.

LITERATURE CITED

ABE. (1985). *Towards a Perspective on Energy Demand and Supply in India in 2004/05.* Advisory Board on Energy, Govt of India, New Delhi.

Arun, L., Chopra, G., and Gupta, S.R. (1988). Foraging activity and soil turn-over by ants and termites in an agroforestry system. *Proceeding of Xth International Soil Zoology Colloquium*, 103 pp.

Berthet, P. (1963). Measure de la consommation d'oxygene des oribatides (Acarins) de la litiere des forests. pp 18-31. In : Soil Organisms. (Eds. J. Doeksen and J. Van der Dift), North-Holland Publ. Co., Amesterdam, Holland.

Brandy, N.C. (1990). *The Nature and Properties of Soil*, Macmillan Publ. Co., New York, USA, 621 pp.

Brown, L.R. and Wolf, E.C. (1984). *Soil Erosion : Quiet Crisis in the World Economy.* World Watch Paper, 60 pp.

Butcher, J.M., Snider, R. and Snider, R.J. (1971). Biology of edaphic collembola and acarina, *Annual Review of Entomology*, 16 : 249-288.

Coleman, D.C., Anderson, R.V., Cole, C.V., McClellan, J.F., Woods, L. E., Trofymov, J.A. and Elliot, T. (1984). Role of protozoa and nematodes in nutrients cycling, pages 17-28. In : Todd, R.L. and Giddens, J. L. (Editors) *Microbial–Plant Interactions.* Soil Science Society of America Society of Agronomy, Madison, USA.

Cornaby, B.W. (1977). Saprophagous organisms and problems in applied resources partitioning. pages 96-100. In : Mattson, M.J. (Editor) *The Role of Athropods in Forest Ecosystems.* Springer Verlag, Berlin, Germany.

Dar, G.H. (1995). Effect of heavy metals (Cd, Cr, Ni and Pb) on soil microbial biomass, C mineralization and enzyme activities. *International Journal of Ecology and environmental Science*, 21 : 87-95.

Dar, G.H., Zargar, M.Y. and Beigh, G.M. (1997). Biocontrol of *Fusarium* root rot in common bean (Phaseolus vulgaris L.) by using symbiotic *Glomus mosseae and Rhizobium leguminesarum. Microbial Ecology*, 34: 74-80.

Dar, G.h. and Ganaie, M.R. (1990). Effect of Cyanobacterial inoculation on soil fertility and yield of paddy under temperate conditions. *Phykos*, 29 : 31-38.

Dar, G.H. and Tandon, S.M. (1987). Recycling potential of waste plant residues for biogas and enriched manure production. *Science and Culture*, 53 : 93-95.

Dash, M.C. and Senapati, B.K. (1986). Vermitechnology potentiality of Indian earthworms for vermicomposting and vermified. pages 61-70. In : Mishra, M.M. and Kapoor, K.K. (Editors). Soil Biology, Haryana Agricultural University, Haryana, India.

Domsch, K.H. (1988). Effects of pesticides and heavy metal on biological processes in soil. *Plant and Soil*, 78 : 367-378.

Hazra, A.K. and Sanyal, A.K. (1997). On some aspects of conservation and management of soil macrofauna and micro-organisms for sustainable soil ecosystem. Proceeding Golden Jubilee National Symposium on Biology for. Sustainable Development, 30 pages.

Jerrey, D.W. (1987). *Soil Plant Relationship : An Ecological Approach*. Groom Helm Ltd., Kent, U.K.

Krivolutzsky, D.A. and Severtzev, A.N. (1978). Oribatids as indicators of effect of radio-pollution biogecoenese. *Zool. Zh*, 55 : 226-236.

Kuhnelt. W. (1961). *Soil Biology*. Faber and Faber, London., UK, 397 pages.

Kumar, H.D. and Rai, L.C. (1986). *Microbes and Microbial Processes*. Affiliated East-West Press, Delhi, India 212 pages.

Lal, R. (1987). *Tropical Ecology and Physical Edaphology*. John Wiley & Sons, Chichester, U.K. 732 pages.

Lee, H. (1993). Research on the use of composted organic wastes. *Spectrum*, 224 : 5-6.

Mishra, M.M. and Kapoor, K.K. (1986). *Soil Biology*. Department of Microbiology, HAU, Hissar, India, 274 pages.

Nordgren, A., Kauri, T., Baath, E. and Soderstorm, B. (1986). Soil Microbiology, activity, mycelial length and physiological groups of bacteria in a heavy metal polluted area. *Environmental Pollution*, 41 : 89-100.

O'Dell, T.E., Castellane, M.A. and Trappe, J.M. (1992). Biology and application of ectomycorrhizal fungi. pages, 379-416. In : Mething, F.R. (Editor) *Soil Microbial Ecology* Marcel Dekker, New York, USA.

Paul, E.A. and Clark, F.E. (1989). *Soil Microbiology and Biochemistry*. Academic Press, New York, USA. 376 pages.

Rather, A.Q. (1988). Observations on the ecology, behaviour and distribution of some oribatid mites (Acari. oribatei) from Jammu and Kashmir. *Proceedings of Xth International Soil Zoology Colloquim*, 110 pages.

Read, D.J. (1990) Mycorrhizae in ecosystems-Nature's response to the law of minimum. pages 103-130. In : Hawksworth, D.L (Editor). *Frontiers in Mycology* CAB International, Wellingford, U.K.

Rishi, N.D. and Rather, A.Q. (1983). Record of mites and their relationship with the forest habitates in the Kashmir valley. pages 151-153. *In : Proceeding of the Workshop on High Altitude Entomology and Wildlife*. Ecol. Zool. Survey, India.

Stolp, H. (1988). *Microbial Ecology : Organisms, Habitat and Activities*. Cambridge University Press, Cambridge, Great Britain. 308 pages.

Tate, R.L. (1987). *Soil organic matter : Biological and Ecological Effects*. John Wilay & Sons, Toronti, Canada. 291 pages.

Tisdale, S.L. and Nelson, W.L. (1975). *Soil Fertility and Fertilizers*. MacMillan Pub. Co., New York, USA. 126 pages.

Wallwork, J.A. (1971). Some aspects of energetics of soil mites. pages 129-134. In : Daniel, M. and Rosicky B. (Editors) *Proceeding of the 3rd International Congress on Acarology*. The Hague, Junk.

Webb, S. (1996). Reducing the use of chemicals in agriculture. *Spectrum*, 253 : 13-15.

Zargar, M.Y., Dar, G.H. and Zargar, A.H. (1999). Mycorrhiza in relation to ecosystem development. In : Khan, M.A. and Farooqs (Editors). *Environment, Biodiversity and Conservation*, University of Kashmir, Srinagar, J&K.

Chapter - 32

Mycorrhizae in Relation to Ecosystem Development

Mohd. Yousuf Zargar, Gh. Hassan Dar and A.H. Zargar

ABSTRACT

Mycorrhizal fungi through their nutrient mobilization and cycling activities enhance the absorption of relatively immobile elements in soil such as P, Cu and Zn beyond nutrient depletion zone. Their ability to conserve soil and water, and enhance plant-tolerance towards pathogens, heavy metals and drought etc. have given them a selective advantage over other microbial commun ties in the rhizosphere. In general, mycorrhizae increase biodiversity of microflora and fauna through their mutualistic interactions. Plant specie diversity in natural forests, grassland and agricultural ecosystems is usually correlated with the presence of different types of mycorrhizas. The diversity of mycorrhizal species decline from natural ecosystem to intensive agricultural system. The mycorrhizae play pivotal role in the development of environmentally sound edaphic ecosystem through their physical exploration and biochemical activities.

Key Words: *biocontrol, biodiversity, energy conservation, microbial interaction, mycorrhizae, nutrient cycling, soil conservation, soil ecosystem.*

INTRODUCTION

Soil is a dynamic entity supporting the growth of plants, microflora and fauna and is an essential component of human environment. The plants derive their macro-and micro-nutrients, water etc. from the soil. However, the primary growth limiting element in plants is either nitrogen or phosphorus. The soil biota help in providing essential nutrients, in mineralised form, through their biochemical activities. About 85% of the metabolic activities in soil are attributed to fungi and bacteria (Paul and Clark, 1989). The fungal communities universally have the limitation of carbon and so, through their heterotrophic activities, derive carbon from litter. Owing to these limitations, natural selection has probably favoured

Division of Environmental Sciences, S.K. University of Agricultural Sciences & Technology, Shalimar, Srinagar (J&K)-191 121

the development of intimate associations between fungi and plant roots. One of such associations is intimate symbiotic relation developed by plant roots and mycorrhizae.

Most of the mycorrhizal studies have remained concentrated on their role in phosphorus uptake and plant growth. Until recently plant ecologists viewed mycorrhizal fungi to be of great significance in phosphorus-limited ecosystem. However, its role in relation to soil ecosystem development and biological diversity has been ignored. Approximately 70% higher plants are obligately dependent on mycorrhizae, 12% facultatively dependent and 18% remain typically uninfected (Trappe, 1987). Except for the members of brassicae and chenopodiaceae, most agricultural plants exhibit mycorrhizal association. Some such as wheat, nut crops etc. show facultative mycotrophy, whereas most of the forest plants, maize and soybean etc. are strongly dependent on mycorrihzae. The herbaceous and shrubby species are generally vesicular arbuscular mycorrhizal (VAM), whereas vascular plants are mostly ectomycorrhizal (ECM). A few tree crops such as apple and other rosaceous trees form both VAM and ECM. More than 150 species of VAM and over 5,000 species of ECM have been identified so far (Marx, 1991).

Mycorrhizae exhibit a great diversity of associations. For instance, Douglas-fir alone associates with ore than 2,000 mycorrhizal species (Trappe, 1987). Similar estimates have been reported for pine, spruce, cider, fir etc. (Castellano, 1996).

The role of mycorrhizae in the enhancement of immobile nutrients through their physical exploration and biochemical activities, and in overcoming edapho-climatic vagaries has lead microbial ecologists to evaluate their functioning in ecosystem development. The present paper reviews the work done in this behalf and the potential thes microbes have in the maintenance of ecologically balanced environment.

NUTRIENT CYCLING AND ENERGY CONSERVATION

In soil only 0.01% of the total phosphorus is available to the plants and most of the fertilizer phosphorus added (about 85%) gets fixed in soil. Plant roots are capable of extracting phosphorus, a least mobile element, only upto 2 mm distance. The mycorrhizal stands transfer water and nutrients beyond nutrient depletion zones over a distance of 30 cm away from roots (Finlay and Read, 1986). Mycorrhizae are capable of mobilizing phosphorus uptake at extremely low phosphorus concentration (less than 0.001 ppm) in soil. In comparison, roots require greater than 0.1 ppm phosphorus for uptake (Hekstra, 1996). This depicts the potential of mycorrhizae in sequestering the nutrients for plant growth in extremely nutrient-limiting conditions. Besides, mycorrhizae influence the residence time of nutrients within the system through nutrients retention for longer periods in hyphal strands. The extensive proliferation of mycorrhizal hyphae enhances the nutrient mobilization. Though the external hyphae for VAM varies with soil type and the plant community, in general 12-19 m of hyphal length per gram of soil has been observed (Pfleger and Linderman, 1994).

It has been recognised that, underwarmer climates, higher pH can precipitate mineral salts as insoluble residues, phosphorus being the major element. However, under cold climates, low temperature, evapotranspiration and pH combine to inhibit the mineralization of nitrogen. Experimental evidences show that the mycorrhizal fungi provide their hosts with access to nutrient elements (N, P, Fe, Zn, CU, S and Mg etc.), limited availability of which would otherwise impose severe limitations on plant survival and growth. Moreover, the nutrient cycling activities like organic matter decomposition, mineralization and elemental fixation is enhanced through boosting in beneficial micro-organisms in mycorrhizosphere.

Mycorrhizal fungi derive energy from the host in the form of sugars and other products of photosynthesis. More than 50% forest and herbaceous photosynthates are directed to the below ground parts to support root, mycorrhizae and associated microbes (Comeau and Kimmins, 1989; Norton *et al.* 1990). External mycorrhizal hyphae represent about 26% of labelled extra-radical organic carbon pool in soil, so seves as an important source of carbon to soil microbes. The presence of mycorrhizae improves composition and quantity of exudates, thus favouring more energy transfer from one system to another. The ecosystem net primary production that goes towards heterotrophic respiration may range from 24% for an alfalfa field to 94% for a mesic deciduous forest (Coleman *et al.* 1976). Removal of trees or other vegetation changes the quality and quantity of energy transferred to below ground and, thereby alters microbial commun ties and soil properties. Bacterial population is 10-100 times higher in ectomycorrhizo sphere than in the soil away from roots (Rambelli, 1973). The mycorrhizae, through organic acid production, besides enzyme activities and physical exploration, enhance nutrient cycling through accelerated weathering of parent materials and solubility of soil minerals (Sollin et al. 1981). Moreover, the humic fraction of soil which consists of organic material is less susceptible to biological decomposition. This gives mycorrhizae a unique advantage over other heterotrophs in soil ecosystem as they have constant accessibility through symbiotic association with their autotrophic hosts.

SOIL CONSERVATION

Soil and water conservation is considered as backbone for sustainable agricultural production, since their management practices reduce soil erosion and help in maintaining soil fertility. Soil structure influences, apart from physical and chemical processes, the biological processes by affecting the accessibility of food, water, shelter, oxygen and nutrients to soil micro-organisms (Elliot and Coleman, 1988). The extra-radical hyphae of mycorrhizal associations is an important, and much overlooked, contribution to the process of creating a stable soil aggregate structure (Miller and Jastrow, 1992). The hyphae of both mycorrhizal and saprophytic fungi along with fibruous roots, bind soil particles and microaggregates into macroaggregate (Gupta and Germida, 1988). The VAM hyphae from aggregates through physical entanglement as well as by their excretions like polysaccharides and organic compounds (Elliot 1986). Moreover mechanical entanglement by extra-radical hyphae physically bring mineral particles and soil debris together (Emerson *et al.* 1986) thus bringing more stability to mico-and micro-aggregates through cementation by their secretions (Elliot and Coleman, 1988). This allows improved soil aeration and water percolation, thus accelerating the biological activities in soil (Tisdall, 1991).

The mycorrhizosphere influences physical, chemical and biological properties of soil. Eroded soils have been reported to have reduced number of mycorrhizal propagules (Habte, 1989). Since mycorrhizal colonization imparts drought resistance and tolerance towards heavy metals, acidity, extreme temperatures and salinity in plants, thus help in plant survival under extreme conditions, thereby does not allow the soil to be exposed to water and wind erosions. Besides these soil conservatory attributes, mycorrhizae induce improvement in water holding capacity and mobilize more nutrients to plant roots.

MYCORRHIZAL INTERACTION WITH SOIL BIOTA

The utilization of sloughed off root carbon by mycorrhizae becomes important in the context of competition by other microbes. Mycorrhizae produce a selective change in the mycorrhizosphere through their exudations as a result of which availability of nutrients for plants also gets changed, because, the quality and quantity of exudates alters the rhizosphere. Beneficial organisms grow within the mycorrhizal root tips (Amaranthus *et al.*, 1990) and the population of plant and fungal antagonistic actinomyceles are reduced in mycorrhizosphere, (Brochers and Perry, 1990). Secilia and Bagyaraj (1987) demonstrated increased number of urea and starch hydrolyzing bacteria in mycorrhizosphere soil. Marks and Foster (1973) showed that bacteria are closely associated with the sheath of ectomycorrhizae and suggested that specific bacteria might be favoured in rhizosphere. In turn, microorganisms affect the mycorrhizal development and function. Some rhizosphere bacteria produce auxins, hormones, nutrient solubilizing enzymes, siderophores etc. thus directly or indirectly stimulate root growth and branching or mycorrhizal growth (Strzelczyk and Pokojska-Burdzirj, 1984; Dar et al., 1995).

Mycorrhizal presence induces significant changes in host root morphology and physiology. The fungi also produce phytohormones such as indole acetic acid, ethylene, giberellin, cytokinis, auxins etc. that induce feeder root proliferation (Graham and Linderman, 1981). Dar et al. (1997) reported enhanced rhizosphere microbial biomass and phosphatase activity in presence of VAM. Increased microbial activity, changes in hyphal proliferation and mycorrhizal composition not only the plant growth favourably but also influence the soil fertility (Mukherji, 1996). Meyer and Linderman (1986) demonstrated growth enhancement when plant growth promoting rhizobacteria, *Pseudomonas putida* and VAM were used together.

The activities of symbiotic and free living rhizosphere nitrogen fixing or phosphate solubilizing microorganisms substantially affect the plant productivity and mycorrhizosphere. Free living diazotrophs such as *Azotobacter, Azospiirillum, Beijerinkia, Clostndium, Pseudomonas* in combination with VAM improve plant growth and nutrient availability significantly. Nitrogen fixing bacteria that live in some ECM mantles also supply a fair amount of nitrogen to the plants (Li and Castellano, 1987). Nursery studies have revealed improved growth and dry matter yield of forest species by *Azospirillum* inoculations (Swaminathan and Vadiaraj, 1988). The interaction between VAM fungi and *Azotobacter* and *Beijerinkia* had positive impact on crop growth and overall ecosystem (Mohandas, 1987). Involvement of hormonal secretions, increased nutrient recycling, efficient nutrient uptake per unit area, enhanced nutrient absorptive surface and more exudation, enzyme production and siderophore release have been attributed as mechanisms of growth enhancement and ecosystem preservation (Chandra, 1992; MacFall et al., 1992; Dar, 1998). The beneficial interactions also help the plant to overcome stress conditions like drought, heavy metal etc. Some free living N_2-fixers have been isolated from surface disinfected mycorrhizal roots and from surface disinfected spores of *G. fasiculatum, G. intraradices, G. mosseae* and *Gigaspora gilmori* (Tilak et al., 1987). Similarly, *Clostridium, Psudomonas* and *Azospirillum* spp. have been isolated from sporocarp and ectomycorrhizospheres (Amaranthus *et al.*, 1990). This all suggests a close association where in partners enjoy mutual benefits.

Mycorrhizae and *Rhizobium* directly interact in rhizosphere of their host. Generally mycorrhizal colonization favours nodule formation by *Rhizobium* (Dar *et al.*, 1995). However, advanced formation of VAM fungi might inhibit the process of nodulation (Bethlenfalvay *et al.*, 1983), possibly due to competition. The qualitative changes in mycorrhizal root exudations favour competitive abilities of *Rhizobium* resulting in enhanced nodulation.

It is apparent that nearly all legumes with *Rhizobium* nodules grow better if they are also mycorrhizal. As with other VAM situations, the plant and the *Rhizobium* benefit from the presence of VA mycorrhizae by the correction of any deficiency that is limiting the system. Those deficiencies could be of P or other nutrients, but other direct or indirect mechanisms also could be operational. Disturbance of agricultural soils may even minimize the involvement of any VAM fungi in the *Rhizobium* system. Positive interactions between phosphate solubilising bacteria (PSB) and VAM fungi on plants have been reported possibly due to production of plant hormones or vitamins by PSB (Raj *et al.* 1981; Krone *et al.* 1987). This has been demonstrated by applying cell free culture extracts from a *Pseudomoonas* sp (PSB) which enhanced VAM infection, mimicking the effects of pure plant hormones (Azcon et al. 1978). The vitamins produced by PSB exhibit a significant effect on VAM fungi, plant growth and other microorganisms. Soil microbes may also produce phosphates. VAM fungal infection is reported to enhance the number and activity of phosphatase-producing microbes (Krishna et al. 1985).

Mycorrhizae interact with yeasts and microscopic soil animals (Ingham and Molina, 1991). Yeasts stimulated germination of mycorrhizal fungal spores in pure culture (Fries, 1979). Some yeasts have been found associated with mycorrhizal sporocarps and fungal propagules dispersed by animals (Li *et al.* 1986). Mycorrhizal presence on the roots of *Pinus ponderosa, Pseudotsuga menziesii, Picea sitchensis, Tsuga Heterophylla* and *Abies grandis* influenced the protozoan community in the rhizosphere by the amount and types of exudations produced by the roots which, in turn, altered the bacterial community (Ingham and Massicotte, 1994). Further, protozoan community, in turn, influenced the successional dynamics of ECM on different host system either through preying upon individuals and perhaps removing particular bacterial species from mycorrhizosphere or by controlling N mineralization in rhizosphere. Cromack (1988) found more ciliates and amoeba together with greater microbial biomass in mycelial mats of *Hysteranguim and Gautieria* compared to adjacent soil.

The activities and excretions of soil fauna influence both hyphal growth and creation of organic microsites. Soil microfauna graze and disrupt hyphal network vis-a-vis help in their dispersal (McGonigle and Fitter, 1988). Further, grazing of external hyphae stimulates hyphal proliferation by creating nutrient enriched microsites of fecal pellets.

Macroscopic animals like collumbola, mites nematodes and microarthropods populate mycorrhizal fungal mats to a greater extent (Cromack, 1988; Kilronoms and Kendrick, 1995). Mycorrhizal fungi and endoparasitic nematodes are often mutually inhibitory, each reducing the population of other (Singh, 1997). Molluscs, such as slugs and snails and insects commonly graze fruiting bodies of higher fungi. Many mammals and some birds even use mycorrhizal fungi as their diets. Some rodents live exclusively on fungi, particularly the ECM hypogeous.

BIOLOGICAL CONTROL

The soil borne pathogens as well as symbionts share common habitat and show differential influence on host plant. The experimental evidence strongly supports the hypothesis that VAM fungi and their microbial associates may contribute in many ways to the suppression of root diseases caused by major fungal and nematode pathogens. Mycorrhizal symbionts offer increased resistance to certain wilt and root not pathogens such as *Olphidium brassicae, Macrophomina phaseolina, Fusarium oxysporum, F. solani, Verticillum* sp., *Pythium splendens, Phytophthora* sp., and *Rhizoctonia solani* (Sharma *et al.*, 1992; Hwang *et al.*, 1992. Dar *et al.*, 1997).

Establishment and functions of VAM fungi are not affected by fungal or bacterial antagonists (Linderman et al. 1991). A selective pressure is exerted by mycorrhizae on soil microorganisms, some of which canantagonise root pathogens. Meyer and Linderman (1986) showed that sporangium and zoospore production by *Phytophthora cinnamomi* was reduced in presence of the leachates from VAM plants. These experiments indicate that VAM fungi are relatively tolerant to antagonists that inhibit fungal pathogens by one or more mechanisms. Moreover, they are highly rhizosphere-competent and are compatible with such antagonists and even function in concert with them.

Because mycorhizae are major components of the rhizosphere, it is logical that they could affect the incident and severity of root diseases (Caron, 1989; Jalali and Jalali 1991). Authors observed that, in addition to decrease in root rot, mycorrhizae cause substantial increase in rhizosphere microbial biomass (Table 32.1). Many mechanisms are involved in disease suppression which include (a) protection of roots from fungal attack by forming protective physical barrier around feeder roots (b) synthesis of toxic substances analogous to synthesis of phytoalexins in host root interaction (c) antibiosis (d) better competition of rhizosphere nutrients as compared to soil pathogen and (d) oxalic acid production. Moreover, the increase in nutrient uptake (P and others) results in healthier plants better able to ward off or tolerate root diseases. There could also be localized morphological and physiological changes which might decrease the root infections by fungal pathogens. Reports indicate that VAM induces alteration in rhizosphere population of antagonists (Meyer and Linderman, 1986; Secilia and Bagyaraj, 1987). Since formation favours the increase of antagonists against root pathogens, therefore, the agricultural practices that disturb mycorrhizal formation are likely to reduce the levels of antagonists, and that, in turn, could account for increase disease controlling inputs. Reversal of the process could, therefore, contribute to sustainable agriculture and allow reduction in fungicide use.

Table 30.1 : Effects of mycorrhizal inoculants on rhizosphere microbial biomass C and root rot diseases

Inoculant used	*% reduction in root rot*	*% increase in rhizosphere microbial biomass C*
(a) Ectomycorrhiza*		
i. *Pisolithus tinctorius*	56	87
ii. *Cenococcum geophilum*	53	81
iii. Mycorhizosphere soil	42	60
iv. Mycorrhizal roots	32	51
(b) VA Mycorrhiza**		
i. *Glomus mosseae*	77	100
ii. *Rhizobium leguminosarum* + *Glomus mosseae*	85	126

* Ectomycorhiza in *Cedrus deodara* against *Rhizocotina solani.*

** Vam inoculated in *Phaseolus vulgaris* against *Fusarium solani.*

Root infections by pathogenic nematodes are generally less on mycorrhizal plants than on non-mycorrhizal ones but the responses and mechanisms may vary (Ingham, 1988). The evidences strongly indicate that VAM suppress nemdatodal root infections or their effects on plant growth and yield. Moreover, the mechanisms involved depend on the conditions of the test, the host plant, edaphic conditions and the VAM species involved. However, it seems safe to say that mycorrhizae play a role in the biological control of root nematode diseases.

ECOSYSTEM DOMINATED BY ECM

Ecotomycorrhizae help in rehabilitation of forest ecosystems directly by fertilizing the soil (Mukherji and Dixon, 1992). The dominance of ectomycorrhizae in most coniferous forest ecosystems has also profoundly altered the classical concept of the role of exudation and rhizodeposition of cell fragments in the formation of rhizosphere. In forests, 3-5 times more organic matter is returned to soil in the form of root and mycorrhizae than is returned by decomposition of litter. Of this, mycorrhizae account 85-95% and rest is from soluble root exudation, root cap and mucilages (Fogel, 1988). Ectomycorrhizal forests characteristically show a relatively low diversity of host species but a great diversity of fungal associates than do VAM ecosystem. Meier (1991) observed differences among frequencies of morphotypes of ECM in plants subject to environmental stress suggesting alteration in species-diversity of ECM. Maintaining mycorrhizal fungal diversity helps in minimizing site degradation caused by unpredictive environmental changes. Moreover, mycorrhizae influences vegetation pattern within the community and allows nutrient transfer directly from one root system to another through mycorrhizal link.

Ectomycorrhizae, involved in mobilization of the organic resources, have a very limited ability to degrade organic polymers (Leake and Read, 1990). Studies reveal that only when in association with appropraite fungal partner, the hosts have access to organic nitrogen (Auzinadah and Read, 1989). A very wide range of proteolytic capability (due to acid proteinase etc.) bestow genera like *Amanita, Boletus, Paxillus, Piloderma, Suillus* and *Thelephora* with ability to assimilate organic nitrogen directly from polymeric sources, besides potential to capture mineral nutrient ions.

Temporal succession of fungal species is associated with aging of forest stands. 'Early' colonizers are replaced with 'late stage' colonizers with age (Dighton and Mason, 1985). Such succession occur due to the changes in resource quality and because of selection pressure (Last et al. 1987). Since aging leads to a progressive accimulation in organic nutrient pool, therefore, 'late stage' colonizers arise because of selection pressure and all of these exhibit well-developed proteolytic potential. Fungal mycelium by proliferation beyond nutrient depletion zones enhance release and mobilization of phosphate ions. Evidence show that phosphate, ammonium and alanine are absorbed by mycelium from forest soil and translocated through myceelial strands over considerable distance (Finlay *et al.* 1988; Read *et al.* 1989). Fogel and Hunt (1979) observed that mycorrhizal sheaths of tree contributed not less than 50% of the total nitrogen input to a Douglus-fir ecosystem. As these fungi senesce, they constitute a major source of relatively labile organic nitrogen which along with other organic residues are then subject to the attack by proteolytic enzymes secreted by living ECM. Thus besides phosphorus cycling, ECM play vital role in nitrogen cycle in forest ecosystems. It seems that as a consequence of providing carbon to fungal symbiont, the infected trees place mycorrhizae in an advantageous position over other soil microbial communities. In turn, fungus through primary mobilization causes release and assimilation of amino and phosphatic compounds. This subsequently releases p essure on host photosynthates and thereby provides a positive feed back of carbon and nitrogen. Such feed back have significance in carbon economy of trees, particularly when growing under shaded environments and stressed conditions.

ECOYSTEMS DOMINATED BY VAM

In contrast to ectomycorrhizae, the VAM fungi show little ability to mobilize nutrient N from organic sources and are physiologically less diverse. However, their mycelium have superior ability to exploit phosphates from organic sources in soil, through the possibility that VAM fungi are directly involved in mobilization of recalcitrant sources of phosphorus cannot be excluded (Gianinazzi-Pearson and Gianinazzi, 1989). This has made VAM a potential durable biofertilizer for nursery raised tree species (Mukerji *et al.* 1996).

In natural ecosystems, seedling establishment occurs in small gaps within communities of more mature plants. VAM infection takes place at an early stage of seedling development. Early integration of the sedling radicle into the mycelial network provides the plant with an enormous increase (about 10 to fold) in absorptive surface at relatively low energy cost, thus giving easy access to resources otherwise utilised exclusively by established plants (Read, 1991). This depicts vital role played by VAM in ecosystem preservation. The experimental evidences suggest that plants in natural communities are functionally interconnected by their mycorrhizal hyphae, and VAM network serves as a major sink for the assimilates continuously flowing into and through the soil eco-system. The significance of these carbon inputs in maintenance of soil microbial community has been ignored in grassland than in forest ecosystems. The mycorrhizal infection induces a shift in the biomass distribution from dominant to subordinate species (Grima *et al.* 1987). Therefore, herbaceous species are more benefited by VAM in terms of survivorship resulting in greater species diversity. Such high diversity in plant species in communities dominated by VAM is probably due to easy water and nutrient access facilitated by VAM to most vulnerable species at critical juncture of their life.

There are some indications that diversity of mycorrhizal species decline from natural ecosystem to intensive agriculture system. High diversity in natural ecosystem may principally be based on a great variety of plant species (Visalakshi, 1997). In contrast, under high intensive agriculture, crop species are few (Sieverdung, 1989). Experiments have proved that a mycorrhizal infection is of great significance under ecologically stressed field conditions (Allen and Allen, 1986; Dixon *et al.*, 1997).

CONCLUSION

The significance of mycorrhizal habitat in establishment and survival of crop plants and forest tree species is now world-wide acknowledged. Their soil conservation and geochemical nutrient abilities, of course, play vital role in maintaining the environmentally balanced ecosystem. Conservation of nutrients within various ecosystems is further enhanced by the formation of semi-permanent mycelial system which form absorptive networks interlinking plants at both intra and inter specific levels. Mycorrhizae exhibit mutual beneficial interactions with other microfora and fauna to accelerate nutrient flow in the edaphic system. The challenges for the future are to evaluate the full range of attributes shown by the different types of mycorrhizae and to integrate the results of such studies into a more broadly-based and realistic understanding of ecosystem dynamics.

LITERATURE CITED

Abuzinadah, R.A. and Read, D.J. (1989). Carbon transfer associated with assimilation of organic nitrogen sources by silver birch (*Betula pendula* Roth). Trees, 3: 17-23.

Allenn, E.B. and Allen, M.F. (1986). Water relations of xeric grasses in field: Interactions of mycorrhizas and competition. *New Phytologist*, 104: 559-571.

Amaranthus, M.P., Li, C.Y. and Perry, D.A. (1990). Influence of vegetation type and madrone soil inoculum on associative nitrogen fixation in Douglas-fir rhizospheres. *Canadian Journal of Forest Research*, 20 : 368-371.

Azcon, R.C., Azcon-Aguilar, C. and Barea, J.M. (1978). Effects of plant hormones present in bacterial cultures on formatio and responses of VAM endomycorrhiza. *New Phytologist*, 80 : 359-364.

Bethlenfalvay, G.J., Bayne, H.G. and Pacovsky, R.S. (1983). Parasitic and mutualistic associations between a mycorrhizal fungus and soybean: The effect of phosphorus on host plant endophyte interactions. *Physiologia Planarum*, 57: 543-548.

Borchers, S.L. and Perry, D.A. (1990). Growth and ectomycorrhiza formation of Douglas-fir seedlings grown in soils collected at different distances from pioneering hardwoods in southwest Oregon clear-cuts. *Canadian Journal of Forest Research*, 20 : 712-712.

Caron, M. (1989). Potential use of mycorrhizae in control of soil borne diseases. *Canadian Journal of Pathology*, 11: 177-179.

Castellano, M.A. (1996). Out-planting performance of mycorrhizal inoculated seedlings. pages 223-301. In Mukerji, K.G. (Editor) Concepts in Mycorrhizal Research Kluwer Academic Publ. Amesterdam, Netherlands.

Chandra, A. (1992). VA Mycorrhiza-Dimensions of itsapplicataion. *Indian Phytopathology*, 45:391-406.

Coleman, D.C., Andrew, C.R., Ellis, J.E. and Singh, J.S. (1976). Energy flow and partioning in selected man-managed and natural ecosystems. Agro-Ecosystems, 3:45-54.

Comeau, P.G. and Kimmins, J.P. (1989). Above-and below-ground biomass and production of lodgepole pine on sites with differing soil moistureregimes. *Canadian Journal of Forest Research*, 19: 447-454

Cromack, K. (1988). Below-ground processes in forest succession. pp. 140-146. In: West E-C; *Forest Succession concepts and Applications*. H.H. and Botkin, D.B. (Editors). Springer Verlag, New York, USA.

Dar, G.H. (1998) Role of mycorrhizae in phosphorus mobilization in agriculture and forests. In : Zargar, M.Y; Dar. G.H; and Pandit. B.A. (Editors). Directorate of Extension Education, SKUAST, Srinagar, Kashmir.

Dar, G.H., Zargar, M.Y. and Beigh, G.M. (1997). Biocontrol of *Fusarium* root rot in common bean (*Phaseolus vulgaris*) using symbiotic *Glomus mosseae* and *Rhizobium leguminosarum*. *Microbial Ecology*, 34 : 74-80.

Dar, G.H., Charoo, G.M. Zargar, .Y. and Beigh, G.M. (1995). Interaction between vesicular arbuscular mycorrhizae and *Rhizobium* in relation to plant growth, nutrient uptake, root colonization and rhizosphere microbial biomass. In : *Annual Conference and National Symposium on Recent Trends on the Management of Biotic and Abiotic Stresses in Plants*, 2-3 Nov., 1995, Indian Society of Plant Pathologists, H.p. K.V.V., Palmpur, HP (India).

Dighton, J. and Mason, P.A. (1985). Mycorrhizal dynamics during forest tree development. pp. 117-139. In : Moore, L. Casselton, D.A. Wood and Frankland J.C. (Editor) *Developmental Biology of Higher Fungi*, Cambridge University Press, Cambridge, UK.

Dixon, R.K., Mukerji, K.G., Chanda, B.P. and Kaushik, A. (1997). Vesicular arbuscular mycorrhizal symbiosis in relation to forestation in arid land. *Annals of Forestry*, 5: 1-9.

Elliot, E.T. and Coleman, D.C. (1988). Let the soil work for us. *Ecology Bulletin*, 39: 23-32.

Elliot, E.T. (1986). Aggregate structure and carbon, nitrogen and phosphorus in native and cultivated soils. *Soil Science Society of America Journal*, 50 :627-33.

Habte, M. (1989). Impact of simulated erosion on the abundance and activity of indigenous vesicular arbuscular mycorrhizal endophytes in an oxisol. *Biology and Fertility of Soils*, 7 : 164-167.

Hekstra, A. (1996). *Sustainable Nutrient Management in Agriculture*. Foundation of Ecodevelopment, Amesterdam, The Netherlands, p. 57.

Hwang, S.F., Chang, K.F. and Chakravarty, P. (1992). Effects of vesicular arbuscular mycorrhizal fungi on the development of *Verticillium* and *Fusarium* wilts of alfalfa. Plant Disease, 76 : 239-243.

Ingham, E.R. and Massiocotte, H.B. (1994). Protozoan communities around conifer roots colonizedby ectomycorrhizal fungi. Mycorrhiza, 5: 53-61.

Ingham, E.R. and Violina, R. (1991). Interactions among mycorrhizal fungi, rhizosphere organisms and plats. pages. 169-197. In : Barbosa, Kirschik and Jones (Editor), *Multi-trophic Level Interactions Among Microrganisms, Plants and Inects*, John Wiley and Sons, New York, USA.

Ingham, E.R. (1988). Interactions between nematodes and VA mycorrhizae. *Agriculture Ecosystem and Environment*, 24: 169-182.

Jalali, B.L. and Jalali, I. (1991). Mycorrhiza in plantdisease control. pages. 131-154. In : Arora, D.K. (Editor) *Handbook of Applied Mycology, Soil and Plants. Vol. I.* Marcel Dekker, New York, USA.

Kilronoms, J.N. and Kendrick, B. (1995). Relationships among microarthropods, Fungi and their environment: The significance and regulation of soil biodiversity. *Plant and Soil*, 170: 183-197.

Krishna, K.R., Shetty, K.G., Dart, J. and Andrews, D.J. (1985). Genotype dependent variation in mycorrhizal colonization and response to inoculation of pearl millet. Plant and Soil, 86: 113-125.

Emerson, W.W. Foster, R.C. Oades, J.M. (1986). Organo-mineral complexes in relation to soil aggregation and structure. pages 521-48. In : P.M. Hang and M. Schnitzer (Editor), Interactions of Soil Minerals with Natural Organics and Microbes, SSSA Sp. Publ. 17, SSSA, Madison, USA.

Finlay, R.D. and Read, D.J. (1986). The structure and function of the vegetative mycelium of ectomycorrhizal plants. II. The uptake and distribution of phosphorus by mycelial strands interconnecting host plants. *New Phytologist*, 103; 157-165.

Finlay, R.D., Ek, H., Odham, G. and Soderstrom, B. (1988). Mycelial uptake, translocation and assimilation of nitrogenn from 15N-labelled ammonium by *Pinus sylvestris* plants infected with four different ectomycorrhizal fungi. *New Phytologist*, 110: 59-66.

Fogel, R. (1988). Interactions among soil biota in coniferous ecosystem. *Agriculture Ecosystem and Environment*, 24 : 69-85.

Fogel, R. and Hunt, G. (1979). Fungal and arboreal biomass in a western Oregon Douglas-fir ecosystem: Distribution patterns and turnover. Canadian Journal of Forest Research, 9 : 265-276.

Fries, N. (1979). Germination of *Cantharellus cibarius*. Mycologia, 71: 216-219.

Gianinazzi-Pearson, V. and Gianinazzi, S. (1989). Phosphorus metabolism in mycorrhizas. Pages 227-243. In L. Boddy, R. Marchant annd D.J. Read (Editors) *Nitrogen, Phosphorus and Sulphur Utiliztion by Fungi*, Cambridge Universty Press, Cambridge, UK.

Graham, J.H. and Linderman, R.G. (1981). Effect of ethylene on root growth, ectomycorrhiza formation and Fusarium infection of Douglas-fir *Canadian Journal of Botany*, 59: 1340-1347.

Grima, J.P., Mackey, J.M.L., Hiller, S.H. and Read, D.J. (1987). Floristic diversity in a model system using experimental microcosms. *Nature*, 328: 420-422.

Gupta, V.V.S.R. and Germida, J.J. (1988). Distribution of microbial biomass and its activity in different soil aggregate size classes as affected by cultivation. Soil Biology and Biochemistry, 20: 777-786.

Krone, W.B., Bichler, E., Vielerock, E., and Moawad, A.M. (1987). Interaction between VA mycorrhiza and phosphate solubilizing bacteria. pages. 328. In : D.M. Sylvia (Editor), Proceedings of the 7th North American Conference on Mycorrhizae , Gainesville, 3-8 May, 1987. Inst. Food Agric. Sci., University of Flora, Gainesville, USA.

Last, F.T. Dighton, J. and Mason, P.A. (1987). Successions of sheath mycorrhizal fungi. *Trends in Ecology and Evolution*, 2: 157-161.

Leake, J.R. and Read, D.J. (1990). Proteinase activity in mycorrhizal fungi. 1. The effect of extracellular pH on the production and activity of proteinase by ericoid endophytes from soil of contrasted pH. *New phytologists*, 115: 243-250.

Lee, A. and Bagyaraj, D.J. (1986). Effect of soil inoculation with VAM fungi and either phosphate rock dissolving bacteria or thiobacilli on dry matter production and uptake of phosphorus by tomato plants. New Zealand *Journal of Agricultural Research*, 29: 525-531.

Li, C.Y. and Castellano, M.A. (1987). Azospirillum isolated from within sporocarps of the mycorrhizal fungi Hebeloma crustulinforme, *Laccaria laccata* and *Rhizopogon vinicolor. Transanctions of the British Mycological Society*, 88: 563-566.

Li, C.Y., Maser, C. and Fray, H. (1986). Initial survey of acetylene reduction and selected microrganisms in the faeces of 19 species of mammals. *Great Basin Nature*, 46 : 646-650.

Linderman, R.G., Paulitz, T.C., Moiser, N.J., Griffiths, R.P., Loper, J.E., Caldwell, B.A. and Henkels, M.E. (1991). Evaluation of the effects of biocontrol agents on mycorrhizal fungi. pages, 379. In : Keister, D.L. and Cragan, P.B. (Editors), *The Rhizosphere and Plant Growth*, Kluwer Academic Publishers, Dordrecht, Netherlands.

MacFall, J.S., Pfleger, F.L. and Linderman, R.G. (1992). *Effects of Ectomycorrhizae on Biogeochemistry and Soil Structure : Mycorrhizae and Plant Health*. Duke University, Durham, Carolina, USA, 237 pages.

Marks, G.C. and Foster, R.C. (1973). Structure Morphogenesis and Ultrastructure of Ectomycorrhizae: *Their Ecology and Physiology*. Academic Press, New York, USA.

Marx, D.H. (1991). Forest application of the ectomycorrhizal fungus *Pisolithus tinctorius*. Marcus Wallenberg Prize Lecture, Stockholm, Sweden, P. 35.

McGonigle, T.P. and Fitter, A.H. (1988). The ecological consequences of arthropod grazing on VA mycorrhizal fungi. *Proceedings of Royal Society of Edinburg*; 94B: 25-32.

Meier, S. (1991). Quality versus quantity: Optimizing evaluation of ectomycorrhizae for plants under stress. *Environmental Pollution*, 73: 205-216.

Meyer, J.R. and Linderman, R.G. (1986). Selective influence on populations of rhizosphere or rhizosplane bacteria and actinomycetes by mycorrhizas formed by *Glomus fasciculatum. Soil Biology and Biochemistry*, 18: 191-196.

Miller, R.M. and Jastrow, J.D. (1992). The application of VA mycorrhizae to ecosystem restoration and reclamation. pages. 438-467. In : Allen, M.F. (Editor) Mycorrhizal Functioning. Chapman and Hall, New York, USA.

Mohandas, S. (1987). Field response of tomato (*Lycopersicon esculantum* Mill 'Pusa Ruby') to inoculation with a VA mycorrhizal fungus *Glomus fasciculatum* and with *Azotobacter vi.elandii. Plant and Soil*, 98: 295-297.

Mukerji, K.G. (1996). *Handbook of Vegetation Science-Concepts in Mycorrhizal Research.* Kluwer Academic Publishers, Netherlands, p. 374.

Mukerji, K.G. and Dixon, R:K. (1992). Mycorrhizae in reforestation pages. 62-82. In M.M. Majid, A. Malik, A. M. D. Z. Hamzah and K. Jussof (Editors) *Rehabilitation of Tropical Rain-forest Ecosystem : R search and Development Priorities.* University Partaniai, Salangar, Malaysia.

Mukerji, K.G., Chanda, B.P., Kaushik, A. and Sarwar, N. (1996). Vesicular arbuscular mycorrhizae-potential biofertilizer for nusery raised multipurpose tree species in tropical soils. *Annals of Forestry*, 4 : 12-20.

Norton, J.M., Smith, J.L. And Firestone, M.K. (1990). Carbon flow in the rhizosphere of ponderosa in pine seedlings. *Soil Biology and Biochemistry*, 22 : 449-55.

Paul, E.A. and Clark, F.E. (1989). *Soil Microbiology and Biochemistry.* Academic Press, New York, USA, 376 pages.

Pfleger, F.L. and Linderman, G.R. (1994). *Mycorrhizae and Plant Health.* American Phytopathological Society, St. Paul, Minnesota, USA. 381 pages.

Raj, J., Bagyaraj, D.J. and Manjunath, A. (1981). Influence of soil inoculation with vesicular arbusular mycorrhiza and phosphate-dissolving bacterium on plant growth and ^{32}p uptake *Soil Biology and Biochemistry*, 13: 105-108.

Rambelli, A. (1973). The rhizosphere of mycorrhizae pp. 279-343. In : G.C. Marks, and T.T. Kozlowski (Editor), *Ectomycorrhizae: their Ecology and Physiology*, Academic Press, New York, USA.

Read, D.J. (1991). Mycorrhizas in ecosystems. *Experientia*, 47 : 376-91.

Read, D.J., Leake, J.R. and Langdale, A.R. (1989). The nitrogen nutrition of mycorrhizal fungi and their host plants. pages. 181-204. In : *L.L. Boddy, R. Marchant and D.J. Read* (Editors) *Nitrogen, Phosphorus and Sulphur Utilization by Fungi*, Cambridge University Press Cambridge, UK.

Secilia, J. and Bagyaraj, D.J. (1989). Bacteria and Actinomycetes associated with pot cultures of VA mycorrhizas. *Canadian Journal of Microbiology*, 33: 1069-1073.

Sharma, A.K., Johri, B.N. and Gianinazzi, S. (1992). Vesicular arbuscular mycorrhizae in relation to plant disease. *World Journal of Microbiology and Biotechnology*, 8: 559-563.

Sieverdung, E. (1989). Ecology of VAM fungi in tropical agrosystems. *Agriculture Ecosystem and Environment*, 29 : 369-390.

Singh, S. (1997). Interactions of mycorrhizas with nematodes. *Mycorrhiza News*, 9: 1-9.

Sollin, P., Cromack, K., Fogel, R. and Li, C.Y. (1981). Role of low molecular weight organic acids in the inorganic nutrition of fungi and higher plants. pages 527-533. In : Wicklow D.T. and Carroll G.C. (Editors). *The Fungal Community–Its Organization and Role in the Ecosystem*, Marcel Dekker, New York, USA.

Strzelczyk, E. and Pokojska-Burdzirj, A. (1984). Production of auxins and gibberellin-like substances by mycorrhizal fungi, bacteria and actinomycetes isolated from soil and the mycorrhizosphere of pine (*Pinus sylvestris L.*) *Plant and Soil*, 81: 185-195.

Swaminathan, M.H. and Vadiaraj, B.A. (1988). Nursery studies on the influence of Azospirillum biofertilizer on the growth and dry matter production of forestry species. *My Forest*, 24: 289-294.

Tilak, K.V.B.R. and Subba Rao, N.S. (1987). Association of *Azospirillum brasilense* with pearl millet. (*Pennisetum americanum*). *Biology and Fertility of Soils*, 4: 97-102.

Tisdall, J.M. (1991). Fungal hyphae and structural stability of soil. *Australian Journal of Soil Research*, 29 : 729-743.

Trappe, J.M. (1987). Phylogenetic and ecologic aspects of mycotrophy in the angiosperms from an evolutionary standpoint. pages 2-25 In : Safir, G.R. (Editor) *Ecophysiology of VA Mycorrhizal Plants*. CRC Press, Boca Roton.

Visalakshi, N. (1997). Dyanmics of vesicular arbuscular mycorrhizae in two tropical dry evergreen forests of south India. International Journal of Ecology and Environmental Sciences, 25: 25-36.

Chapter - 33

Suitable Agroforestry Models For Degraded and Wastelands of Srinagar District, Kashmir

A.H. Mughal and A.S. Makaya

ABSTRACT

Soil and water are the basic resources of the country and should be appropriately conserved. The increasing pressure of human and animal population and improper land use has resulted in increased run off and soil loss and caused several land degradation problems. The present research work was carried out at Sheep Breeding Farm, Shuhama Srinagar and the aim was to identify suitable agroforestry models along with suitable trees and agricultural crop components for different degraded areas. Silviagriculture and silvipasture models were found suitable for such degraded areas. For southern slopes trees like *Melia azedarach, Ailanthus excelsa, Aesculus indica, Gleditsia inermis. Gleditsia tricanthos, Robinia psuedoacacia* and *Ulmus wallichiana* were found suitable while as Zea mays performed better among agricultural crops. *Ulmus wallichiana, Robinia pseudoacacia, Ailanthus excelsa, Morus alba, Prunus armenica* and *Catalpa bignonoides* along with some shrubs and fodder grasses performed better on degraded lands besides, southern slopes. On areas with high water table Salix tetrasperma took lead over other species.

Key Words : agroforestry, Contour, degraded lands, impeded drainage, Silviagriculture, Silvipasture, Wastelands

INTRODUCTION

The problem of land degradation has not occurred overnight, it had started right from the time man started the practice of cutting of trees, but the impact could not be felt immediately as the population was less and so perfect balance existed between deforestation and regeneration. But now due to increase in human and cattle population there is an

Division of Forestry S.K. University of Agricultural Sciences and Technology Shalimar Campus Srinagar - 191 121

increased pressure on the biological resources, which are becoming scarce and are resulting in degradation of areas harbouring these resources. In order to avert environmental disaster and to make this earth congenial for human survival rehabilitation of degraded and wastelands should receive priority. Rehabilitation should be brought about properly and scientifically and agroforestry is the front line defences on the fight for wastelands and degraded lands. The present research work carried out at Sheep a Breeding Farm Shuhama, Srinagar has revealed that to arrest land degradation and to rehabilitate wastelands Silviagriculture and Silvipasture models constitute suitable approach for such areas.

(i) Degraded Wastelands

The term wasteland refers to degraded land which has deteriorated for want of appropriate soil and water management and which can be brought under vegetative cover with reasonable efforts (NWDB, 1987). Land degradation has been aggravated primarily by the pressure of human and animal population and secondarily due to mismanagement of biophysical environment. The effect of deforestation, mining, construction of dams and roads, industrialisation, over-grazing, unplanned changes in land use pattern, urbanization and land encroachment activities have contributed to this process. Decline in forest cover has led to shortage of fuel wood timber. In India deficit of feed, fodder (green and dry) and fuel wood would be to the tune of 75 11 00 and 140 million tonnes by 2000 A.D. (Sood et al. 1995). Besides. conversion of forest lands into agricultural lands, use by other development projects has led to soil erosion, ground water depletion, frequent floods and drought. In the country out of total geographical area of 329 million hectares uncultivated wasteland cover is about 85 to 90 million hectares (World Bank, 1988). The land use data indicates that about half of the geographical area (158.06 million hectares) of the country is wasteland or degraded land yielding less than 20% of its biological productivity (NWDB 1986).

(ii) Agroforestry

Agroforestry may be defined as a collective term for multiple land use (forestry, horticulture, agriculture, animal husbandry, fisheries, apiculture or lac culture) of a unit area simultaneously or sequentially and the aim is to meet the diverse needs of a farmer by getting multiple outputs from the unit area in a sustainable manner. Outputs may include food, fodder, fruits, fuel wood and timber.

Studies conducted on agroforestry systems in different places in our country have revealed that the best and suitable way for rehabilitating degraded and wastelands is by adopting agroforestry on such degraded lands (Rai, 1997). Elsewhere also, emphasis has been laid on the utilization of wasteland/forest land for *in situ* conservation of endangered and economically important species.

In Kashmir, a valley surrounded by lofty mountains, huge tracts are denuded due to deforestation and heavy human and animal pressure (Fig. 33.1). Besides, the southern slopes of mountains are completely barren and no vegetation worth the name exists on these slopes, thus all these areas are exposed to scorching sun during summer and soil erosion during heavy rains. (Fig. 33.2)

MATERIAL AND METHODS

The three sites selected in the year 1990 to identify suitable agroforestry models are (1) Southern slopes of the hills (2) Degraded lands prone to soil erosion besides southern slopes (3) Sites with high water table in the plans of the valley.

Fig. 33.1 : Degraded hills, need for rehabilitation through agroforestry

Following multipurpose tree species were planted on all these sites in blocks, each comprising of 100 trees.

1. *Melia azedarach*
2. *Robinia pseudoacacia*
3. *Ailanthus excelsa*
4. *Morus alba*
5. *Aesculus indica*
6. *Ulmus wallichiana*
7. *Gleditsia inermis*
8. *Gleditsia tricanthos*
9. *Catalpa bignonoides*
10. *Salix alba*
11. *Salix tetresperma*
12. *Alnus nitida*
13. *Populus deltoides*
14. *Populus nigra*
15. *Prunus armenica*

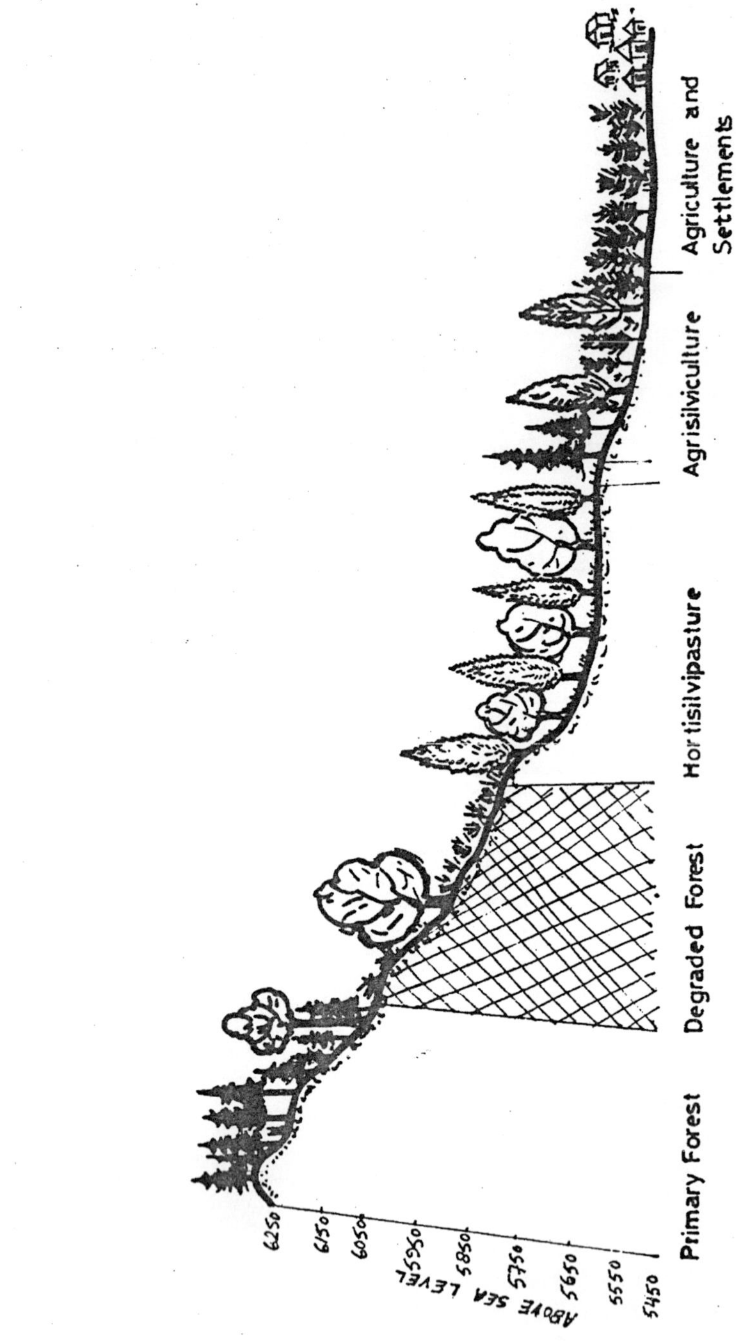

Fig. 33.2: Ecological transect of hills surrounding the villages of Kashmir

Besides, *Amorpha fruticosa* and *Cytisus scorparious* formed the shrub components while as fodder grasses like *Andropogon verginicus, Festuca pretense, Dactylis glomerata & Trifolium repans* were also tried in between the rows of trees.

Zea mays, Triticum spps. *Pisium sativum* and species of *Brassica* were raised as agricultural crops. Yearly data for survival percentage and growth in terms of height and diameter was recorded continuously for seven years. Yield of agricultural and fodder crops was also recorded both in Kharif and Rabi seasons.

RESULTS AND DISCUSSION

(i) Silviagricultural model for Southern slopes

There exists a tremendous scope for silviagricultural system in the areas located on the southern slopes. The system will not only help to reduce scorching heat but at the same time will increase moisture retention capacity and reduce soil erosion which in turn will increase agricultural production. Besides, the system will increase the fuel wood, timber and fodder production which are deficit in the area and people have to depend on alternate source of fuel and have to purchase fodder for the animals from the market in the form of the rice and wheat husk which is imported by the State from other parts of the country.

Experiments conducted on the southern slopes of these hills (Fig. 33.3) with no irrigation facility during the last seven years (1990-97) have revealed that following multipurpose tree species are best suited for these southern slopes in order of merit when planted along contour.

1. *Melia azedarach*
2. *Ailanthus excelsa*
3. *Aesculus indica*
4. *Gleditsia inermis*
5. *Robinia pseudoacacia*
6. *Ulmus wallichiana*

Maize crop was found suitable for the area. Trees should be closely spaced and should be planted along the contour whose distance should be maintained at 3.7m.

This type of Silviagricultural model will go a long way in increasing the agricultural production, meet requirements of fodder, fuel wood and timber and at the same time stabilize the slopes.

(ii) Silvipasture model for degraded lands besides southern slopes

Of the total geographical area, pasture and grazing lands constitute a mere 8.05% for an estimated 21.6 lakh cattle population in Kashmir province of which 2.499 lakh cattle are in Srinagar District only (J&K Live stock census, 1988). Of the total geographical area 4163.62 km is under pastures and grazing lands (Misri, 1995).

It is only from late eighties that the farmers are raising *Avena sativa* (oats) in their ochards, otherwise fodder production hardly finds any place in the common cropping pattern, consequently there is acute shortage of fodder in these areas. At present 50,000 hectare is under oat production in winter (Kachroo, 1992). It is well known that in Kashmir live stock depends only on grazing on the forest areas and pastures which are in a degraded state due to mismanagements. The problem is further compounded by the nomadic nature of Bakerwals and Gujjars who leave their cattle to graze and browse without any regulation. In the wake of large scale afforestation of degraded lands and other available areas, the lands have been closed for grazing to allow new plants to attain height beyond the reach of cattle, but this has given rise to conflict between the people and the concerned departments. So it is necessary to formulate silvipasture models on degraded lands belonging to the individuals as well as to various departments so as to tide over the scarcity of fodder and at the same time stabilize the degraded lands. Species which were found suitable during seven years of experimentation on these sites (Fig. 33.4);

1. *Ulmus wallichiana*
2. *Robinia pseudoacacia*
3. *Ailanthus excelsa*
4. *Aesculus indica*
5. *Morus alba*
6. *Prunus armenica*
7. *Catalpa bignonoides*

The trees were planted along the contour at a close spacing of 1.25m and trenches were dug just above the contour to preserve moisture and make it available to the plants. Contour to contour distance maintained at 3.7m.

Shrubs and grasses found suitable were as under:

1. *Amorpha fruticosa*
2. *Cytisus scorparious*
3. *Andropogan verginicus*
4. *Festuca pratense*
5. *Dactylis glomerata*

(iii) Rehabilitation of areas with high water table

Kashmir being located in the valley surrounded by lofty mountains on all its sides which are generally barren and all the rain and snow water flows into the valley with the result water table in the areas located just at the foot hills is very high and these areas also suffer from impeded drainage and as such are not suitable for agriculture and horticulture.

These areas can be rehabilitated by planting suitable forest tree species. Some of the trees found suitable for such areas are ;

1. *Salix alba*
2. *Salix tetrasperma*
3. *Ulmus wallichiana*
4. *Alnus nitida* and other species of Salix

Fig. 33.3 : Silviagricultural model for southern slopes. Tree component : *Melia azedarach, Ailanthus excelsa, Aesculus indica, Gleditsia tricanthos, Robinia pseudoaccacia, Ulmus wallichiana.* Crop component : *Zea mays, Triticum* spp. *Pisum sativum* and *Brassica* spp.

Fig. 33. 4 : Silvipastoral model for degraded lands. Tree component : *Ulmus wallichiana Ailanthus excelsa, Robinia pseudoacacia, Aesculus indica, Catalpa bignonoides, Prunus armenica* and *Morus alba*. Grass/shrub component : *Amorpha fruticosa, Cytisus scorparious, Andropogan werginicus, Festuca pratense* and *Dactylis glomerata*.

ACKNOWLEDGEMENTS

Authors are thankful to Hon'ble Vice Chancellor of the University for providing necessary facilities for the smooth conduct of research programme. Thanks are also due to the Ministry of Non-Conventional Energy Sources Govt. of India for financial assistance under Energy Plantation programme.

LITERATURE CITED

Census (1988). Live stock census of J&K. Directorate of Economics and Statistics. Planning and Development Department J&K.

Kachroo, V.K. (1992). Problems and prospects of Forage production in Kashmir. Paper presented at the workshop on Grazing policy held at Jammu, 19-20 March, 1992.

Misri, B (1995), Forage production in temperate Himalaya. pages 212-220. In : C.R. Hazara and Bimal Misri (Editors) New vistas in Forage production Publication & Information Directorate, New Delhi.

NWDB (1986). Wasteland challenges and resources. National Wasteland Development Board, New Delhi.

NWDB (1987), Report of committee on fodder and grasses National Wasteland Development Board, New Delhi.

Rai. P and Shukla S.K. (1997). Greening of wastelands through hortisilvipostoral system in India. pages 54-57. In : Proceedings on recent advances in agroforestry and its role in conservation of soil, water and biodiversity. 4-13th June 1997. Organizer NRCAF, Jhansi.

Sood, B.R., Singh, C.M. and Premi O.P. (1995). Forage production systems in humid, western Himalayan Region pages 184-196. In : C.R. Hazara and Bimal Misri. (Editors). New Vista in Forage Production.

World Bank (1988) India : Wasteland Development Review. Washington. The World Bank.

SECTION - IX

ENVIRONMENTAL LEGISLATION

Chapter - 34

Legislative and Judicial Response to Environmental Issues

A.M. Matta

ABSTRACT

This communication attempts to analyse the legislative and administrative measures for the protection of environment together with the role of judiciary. Despite some progress made in the area of environmental safeguards, apathy for environment, ecology and biodiversity is still pervasive, Realizing its responsibility judiciary has invoked its activism to curb the growing menace of environmental pollution. Besides making plea for strengthening the social action groups, the paper discusses suggestions aimed at enforcing the legal machinery for the environmental protection.

Key Words : critical appraisal, environmental issues, Indian scenario, international efforts, legislative measures.

INTRODUCTION

Environment has assumed a place of priority in human agenda. Environmental pollution has awakened man from his materialistic slumber. It has become the issue of concern for all-the scientist, the planner, the lawyer and the social activist. The scientists attempt to comprehend the scope of environment, the planner tries to accommodate environment in his policies, the social activist tries to educate people about its importance and the lawyer lays down the laws for safeguarding the environment. All have come to agree that little attention has so far been paid to be environment, its importance for mankind has not been fully realised and no steps have been taken for its safeguard. The post world war period depicts obsession with industrialisation which was considered to be a means of development. This path of industrialisation was chosen by man according to his own perception of development unmindful of its pernicious effects on environment. The result is that tones of industrial waste containing toxic substances finds its way into rivers and other water bodies which is being used not only for animals and plants but also for human consumption, causing health hazards.

*Department of Law, University of Kashmir, Srinagar - 190 006, India

Air pollution is used by emission of poisonous substances such as carbon dioxide, sulphur dioxide, hydrogen sulphide, methane etc. in the form of vehicular exhaust besides the smoke and gases billowing from the chimneys of big and small industrial units. Mining industry is yet another source causing atmospheric pollution.

Noise pollution is the result of unplanned location of industrial units, ever increasing number of vehicles, aviation and other similar activities. Besides unmindful dumping of garbage on open spaces, disposal of human refuse and domestic waste in natural water resources such as rivers, ponds and lakes results in exposure to pollution causing grave threat to human, plant and animal life. It has started being realised that the unabated industrialisation unmindful of environmental requirements is fatal to the health and prosperity of the people. Efforts have been set afoot in order to arrest the process of degradation of environment. International agencies and organisations are busy in identifying the causes of environmental pollution and evolving measures for its prevention and control. At national level also the legislative and administrative measures have been taken in this direction. The judiciary has also realised its responsibility and has come down heavily on the violators of environment. It is now being felt that economic development cannot be pursued with a blind eye towards the environmental requirements of human, animal and plant life. Economic development has to be compatible with environment which is essential for the sustenance of life on the universe.

An attempt is made in this paper to find the nature and scope of environment, to analyse the legislative and other measures taken for the safeguard of environment and to evaluate the role of judiciary in this direction. In the end some suggestions are made with the object of strengthening the legal machinery for the safeguard of environment and take timely action in order to punish the violators so that the propensity of violation is curbed with deterrence.

ENVIRONMENT : LEGAL PERCEPTION

Although it is primarily the job of a scientist to define environment but the essence of the matter has awakened mankind as a whole to the occasion and attempts have been made to define environment from a common man's perspective. The courts have also ventured in adopting some definitions of environment. Environment is a concept of composite nature. The meaning and scope of environment cannot be restricted. According to the Supreme Court, the word "Environment" is of broad spectrum which brings within its ambit "*hygienic atmosphere and ecological balance*"[1]. It is a wide term which could be used to encompass anything from the whole biosphere to the habitat of the smallest creature or organism. It is difficult to define this term and arrest the scope of its meaning within artificial boundaries. It is sometimes defined as the aggregate of social and cultural conditions that influence the life of an individual or any community. It has also been defined as the aggregate of all the external conditions and influences affecting the life and development of an organism[2]. It is said to include whole surrounding objects, regions or circumstances. Thus, the whole range of perceptible surroundings that influence the human life consciously or unconsciously would qualify as environment. Air, water, land, forests, vegetation, population, social values, moral practices, education, economic development, human needs and desires could easily be brought within the ambit of environment. The definition of environment under the Environment (Protection) Act, 1986 is also not exhaustive. Environment is defined to include "water, air, land and the inter-relationship that exists among and between water, air and land and human beings, other living creatures, plants, micro-organisms and property"[3].

Some of the well known causes of environmental pollution are soil erosion, deforestation, industrialisation, over exploitation of natural resources, and urbanisation. Besides these, big dams, oil slicks and nuclear industry have high propensity of pollution causing grave threat to human, animal and plant life.

INTERNATIONAL EFFORTS IN SAFEGUARDING ENVIRONMENT

The rapid environmental degradation alarmed people at all levels. At international level various measures were adopted to arrest the trend of pollution and lay down rules and regulations for its prevention and control. Some specialised agencies of the United Nations Organisation such as WHO (World Health Organization), UNESCO (United Nations Economic Social and Cultural Organisation), IMO (International Marine Organisation), OECD) (Organisation of Economic Cooperation and Development) had environmental concerns on their agenda but they simply issued guidelines for safeguarding environment. The role of most organisation was confined to the production of soft law through non-binding codes and guidelines, negotiation processes etc. There was to specific mandate for environmental protection given to these agencies. They exercised these powers by broadly interpreting their constituent documents considering environmental protection as their social, economic and humanitarian responsibility. However, it was felt, generally, that the traditional legal order of environment is essentially a laissez faire system oriented towards the unfettered freedom of states[4].

In 1972 the United Nations Conference on Human Environment (UNNCHE) was held at Stockholm with large participation. This Conference is considered to be a landmark development in the field of environmental protection. In this conference, a perception was evolved that environmental protection has a direct bearing on the economic development, especially of developing nations. For protecting environment from pollution and preserving the natural resources there has to be a restructuring of industrial, agricultural, fisheries and other development activities. Since these activities were essential for them the new approach had serious economic and political implications for developing states[5]. This was a cause of concern for them. The result was that a quick note of these implications was taken and in the following years attempts were made to seek redressal through a New International Economic Order based on a charter of economic rights and duties of states. It was advocated by the developing states that the developed countries through their process of industrialisation and exploitation of natural resources were mainly responsible for environmental pollution. It were those countries that had to bear the major cost of environmental protection. This approach was given went in the concept of New International Economic Order. Thus, a growing concern was evinced in the years to come.

During 1980's it was realised that environment could not sustain unlimited development in all cases. Efforts were made for conservation of and protection of natural resources. This required a check on the unlimited industrialisation and productivity. Thus, optional sustainable productivity and sustainable development were evolved as the key words as a concern for environmental protection.

In the World Commission on Environment and Development (WCED) in 1987, varying concerns that were put forth were analysed and it was realised that "sustainable development" had to be ensured in order to prevent environmental degradation. Sustainable development was defined as the "development that meets the needs of the present without compromising the ability of future generations to meet their own needs[6]. Thus the right of future generations to the natural resources was recognised and sought to be protected.

Subsequently, it has been stressed that sustainable development requires the maintenance, rational use and enhancement of the natural resources base that underpins ecological resilience and economic growth[7]. Sustainable development must not endanger natural systems that support life on earth, atmosphere, water, soil and the living bodies and living beings. All human beings have the fundamental right to an environment adequate for their health and well being. Conservation needs to be treated as an integral part of planning. Ecosystem and ecological processes essential for functioning of bio-sphere are to be maintained and biological diversity is to be preserved. Measures to limit the risk in carrying out dangerous beneficial activities have to be adopted.

The international organisations have thus identified sustainability which should constitute the prime objectives in our goals of economic and social development. The rights and interests of future generations in equitable utilization of natural resources have been acknowledged. The demarcation of the rights and duties of successive generations, their claims and concerns for development and sustainability in order to protect and improve environment, is, therefore, a complicated job which the law has to perform. The conventional role of the international law in this sphere in evolving non-obligatory guidelines and suggestions is making way for discemible imperatives which may even have punitive consequences for the erring states. Sustainable development is on its way to become a norm of international law. It is a notion around which legally significant expectations regarding environmental conduct have begun to crystalise. Expectations have arisen. Competing interests are being projected. The need for sustained efforts towards an amicable understanding of the issues and their resolution has been recognised. It was in this context in order to secure sustainable development in a wide spectrum of environmental issues that the United Nations Conference on Environment and Development (UNCED) was held at Rio de Jeniero in 1992. The issues of climate change, biodiversity and land based sources of pollution constituted the top agenda of the Conference. It aimed at establishing and improving co-ordination between those organisations whose role and programmes conflict or overlap in areas of sustainable development. It attempted to promulgate an "Earth Charter" as a consensus document for the world community. Although it was a sincere endeavour but the complex and insurmountable economic and social problems involved were such that doubts are raised about the success of this conference.

The post Stockholm period has seen a remarkable growth in the international treaties for environmental protection. New legal principles and concepts have been evolved with a view to remove the traditional hurdles of jurisdiction and enforceability. Detailed and concrete rules of environmental law have been framed. Multilateral treaties on subjects like Marine Pollution, Protection of the Atmosphere, Wild life Conservation, International Water Courses etc., have been arrived at. Multidirectional efforts are focused on the prevention of environmental harm and conservation of sustainable development of natural resources and ecosystem. The 1979 Geneva Convention on Long Range Transboundry Air Pollution, the 1989 Basel Convention on the Control of Transboundry Movement of hazardous Wastes, the 1985 Vienna Convention for the protection of Ozone layer, the 1979 Bame Convention on the Conservation of European Wild Life and Natural Habitat and the 1991 Protocol to the Antarctica Treaty on Environmental Protection are only a few cases showing the steady progress of environmental law at the international level.

INDIAN SCENARIO

Under the influence of International efforts for the prevention and control of environment pollution the Indian mind has also awakened to the situation. In the overall planning and development of the country the environmental issues have been taken note of. In its successive five year plans (particularly fourth plan onwards) environmental preservation and protection has received major place and stress[8]. The need for proper balance between development, planning and environmental management has been acknowledged[9]. The sound environmental and ecological principles in land use, agriculture, forestry, wild life, water, air, marine environment, minerals, fisheries, renewable resources, energy and human settlements have been emphasised[10].

The Seventh plan saw significant progress in environment and ecology. Ganga action Plan, forestry and wildlife, wasteland development and Island development subsectors constituted an integral part of the plan. Two hundred water quality monitoring stations, 85 air quality monitoring stations and 173 coastal monitoring stations were established. More than 50% of the major and medium industries have installed pollution treatment plants. Environmental impact assessment of major river valley and Hydro-electric, mining, industrial and thermal power projects were carried out through Environment Appraisal Committees. A multi departmental National Natural Resources Management System (NNRMS) has been set up. Eco-task forces have been deployed in some states for ecological restoration and land reclamation[11].

Implementation of the 10 point National Wildlife Action Plan (NWAP) was started: Thirty National Parks and 75 Sanctuaries are assisted financially by the central government. The number of tiger reserves rose from 15- 18 in the seventh plan. A National Wasteland Development Board (NWDB) has been set up in 1985. Social forestry projects were continued, envisaging a tree planting and afforestation of 1984600 hectares of wasteland during seventh plan. These projects were assisted by foreign agencies like the World Bank, United States Agency for International Development and Overseas Development Agency of U.K.[12]

In the eight plan (1992-97) an outlay of Rs. 525.00 crores was provided for forestry and wildlife sub sector. For ecology and environment sub-sector an outlay of Rs. 675.00 crores has been made. This is in addition to the state plan outlay of Rs. 3556.37 crores and Rs. 153.11 crores for these sub-sectors respectively.[13]

CRITICAL APPRAISAL OF LEGISLATIVE MEASURES

There are three main statues enacted by the parliament of India for the prevention and control of pollution. The Water (Prevention and Control of Pollution) Act, 1974 aims at the prevention and control of water pollution and the maintenance and restoration of the wholesomeness of water. Its protection extends to streams, rivers, water courses, inland waters, subterranean water, sea and tidal water.

The Air (Prevention and Control of Pollution) Act, 1981 is a post Stockholm statute aiming at the prevention and control of air pollution. It is direct towards the preservation of the quality of air.

The Environment (Protection) At, 1986 is a direct consequence of the Stockholm conference and intends to implement the decisions taken therein. It is the latest direct legislation on environmental pollution and seeks to preserve environment by establishing high powered administrative bodies and confers enhanced powers of punishment on the state.

Under these anti pollution laws Pollution Control Boards are envisaged at the central and state levels[14]. It is these boards which lay down, modify or annul the standards of water and air quality, and of pollutants for emission in the water and air. They have powers of inspection of industrial plants to ensure the execution of the intended aims under these statutes. The Environment (Protection) Act, 1986 confers wide powers on the central government. These powers are required for the enforcement of the provisions of the statute which aim at the prevention and control of water, air, soil and noise pollution as also the treatment of hazardous materials. The central government can take all necessary measures which it deems necessary or expedient for the purpose of protecting and improving the quality of environment. It has power to lay down standard of air, water and soil quality and the standards for discharge of pollutants and their maximum permissible concentration in the environment. It can investigate the causes of environmental pollution and restricted the location of industries by exercising all necessary powers such as inspection, examination, directions for closures, prohibition or regulation of any industry including the stoppage or regulation of electricity, water and other supply to the industry.

These anti-pollution statutes have created offences and prescribe punishments. The punishment prescribed under the Environment (Protection) Act, 1986 is harsher than under the Water Act and Air Act. This is understandable because the importance of clean environment is felt much more than before and its violation is sought to be prevented by prescribing stringent punishments under the Environment (Protection) Act, 1986. However, it is baffling to note that the Act of 1986 leans towards the offender when it gives him an option to be governed under the Water Act or the Air Act in case the offense falls under all the three Acts[15]. Thus the offender would obviously opt for the Act other than the Environment (Protection) Act, 1986 as the would thereby attract lesser punishment. This optional provision does not reconcile with the object of anti-pollution legislations. The matter deserves reconsideration.

The time consuming and expensive nature of litigation, the diatory and technical procedures and rules of evidence frequently come in the way of timely and purposeful control of pollution[16].

Now-a-days, the inaction on the part of the governmental agencies and the statutory authorities has been noticed with concern. The courts have invoked their inherent and writ jurisdiction and passed regulatory and other kinds of orders through the process of judicial activism. The courts are becoming more and more conscious about their activist role in preventing the pollutant state of affairs. The reasons and occasion for such a judicial role has been furnished by the disdainful violation of statutes in a callous and derogatory manner at all levels. The political protection to environmental polluters and the administrative inaction in carrying out the objectives of anti-pollution laws, left no choice with the judiciary but to adopt the activist approach in its endeavour to retrive the situation. It is in these circumstances that Supreme Court ordered the issuance of show cause notices to nine thousand and thirty eight industrial units in the capital for relocation[17]. The court has recently ordered the closure/shifting of one hundred and sixty eight industrial units from Delhi. The tanneries, foundries and other industries in the vicinity of historic Taj Mahal of Agra have been ordered to be closed down. The Supreme Court which is seized of a large number of petitions relating to environmental pollution ordered the closure of all stone crushers of Delhi and Haryana. The court even disbanded the Pollution Control Board of U.P. and threatened to disband the Board of Haryana on the ground of falsehood of its report[18]. The Supreme Court was critical recently when it took a strong note of the inaction on the part of the government in matters of environment. It remarked that "unfortunately the job which under the law was to be done by the government is left to the courts."[18a]

The constitution of the State Pollution Control Boards (SPCBs) and the Central Pollution Board (CPCB) under the Water Act and the Air Act is defective to the extent that no statutory qualifications have been prescribed for the members of these boards. The government generally nominates unqualified persons on these boards on political considerations with the result that the inexperienced and unskilled people come on these boards rendering them ineffective and jeopardising the intended objectives of the statutes. This is more so when these statutory authorities are made subservient to the general control and direction of the respective governments[19]. It is heartening to note that courts in India have widened the scope of locus standing in environmental issues otherwise the public spirited individuals have to comply with the requirement of sixty days statutory notice to the competent authority before making a complaint against a polluter. In view of the insensitiveness of the state authorities towards environmental issues, such individuals should be encouraged with all kinds of cooperation rather than creating obstacles in their way. Recently, the Environmental Clearance Notification 1993 has been issued by the central government under the powers given to it vide Environment (Protection) Act, 1986 [S. 3 (1) and 3 (2)(v)]. Under the notification, for the setting up of new projects or the expansion and modernization of the existing ones, a clearance certificate is required from the central or state government as the case may be. This has been done with the object of preventing environmental degradation with excessive soil erosion, water and air pollution on account of certain developmental activities engendering destruction of natural resources like forests, mangrove, wetlands, rivers, lakes, genepool reserves and vegetation covers. The clearance certificate is to be issued by the government agency after careful assessment of the project and its impact on environment.

The Wildlife (Protection) Act, 1972 and the Forest Act, 1927 are two other statutes which aim at the preservation and protection of environment through the protection of biodiversity for the conservation of wildlife habitat. The survival of wildlife is inextricably linked to the health of the ecosystem in which they live. It is with this view that certain areas have been declared as the protected areas, under the Wildlife (Protection) Act, 1972 and the Forest Act, 1927. Initially it was the state legislature which had the power to legislate in respect of wildlife before the transfer of this subject from the state list to the concurrent list by the Constitution (42nd Amendment) Act, 1976. The operation of the Wildlife Protection Act, 1972 now extends to the whole of India except the State of Jammu and Kashmir[20]. The Act envisages the establishment of protected areas in the form of "Sanctuaries", "National Parks" and "close areas". The entry, movement and residence in such areas have been regulated. Similarly, under the Forest Act, 1927 also three categories of protected areas have been constituted. These are the "reserved forest" the "protected forest" and the "village forest". Under the Act, the forest use such as grazing, felling trees, removing any forest produce, cultivation, quarrying, burning lime or charcoal, hunting and fishing is under the authority of the government which has to regulate it in the national interest.

It is, however, seen that the protected areas of forests have degraded and the forest wealth is being exploited by vested interests in an unabashed manner. Indiscriminate access into the protected areas and the extinction of wild animals and birds is a clear violation of the mandated purpose of protecting, propagating or developing wildlife or its environment and the regulated utilization of natural resources as envisaged under the Wildlife (Protection) Act, 1972 and Forest Act, 1927.

Artificial management of natural habitat through human efforts is an external interference in the course of nature and is likely to produce adverse genetical pressures. The natural state of habitat is disturbed by artificial conservation management and may be counterproductive. Further, the conservationists are divided on the evolution of methods of conservation management. The human knowledge is insufficient to prescribe certain methods for achieving conservation objectives and the natural state of habitat. Thus, it is advisable to report to minimum of human interference in natural habitat for preserving the self perpetuating ecosystem for the protection of wild life.

The Public Liability Insurance (Amendment) Act, 1992 has improved upon the Principal Act of 1991 with the object of providing immediate relief to the victims of accidents in handling hazardous substances. It is based on the principle, of no fault liability. Under the amendment Act the owners liability is unlimited. The Environmental Relief Fund has been created under the Act for providing immediate relief to the victims.

ROLE OF JUDICIARY

In the event of offensive inaction on the part of the executive in matters relating to environment the judiciary has exercised its inherent jurisdiction and intervened to prevent environmental degradation. The courts initially relied on the specific directive principles of state policy for enforcing environmental protection. Article 48A of these principles provides that the state should endeavour to protect and improve the environment and to safeguard the forests and wildlife. Article 51A *inter alia* casts a duty on every citizen to protect and improve natural environment including forest, lakes, rivers and wild life. The courts have heavily relied on these directive principles in issuing writs and directions ignoring their non justifiable nature.[21]

Now the courts derive authority from the Fundamental rights contained in article 21 for invoking powers to curb environmental pollution. According to the Supreme Court, Article 21 of the Indian Constitution protects right to life as a fundamental right. Enjoyment of life and its attainment including the right to live with human dignity encompasses the protection and preservation of environment, ecological balance free from pollution of water and sanitation without which life cannot be enjoyed[22]. Thus, human life is being valued high. The essential of human life are being spelt out in its true sense. It has rightly been said:

> "The expression "life" no longer means mere animal existence or continued drudgery but includes the finger graces of human civilization."[23]

It is on these considerations of human values that in 1985 some lime-stone quarries in Musori which resulted in degradation of environment affecting ecological balance were ordered to be closed.[24] Stressing the need for evolving a balance between ecology and development in *M.C. Mehta* v *Union of India & others*[25]. Justice K.N. Singh of the Supreme court observed :

> "We are conscious that closure of tanneries may bring unemployment (and) loss of revenue, but life health and ecology have greater importance to people".

It was perhaps for the first time the Supreme Court explicitly recognised the right of a citizen to healthful environment. The judicial concern for health, environment, ecology and habitat can be gauged by a cursory look into some orders that have been handed down recently by courts. For example, the practice of throwing burnt and semi-burnt corpses into river Ganga was ordered to be stopped as it was a source of water pollution.[26]

The Kanpur municipal council was directed to improve the sewerage system in order to preserve ecological balance[27]. The state government was restrained from reclining any further wetlands on the marshes along Hubli and Budyadhari rivers in orders to avoid threat to natural habitat[28]. The government was restrained from proceeding ahead with its public construction project on the ground that it would result in conversion of wetland which was harmful to be healthy environment[29]. The people resisted the plan of conversion of a park site into a hospital on the grounds that it would disturb ecosystem[30]. The open space of land meant for public use was ordered to be maintained disallowing the construction of dharamshala on it[31]. The tobacco manufacturing unit generating highly polluted effluent in violation of the Water (Prevention and Control of Pollution) Act, 1974 was ordered to be closed down and its electricity and water supply disconnected[32].

The court has said that the right to life under article 21 of the Constitution includes the right to enjoy pollution free water and air. Thus right to pollution free environment was brought within the ambit of fundamental rights[33].

Entitlement to a clean environment is recognised as one of the basic human rights. Any legal formation which impedes the dispensation of social justice is given a goodbye. In 1980 Krishna Iyer, J. of the Supreme Court in acknowledgement of public interest litigation bypassed the traditional rules such as "locus standi" when he said:

> "An issue is coming of age of that branch of public law bearing on community actions and the court powers to force public bodies, under public duties to implement specific plans in response to public grievances."[34]
>
> Justice Iyer continued to say:
>
> "Decency and dignity are non-negotiable facets of human rights and are the first charge on local self governing bodies".[35]

The Kerala High Court giving due recognition to human rights dispelled formalism. It said:

> "The human rights jurisprudence cannot be permitted to be thwarted by status quoism on the basis of unfounded apprehensions".[36]

Thus, the courts broke the shell of traditional technicalities and came out with dedication and determination for environmental protection. Recognising pure environment as basic to human life the judiciary realised that it had to innovate new methods and devise new strategies for purpose of providing access to justice to large masses of people who are denied basic human rights and to whom freedom and liberty have no meaning[37].

Economic development in terms of industry is the main consideration in both developed as well as the developing countries. The policies of development are pursued without any consideration for protection and safeguard of human values through environment. However, the courts have circumscribed development so as to be in closest possible harmony with environment. The courts have reiterated their duty-fullness towards society in carrying out the mandate of legislation and have said that there should be proper balance between development and environment.[38] If balance was not struck between the needs of development and that of protection of ecology it would result in violation of fundamental rights under articles 14 and 21.[39]

Recently the Supreme Court[40] has taken a strong note of the failure of the central government in not constituting the statutory authority under section 3(3) of the Environment Protection Act, 1986, for controlling pollution and protecting the environment. It issued comprehensive directions for maintaining standards stipulated by Pollution Control Board and requested the High Court of Madras to constitute Special Bench "Green Bench" to deal with he instant case of pollution of water courses by discharge of effluent from tanneries and other environmental matters. It also reiterated the precautionary principle and the pollutor pays principle as parts of the environmental law of the country, under which the pollutor has not only to compensate the victims but also the cost of restoring the environmental degradation.

In yet another recent case[41] pollution caused by mining operations near Badkal Lake and Surejkund in Haryana was sought to be prevented by prohibiting such activities within 2 Km range of these lakes. The Supreme Court directed that mining leases within 2-5 km of these lakes should not be renewed unless no objection of the State and the Central Pollution Control Board was obtained. The tourist resort camps set up by the state government of U.P. in the open pastures and meadows of Garhwal area were ordered to be closed by the court as it was considered to cause ecological imbalance.[42] Similarly, the setting up of Shrimp Culture Industry within the prohibited area in ecology fragile coastal area was said to have adverse effect on environment and coastal ecology and economics. The Supreme Court did not permit it to operate. The retrenched employees were entitled to compensation. The court ordered the constitution of a High Power Authority to consider grant of permission for installation of Shrimp industry.[43]

Locus standi could not come in the way of the Judiciary's resolve to preserve environment and ensure ecological balance. The courts consider it their duty to create innovative methods of administration of justice in human rights jurisprudence. The public interest litigation (PIL) has been given due consideration and writ petitions have been entertained from public spirited people in the cases of environmental pollution. Bare letters have been entertained as petitions and judicial activism invoked. If the administrator shows indifference to the principle of accountability, law will become a dead letter on the statute book and public interest will be the casualty. With this perception the technical locus standi was given wider interpretation.[44]

The Supreme Court through an interim order in *Indian Council for Enviro-legal Action v Union of India*[45] has directed that the restrictions, prohibitions regarding the construction and setting up of industries or for any other purpose should be meticulously followed by all concerned states. The activities declared prohibited within the Coastal Regulations Zone under the Coastal Areas Classification and Development Regulations should be strictly followed by state. The Supreme Court has gone further ahead and directed the concerned authorities to spare time slots on T.V. and Radio as also in cinema and video parlours for environmental education. Environment as a subject to be taught in under-graduate classes was made obligatory on UGC.[46]

CONCLUSION

Environmental protection today, is everybody's concern. It has come to occupy a place in the agenda of human values. Processes and programmes have been initiated at international level for the preservation of environment, ecology and natural habitat. At national level in India administrative as well as legislative measures have been taken towards the prevention and control of environmental pollution and the maintenance of biodiversity. However, like all other statutes, the non-implementation and laxity on the part of the administrative machinery for unexplainable reasons has been the major cause of our failure. The judiciary has taken strong note of inaction on the part of government. The people responsible for such laxity or inaction should be identified for deterrent punishment. It is the judiciary which has realised its responsibility and invoked its activism in its endeavour to curb the menace of environmental pollution. Orders of closure of industrial units, stoppage of electricity and water supply, satisfactory treatment of hazardous effluent by industry, relocation of pollutant industries etc., have been passed. Social action groups have demonstrated their will to fight all measures which are detrimental to human health and environment.

Much advance appears to have been made in the area of environmental protection. However, it is observed that disregard for environment, ecology, and biodiversity still persists. The will to implement the anti pollution laws and maintain ecological balance has not been fully generated. Administrative machinery has not shown the determination to carry out the mandate of environmental protection. Although the judiciary has broken the traditional barrier of locus standi and provided the desired lead in the sphere of environmental protection, the message has not fully percolated to the administrative rung of the state machinery. Environmental education has not reached the common man. People take environmental violations casually. They hesitate in entering the arena of litigation for social causes of environmental protection, for, the experience has shown it to be dilatory, technical and burdensome. Again, people hesitate also because they do not want to take action at the cost of social and neighbourly relations. In these circumstances the social action groups need to be strengthened so that they take up the task of social action. This would save individual exposure of people to the risk of spoilling relations or becoming targets of victimisation. For the speedy redressal of environmental grievances separate courts need to be set up with expert personnel constituting a part of the system. Last but not the least is the need to spread environmental education and inculcate a sense of human rights and human values in the minds of people so that a sense of self-realization is generated. This does not, however., mean that there should be any laxity in punishing the violator. For, heavy sanctions alone mend the deliquent.

LITERATURE CITED

1. *Virender Gaur and Others* V *State of Haryana and others* (1995) 2 SCC 577.
2. International Encyclopaedia of Social Sciences.
3. Sec. 2 (a) of Environment Protection Act, 1986.
4. Schneider, World Public Order of the Environment ; Towards an Ecological Law and Organization, Toronto 1979, 30 cited in Bernie, Patricia W. and Boyle, Allen E. International Law and the Environment, Reprint 1994 P. 1.
5. In this Stockholm Conference 113 states, both developed and developing participated. The USSR and other Eastern Block states abstained for the reasons that East Germany was not invited to the Conference. However in the year 1989, USSR participated in UNEP.
6. See WCED Document, Our Common Future, P. 43.

7. This was emphasized by the United Nations Environment Programme. See UNEP Governing Council Decision 15/2 May, 1989. Annexure II Session Supply No. 25 (A/44/25).

8. Planning Commission, Govt. of India, Fourth Five Year Plan 1969-70 to 1973-73.

9. Planning Commisssion, Govt. of India, Fifth Five Year Plan 1974-1979.

10. Planning Commisssion, Govt. of India, Fifth Five Year Plan 1980-85 p. 343.

11. Planning Commisssion, Govt. of India, Fifth Five Year Plan 1992-97, Vol. 11, 95-96.

12. Ibid. at 96.

13. Ibid. At 96-100

14. The Air and Water Pollution (Prevention and Control) Act, confer vast powers on the State Pollution Control Boards while as the Environment (Protection) Act, 1986, confers such powers (even wider) on the Central Government.

15. The same is true of other Environmental Acts also, under which the offence simultaneously falls e.g. The Mines and Minerals (Regulation and Development) Amendment Act, 1986 under which the maximum penalty is two years imprisonment of the fine of Rs. 5000/- whereas under the Environment (Protection) Act, 1986, it is seven years imprisonment and a fine of Rs. 1.00 lac and Rs. 5000/- per day of continuation of offence.

16. This is also borne out by the Annual Reports of the Central Boards for Prevention & Control of Pollution. See Vir Singh, "Judicial Activism and Environmental Pollution" in New World Environment Series--New Environmental Age ed. Saproo R.K. and Bhardwaj Shyama (1990) PP. 37-47.

17. See Indian Express, of 25th March, 1995. The order was passed on 24th March, 1995

18. See the Times of India, Nov. 22, 1995, P. 7, The Pioneer, Sept, 28, 1995.

18a. Vellore Citizens Welfare Forum v Union of India and others AIR 1996 SC. 2715

19. For further details, see Vibhute K.I. Environment, Development and Law Indian Perspective JILI, Vol. 37, No. 2, 1995 PP. 182-194.

20. This was done through Wildlife (Protection) Amendment Act, 1991.

21. In *M.C. Mehta* v *Union of India & others* AIR 1991 S.C. 813. The Supreme Court ordered the state to appoint a higher power committee to look into the problem of vehicular pollution.

22. *Virendra Gaur* v *State of Haryana*, 1995, 2 SCC 577 at 578.

23. *Autonv* v *Commercial Corporation of Cochin*, 1994 (1) KLT 169.

24. *RL & E Kendra Dehradun* v *State of UP* AIR 1985 SC 652.

25. 1987 SCC 463 at 482

26. *M.C. Mehta* v *Union of India & others* A I R 1988 SC 1115

27. Ibid.

28. *People United for Better Living in Calcutta* v *State of West Bengal* AIR 1993, Cal. 215

29. *Consumer Protection Group* v *Union of India Mgd.* (Unreported)

30. *Bangalore Medical Trust* v *Mudappa and others*, AIR, 1991, SC, 1992

31. *Virender Gaur and Others* v *State of Haryana and others* 1995, SCC, 577.

32. *M/s Mahabir Soap and Gudakhu Factory* v *Union of India* AIR 1995, Ori. 218.

33. *Subash Kumar* v *State of Bihar* AIR 1991 SC 420.

34. *Ratlam Municipality* v *Vardchichand* AIR 1980 SC 1622.

35. Ibid. At 1623.

36. *V. Laxmipatty and others* v *State of Karnataka and others* 1991 (2). Ker L J 453.

37. *S.P. Gupta and others* v *President of India and others* AIR 1982 S.C. 149

38. *Consumer Action Group* v *Union of India, Madras* (Unreported).

39. *Kinki Devi and Another* v *State of H.P. and Others* AIR 1988 H.P. 4; *L. K. Koolwal* v *State of Rajasthan and others,* AIR 1988 Raj. 2; *People United for Better living in Calcutta* v *State of West Bengal* AIR 1993 Cal. 215.

40. *Villore Citizens Welfare Forum* v *Union of India and others* AIR 1996 SC 2715.

41. *M.C. Mehta* V *Union of India,* AIR 1996 SC 1977.

42. *Om Prakash Bhatt* V *State of U.P.* AIR 1997 All. 259.

43. *S. Jagannath* V *Union of India* AIR 1997 SC 811.

44. *V. Laxmipatty & others* v *State of Karnataka and others* 1991 (2) Ker. L.J. 453.

45. AIR 1995 SC 2252.

46. *M.C. Mehta* v *Union of India* AIR 1992 S.C. 382.

Index

B

C

D

G

H

J

K

N

O

P